Monitoring, Simulation, and Management of Visitor Landscapes

Monitoring, Simulation, and Management of Visitor Landscapes

Edited by

Randy Gimblett and Hans Skov-Petersen

THE UNIVERSITY OF ARIZONA PRESS, TUCSON

The University of Arizona Press
© 2008 The Arizona Board of Regents
All rights reserved

ISBN 978-0-8165-2729-8 (paperback edition)
ISBN 978-0-8165-2728-1 (electronic edition)

Library of Congress Control Number: 2007940043

This book was published from formatted electronic copy that was edited and prepared by the volume editors.

Cover photos: Hans Skov-Petersen, Laura Kennedy, and Ole Malling

Typesetting and layout by Inger Grønkjær Ulrich

Manufactured in the United States of America on acid-free, archival-quality paper containing a minimum of 30% post-consumer waste and processed chlorine free.

13 12 11 10 09 08   6 5 4 3 2 1

CONTENTS

Preface — vii

Chapter 1 — 1
Introduction: Monitoring, Simulation and Management of Visitor Landscapes
Hans Skov-Petersen and Randy Gimblett

Wayfinding And Prediction Of Behavior In Visitor Landscapes

Chapter 2 — 13
What Visitors "do" in Recreational Landscapes: Using Categories of Affordances for Evaluation, Simulation and Landscape Design
Dennis Doxtater

Chapter 3 — 37
Sensemaking Experiences on the Bardedjilidji Sandstone Track
Michael L. Steven

Chapter 4 — 59
Keep an Eye on Nature Experiences: Implications for Management and Simulation
Birgit Elands and Ramona van Marwijk

Monitoring Systems For Capturing Visitor Patterns of Use

Chapter 5 — 85
Techniques for Spatial and Temporal Simulation Modeling: Evaluating both Macro and Micro Scales
Jianhong (Cecilia) Xia and Colin Arrowsmith.

Chapter 6 — 107
Simulation, Calibration and Validation of Recreational Agents in an Urban Park Environment
Daniel Loiterton and Ian Bishop

Chapter 7 — 123
Applying Data from Automatic Visitor Counters to Agent-Based Models
Hans Skov-Petersen, Henrik Meilby and Frank Søndergaard Jensen

Chapter 8 — 143
Probability of Visual Encounters
Hans Skov-Petersen and Bernhard Snizek

Chapter 9 — 159
Exploring Spatial Behavior of Individual Visitors as Background for Agent-Based Simulation
Karolina Taczanowska, Andreas Muhar and Arne Arnberger

Computer Simulation And Management of Visitor Landscapes

Chapter 10 — 175
Measuring, Monitoring and Managing Visitor Use in Parks and Protected Areas Using Computer-Based Simulation Modeling
Steven R. Lawson, Jeffrey C. Hallo and Robert E. Manning

Chapter 11 — 189
Choice models: Estimation, Evolution, Limitations and Potential Application for Simulation Modeling
Len M. Hunt

Chapter 12   213
Modeling Indigenous Cultural Change in Protected Areas
Laura Sinay, Carl Smith and Bill Carter

Chapter 13   239
Agent-based Modelling: Views from the Management Perspective
Ulrike Pröbstl, Peter Visschedijk and Hans Skov-Petersen

Chapter 14   253
Linking Ecological and Recreation Models for Management and Planning
Rogier Pouwels, Claire Vos and Paul Opdam

Chapter 15   269
MASOOR: Modeling the Transaction of People and Environment on Dense Trail Networks in Natural Resource Settings
Rene Jochem, Ramona van Marwijk, Rogier Pouwel and David Pitt

Chapter 16   295
Applying Agent-Based Modeling for Simulating Spring Black Bear Hunting Activities in Prince William Sound, Alaska
Spencer Lace, Randy Gimblett, Aaron Poe and Dave Crowley

Chapter 17   315
Balancing Increased Visitor Use and Wilderness Character using Simulation Modeling in Misty Fjords National Monument
Rachel Nelson, Chris Sharp, Chris Prew and Randy Gimblett

Chapter 18   331
Level of Sustainable Activity: Moving Visitor Simulation from Description to Management for an Urban Waterway in Australia
Robert Itami

Chapter 19   349
Monitoring and Simulating Recreation and Subsistence Use in Prince William Sound, Alaska
Phillip Wolf, Randy Gimblett, Laura Kennedy, Robert Itami and Brian Garber-Yonts

Chapter 20   371
Assessing the Reliability of Computer Simulation for Modeling Low Use Visitor Landscapes
Brett Kiser, Steven Lawson and Robert Itami

Chapter 21   389
Replication, Reliability and Sampling Issues in Spatial Simulation Modeling: A Case Study Exploring Patterns of Human Visitation in Canada's Mountain Parks
Robert Itami, Randy Gimblett, Frank Grigel, Darrel Zelland and David McVetty

Conclusions and perspectives

Chapter 22   411
What Can Been Learned from Multi-Agent System Modeling?
Danielle J. Marceau

Chapter 23   425
Conclusion: Monitoring, Simulation and Management of Visitor Landscapes.
Randy Gimblett and Hans Skov-Petersen

List of Contributors   439

Index   445

PREFACE

There is a growing body of research focused within the context of human-environment interactions. This book "*Monitoring, Simulation and Management of Visitor Landscapes*" was conceived as a beginning point to explore the current state of spatial/temporal simulations that integrate human behavior and environmental factors as applied to decision-making in spatially referenced dynamic environments. The term *"Visitor Landscape"* is an extremely important part of the title of this book. Visitor landscapes are those that humans inhabit temporarily and seek experience opportunities in a natural outdoor setting. These experience opportunities such as solitude or adventure, come with the potential to destroy the very nature of the experience sought through crowding, overuse and degradation to the landscape. Frequency of use, duration of stay, the recreation activity visitors are engaged in, season of use, numbers of visitors in the landscape setting and the fragility of the landscape, all have a significant impact on social and ecological values that require extensive management.

This idea for this book grew out of a series of conferences with a focus on *Monitoring and Management of Visitor Flows in Recreational and Protected Areas*. These conferences have been held every other year starting at University of Natural Resources and Applied Life Sciences, Vienna, Austria in 2002, Rovaniemi, Finland in 2004 and in Rapperswill, Switzerland in 2006. These conferences have brought together scientists, managers and decision-makers from around the world. While this book will be of interest to this enlightened group, the intended audience of this book is the professional ranging from computer scientists, biologists, social scientists, to field technicians, resource managers and decision makers. Students of environmental studies or those engaging in building models of human use in spatially referenced landscapes should embrace the lessons learned by the chapter authors. While there are many important points that you will encounter throughout this book, it's purpose is to provide a glimpse of field methods and tools that can aid decision makers in planning and managing lands to accommodate increasing human use while maintaining the ecological integrity of those lands. Conventional methods used in planning and management of human- landscape interactions fall far short of the needs of decision makers who need to evaluate the cascading impacts of humans in visitor landscapes. Many public land agencies, local governments and international organizations are exploring tools such as multi-agent simulations coupled with social science data for developing long-term strategies for evaluating human-landscape interactions. In particular, spatial agent-based models are being explored with some success to provide a better understanding of the spatial and temporal patterns of human-landscape interactions and to predict how distributions of this use are likely to change in response to policy implementation and proactive management actions. While the application of simulation to study human-landscape interactions is in its infancy, there is need to develop a comprehensive and empirically based framework for linking the social, biophysical and geographic disciplines across space and time.

All chapters in this book have gone through a rigorous peer review process. A list of peer reviewers can be found in the end of this book. Each chapter in the book has been reviewed by at least two external reviewers and by both of the editors. Chapters have been organized in such a way as to demonstrate the necessity of careful integration of monitoring data using statistically valid sampling and data collection techniques, with a growing number of modeling platforms (ie. Agent-based, State preference/choice models and Baysian belief systems) to specifically answer resource based management questions. While it seems almost trite to say but worth repeating is that the most critical lesson learned is that all of these issues need to be solely directed by the management. There are many questions at hand. Several chapters in this book that provide good examples of how decision makers can help modelers and field staff articulate management questions and subsequently develop monitoring and data collection techniques, coupled with the appropriate modeling platform to directly answer those questions and implement model results to manage visitor use. Without such a planning framework, a model can be only minimally useful, under utilized in predicting future trends and lead to misrepresentative or speculative results. If decision makers are our intended audience, who use data and model outputs to aid decision making, then it is our ethical responsibility as professionals to ensure that current monitoring techniques are consistently explored and statistically representative and valid methods are used to capture spatial and temporal patterns of human use that meet social science standards. In addition, spatial simulation models need to be constructed using appropriate simulation standards including degrees of reliability and confidence intervals. This volume explores all of these issues and provides excellent examples of where coupling of monitoring techniques with spatial simulations and management questions can lead to policies that dramatically improves decision-making in visitor landscapes.

Randy Gimblett
Hans Skov-Petersen
September, 2007

CHAPTER 1

INTRODUCTION: MONITORING, SIMULATION AND MANAGEMENT OF VISITOR LANDSCAPES

Hans Skov-Petersen
Randy Gimblett

The bonds between people and the environment are inherited, tight and ever lasting. We have developed in and continuously depend on the primary resources afforded by the environment we inhabit; such as fresh air, water, and food. In a modern stressful world we need to remind ourselves of our inheritance: Once our primarily needs are fulfilled, we need nature for exercise, to relax in, to learn about, and to enjoy. This way we recreate ourselves. To many people - a still increasing number - recreation in nature is considered an irreplaceable part of living. Accordingly the pressure on our visitor landscapes is increasing and is likely to continue to increase to or beyond the carrying capacity of the areas involved; leading to the 'tragedy of the commons' (Manning, 2007) where our wish to ensure public access to nature will led to biological, physical or social overuse and degradation. Therefore a major objective for planning and managing of visitor landscapes that includes parks, protected areas, and urban forests is to avoid the negative effects of recreational use and to ensure that expectations of visitors can be afforded (Kajala et al, 2007 and Hornback and Eagles, 1999).

Management and planning recreation in nature areas needs adequate and reliable data about visitor use before effective decisions can be made. Monitoring schemes in relation to recreation often aims at measurement of the spatial and temporal distribution of visitors. This can include the number of visitors entering a park at different points at different times of the year, week and day. Further, information about the visitor's expectations, activities, expectations, motivations, duration of stay, trip itineraries inside the park, etc. can be of interest. Methods for visitor monitoring include:

- Automatic traffic counters (see for instance Arrowsmith and Xia chapter 5 and Skov-Petersen et al. chapter 7)
- Assessment of the number of tickets or licenses granted (see for instance van Wagtendonk 1978, Gimblett et al. 2002, and Lace et al. chapter 16)
- Interviews with visitors which are frequently based on structured questionnaires (see for instance Taczanowska et al. chapter 9)
- Evaluation of preferences, norms, experiences, and choices including ranking

statements and photographs (e.g. Jensen 1999), manipulated photos displaying different levels of visitation (Manning 2007), and in situ self-registration of the linkage between stimuli/thoughts/feeling (Steven chapter 3)
- Preparation of sketch maps (Taczanowska et al. chapter 9), sports timing devices, and application of GPS technology for registrations itineraries (see for instance Jochem et al. (chapter 15), Arrowsmith and Xia (chapter 5), and Loiterton and Bishop (chapter 6)).

In cases where the purpose is general physical planning household questionnaires can be carried out. This way important information about people that do not participate in recreation in nature can be obtained. Often questions related to recreation will be posed with reference to the respondent's last visit to a recreative area (Jensen 1999). Objectives for investigations based on household questionnaires include levels of participation in recreation given different socio-economic premises, urban structures, and the relation between site choice and transport distance (e.g. Jensen 1999 and Termansen et al. 2004).

FROM MONITORING TO SIMULATION

Major motivations for monitoring in relation to recreation include the need to get a status of or follow the development in use levels, or assessment of the ecological or social effects of recreation might have on nature. In cases where monitoring is impracticable or expensive, for instance if a nature setting is vast and includes remote points of interest, information has to be extrapolated from the available data by means of simulation modeling. Likewise in situations where future settings have to be evaluated, simulation models can be an appreciated option. Often generic data from monitoring schemes are used as baseline for simulation models, but most frequently they have to be supplemented by additional, case-specific collection of data.

Reasons to simulate include:
- Provision of proxies for expensive and impossible monitoring schemes (Manning 2007) including identification of potentially problematic 'hot spots' (see for instance Nelson et al. chapter 17).
- Evaluation of existing management policies and standards (Nelson et al. chapter 17)
- Obtaining a better understanding of spatial and temporal patterns of visitor use (Cole 2005) e.g. through hypothesis testing.
- Prediction visitor/landscape systems' response to factors not subject to managers control (Cole 2005) e.g. a general increase in the number of visitors due to demographic changes, changes in population mobility, regional infrastructure etc.
- Evaluation of effects of management actions, i.e. generation of 'what if' scenarios (Castle and Crooks 2006). Including changes in access (parking lots, path network, temporal closure, etc.)
- Estimation of maximum visitor use levels (Manning 2007)
- Provision of communicative means in participatory processes (Cole 2005 and Jochem et al. chapter 15) or as a learning tool (Manning and Potter 1984 and Marceau chapter 22).

In the present volume Lawson et al. (chapter 10) provide a systematic discussion of various approaches to application of simulation models in addition to or alternating monitoring schemes. The chapter addresses indicators of a) description of existing visitor use conditions, b) monitoring 'hard to measure' indicators, c) 'proactive' managing carrying capacity, d) test of alternative visitor management practices and e) research in public attitudes.

Indicators revealed from simulation models typically address visitor loads distributed over space, time, state-of-mind or type of activity, and type of visitor. Most frequently in relation to models of recreational behavior space is represented by locations (typically as network nodes) or stretches of infrastructure (network edges). Space based indicators can address the overall number of persons traversing given stretches of the path network (Jochem et al. in chapter 15) or the number of cars at a parking lot (Itami 2005) or a site of special interest in the park. A temporal distribution addresses the number of visitors over defined durations of time. It can for instance be per hour over the day, per day of the week or monthly over the span of a year. State-of-mind or activity type describes the present mental state of the agent (e.g. satisfied, sad, happy etc.) or activity (e.g. walking, eating, resting etc.) at a given time. Taking that as a point of departure an indicator can address the relative amount of time visitors were satisfied or felt annoyed, for instance by over crowding. In models where different types of agents are acting simultaneously indicators can of course be split into user type.

Often the grouping categories are combined into complex indicators, for instance assessment of the relative amount of time where crowding norms in terms of People At One Time (PAOT) are exceeded (Manning 2007).

SIMULATION MODELS OF RECREATIONAL BEHAVIOR

Models of recreation vary in their level of aggregation. Very aggregated models attempt to predict for instance the number of visitors to a give nature area only by means of the number of people having access to it, given different conceptions of accessibility and mobility. More advanced aggregate models include also local characteristics of the given forests, different types of users and the presence of alternative opportunities (Skov-Petersen 2002 and Termansen et al. 2006). Among the most disaggregated of these approaches are the Random Utility Models (Romano et al. 2000 and Len Hunt in chapter 11). Here the probability of people of certain characteristics (socially, environmentally etc.) would visit destinations of certain characteristics (distance, nature-type, etc.) is included. Whereas these types of models are potentially useful for assessment of regional effects of changes in e.g. forest facilities, they are not suited for situations where individual behaviors, choices and preferences change as a consequence of e.g. the infrastructure network or other visitors. Complete disaggregated simulation of individual behavior can be accomplished by models based on Autonomous Agents (Batty 2001 and 2003). Agent-Based Models (ABM) and Individually-Based Models (IBM) are synonymous terms. In the present volume ABM´s are addressed entirely in the context of simulation of recreational behavior. Obviously this type of models is applied to a much broader range of phenomena. Readers interested in generic, computer scientific aspects of multiagent systems (MAS) are advised to consult the extensive volume of literature on the topic. See for instance Ferber (1999).

Agent-based simulations are applying object oriented programming technology. The agents are autonomous because they are programmed to move around the landscape like software robots. The agents can gather data from their environment make decisions based on the information and make decisions in accordance. Each individual agent has it's own physical mobility, sensory, and cognitive capabilities. This results in behavior that mimics real animals and humans. The process of building an agent is iterative and combines knowledge derived from empirical data with the theoretic models of behavior. By continuing to program knowledge and rules into the agent, watching the behavior resulting from these rules and comparing it to what is known about actual behavior, a rich and complex set of behaviors emerge. This type of simulation is compelling because it is impossible to predict the behavior of any single agent in the simulation, but by observing the interactions between agents it is possible to draw conclusions that are impossible using any other analytical process (Gimblett 2002).

The existence of a spatial component in recreational modeling seems quite obvious. Accordingly, there seems to be ample opportunity to use Geographical Information Systems (GIS) and its capability of automation, management, analysis, and presentation of geographic information. As the technology evolves both in terms of analytical and cartographic capabilities as well as user-friendliness, applications of GIS to planning and management will become more abundant (see for instance Bahaire and Elliot-White 1999). A range of studies from the last decade is addressing the advantages of using GIS in relation to studies and management of recreation. See for instance de Vries and Goossen (2001), Powe et al. (1997), Lovett et al. (1997) and Boxall et al. (1996). A widely recognized limitation of most contemporary GIS's and geo-databases is their lack of ability to handle time (see for instance Gimblett 2002 or Geertmann 1999). Phenomena where individuals are interacting in space obviously require this ability.

The recognition of simulation models as a tool for management of recreation in nature has been present since the 70's. Running on mainframe computers, a simulation program written in the General Purpose Simulation System (GPSS) was developed as early as 1973 (van Wagtendonk and Cole 2005). In the following years the program was successfully applied to a range of cases mainly focused on relations between frequency of visitor encounters and spatial and temporal distribution of visitor loads. Cases addressed include wilderness areas, the Colorado River, and the Appalachian Trail. Despite of the simulation endeavors in general being well received by managers a main problem of their further use appeared to be the difficulty and monetary cost of applying and running them. Yet it appears that these early attempts to develop simulation models have played a major role in the development of present day models. For a more comprehensive story of the early development of recreation simulation models, refer to van Wagtendonk and Cole (2005).

Three approaches to simulation of recreational behavior appear; Trace, probabilistic and rule-based models which are applying an increasing degree of model complexity in terms of agent behavior (Manning et al. 2005). In trace-based models agents are programmed to follow entire travel itineraries as registered in the field. Probabilistic models extend this to allow agents to make spatial 'choices' at appropriate points along a trip (typically at network nodes). The choices are based on preprogrammed probabilities of choosing one or the other path option, and are not based on perceptions or other type of sensory information. The early models mentioned above were based on track or probabilistic simulation, where the behavior of the individual agent was defined and parameters assigned at simulation initiation. Accordingly agents are able to move and observe, but not make decisions and change attitude according to what they experience. This approach is further developed into the rule-based models, where the behavior of agents is based on decisions according to behavioral rules applied during the model run (Gimblett et al. 2001). Rules can relate to spatial decisions (e.g. which way to go), changes in attitude (e.g. whether to camp or not), and changes in state-of-mind (See Lace et al. chapter 16).

Typical components of an ABM include: A scheduler, a list of agents, an environment, and a log module. Simulations are performed in discrete time steps. At each time step the scheduler iterates through all agents on the agent list making them perform as they are programmed, for instance they can move, sense their surroundings, and for rule-based simulation systems change activity. At time steps the scheduler further evaluates if agents have reached their goal and therefore can be removed from the model and whether new agents have to be created and can be added to the agent list. Finally the scheduler controls if the overall duration of the model is up and it has to be halted. Creation of new agents takes place at discrete entry points such as park gates, public transportation stops, or parking lots. In rule-based systems the agent performance can include: a) Perception where the agent senses its surroundings (including the infrastructural options, the landscape and other agents), b) comprehension where the perceived information is compared to the agents' goals and abilities leading to a decision, and c) reaction, which typically can be movement and transitions (e.g. changes in state-of-mind or activity). The physical environment of the model can include: infrastructure (roads and paths), location of entry points and points of interest (which can be parts of the infrastructure), land cover/land use, and terrain. In most models of recreational behavior, visitor movement is restricted to a digital network that is composed of edges (connections) and nodes (junctions or locations). In cases where movement is not restricted to network (roads or paths) space is represented as a grid of (raster) cells. Whereas models of visitor movement rarely are based on movement in rasters, models spatial behavior of animals (see e.g. Topping et al. 2003) and pedestrian movement in build-up environment, e.g. evacuation models, often are (see e.g. Castle 2006). Hybrid models where agents are able to switch between on- and off-network behaviors are yet to be developed. The logging module of an ABM keeps track of location and state-of-mind of the agent during the model run. Logging can be performed at a high level of detail for each individual agent at each time step or more coarsely for instance by storing the number of agents present at specified locations at given time intervals. An idealized flow chart of rule-based ABM is shown in figure 1.

Most probabilistic and rule-based simulation models generate input variables based on randomness. Arrival times, journey duration, speed, travel routes etc. are set on the basis of a well described distribution (typically Gaussian) or selected from a set of options (trip itineraries, visitor types etc.). Accordingly the output of consecutive model runs will not be identical. An obvious question will therefore be if a single output from a model is reliable and if not how many runs it will take to reveal a trustworthy, central estimate. Kiser et al. in chapter 20 pose this question and demonstrate a method, which is applied to a model simulating trail encounters in Great Smokey National Park, US. Subsequently the authors discuss the possible difference in precision of model outputs at different locations in the park and for different visitor groups. In chapter 21 Itami et al. also address the required number of replications to obtain an acceptable level of confidence of model results. Further, the spatial and temporal sampling strategies that are required to obtain representative parameters on visitor behavior.

The majority of software options for ABM that have been applied to the cases of the present volume are specifically developed for recreational modeling including RBSim (Lace et al chapter 16, Nelson et al. chapter 17, Itami chapter 18, Wolf et al. chapter 19, Kiser et al chapter 20 and Itami et al chapter 21, MASOOR Jochem et al. chapter 15, Taczanowska et al. chapter 9, Pouwels et al. chapter 14, Elands and van Marwijk chapter 4), and iRAS (Loiterton and Bishop chapter 6). A number of generic systems are available too including commercial packages like Extend (Lawson et al. chapter 10 and Manning 2007) or open source systems like SWARM (http://www.swarm.org) and Repast (http://repast.sourceforge.net). The methods developed

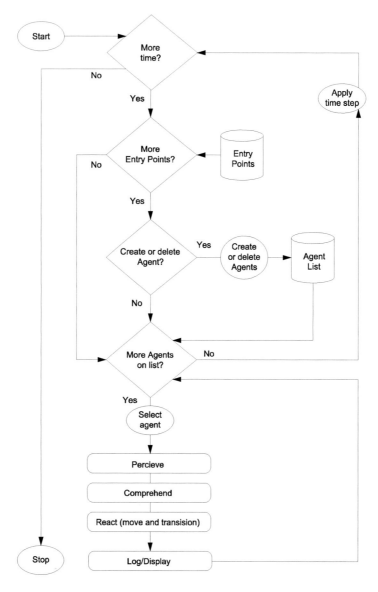

Figure 1. A generic model of a rule-based agent-based simulation model with multiple entry points. Refer to the text for further description.

for studies of probabilistic visibility described in Skov-Petersen and Snizek (chapter 8) are applied in a REPAST-based simulation model.

LEVEL OF DETAIL AND RESOLUTION

Looking at the different recreational simulation models applied in this volume it appears that there are big differences in terms of scale (both in space and time) and the visitor types anticipated. In some cases the setting assessed is vast with relatively low visitation levels; in others areas are small with a high amount of visitors. Likewise trail networks can be very extensive, like in backcountry settings, where as in areas with a high number of visitors, the network is more likely to be dense affording a lot of local spatial choices to the visitor.

In the present book cases ranges from a regional scale like the model of hunters' behavior in Prince William Sound, Alaska (5.513 Km^2 chapter 18 and 20), Juatinga Ecological Reserve, Brazil (800 km^2), and of visitor use in Misty Fjords National Monument, Alaska (615 km^2, chapter 17), over models of smaller densely visited forest areas in Holland (35 km^2, chapter 3, 13, 14 and 15) and Austria (15 km^2, chapter 9), down to a botanical garden (chapter 6) and a harbor area (chapter 18) in Australia.

It appears that the level of detail of the models seems to be higher in smaller case areas. In extensive areas the input parameters to the models includes only e.g. entrance schedules, itineraries, and to some extent activity types whereas in smaller areas with a high visitation level (relative to area) more focus is put on details of perceptive (Skov-Petersen and Snizek chapter 8) and cognitive processes (Elands and van Marwijk chapter 4, Taczanowska et al. chapter 9 and Loiterton and Bishop chapter 6), and mechanisms behind local, spatial choices (Jochem et al. chapter 15). The finer the degree of detail of the phenomena modeled the higher the need for application of approaches related to Artificial Intelligence (AI). To some extent it is sufficient for agents of extensive models to be active, not reactive. The movement patterns can be preset at agent-instantiation and need only adjustment at a fairly low frequency, during an agent's lifetime, i.e. the agent's overall, spatial objectives are fixed to a higher extent compared to smaller scale situations where visitors tend to be more 'browsing' (or rule-based), pursuing their recreational goals in a less stringent manner. Examples of agents that makes spatial choices at every node of the traversed trail network, as opposed to application of agent's following fixed, preset itineraries, is provided by Jochem, et al. (chapter 15). In this respect the techniques and rationales of high detailed agent-based models approaches more and more those applied in relation to pedestrian models in urban systems (see for instance Castle 2007, Batty 2006, Castle and Crooks 2006, and Benenson and Torrens 2004).

Most of the chapters of the present volume are primarily addressing visitor behavior and visitor-visitor interaction and encounters. Protection of biodiversity is a frequent quoted motivation for planning and management visitor landscapes models aimed at visitor-wildlife too. In Pouwels et al. (chapter 14) an ABM of visitor behavior is used to evaluate the disturbance pressure in corridors along trail edges. The level of disturbance is loosely coupled to another ABM aimed at population development of Skylark. This way, areas particularly sensitive to disturbance are identified. A similar approach is employed by Nelson et al. (Chapter 17) where an ABM of visitor behavior in Misty Fjords National Monument (Alaska) is used to evaluate wildlife disturbance zones along frequently used trail segments. Wildlife behavior is not part of the ABM as such. In a different way, visitor-wildlife interaction is a serious issue examined by Lace et al. (Chapter 16). Here an ABM is used to model success rates of black bear hunting in Prince William Sound (Alaska). It is the hunters' movement (in boats on the sound) that is in focus, rather than the bears' movement.

Smith et al. (2002) have worked with another agent-based framework that is based upon the Belief, Desire, and Intention (BDI) theory of intelligent agents and implemented in Visual Basic. Agent´s are initialized with a set of goals they want to achieve and a library of plans that stores knowledge about landscape management actions. Each plan is like a 'if-then' rule where the 'if' is a set of conditions that need to be matched to agents beliefs for execution and the 'then' part, which describes an outcome when the plan is executed. Agents act by looking for plans that match their goals and their current beliefs. At each time step, for each outstanding goal they examine their plan library and choose the best plan, if any, which satisfies the goal. Plan choice is influenced by the agent's current belief state and values. Plans can be linked together in a hierarchy providing flexible and responsive reasoning to the agents. Agent-based models include agents that are able to communicate both by expressing its own knowledge and by storing and applying received information (See for instance Weiss 1999 and Ferber 1999).

FROM THE REAL WORLD, TO DATA, AND FURTHER ON TO SIMULATION

Simulations are built on models that are ideal representations of real-world phenomena. The ideals of the models applied have to be based on deductive theories, which again must be supported by inductive arguments obtained through measurements in situ. Accordingly, theories are important because they facilitate both the formulation of the rationale behind the models, and the design of the monitoring and surveying schemes from which model parameters are obtained. Itami et al. (chapter 21) explore the spatial and temporal sampling of visitor behavior required to develop probabilistic simulations for accurate forecasting and decision-making.

In the context of simulation of individual human behavior (including recreation) it has often been acknowledged that Geographical Information Systems (GIS) are obvious data sources and to some extent platforms for spatial modeling and analysis (see for instance Castle and Crooks, 2006, Manning et. al., 2005 and Gimblett, 2002). To a vast extent simulation models are founded on data stored as digital maps; either as raster grids (where the mapped area is divided into evenly sized square cells) or features (where real-world objects are represented as points, lines or polygons). In both cases thematic characteristics of the objects are stored as attributes to the raster cells, groups of cells, or to individual features. A generally recognized problem is that cartographic data are expected to be generic and objective. Accordingly data are frequently applied outside their scope - for purposes they were not originally meant for – neglecting the fact that they are only representations of the World, created by a specific organization, for a specific reason (Longley et al. 2005 and Monmonier 1996). In relation to simulation of recreational behavior geographic base data applied often include land cover (or land use), infrastructure (trails or roads) and terrain. A regularly used data source is 'generic', digital topographic maps made for other purposes (for instance strategic applications) which accordingly need further refinement before they can be used for individual-based simulations. The recoding of 'raw' base-maps to facilitate recreational simulation requires a thorough, theoretic background and often further field reconnaissance and surveying. As examples of base data being refined for agent-based simulations in the present volume (Taczanowska et al. Chapter 9, Jochem et al. Chapter 15, and Skov-Petersen and Snizek Chapter 8).

In chapter 2 Doxtater argues, on a theoretic basics, that much more consistence categories of the affordances of nature settings are required to enable evaluation of recreational qualities and thereby the activities that can take place, and the level of appreciation of the visitors. Natures' affordance is discussed in categories of Wayfinding, Task-performance, Social territories, Cultural expressions, and Visual and non-visual aesthetics.

Between perception of objects of the natural setting and the visitors' sense making is an informal learning process. In chapter 3 Steven applies a cognitive mapping instrument, called 'Environmental N' to track the linkage between visitors' observations (stimuli) to the consequential thoughts and feelings in Australias Kakadu National Park. By locating recordings of stimuli/thoughts/feeling on sketch maps a typology of recreational experiences is established. Assessment of cultural impact and predicting the efficiency of planned or un-planned cultural changes has received little attention in relation to human/landscape interaction. This is partly due to the lack of monitoring and modeling techniques applicable in such situations. In chapter 12 Sinay et al. demonstrate and discusses how application of a Bayesian Belief Network approach based on an empirical study in Juatinga Ecological Reserve, Brazil can be used to study and understand cultural changes in an ethnic, cultural, and social heterogeneous region.

Individual-based models are constituted by types of user- or activity-groups, which perform different activities, prefer and perceive the environment differently, and sped different amounts of time while recreating in nature. The delineation of activity-groups applicable to the simulation at hand is of course of prime interest. In chapter 4 Elands and van Marwijk study the interrelation between nature experience, environmental values and visitor behavior based on theoretical reflections. A simulation model based on the resulting four activity groups is demonstrated.

While simulation models are often claimed to be highly successful in management and planning of visitor landscapes, studies and discussion of the actual effect in terms of better management strategies and better plans are rare. A major requirement for pursuing this is formulation of a typology of motivations and norms stakeholders might have – given the theme, process and type of participants of the planning or management situation in question. In chapter 13 Pröbstl et al. set up a framework for such a typology, which is applied to three case studies of the use of simulation models in Austria, England and France.

This book has been strategically organized to provide the reader with background to Wayfinding and Prediction of Human Behavior in Visitor Landscapes. This section is imperative as it provides a brief background to work that examines both the theory of wayfinding and psychological responses of individuals in visitor landscapes. This book links monitoring to simulation to the underlying reasons of management and protection of visitor landscapes. Finally the concluding section of this book provides insight into what has been learned in applying monitoring systems and data collection to simulation modeling to resolve natural resource based planning and management issues in visitor landscapes.

LITERATURE CITID

Bahaire, T. and Elliot-White, M., 1999. The application of Geographical Information Systems (GIS) in sustainable tourism planning: A review. J. of sustainable tourism, Vol. 7 nr. 2 1999.

Batty, M. 2001. Agent-based pedestrian modelling. Editorial. Environment and Planning B. Vol. 28, pp 321-326.

Batty, M. 2003. Agent-based pedestrian modelling. In Longley, P. and Batty, M. 2003. Advanced Spatial Analysis. ESRI Press.

Benenson, I. and Torrens, P., M. 2004. Geosimulations – Automata-based modeling og urban phenomena. Jonhn Wiley & Sons, ltd. 287 pp.

Boxall, P.C., McFarlane, B., and Gartell, M. 1996. An aggregate travel cost approach to valuing forest recreation at management sites. The forest chronicle. Vol. 72, No 6. Pp. 615-621.

Castle, C.J.E. 2006. Developing a prototype agent-based pedestrian evacuation model to explore the evacuation of King's Cross St. Pancras Underground Station. CASA working paper nr. 108. URL: http://www.casa.ucl.ac.uk/working_papers/paper108.pdf. Last accessed July 2007.

Castle, C.J.E. 2007. Guidelines for assessing pedestrian evaluation software applications. CASA, University college London. 36 pp. URL: http://www.casa.ucl.ac.uk/publications/workingPaperDetail.asp?ID=115. Last access July 2007.

Castle, C.J.,E. and Crooks, A.,T. 2006. Principles and concepts of agent-based modeling for developing geospatial simulations. CASA, University college London. 59 pp. URL: http://www.casa.ucl.ac.uk/publications/workingPaperDetail.asp?ID=110. Last access July 2007.

Cole, D. 2005. Why model recreation use? In Cole, D. (compiler). 2005. Computer simulation modeling of recreation use: Current status, case studies, and future directions. USDA, General technical report RMRS-GTR-143.

de Vries, S and Goossen, M., 2001. Planning tools for the recreational function of forests and nature areas. Paper presentes at the 12'th Euro Leisure-congress "Cultural events and the leisure systems". Amsterdam, The Netherlands.

Ferber, J. 1999. Multi-agent systems. An introduction to distributed artificial intelligence. Addison-Wesley. 507 pp.

Geertmann, S. 1999. Geographical information technology and strategic physical planning. In Stillwell, J., Geertmann, S., and Openshaw, S. 1999. Geographical information and planning. Springer.

Gimblett, H.R. 2002. Integrating Geographic Information Systems and Agent-based Modeling Techniques for Simulating Social and Ecological Processes. In Gimblett, H. R. (ed.) Integrating Geographic Information Systems and Agent-based Modeling Techniques for Simulating Social and Ecological Processes, Oxford University Press. 327 pp.

Gimblett, H.R., Richards, M.T. and Itami, R.M. 2001. RBSim: Geographic simulation of wilderness recreation behavior. Journal of forestry. 99(4): 36-42.

Gimblett, R., Roberts, C., Daniel, T., Ratcliff, M., Meitner, M., Cherry, S., Stallman, D., Bogle, R., Allred, R., Kilbourne, D., and Bieri, J. 2002. An intelligent agent-based model for simulating and evaluating river trips scenarios along the Colorado River in Grand Canyon National Park. In Gimblett, R (ed). Integrating Geographic Information Systems and Agent-based Modeling Techniques for simulating social and ecological processes. Oxford university press. ISBN0-19-514336-1. 327 pages.

Hornback, K., E., and Eagles, P., F., J. 1999. Guidelines for public use measurement and reporting at parks and protected areas. IUCN, Gland, Switzerland and Cambridge, UK. 90 pp.

Itami, R.,M. 2005. Port Campell national park, Australia: Predicting effects of changes in park infrastructure and increasing use. In Cole, D. (compiler), 2005. Computer simulation modeling of recreation use: Current status, case studies, and future directions. USDA. Forest service. Rockey mountain research station. General technical report RMRS-GTR-143. 75 pp.

Jensen, F., S., 1999. Forest recreation in Denmark from the 1970s to the 1990s. The research series, vol 26. Danish forest and landscape research institute. Hoersholm, Denmark. 166 pp.

Kajala, L., Almik, A., Dahl, R., Dikšait, L, Erkkonen, J., Fredman, P., Jensen, F. Søndergaard, Karoles, K., Sievänen, T., Skov-Petersen, H., Vistad, O. I. and Wallsten, P. 2007. Visitor monitoring in nature areas – a manual based on experiences from the Nordic and Baltic countries. TemaNord 2007:534. Editor: Liisa Kajala. ISBN 91-620-1258-4. 202 pp.

Longley, P., Goodchild, M., Maguire, D., J., and rind, D.,W., 2005. Geographical Information Systems and Sciences (2'nd ed.). John Wiley & Sons ltd. 517 pp.

Lovett, A.A., Brainard, J., and Bateman, I.J. 1997. Improving benefit transfer demand functions: A GIS approach. J. of Environmental Management, 51, pp 373-389.

Manning, R.E., Itami, R.M, Cole, D.N., and Gimlett, R.H. 2005. Overview of computer simulation modeling approaches ad methods. In Cole, D. (compiler), 2005. Computer simulation modeling of recreation use: Current status, case studies, and future directions. USDA. Forest service. Rockey mountain research station. General technical report RMRS-GTR-143. 75 pp.

Manning, R.E., 2007. Parks and Carrying Capacity. Commons without tragedy. Island Press. 313 pp.

Manning, R.E. and Potter, F.I. 1984. Computer simulation as a tool in teaching park and wilderness management. Journal of environmental education. 15(3): 3-9.

Monmonier, M. 1996. How to lye with maps (2'nd ed.). University of Chicago Press. 207 pp.

Powe, N.A., Garrod, G.D., Brunsdon, C.F., and Willis, K.G. 1997. Using geographical information system to estimate an hedonic price model of the benefits of woodland access. Forestry, Vol. 70, No. 2. Pp 139-149.

Romano, D. Scarpa, R., Spalatro, F. and Viganò, L (2000). Modelling determinants of participation, number of trips, and site choice for outdoor recreation in protected areas. Journal of Agricultural Economics, 51(2):224-238.

Smith, L., Itami, R.M. and Bishop, I.D. (2002). An architecture for modelling individual behaviour and landscape scale outcomes in an intelligent agent-based simulation of environmental management. IEMSs 2002 - Integrated Assessment and Decision Support, Lugano, Switzerland, International Environmental Modelling and Software Society (iEMSs), 2:214-219.

Termansen, M., McClean, C. and Skov-Petersen, H. 2004. Recreational Site Choice Modelling Using High Resolution Spatial Data. Environment and Planning A 2004, volume 36(6) June, pages 1085 – 1099.

Topping, C.J., T.S. Hansen, T.S. Jensen, J.U. Jepsen, F. Nikolajsen and P. Odderskær. 2003. ALMaSS, an agent-based model for animals in temperate European landscapes. Ecological Modelling 167: 65–82.

van Wagtendonk, J.W. 1978. Use of a wilderness simulator for management decisions. In Shechter, M., Lucas, R.,C. (eds). Simulation of recreational use for park and wilderness management. Baltimore, MD. John Hopkins University Press. 197-203.

Van Wagtendonk, J.W. and Cole, D.N. 2005. Historical development of simulation models of recreation use. In Cole, D. (compiler), 2005. Computer simulation modeling of recreation use: Current status, case studies, and future directions. USDA. Forest service. Rockey mountain research station. General technical report RMRS-GTR-143. 75 pp.

Weiss, G. (ed). 1999. Multiagent systems. A modern approach to distributed artificial intelligence. The MIT press. 619 pp.

CHAPTER 2

WHAT VISITORS "DO" IN RECREATIONAL LANDSCAPES: USING CATEGORIES OF AFFORDANCES FOR EVALUATION, DESIGN AND SIMULATION

Dennis Doxtater

Abstract: Contrasting with frequently more normative and holistic views of landscape meaning, and omitted in published approach summaries are recent appropriations of J.J. Gibson's affordance theories in applied fields of design even including larger settings at the scale of landscape. Rather than assume a holistic complexity, one can begin with specific events or situations where people are actively involved in "doing" something as a transactional interaction between themselves and the physical setting. The psychological ability to comfortably dissociate one event from another diminishes the need to more holistically perceive or use the landscape. Object, space, action are the necessary ingredients to describe and eventually simulate affordances. Instead of considering holistic places or scales of places as composites of affordances, social science literature - from psychology to culture - suggests that certain kinds of affordances tend to experientially string together at larger scales. The categories of wayfinding and task performance are immediately understandable in their function. Visual and non-visual aesthetics contrasts the intrinsic or even innate aspects of natural form with socially motivated experiences, which themselves can be distinguished as simple territoriality and symbolic or cultural expression. While significantly reduced in apparent complexity from normative and holistic approaches to landscape, a reasonable assessment and simulation of all categories of specific "doings" nonetheless can achieve a rich understanding of visitor experience for practical uses by designers, simulators, and managers.

Key Words: Affordances, Wayfinding, Task-performance, Social Territories, Cultural Expression, Visual/Non-Visual Aesthetics

THE AFFORDANCE APPROACH

Even though the physical scale of recreational landscapes can obviously be quite large with an equally large set of possibilities of visitor activities, the idea of "affordances" may allow us to focus more efficiently on specific realities of experience. J. J. Gibson's (1966) concept of affordances is quite straightforward as used in applied processes. As an example of more typical small scale studies of affordances, yet with some complexity, one can cite Boschker et.

Figure 1. Climbing a rock wall. (Photo by Steve Hawkshaw)

al.'s (2002) comparison of the way experienced and inexperienced rock climbers understand physical form (holds on a climbing wall) as opportunities for, and dangers of, action. They speak of abilities to use "fine" or "course" grain affordances, "chunking" and "nesting" of information as they think about and execute a route up the wall. The inexperienced climbers attended to relatively superficial, structural characteristics of the climbing wall, rather than to the specifically useful details and routes necessary for a successful climb. Moving to a larger scale than that of rock climbers, and coming out of a center for urban and regional studies is a comparison of children's affordances in a continuum of settings from the rural to the urban (Kytta 2002). Here the researcher asks children about thirty-four different kinds of physical activities, such as running on flat surfaces, jumping on certain things, swinging, playing with water, and the like. Each is referred to in the manner of "affords skipping" or "affords skateboarding", for example.

While somewhat predictable, more affordances were possible in the rural settings with their greater landscape attributes.

At a similar scale, but addressing the largely unresearched assertions of New Urbanism, we find a recent planning dissertation using Gibson's ideas of affordances, along with place theory of Canter (1991). This comparison of New Urban town centers (Disney's new town, Celebration) and conventional shopping centers (Bohl 2004), focuses on twenty-five affordances that fuel the statistical study. While some of these kinds of setting integrated activities seem vague as affordances, e.g. "this is a center of activity and meeting place for the community", others are much more to the point, e.g. "this would be a good place to sit and watch people", or "this would be a good place to ask for signatures in support of a new local law" (ibid:127). Often when researchers ask questions of landscape users, the labels of experience tend to be more implicit than specific. In a study of greenway trail users in Chicago (Gobster 1995), for example, informants were queried about positive or negative effects of physical elements, e.g. "scenic beauty", "smooth trail", "safe", "close to home" etc. While some actual affordance or experience can be connected to these typological labels, one cannot be certain about just when or how long specific aspects of the physical setting were actually being used. A more recent Danish study of recreation in forests (Jenson & Koch 2004) also raises questions about specificity; it asks respondents, via mail, about nineteen nominally separate experiences, e.g. "going for a walk", "enjoying nature", "studying nature", "going for a drive", "sitting still", "exercising". While users might rely on these terms to define their reason to go to the forest, their actual experiences there might well be much more specific.

People are usually aware of what their

Figure 2. Reading while walking (Sabino Canyon, Arizona)

attention is currently focused on. Holists would probably argue that a good deal of environmental attention is being paid to things not immediately connected to what people report doing at specific times and places. Nevertheless, there may be little evidence that this peripheral or background use of the physical setting is very important. Even though people are always in one sort of physical setting or another, a good deal of the time the environment plays only a minor background role compared to a more purely "medial" exchange of information. We spend a lot of time talking or listening to other people talk, both in real encounters and in those extended by electronic media. We also may be reading a book, watching a T.V. program, on the internet, or just thinking about something, none of which again seriously involves larger physical settings and their affordances. It is also true that we spend about a third of our time sleeping and dreaming.

In this writer's opinion most designers and many landscape theorists radically overestimate the actual amount of time that the physical environment significantly holds the attention of human beings. Even though managed landscapes are billed to be much more interesting, in this regard, it is a good bet that a not inconsequential portion of the time visitors are doing things that do not heavily depend on physical environmental affordances.

CATEGORIZING WHAT PEOPLE DO IN VISITOR LANDSCAPES

As one attempts to operationalize important environmental affordances in evaluation, design or simulation, there appear to be certain logical or informational differences between kinds of things people do with physical settings. Typically when professionals program or evaluate settings they begin with very open ended observations or questions. Often these inquiries are more in the vein of a "walk-through" where users are encouraged to report about various aspects of the setting that they think important, for whatever reason. If one attempted to frame the evaluation in terms of what people "do", where would one start? When walking around a setting, or viewing a film, one could ask what users do or have done in certain part of the scene. Yet here too a bias would exist toward a sort of site determinism. Again much of the time visitors or users won't be using the setting very intensely at all, and it follows that any recollection of setting specific episodes in the context of the overall stream of experience might be difficult. How can one be sure that all the significant, shared environmentally based experiences are recorded?

It is also true that most evaluations of physical settings are in the form of case studies often with resulting guidelines (Francis 2001, Cooper-Marcus and Sarkissian 1986). The overall physical setting is again the primary point of departure. While case studies obviously generate good usable information about how people use settings, nevertheless, their somewhat holistic or at least *ad hoc* nature tends to make cross study comparison more difficult. More importantly, without some overriding understanding of potential kinds of user experiences, risks of omission could be high.

This paper introduces the possibility of categorizing different kinds of environmentally important affordances as a means of facilitating important processes of design, simulation, and eventually new significant research. The distinction of affordances for purposes of *wayfinding, task-performance, social territories, cultural expression* and *visual/non-visual aesthetics* can initially be useful in prompting people to recall all the different kinds of experiences one did or might do in a setting. This taxonomy can then be used to structure information about environmental experiences in subsequent processes of design schematics and simulation, most of which involve the management of input from multiple user groups.

This way of thinking is not at all dissimilar to McHarg's (1994) layers of independently considered components of landscape. His layers, however, are ecologically or infrastructurally constituted. These are quasi-systematic physical aspects of the environment operating somewhat independently of each other, competing and sometimes cooperating in the appropriation of the landscape; see for example Schrijnen's (2000) discussion of how urban infrastructure relates to more natural or green landscapes in the Netherlands. Most aspects of natural ecosystems or even built infrastructure, e.g. engineered structures or mechanical systems, are not primarily created as active affordances for human beings. Even though these processes eventually affect every person and animal on the planet, a very large part of our physical settings are not actively attended to as part of what we "do". The more ethical interest in making ecology and infrastructure "affordable" or bringing aspects to a level of actual experience is a relatively minor subset of mapping significant setting based experience, even though some perceptual and cultural studies of the landscape work in this dimension.

While McHarg's layers are primarily ecological and infrastructural, adaptations of this methodology often include a more clearly human dimension, ranging from the psychological to historical or cultural contents. While not really developed as affordances *per se*, this essentially experiential layer might itself be composed of sub layers of categorical affordances potentially stringing across the same overall landscape setting. They may or may not actively interact with aspects of the ecological or infrastructural layers, depending again on what people are doing.

WAYFINDING

Perhaps the most fundamental kind of affordance helps the individual maintain his or her orientation in physical world. Following Lynch's (1960) and Appleyard's (1970) early work, this has been one of the

Figure 3. Planning the route" (Sabino Canyon, Arizona)

most easily understood, and most specific kind of environmental information. It is also true that understanding the objects (landmarks), spatial relationships (paths or routes), and action of *wayfinding* exemplifies the way smaller affordances can string together in mapable structures across the larger landscape, both urban and natural. Much of the *wayfinding* application has naturally occurred in regard to complex, intensely used architectural settings, e.g. Authur and Passini (1992). Case study examples appear in Baskaya et. al.'s (2004) mapping of user polyclinic affordances in *wayfinding* critical medical facilities. More recently, Space Syntax has also been adapted to architectural settings such as hospitals (e.g. Haq & Zimring 2003), illustrating how this often larger scale methodology might be adapted to visitor landscapes.

Coming out of Space Syntax methodology as well are studies of "least-angle" strategies of *wayfinding*, where users tend to rely upon the spatial affordances of straightest lines between known aspects of a setting (Hochmair 2005, Dalton 2003). Differences in individual strategies of *wayfinding* by different kinds of users, as emphasized earlier in Appleyard and Passini are also maintained in recent studies. Lawton and Kallai (2002) combine both cross-cultural and cross-gender characteristics in their study of men and women in Hungary and the United States. Men in both countries preferred global reference points, while women were more likely to use strategies of route information. Cultural differences were less gender specific and clearly involve social meanings of environmental experience, such as personal safety, and fall into other *non-wayfinding* categories. This sorting out of distinct affordance categories was also problematic much earlier for Lynch and other cognitive mappers.

In a similar vein, but from the business

Figure 4. Checking the map (Sabino Canyon, Arizona)

and management field, we find a study of men's use of landmarks and women's use of verbal messages to find destinations in a shopping mall (Chebat, Chebat, & Therrien 2005). This is a relatively clean evaluation of *wayfinding* information as a specific category, although some meaning and preferences of consumer products and advertising enter into the researcher's agenda. As will be illustrated below, one can independently distinguish such object, spatial and action components of distinct kinds of affordance as *cultural expression*, not *wayfinding*.

Even though most *wayfinding* studies target more intensely used and perhaps more critical building and urban settings, some work moves in the direction of visitor landscapes. In spite of occurring in a small-scale totally created artificial setting, Vosmik & Presson's (2004) research on children's map-reading abilities has implications for any setting, built or natural. Soh & Smith-Jackson (2004) look at map reading in relation to individual differences and culture as users find their way along trails in recreational landscapes. Many visitor landscapes involve the use of maps by not only children but others with perhaps different information processing preferences.

Signs are often an important part of *wayfinding* for first-time visitors, whether in hospitals or landscapes. Thompson et. al. (2004) focuses primarily on signage in her evaluation of how visitors know where to go on arrival to a forest recreation area in Britain, and how they map routes to various sites in the larger setting. This work particularly fits the present context because of their definition of specific situations associated with: "pre-arrival information (Where do I want to go? What is the site called")"; "approach routes (How do I get there?)"; "finding the site entrance (Is this the right place?)"; and "arrival on site (What can I do)" (ibid 40).

Returning to affordance literature, we first see the detailed example what the rock climbers (Boschker, Bakker & Michaels 2002) appear to be doing as *wayfinding*. Yet if we look closer, what connects the handholds across the climbing wall is not their orientation to each other *per se*, but how they are used to physically support the movement of the climber. This is an example of *task performance* or even *non-visual aesthetics*. It is also true that neither Kytta's (2002) listing of children's activities, nor Bohl's (2004) indices of New Urban affordances contain *wayfinding* experiences.

The only possible *wayfinding* experience included in Jensen & Koch's (2004) survey of Danish forest users might be found in their "followed marked paths". Presumably this is a *wayfinding* experience based on reading signs, although other kinds of content, e.g. such as an interpretative time line, might be communicated by marked paths. The researchers do not include an experience such as "got lost", which even if reported would not specifically define their attempted affordance use of the physical setting at the time. Gobster's (1995) list of attributes of Chicago greenway experience, not dissimilarly from the Danish forest study, only includes the negative issue of "poor signage". Again this may or may not be *wayfinding* affordance.

Clearly much work is needed to develop good, specific descriptions of object, space and action relative to those instances where visitors are "doing" *wayfinding*. What are the specific object features in visitor landscapes being used as landmarks, including aspects of the natural and built setting, and signage? What are the specific spatial strategies the user is aware of in trying to cognitively map orientations between landmarks? How does the user move around in the setting in a focused effort to better understand landmarks and spatial orientations?

TASK-PERFORMANCE

In considering the work that people do, it is clear, for example, that affordances of the physical setting may be more important in the case of a construction worker than for an accountant. In experience where the physical setting is actively involved, not just as background, we can describe object, space and action characteristics that appear to be unique to the category of *task performance*. Objects being used for work are primarily being experienced in terms of some physical manipulation, accommodation or capacity, or involvement of measurable energy. In terms of spatial relationships, cognitive work maps are abstract understandings of patterns of efficient work sequence such as in the case of the experienced rock climbers. The way a systems engineer or a construction worker plans the most efficient route of moving materials around a warehouse or building site are clear exam-

Figure 5. Getting the kids ready (Sabino Canyon Arizona)

ples here. The actual movement or conduct of work among objects in a mapped space fills out the affordance characteristics. Here adjacency (distance) and movement using both object and spatial characteristics has the potential to come to one's attention usually as fatigue or production inefficiency.

Park planners do not obviously intend for visitors to be frequently aware of doing work in managed landscapes. Nevertheless if a path is too steep, in its object characteristics, people will shift their attention to this experience and likely perceive it negatively. More spatially, if the path system is not efficiently laid out, people could become aware of expending too much energy to do the other things they want to do. If the adjacency between any two points of interest is too far, the actual movement might be more immediately attended to as work. In an architectural example of a large museum, Jeong & Lee (2006) identify physical fatigue along with other experiences.

Certainly one of the most frequent kinds of task performance visitors do in recreational landscapes is "using the restroom" (physical exercise will be discussed below as a non-visual aesthetic). How far one has to walk under various kinds of restroom situations (adjacency), and waiting for a stall or urinal (capacity) of this work are different than the *wayfinding* affordance of "knowing where the restrooms are". Smaller scale, more ergonomic or workstation affordances include various experiences using the toilets, stalls, diaper changing tables, washbasins, and more. As Aoki (2005) points out in the case of public park restrooms, the daily maintenance of these facilities is a major *task performance* demand for staff. Her article speaks as well to issues of personal safety in and around restrooms from social deviants in secluded and/or poorly lit areas of parks. Here one distinguishes two kinds of safety in landscape settings. We will wait until the category on *social territories* to include getting mugged or raped as negative affordances. In *task performance*, if paths, stairs or railings cause or allow people to fall, or if people slip on flat surfaces when wet, these are negative experiences in this category. Essentially "aesthetic" experiences like those done by

Figure 6. Going off the bridge (Sabino Canyon, Arizona)

Finnish children (Kytta 2002), including skateboarding, and activities like mountain biking or skiing among other populations, all include the high probability of some negative *task performance* event, i.e. "falling" or "crashing", which are of concern to recreational users and managers. Jones & Graves (2000) evaluate skate board parks, but as we will see in the *social territories* section, the primary issues are social relationships between skate boarders and neighbors, rather than *task performance* safety.

What, however, would one call the safety issue of separating visitors from grizzlies at campgrounds, as in Creachbaum, et. al (1998)? Unlike the situation where muggings or rapes are being caused by social deviants at park restrooms and the lack of social, territorial surveillance we can't expect grizzlies to be on their good behavior when coming into contact with visitors, or to respect the surveillance of groups of visitors. It needs to be physically insured that the two users don't wander into each other's paths, or in the last resort, such as in zoos, impenetrable barriers need to be created between the two. These *task performance* strategies of safety also occur, of course, in separating some humans from others where the social control of public surveillance isn't effective. Preventing theft in visitor landscapes might rely on both kinds of affordances, one the *social territorial* public presence of others, or two the *task performance* of locking up valuable objects. In the latter case, one would to want to understand the object (cutting a lock), spatial relationships (using a route that hid the presence of the thief), and movement (actually avoiding discovery during the event) characteristics of the affordance. One should distinguish using a clandestine route from a more general knowing where things are in *wayfinding*. Additional task performance aspects may also be important as the thief moves the goods, heavy or awkward as they might be, some distance in the act of getting away.

Some accessibility for the physically disabled is of course mandated in most landscape recreational settings. *Task performance* aspects of these affordances clearly focus on the way the physical chair, whether motorized or manually powered, works with paths and ramps as objects (see for example Longmuir, et. al. 2003). Are they too steep? Are they wide enough? Does the surface allow adequate traction? How can chairs fit into auditorium seating? Use of restrooms, of course, has more universal conditions. In terms of spatial structures, one must again distinguish work patterns from *wayfinding*. "Knowing where the accessibility route is" involves different kinds of information from "understanding the easiest route to manipulate the wheelchair or scoter".

Adjacency and capacity attributes are always present when numbers of visitors arrive in or use their cars in relation to visitor landscapes. While "knowing where the parking lots are" is *wayfinding*, being able to find a spot (capacity) close by (adjacency) are clearly *task performance*. Certainly much is known about parking lot layout and capacities in public ordinances or guidelines. Yet McPerson's (2001) study of park-

Figure 7. Keeping clear of the mountain lions (Sabino Canyon, Arizona)

ing in the city of Sacramento questions the accuracy of these conventions, citing an overbuilding of capacity, and even proportionally greater under use during peak periods. This report, however, is really about shade from trees in parking areas. Thus two categories of affordance are potentially at play here, one *task performance* ("having an adjacent space to park in"), the other *visual and non-visual aesthetics* ("enjoying the beauty of green areas", or "keeping cool on a hot day"). These are separate affordance experiences within the larger gloss of "parking", and might well be understood for their own realities.

Then of course is the whole efficiency and safety of traffic movement, pedestrian separation, and the like. Crankshaw (2001) outlines convenient parking locations and adjacent pedestrian route issues in historic downtowns, ideals not unrelated to those in New Urbanism. In landscape settings one does find occasional pieces like Widmer and Underwood's study of recreational boat traffic and moorage patterns in busy Sydney Harbor (2004). Maneuvering boats under the conditions described may be even more demanding on affordance attention than auto driving in urban streets.

Finally, it seems true that *task performance* affordances are also relatively easy to document in their dimensions of object, space, and movement. People, it seems, are usually aware of and can describe their work experiences. But as discussed above, this does not mean that evaluators, designers and managers don't often inappropriately lump specific task performance affordances in with other kinds of experience.

SOCIAL TERRITORIES

The simplest kinds of social spaces in human behavior are not that different from territoriality in animals generally. Yet some researchers conflate these more immediate, occupation based social uses of space with symbolic meanings people attach to their settings, e.g. using terms like "symbolic territories". For present purposes these more complex symbolic phenomena are included in the final affordance category of *cultural expression*. In terms of the simplest, non-symbolic uses of space, one refers to writings on concepts like privacy, personal space, human territoriality, defensible space, or proxemics ideas developed in 70's and 80's environmental psychology. In design related practice, however, one needs to distill these often broad concepts down to more specific affordances or subcategories.

Dangerous space

The easiest to define is Oscar Newman's "defensible space" (Newman 1996, Tijerino 1998), already introduced as surveillance control of deviant behavior in the restroom example above. Defensible space, when effective, eliminates the negative affordance or experience of being raped or mugged. In the best of all worlds, people will be largely unaware of defensible space and will be doing other things in the setting. The true affordance focus is really on "dangerous" space, or those experiences where people think about or actually are unfortunate participants in these events. Most current defensible space literature involving landscape spaces exists in urban settings. Freestone & Nichols (2003) describe the "reserve community space at the rear of residential lots" initially intended for recreation, but eventually perceived as dangerous through disuse and lack of maintenance. One interesting issue here is the distinction between perception and real activity of the negative affordance. If in fact people are not actually involved in deviant acts in the setting, is this really an "affordance"? Might not these images of possible affordances be better subsumed in the symbolic category of *cultural expression*. As discussed in the author's evaluation of La Paz residence hall on the University of Arizona campus (Doxtater 2005), a published example of New Urbanism in Katz (1994), much of these places' apparent sociability may function as a sort of fictive surveillance an opposite symbolism to Freestone & Nichol's example.

We know from criminal justice literature that a fair amount of crime happens in large recreational landscape settings such as national parks. Gilbert (2000) cites data from the year 1995 in the United States showing that 4,700 felonies and 31,000 misdemeanors occurred in our national parks, among which were 13 homicides, 34 rapes, and 164 aggravated assaults. This article, however, does not break out those crimes that may be due to *dangerous space*, rather than to non-environmentally facilitated situations between drinking buddies, feuding biker groups, marital partners, and the like.

Exclusive Space

Being exclusive or private forms an important variable in Bishop & Gimblett's (2000) analysis and simulation of certain kinds of recreation visitors in Sedona, and the Grand Canyon Arizona, Gimblett, et. al. (2001). The affordance is negative when a visitor's exclusivity is threatened by others. When actually alone or enjoying relative exclusivity in a setting, individuals or small social groups will be doing other things, many of which may not be very dependent upon the physical setting. Probably the visitor who seeks the most exclusion is the solitary wilderness hiker. In this case, the non-social affordances, once other people are avoided, may well be linked to *visual & non-visual aesthetics* or symbolic, ecological, or religious meanings as *cultural expression*.

At the opposite end of the spectrum of "avoiding other people or groups", and achieving seclusion, are negative affordances of "feeling crowded" in close

Figure 8. Being alone

quarters with other people. Most visitor experiences in landscapes are comparatively spacious in comparison to being in elevators or corridors in larger buildings or busy urban sidewalks. Nonetheless, situations might occur where people in busy nodes of focused experience, such as interpretative presentations, might become socially uncomfortable by the too close proximity of too many other people.

Reports of these negative affordances might commonly include "there are too many people", "you can't turn around", or of course "it's too crowded". Here again the gloss of "crowding" may conceal a finer discrimination of more specific affordance experience. If at a viewing platform at a scenic overlook some people cannot get an adequate view because they can't see over taller folks at the rail, then this is *task performance, cultural expression or visual & non-visual aesthetics*. The design solution is not necessarily to thin out the numbers of people at the overlook at any one time, but to devise a way for viewing from elevated positions away from the rail. Even though users might report the overlook problem as "being too crowded" the experience isn't really social exclusion or spacing, but "being able to see the view".

Ranked Space

In addition to being exclusive, some ani-

Figure 9. Its too crowded (Sabino Canyon, Arizona)

mals and certainly humans are also capable of using simple territorial space to socially distinguish members along dominant and subordinate scales. In Calhoun's (1962) classic experiment with rats, a limited rectangular nesting box created dominant male rats in the four corners, while subordinate others could not nest or mate. They became socially deviant. In contemporary human settings a not dissimilar ranking frequently occurs in those intense social settings called "offices" (Mazumdar 1992, Doxtater 1994:338). Most landscape settings, whether urban or national parks, may be significantly valued as places where ranking is much less likely to occur a refuge from negative territoriality in urban buildings and layouts, e.g. in Shin et. al.'s (2005) evaluation of "psychosocial" effects of urban forest park use in South Korea. In this consideration, unlike the city, powerful people cannot control areas of space, most use is relatively transient, and even desired exclusivity often for socializing in small family or friendship groups is generally available on a first come, first serve basis.

It is perhaps this sense of an egalitarian right to be exclusive that most bothers hikers when they have to step off the trail for pack trains of horses or mules. Watson and Niccolucci (1994) describe hiker's feelings of lower status in these affordance experiences of "getting off the trail for stock groups", a clear sense of social ranking. Landscape recreation literature seems almost nonexistent, however, about possible ranking in designated spaces of parking lots, more expensive lodging within the same facility, more expensive seats in auditoria events or status restaurants along side of fast food. Nevertheless, some of these socially differentiating negative or positive affordances probably exist for visitors, e.g. "we can't afford that", or "we don't belong there".

Strategies of moving through the landscape involve considerations of social effects as the user moves in either dominant or subordinate directions of the spatial structure. People of lower status will be expected to move in the direction of their superiors, e.g. when the boss wants to see an employee, the employee usually goes to the boss's office. The primary interest in less socially structured visitor experiences should not keep evaluators, designers, and simulators from attending to the way employees or staff use the physical setting to socially influence. These things are always more important in work groups where people have to co-exist with each other in the same space often over considerable periods of time.

Figure 10. A lot of money in those R.V.s. (Photo by Lauren Hillquist)

Spontaneous Space

The best research based example of these

kinds of experiences is widely seen in Whyte's film and book on the "public life of small urban spaces" (1980). Even though some evidence of exclusion occurs, e.g. in sitting to the back of a public space under trees (Appleton's "prospect and refuge"), people are clearly interested in seeing and being seeing in this modern "I-thou" or "communitas" condition (Turner 1974). Some of the social things that people do are: "watching people", "talking with friends", "eating in an active outside place", "watching street performances", "running into friends", or "necking in public". Of these, "talking with friends" as an affordance might have little to do with the physical setting, in contrast with the others. Certainly Space Syntax methodology has paved the way in their analyses of how people move through public space, assembling in greater densities here or there. As useful as these more abstract contrasts to structured space are in predicting the general success of a design, however, it remains necessary to more fully define experiences in terms of specific affordances like those detailed by Whyte.

One good literature example of the tension between spontaneous and structured social space occurs in the case of skateboarding (Jones & Graves 2000). Of the six case studies the authors evaluate, one

Figure 11. Talking with friends (Sabino Canyon, Arizona)

appears to be the ideal and is specifically compared to Whyte's work. The case is a park under a bridge in Portland, Oregon, where skateboarding can occur in a more integrated social context, not unlike some very active inner city streets. The other sites studied create exclusive spaces for skateboarding away from publicly active areas. Certainly the primary affordance for skateboarders is a kinesthetic (*visual & non-visual aesthetic*) with the occasional negative *task performance* experience of being physically injured. Nonetheless, at certain times boarders will be quite aware of situational, territorial issues of social exclusion from or inclusion in spontaneous settings, e.g. "being ticketed for skateboarding in illegal areas", or "being able to hang out with other skateboarders in public places".

As in the other social, territorial situations described above, other people will be the primary "objects" in describing affordances of spontaneous space. Physical appearance, dress, behavior, contribute an endless variety of interesting people we are likely to watch, think about, or perhaps engage in conversation with in active public spaces. But what about the physical setting, particularly Whyte's ingredients for a successful plaza, i.e. sittable space, adjacency to street, water and trees for shade? One could look at each of these from other category points of view. Sitting and adjacency to the street could be *task performance* as well; water and trees are *visual and non-visual aesthetics*. Yet much of the time these must be background affordances, with the focused attention on spontaneous socializing.

Only Whyte's term of "triangulation" an admittedly vague notion about positioning between most interesting people or other behavioral events, such as street artists begins to capture spatial aspects of the setting. To a large extent, if the setting is safe, with large numbers of people, the object characteristics of the physical setting will be less interesting or mostly background. Spatial mapping of where we are likely to find such numbers of people seems a likely flip side to exclusive space, i.e. where one is unlikely to run in to people. Strategies for movement may also involve times when numbers of people are present at certain locations.

CULTURAL EXPRESSION

The reader at this point may perceive a certain scale of importance in the order of kinds of categorical affordances presented herein. We have yet to cover the two most ostensibly important experiences visitors have in recreational landscapes. First, people are likely to attach shared, symbolic meaning to landscapes in an essentially extrinsic manner somewhat independent of physical form. Cultural meanings, aside from learned task-performance ways of doing things, will be social in purpose, including the full range of belief religious, educational, ethical, etc. Intrinsically, however, landscape form may hold more universal interests not dedicated to any of the other more purposeful affordance categories, but to the pure delights of the senses described by Stolnitz in his philosophy of aesthetics (1992).

The term *cultural expression* includes numerous specific experiences that can be religious, commemorative, or educational. It is true that one cannot expect to map all the personal, idiosyncratic associations today's diverse individuals attach to visited landscapes. In traditional cultures, however, most symbolic meanings, ritual frameworks of space, and ceremonial movement in the landscape were shared and therefore more available to ethnographic inquiry. Some of this meaning still exists in Native American groups living in a larger landscape that for most others has become more recreational than religious. Applied anthropologists such as Stoffel,

Figure 12. Pilgrimage to the mountain shrine (Source: www. sacredsites.com).

Zedeno & Halmo (2001) provide an excellent example of how to record what objects mean, their location in space, and movements in the larger landscape. They rely upon onsite information from native informants, e.g. with the Paiute Tribe in areas around Las Vegas. Spiritual "power" from these culturally defined natural landscapes still provides the basis for healing and social ceremonies such as rites of passage. Most managed landscapes, however, are not in the business of providing religious experiences, at least as shared by any institutionalized group. One does find, however, designed commemorative experiences in the landscape with spiritual overtones, many influenced in one way or another by the precedent of the Viet Nam Memorial. One of the best comprehensive overviews of these kinds of landscapes can be found in Wasserman (1998) who identifies several kinds of experience: the "sacred" ("making contact with spirits" in affordance terms), "memory" evocations of particular people or events, "mourning", "reflection", "healing", "ritual" and "collective action". This study is more overview than applied methodology.

In spite of being one of the most significant and influential design projects in decades, no one has documented what people actually do at the Viet Nam Memorial. True, several articles, e.g. Griswold (1986), speak of normative symbolic, poetic meanings of the physical forms in their mall context, particularly the contrast with the rhetorical architecture of buildings like the Lincoln Monument. But most visitors will know more specifically what they and others actually do: "looking up a name", "figuring out the time sequences", "thinking about a lost love one", "leaving a personal object", "being in a grave", "watching other people's emotions", "seeing one's face reflected along with the names", "remembering a particular event", "gathering on the lawn as part of a visiting group" (e.g. biker groups on Memorial Day). Spatial movements as well might be mapped, although there appears to be little formal ritual-like framework to these affordances, as in more traditional settings or churches.

Even though published studies of visitor attention or interests in museums or zoos are few and far between, many interpretative landscape organizations will undoubtedly have conducted some sort of internal evaluation of signage, exhibit appeal and the like. To what degree is this information dependent on exit surveys with general questions of preference rather than real time data about what a person is actually doing at particular points along an interpretative sequence? How possible is it to map these actual experiences of *cultural expression* among the other things people do while in these settings? How specific can one practically get? In the case of historical or archaeological settings, for example, the National Register's criteria for evaluation include "association to events or persons important in the past", "properties significant as representatives of the manmade expression of culture or technology", or "properties significant for their ability to yield important information about prehistory or history" (National Register Bulletin nd). But while we might relatively easily evaluate the presence of these contents in the physical setting, the affordance approach requires also knowing if and when people attend to them. In addition to ideally knowing which object content people actually attended to and internalized to some degree, we also must consider the unique information that spatial structure and movement can add to affordances. The most obvious of these will be "time lines" as an associational, educational device, again for history, archaeology, geology, ecology, etc.

An example of a more rhetorical (object based), less highly structured cultural space as visitor landscape can be found in Northern Italy where a local group was determined to attract tourists by creating a map of things to do and see in their landscape region Grasseni (2004). While, again, the source of information is not the actual affordances the tourists eventually had, nevertheless, the knowledge about land use, local history, identity, and conservation is quite specific and seems to capture *their* experiences as remembered or even reconstituted. Grasseni's article is also interesting in her effort to reconcile more initial, holistic, phenomenological impressions of landscape, that are difficult to specifically define, with the "skilled vision" of locals which at times approaches the specificity of affordances. It should also be noted that as an anthropologist, Grasseni measures primarily social meanings and experiences. She speaks of a "reterritorialization", or an attempt to add symbols to a defined or exclusive space (the local landscape) which

Figure 13. Reflecting on the Korean War (Washington, D. C.)

creates tourism identity in relationship to the outside world.

At the much smaller landscape scale of front or back yards in Phoenix, Larsen & Harlan (2006) associate four different kinds of landscape styles with social status and Goffman's presentation of self. Having "lawn" is preferred by lower income people, "desert" connects with middle income, and "desert" and "oasis" associates with upper income (even though it is the same house plan and elevation). Here front yards seem to have much less to do with intrinsic *visual* or *non-visual aesthetics* than the cultural or educational awareness that "desert" is ecologically preferred over "lawn". This doesn't of course mean that the front yards cannot at other times have other affordance properties in other categories. A similar discussion of front yard status in New York, this time with racial overtones, can be found in Duncan and Duncan (2003). While these examples are from private residences, not dissimilar affordance experiences might well be possible with stylistic aspects of landscape or built form in visitor landscapes, where they will usually be associated with some sort of spatial exclusivity or ranking.

These cultural views of nominally *visually and non-visually aesthetic* landscapes again raises the question of which may be more important as affordances. Perhaps the most documented example of this question is Gobster & Hull's (2000) anthology dedicated to the controversy over proposed restoration of forest lands in the Chicago region. Should governmental agencies restore the lands to their aboriginal prairie conditions, or leave the more park-like character created by many imported trees? The volume speaks most clearly about the propensity of culture to interpret particularly social importance of landscapes, perhaps both pristine and park-like, at least at the level of discourse. No theoretical distinction between *cultural expression* (extrinsic) and *visual and non-visual aesthetics* (intrinsic) surfaces.

Figure 14. Learning about Mt. St. Helens". (Photo courtesy: USDA)

VISUAL AND NON-VISUAL AESTHETICS

Finally, we define those interesting moments of our sensory experiences when some intrinsic aspect of the physical setting captures our attention, away, as it were, from all the other obviously "useful" things we do (Terry Daniels: class presentation), some dependent upon the setting, some not. In visitor landscapes, when will people report "watching a sunset" or "taking a picture of a rugged canyon", (visual); "smelling the flowers" (olfactory); "seeking shade on a hot day" (thermal); "playing in the water" (tactile); or "riding a big wave on a surf board" (kinesthetic)?

The term "visual aesthetics" comes primarily from a considerable literature assessing human experiences of natural landscapes, e.g. Taylor, et. al. (1986), Ulrich (1983), Kaplan et. al. (1998), Clay & Daniel (2000). These evaluations run the range from precise calculations of the scenic in unique natural features to more holistic, phenomenological feelings about usually larger scale perceptions of "being" in a landscape. Such studies are typically dependent upon preferences subjects give to photographs of various kinds of natural form. Some recent work extends this kind of process using film and movement (Heft & Nassar 2000). At the more specific end of the evaluation spectrum, where perceptual details of natural objects are clearly identified, photo judging done by subjects might actually correspond to reportable affordance experiences in the field. Nevertheless, even here too, there exists a holistic assumption about being interested in relatively large, diverse swatches of settings framed by photographs. Because subjects are not typically asked to imagine "doing"

Figure 15. Looking at the Bryce Canyon. (Photo by Tina Mattsson)

something in the photographed setting, it is not clear, from an affordance point of view, what preferences really mean. When subjects prefer photos with water in them, or preferentially distinguish different kinds of landscape character or types but have no notion of actually doing anything does this suggest that these meanings are attended to in some sort of less conscious, pre-affordance process?

Many architects very much believe that not dissimilar kinds of holistic meanings are the primary aesthetic force operating separate from or in spite of all of the other more measurable things the built setting can do (*wayfinding, task-performance, social territories, cultural expression*). Bachelard (1969) and Tzonis & Lefaivre (1986) speak of poetic structures of ordinary dwellings or Greek temples, form not at all traceable to any symbolic, socio-cultural intent. Benedikt (1991) uses the term "archetype", not unlike Bachelard, in describing fundamental formal and volumetric attributes that create "difference", al la Derrida, in the work of famous architects like Frank Lloyd Wright or Louis Kahn. Peter Smith attempts to associate certain formal properties of silhouette, mass, pattern and the like to different parts of the human brain (1979). Not unrelated to the mentioned work that measures preferences of moving images or film, Weber (1995) tracks the movement of the eye over a photograph in an attempt to determine aesthetic preferences. Again, however, subjects are not "doing" anything, presumably just attending more subconsciously to visual form.

Our present question asks to what extent are *visual and non-visual* experiences more reportable, and conscious perhaps, like the affordances of other categories. This issue emerges in Palmer's (2000) recent discussion of positivist vs. post-positivist (phenomenological) evaluation of visible landscape quality. After testing differences between denotative (fact-like) and

Figure 16. I wonder where this path goes

connotative (subjective) qualities of evaluation processes, his conclusion suggests the possibility that "photographs are inadequate to effectively trigger our information-seeking instincts", and that "more attention needs to be paid to how movement through the landscape influences both landscape preference and information-seeking preference" (ibid: 175). In spite of the ambiguity created by the term "instincts", nevertheless, he seems to point to more affordance oriented approaches in the tricky process of determining how often people really "do" visual and non-visual aesthetics.

In addition to the frequency issue, in context with all the other setting related and setting unrelated things people do, there appears to be clear differences in aesthetic preferences between age groups. Returning to Kytta's work with children (2002), we see a preponderance of interests in the kinesthetic, e.g. skipping, jumping, swinging, skate-boarding, etc. The tactile as well may afford more to children. We all know how kids love to play in sand and water. Some indication of age differences resulted in a study of photographs taken by the full age range of visitors to a recreational forest outside of Osaka, Japan (Oko & Fukamachi 2006). Part affordance, part phenomenology, the photographs were ultimately organized into categories that emphasized "object", "event", or "place".

One conclusion was that the youngest visitors were more interested in objects like flowers or waterfalls, while the more solitary oldest participants had a keener view for "total scenery" with a wider interest in nature. One assumes that the latter is part of more passive visual affordance experiences while waterfalls, at least, might well involve the tactile and even kinesthetic. The color of flowers, on the other hand, might well be interesting to all ages, even though their aromas might be best sensed by younger noses.

The degree to which the Japanese study captures any specifically aesthetic spatial structure in the experiences is unclear. Their "events" and "places" seem more node-like, in addition to "objects", and offer little information about possible aesthetic enhancement from structured spatial relationships between them. This raises the interesting question about whether *visual & non-visual aesthetic* affordances, because they tend to exist at a more perceptual than cognitive level perhaps, frequently do not string together in space and movement as a means of enhancement. Certainly one can quickly imagine a skier thinking about the spatial strategy of a run, one that would include object or nodal highlights of imagined high speeds, turns or jumps. But the question is whether the spatial structure of the run actually contributes to some heightened kinesthetic

Figure 17. Like frozen music (Photo by Kelly Drawn)

experience, beyond a simple orientation between somewhat disconnected focal events.

CATEGORICAL COMPETITION, MEDIA, AND FUTURE METHODOLOGY

Hopefully the reader will be left with a less holistic, less normative view of possible experiences of visitors in recreational settings. Persuasive assumptions of an overwhelming complexity of these large-scaled settings are first diminished by the likelihood that for some appreciable time in these settings, visitors are doing things that are not that dependent upon the physical setting. Most important, however, is the need to record, design and evaluate in terms of very specific affordances, each with its psychologically unique, practically describable characteristics of object, space and movement. Within this list of affordances, then, there appear to be unique clusters that may tend to be understood by the visitor as related or strung together across the setting: *wayfinding, task-performance, social territories, cultural expression, and visual and non-visual aesthetics*. We do not, either as visitors or as practitioners, need to understand these potential layers of experience as any whole. While any final design is a single form that ideally will accommodate multiple experiences strung together categorically perhaps the best definition of "place" complexity can be reduced by understanding each layer. This does not, however, mean that visitors will be in agreement about which kinds of experiences are most important to them personally. But ideally, most visitor experiences can be maximized with some ultimate design solution.

We must avoid terminological glosses of setting behavior that may hide multiple affordances in different categories. Many examples can be found where research, case studies or other reports focus on a problem in the physical setting without clearly identifying the actual affordance meaning. In Lynn & Brown (2003), the effects of recreational "trail erosion, extension and widening, muddiness, tree and plant damage, fire rings and litter" were measured in terms of "solitude, remoteness, naturalness, and artifactualism". The focus of the study is largely on scenic or *visual aesthetics*, which here includes solitude in a more intrinsic less social sense. Yet the physical aspects of the trails might well involve *task-performance* ("getting wet" or "slipping"), *social territories* ("feeling crowded"), and *cultural expression* ("looking down on people that vandalize or don't adequately maintain their settings").

The present paper advocates a more consistent specificity and organization of these experiences in applied processes. As a more disciplinary basis develops, comparative research may lead to a much greater theoretical understanding of how categorical affordances trade-off or compete with each other, within a particular setting or between settings. An interesting example in this direction compares the experiences of people in a naturalistic Olmstead designed suburb with those in a more conventional gridded layout (Crow, et. al. 2005). Even though the perceived attributes by inhabitants are not really specifically stated at the level of affordances or categories, nevertheless, it is reported that the Olmstead people value their naturalness, while those living in the grid hold function and wayfinding as most important. Thus we have the possibility that naturalness, either for visual and *non-visual aesthetics* or *cultural expression*, may diminish *wayfinding* and *task-performance*. What are the consequences of these trade-offs?

One final example of competition between affordances can be found in the film *In the Light of Reverence* (McLeod 2001). One of its three case studies compares the experiences and values of two groups who differently use the remarkable natural feature Devil's Tower in Eastern Wyoming. The Native American Lakota use it as part of important ceremonial or *culturally expressive* experiences. These can be described in terms of key symbols, spatial structure in the landscape, and ritual movement. Then there are the non-native rock climbers who satisfy their kinesthetic (*non-visual aesthetic*) needs by focusing on hand hold features, routes up a particular face, and of course the timing or actual movement of the climb. In this case each of the two groups sees the activities of the other as an infringement on their particular affordance orientation. The National Park Service attempts to solve the conflict by asking climbers to voluntarily stay off the feature during major native ceremonial times of the year. While it is unlikely that these two groups would ever be comfortable with the concept of alternating affordance for a particular landscape feature or setting nevertheless in less politicized contexts visitors may well understand that this is the typical way we use our physical environments.

Media

The research question of how media, particularly film and printed brochures, can influence our actual affordance experiences in real physical settings is relatively unexplored. We do find comparisons of visual representations of natural settings with the perception and preference for the real thing (Daniel & Meitner 2001). But again, these efforts rely on views of scenic settings shown to subjects without affordance contexts of actually "doing" something. More suggestive of specific affordances were the pictures taken by visitors to the Japanese forest park mentioned previously (Oku & Fukamachi 2006). But neither of these examples is used in the manner of numerous film and printed promotions illustrating potential affordances in some to-be-visited landscape, such as when the Australian Tourist Commission sent out a call for photographs of the "real Australia" by ordinary citizens (Evans 2003). The photos were to be used in an advertising campaign to attract more tourists from abroad. Presumably some of these photos will document a specific affordance in a physical setting that will be previewed by a visitor prior to his or her actual re-creation of the experience on site.

Seeing potential affordances in films of brochures, particularly for *culturally expressive* and *visual-non-visual* experiences, are quite common at least for the more popular national and international recreational settings. Many such landscapes have representations of potential experiences as an introduction when visitors first arrive on the site. These and more distant previews must have an effect in planning one's experiences in the settings, and perhaps even heightening them when they actually occur. We return to the question of whether representation of potential *visual aesthetics* in effect casts a *culturally expressive* definition on the experience, complete with felt social obligations to participate in the affordance. Does representing things that ideally should be unrepresented, intrinsic and inherent, diminish any truly aesthetic experience?

It is also possible that some portion of tourist experiences are staged, either by visitors themselves or setting managers, in part to be captured in photography and film and to be post-viewed after the visit. Iles (2006) describes this "performing tourism" as a central experience in visiting the WWI battlefields of the Western Front. How does this staging of cultural expressive affordances depart from nominally self-initiated affordance seeking in recreational or educational landscapes? At an extreme, guided tours provide the ultimate in prestructuring affordances in visitor landscapes, in most cases using verbal communication by docents as the amplifying medium. How does one compare the effectiveness of these experiences with those "discovered" more by visitors themselves? How effectively will cell phones and GPS identify and structure affordance experiences?

As in the Vietnam Veterans Memorial, as well as the Western Front battlefields, visitors to some "pilgrimage" settings may bring a great deal of previous, media-based experience to the actual memorial or historical site. The relationship between pervasive mass-media "myth" and real setting experience is explored by Dickinson, et. al. (2005) in the setting of the Buffalo Bill Museum in Cody Wyoming. According to the authors, the physical affordances reify the images of masculinity and Whiteness of the myth, "carnivalizing" the violent struggle between Anglo Americans and Indians. They imply, alternatively, that a constructive cognitive dissonance might have been created in the physical setting by confronting long held biases of media.

Research is also needed to determine if film and photography can adequately capture or preview all characteristics of affordances, in all categories. Photography is certainly biased toward perceptions of objects or images, and lacks information about spatial structures and movement. How adequately then, does film represent experienced cognitive space in its different categories of affordance? We know that film does create a structure that connects various scenes, yet these images really are not mappable to any real space as more normally cognized. And if film space is not real space, as it were, how similar are its impressions of movement and action to the strategizing and timing that normally depends on cognitive mapping of real settings?

Future Methodology

Most applications of categories of affordances in visitor landscapes will begin with a close evaluation of what people are actual doing in the setting. Instead of more open ended methods of gathering information about experiences, one can envision separate mapping of the five different kinds of things people do. Considering individual (demographic) differences between kinds of visitors, methods must enable the researcher to capture information about object, space and movement relevant to each category of experience. Once this layered baseline data is gathered, simulation models can be developed by which to manage new pre-design and design schemes which will alter the physical setting. The author is fully confident from experience in architectural post occupancy evaluation and design that good category information can also be collected from recreational landscape visitors using a combination of spatial tracking, reporting devices, and observation.

At present, category information in architectural processes has been used primarily to develop a manageable list of things people do in a particular kind of building, usually as part of a post occupation evaluation (e.g. in the evaluation web site linked to Doxtater 2005). Such information can be useful in maintaining a dialogue with users and professionals during pre-design (programming) and design processes. This potential, however, has yet to be integrated into high-end computer modeling or simulation of architectural settings, where typically viewers "fly though" presentations without really "doing" anything. The development of simulation models of experience has occurred first in landscape oriented adjacent fields. It is here that ideas about categories of experience might be expected to bear first fruits as part of applied simulation to make better decisions about changes in these physical settings.

LITERATURE CITED

Aoki, Kris. 2005. Dealing With a Dirty Little Secret. In *Planning*, Vol. 71, Iss. 4, pg. 36-40.

Appleyard, Donald. 1970. Styles and methods of structuring a city. In *Environment & Behavior*, June, v. 2, n. 1, pp. 100-117.

Arthur, Paul and Passini, Romedi. 1992. *Wayfinding: people, signs, and architecture*. Toronto: McGraw-Hill Ryerson.

Bachelard, Gaston. 1969. The poetics of space. Boston: Beacon Press.

Baskaya, Aysu; Wilson, Christopher; Ozcan, Yusufziya. 2004. Wayfinding in an Unfamilian Environment: Different Spatial Settings of Two Polyclinics. In *Environment & Behavior*, Vol. 36 No. 6, November, pp. 839-867.

Benedikt, Michael. 1991. *Deconstructing the Kimbell: an essay on meaning and architecture*. NY: Sites Books.

Bishop, I. D. and Gimblett, H. R. 2000. Management of recreational areas: GIS, autonomous agents, and virtual reality. In *Environment and Planning B:Planning and Design*, volume 27, pp. 423-435.

Bohl, Charles. 2004. *The Social, Civic and Symbolic Functions of the Public Realm: A Comparative Analysis of New Urbanist Town Centers and Conventional Shopping Centers*. Dissertation: University of North Carolina at Chapel Hill.

Boschker, Marc; Bakker, Frank; and Michaels, Claire. 2002. Memory for the Functional Characteristics of Climbing Walls: Perceiving Affordances, in *Journal of Motor Behavior*, Vol. 34, No. 1, pp. 25-36.

Calhoun, John B. 1962. Population density and social pathology. In *Scientific American*, 206(2), pp. 139-150.

Canter, David. 1991. Understanding, assessing and acting in places: Is an integrative framework possible?. In *Environment, cognition, and action: An integrative approach*, edited by G. W. Evans. NY: Oxford University Press.

Chebat, Jean-Charles; Gelinas-Chebat, Claire; Therrien, Karina. 2005. Lost in a mall, the effects of gender, familiarity with the shopping mall and the shopping values on shoppers' way finding processes. In *Journal of Business Research* 58, pp. 1590-1598.

Clay, Gary R. and Daniel, Terry C. 2000. Scenic landscape assessment: the effects of land management jurisdiction on public perception of scenic beauty. In *Landscape and Urban Planning* 49, pp. 1-13.

Cooper-Marcus, Claire and Sarkissian, Wendy. 1986. *Housing as if people mattered: Site design guidelines for medium density housing*. Berkeley: University of California Press.

Crankshaw, Ned. 2001. Spatial Models for Parking and Pedestrian Access in Historic Downtowns: A Preservation and Design Perspective. In *Landscape Journal*, 20:1, pp. 77-89.

Creachbaum, M. S.; Johnson, C.; and Schmidt, R. H. 1998. Living on the edge: a process for redesigning campgrounds in grizzly bear habitat. In *Landscape and Urban Planning* 42, pp. 269-286.

Crow, Thomas; Brown, Terry; and De Young, Raymond. 2006 The Riverside and Berwyn experience: Contrasts in landscape structure, perceptions of the urban landscape, and their effects on people. In *Landscape and Urban Planning* 75, pp. 282-299.

Dalton, Ruth Conroy. 2003. The Secret is to Follow Your Nose: Route Path Selection and Angularity. In *Environment & Behavior*, Vol. 35, No. 1, January, pp. 107-131.

Daniel, Terry C. and Meitner, Michael M. 2001. Representational Validity of Landscape Visualizations: The Effects of Graphical Realism on Perceived Scenic Beauty of Forest Vistas. In *Journal of Environmental Psychology* 21, pp. 61-72.

Dickinson, Greg; Ott, Brian L.; and Aoki, Eric. 2005. Memory and Myth at the Buffalo Bill Museum. In *Western Journal of Communication* Vol. 69, No. 2, April, pp. 85-108.

Doxtater , Dennis 1994. *Architecture, Ritual Practice and Co-Determination in the Swedish Office*. Aldershot: Avebury.

Doxtater, Dennis. 2005. Living in La Paz: An Ethnographic Evaluation of Categories of Experience in a "New Urban" Residence Hall. In *Journal of Architectural and Planning Research*, Vol. 22, Number 1, Spring, pp. 30-50.

Duncan, James and Duncan, Nancy. 2003. Can't live with them - can't landscape without them: racism and the pastoral aesthetic in suburban New York. In *Landscape Journal*, 22-2, pp. 88-98.

Evans, David. 2003. ATC showcases 'Real Australia'. In *Media Asia* 10/17.

Francis, Mark. 2001. A Case Study Method for Landscape Architecture. In *Landscape Journal* 20: 1, pp. 15-29.

Freestone, Robert and Nichols, David. 2004. Realising new leisure opportunities for old urban parks: the internal reserve in Australia. In *Landscape and Urban Planning* 68, pp. 109-120.

Gibson, J. J. 1966. *The senses considered as a perceptual system*. Boston: Houghton Mifflin.

Gilbert, James N. 2000. Crime in the National Parks: An Analysis of Actual and Perceived Crime within Gettysburg National Military Park. In *The Justice Professional,* Vol. 12, pp. 471-485.

Gimblett, Randy; Daniel, Terry; Cherry, Susan; and Meitner, Michael J. 2001. The simulation and visualization of complex human-environment interactions. In *Landscape and Urban Planning* 54, pp. 63-78.

Gobster, Paul H. 1995. Perception and use of a metropolitan greenway system for recreation. In *Landscape and Urban Planning* 33, pp. 401-413.

Gobster, Paul H. and Hull, Bruce R.(eds.) 2000. *Restoring Nature: Perspectives from the Social Sciences and Humanities*. Washington, D. C.: Island Press.

Grasseni, Cristina. 2004. Skilled landscapes: mapping practices of locality. In *Environment and Planning D*: Society and Space 2004, Vol. 22, pp. 699-717.

Griswold, Charles L. 1986. The Vietnam Veterans Memorial and the Washington Mall: Philosophical Thoughts on Political Iconography. In *Critical Inquiry* 12, Summer, pp. 688-719.

Haq, Saif and Zimring, Craig. 2003. Just Down the Road a Piece: The Development of Topological Knowledge of Building Layouts. In *Environment & Behavior*, Vol. 35 No. 1, January, pp. 132-160.

Heft, Harry and Nasar, Jack L. 2000. Evaluating Environmental Scenes Using Dynamic Versus Static Displays. In *Environment & Behavior*, Vol. 32 No. 3, May, pp. 301-322.

Iles, Jennifer. 2006. Recalling the Ghosts of War: Performing Tourism on the Battlefields of the Western Front. In *Text and Performance Quarterly*, Vol. 26, No. 2, April, pp. 162-180.

Hochmair, Hartwig H. 2005. Investigating the effectiveness of the least-angle strategy for wayfinding in unknown street networks. In *Environment and Planning B: Planning and Design*, volume 32, pp. 673-691.

Jensen, Frank Sondergaard and Koch, Niels Elers. 2004. Twenty-five Years of Forest Recreation Research in Denmark and its Influence on Forest Policy. In *Scandinavian Journal of Forest Research* 19 (Suppl. 4), pp. 93-102.

Jeong, Jae-Hoon and Lee, Kyung-Hoon. 2006. The physical environment in museums and its effects on visitor's satisfaction. In *Building and Environment* 41, pp. 963-969.

Jones, Stanton and Graves, Arthur. 2000. Power Plays in Public Space: Skateboard Parks as Battlegrounds, Gifts, and Expressions of Self. In *Landscape Journal* pp. 136-148.

Kaplan, Rachel; Kaplan, Stephen; and Ryan, Robert L. 1998. *With people in mind: design and management of everyday nature*. Washington D. C.: Island Press.

Katz, Peter. 1994. *The new urbanism: toward an architecture of community*. NY: McGraw-Hill.

Kytta, Marketta. Affordances of Children's Environments in the Context of Cities, Small Towns, Suburbs and Rural Villages in Finland and Belarus, in *Journal of Environmental Psychology* 22, pp. 109-123.

Larsen, Larissa and Harlan, Sharon. 2006. Desert dreamscapes: Residential landscape preference and behavior. In *Landscape and Urban Planning* 78, pp. 85-100.

Lawton, Carol A. and Kallai, Janos. 2002. Gender Differences in Wayfinding Strategies and Anxiety About Wayfinding: A Cross-Cultural Comparison. In *Sex Roles,* Vol. 47, Nos. 9/10, November, pp. 389-401.

Longmuir, Patricia E.; Freeland, Michelle G.; Fitzgerald, Shirley G.; Yamada, Denise A.; and Axelson, Peter W. 2003. Impact of Running Slope and Cross Slope on the Difficulty Level of Outdoor Pathways. In *Environment & Behavior*, Vol. 35, No. 3, May, pp. 376-399.

Lynch, Kevin. 1960. *The Image of the City*. Cambridge, Mass.: The M.I.T. Press.

Lynn, Natasha A. and Brown, Robert D. 2003. Effects of recreational use impacts on hiking experiences in natural areas. In *Landscape and Urban Planning* 64, pp. 77-87.

Mazumdar, Sanjoy 1992. "Sir, please do not take away my cubicle": the phenomena of environmental deprivation. In *Environment & Behavior*, Vol. 24, n.6, pp. 691-722.

McHarg, Ian L. 1994. *Design with Nature*. NY: J. Wiley.

McLeod, Christopher. 2001. In the light of reverence (videorecording). The Independent Television Service and Native American Public Telecommunications.

McPherson, Gregory E. 2001. Sacramento's parking lot shading ordinance: environmental and economic costs of compliance. In *Landscape and Urban Planning* 57, pp. 105-123.

Newman, Oscar. 1996. *Creating Defensible Space*. Washington D.C.: HUD USER Publications.

Oku, Hirokazu and Fukamachi, Katsue. 2006. The differences in scenic perception of forest visitors through their attributes and recreational activity. In Landscape and Urban Planning 75, pp. 34-42.

Palmer, James F. 2000 Reliability of Rating Visible Landscape Qualities. In Landscape Journal Vol.19, n. 1-2, pp. 166-178.

Schrijnen, Pieter M. 2000. Infrastructure networks and red-green patterns in city regions. In *Landscape and Urban Planning* 48, pp. 191-204.

Shin, Won Sop; Kwon, Hon Gyo; Hammitt, Willian E.; and Kim, Bum Soo. 2005. Urban forest park use and psychosocial outcomes: A case study in six cities across South Korea. In *Scandinavian Journal of Forest Research*, 20, pp. 441-447.

Smith, Peter. 1979. Architecture and the human dimension. London: G. Godwin.

Soh, Boon Kee and Smith-Jackson, Tonya L. 2004. Influence of Map Design, Individual Differences, and Environmental Cues on Wayfinding Performance. In *Spatial Cognition and Computation*, 4(2), pp. 137-165.

Stoffle, Richard W.; Zedeno, Maria Nieves; and Halmo, David B. (eds). 2001. *American Indians and the Nevada Test Site: A Model of Research and Consultation*. Washington, D.C.: U.S. Government Printing Office.

Stolnitz, Jerome. 1992. The Aesthetic Attitude. In Alperson, Philip (ed) *The Philosophy of the Visual Arts*. NY: Oxford University Press, pp. 7-14.

Taylor, Jonathan G.; Zube, Ervin H.; and Sell, James L. 1986. Landscape Assessment and Perception Research Methods. In Bechtel, Marans & Michelson (eds) *Methods in Environment Behavior Research*. NY: Van Nostrand Reinhold.

Thompson, Catharine Ward; Findlay, Catherine; and Southwell, Katherine. 2004. Lost in the Countryside: Developing a Toolkit to Address Wayfinding Problems. In Martens, Bob & Keul, Alexander G. (eds.) *Designing Social Innovation: Planning, Building, Evaluating*. Bern: Hogrefe & Huber Publishers, pp. 37-46.

Tijerino, Roger. 1998. Civil Spaces: A Critical Perspective of Defensible Space. In *Journal of Architectural and Planning Research* 15:4, pp. 321-337.

Turner, Victor W. 1974 *The Ritual Process*. Harmondsworth, Middlesex: Penguin Books Ltd.

Tzonis, Alexander and Lefaivre, Liane. 1986. *Classical Architecture, the Poetics of Order*. Cambridge: Cambridge University Press.

Ulrich, Roger S. 1983. Aesthetic and affective response to natural environment. In Human Behavior & Environment: Advances in Theory & Research, Vol. 6, pp. 85-125.

Vosmik, Jordan R. and Presson, Clark C. 2004. Children's Response to Natural Map Misalignment During Wayfinding. In *Journal of Cognition and Development*, 5(3), pp. 317-336.

Wasserman, Judith. 1998. To trace the shifting sands: community, ritual, and the memorial landscape. In Landscape Journal 17-1, pp. 42-61.

Watson, Alan E. and Niccolucci, Michael J. 1994. The Nature of Conflict Between Hikers and Recreational Stock Users in the John Muir Wilderness. In *Journal of Leisure Research*, Vol. 26, No. 4, pp. 372-385.

Weber, Ralf. 1995. *On the Aesthetics of Architecture*. Aldershot: Avebury. *Behavior*, Vol. 36 No. 4, July, pp. 461-482.

Whyte, William H. 1980. *The social life of small urban spaces*. Washington, D.C.: Conservation Foundation.

Widmer, W. M. and Underwood, A.J. 2004. Factors affecting traffic and anchoring patterns of recreational boats in Syndey Harbour, Australia. In *Landscape and Urban Planning* 66, pp. 173-183.

CHAPTER 3

SENSEMAKING EXPERIENCES ON THE BARDEDJILIDJI SANDSTONE TRACK

Michael L Steven

Abstract. Little is known of the capacity for informal learning, or *sensemaking*, that attends recreational transactions with natural environments. What stimuli are encountered in the biophysical landscape, and how do cognitive and affective responses to those stimuli colour recreational experiences? Using a multi-layer cognitive mapping instrument, *Environmental N*, recreationists experiences on the Bardedjilidji Sandstone Track in Australias Kakadu National Park were recorded on site. A typology of recreational experiences grounded in participants responses provided the basis for characterising sensemaking experiences at Kakadu. Three core experience themes encapsulating fifteen sub-themes emerged from the data: *Wayfaring; Seeking, discovering, encountering*; and *Making sense of place*.

Key words: Sensemaking, recreation experience, cognitive map, stimuli, affective tone

INTRODUCTION

Very little is known about the environmental stimuli that contribute towards recreational experiences, but it is generally accepted that lay and professional understandings of natural landscapes likely vary in many respects. As Knopf (1983), p. 223) has observed, "The environments people actually 'see' are not the same environments that recreation planners strive objectively to define." A cognitive gap between lay and professional understandings of natural landscapes has implications for natural area management, particularly within landscapes intensively used for recreation.

A better understanding of the sense that recreationists make of the natural environments within which they recreate could have particular implications for the promotion of learning as an integral aspect of the recreation experience in protected natural areas. Roggenbuck, Loomis and Dagostino (Roggenbuck et al., 1990), in discussing the types of learning that occur in leisure contexts, comment that the learning benefits of leisure are often overlooked by natural area planners and managers. While formal interpretation programs are commonly encountered in Australian national parks, little is know regarding the extent to which informal, unmediated learning takes place in the course of casual encounters with the natural environment, nor how unmediated approaches to discovery and learning might be promoted by park management.

A call for alternative, experiential approaches to interpretation and learning has been made by Uzzell (1989) who proposes that interpreters should be giving away the skills of revelation rather than information. Beck (1993; p.30) asks how visitors might be encouraged to "be curious about, to speculate upon, and to seek their own answers", adding that "the most rewarding meanings may well be those prompted by the managing agency, but ultimately discovered by the visitors themselves". Within Australia there is a growing interest within the discipline of interpretation for greater reliance upon experiential learning, whereby visitors make their own discoveries and construct their own meanings within natural environments (Markwell 1996). Of the actual process of discovery and learning, however, little is known. Specifically, what stimuli do recreationists perceive, and what cognitive and affective responses accompany the perception of stimuli in the landscape?

In terms of Uzzell's "skills of revelation", to what extent are nature walk experiences "revelatory", and how might heightened levels of revelation be fostered? If the practice of experiential learning is to be promoted in the field of natural heritage interpretation, then gaps in the knowledge of recreationists' affective and cognitive responses to environmental stimuli must be addressed.

MAKING SENSE OF THE LAND

The doctoral research project, *Making Sense of the Land: A Sensemaking Approach to Environmental Knowing* (Steven, 2004) sought to investigate environmental knowing associated with recreational experience through an investigation into the cognitive and affective responses that attend the perception of landscape stimuli during short duration nature walks. Recreational experience was investigated from a sensemaking perspective, the intention being to gain an understanding of recreationists' capacity to 'make sense of' the environments within which they engaged in recreational activities.

Landscape sensemaking, as conceptualised in this study, can be regarded as being synonymous with processes of informal learning, or the unmediated, unprogrammed interpretation of the biophysical and cultural environment. The concept of sensemaking draws upon the work of organisational behaviourist Karl Weick (1995; 1979). Weick argues that sense is not given by the environment, but emerges as a result of the process of construction, or authoring; sensemaking addresses the act of constructing the sense, as well as its interpretation. The process involves the observation of cues or stimuli, selected by the individual sensemaker from within the environment and organised within a framework (Starbuck & Milliken, 1988, cited in Weick, 1995). Sensemaking is also described as "the reciprocal interaction of information seeking, meaning ascription and action" (Thomas *et al.*, 1993, cited in Weick, 1995, p. 5), while Weick (1979) defines sensemaking as a process involving perceiving, believing, interpreting, explaining, predicting and acting, both individually and collectively, in a given organisational setting.

Wicker (1992) has applied the sensemaking concept to the area of environment-behaviour studies, particularly to the study of behaviour settings (Barker, 1968). Wicker (1992) defines sensemaking as the cyclical process of attending to, interpreting and acting upon environmental stimuli in the course of transactional experiences in behaviour settings. Wicker claims the concept is readily applicable and complementary to the notion of the behaviour setting. He proposes that transactions undertaken within the behavioural setting, and in support of the setting program, "can be understood in terms of people's attempts to make sense of environmental events" (Wicker 1992, p. 173).

GEOGRAPHIC CONTEXT

Data for the project *Making Sense of the Land* was collected from four geographically distinct national park settings across Australia. This paper deals with the findings to emerge from just one of these sites; the Bardedjilidji Sandstone Track and the Rockholes Sandstone and Riverside track in Kakadu National Park. Located within the Northern Territory of Australia, between 12° and 14° degrees south, Kakadu National Park is designated a World Heritage site, and is one of six national parks administered by the federal Department of the Environment and Heritage. Kakadu NP is widely known and appreciated for its landscape of tropical savanna grasslands and woodlands, eroded sandstone formations and dramatic escarpments, meandering rivers, extensive wetlands, wildlife, and Aboriginal heritage.

The walks surveyed were the Bardedjilidji Sandstone Track and the Rockholes Sandstone and Riverside track adjacent to the Merl campground, a short distance from Cahills Crossing on the East Alligator River (AUSLIG, 1996). The location of the walks adjacent to a popular camping area ensured a steady flow of participants, while the eroded sandstone, wetland and riverine landscape is characteristic of the park. The walks surveyed are contiguous with each other and provide a short (2.5 km) option in the case of the Bardedjilidji Sandstone Track or a longer 6.5 km) walk in the case of the Rockholes Sandstone and Riverside tracks. The short walk passes relatively flat sandy terrain within sandstone formations of pancake rocks sculpted into pillars, arches and caves. The longer walk option involves a section along a dry, sandy riverbed, and a walk along the levee bank of the East Alligator River. The walks are regarded as dry season walks, and the area in inundated during the wet season and inaccessible to recreationists and tourists. For convenience, both walks will be referred to by the name, 'Bardedjilidji Sandstone Track'.

Evidence of Aboriginal occupation, believed to extend back as far as 65,000 years, is apparent along sections of the track in the form of rock art within caves, and tool sharpening grooves in the sandstone rocks. Wildlife is abundant, and of particular interest to visitors to the locality are the saltwater crocodiles (*Crocodylus porosus*), that are found in the East Alligator River and adjacent billabongs. Prominent signage at the commencement of the walk and in all national park literature relating to recreation within the park, warns of the danger posed by these reptiles.

At the time of the fieldwork in Kakadu, daily temperatures were in the range 32-34°c. The high temperatures, combined with sparse vegetation that affords little shade and the ubiquitous Australian bushfly, creates an environment for walking that many find physically and mentally challenging, despite the flat terrain. Nevertheless, it is one of the park's most popular locations and walks.

The Kakadu walks were the only walks surveyed in the study to be formally interpreted by park management. On the Bardedjilidji Sandstone Track, a series of eleven features were highlighted by marker posts, the numbers on which corresponded to notes on interpretation handout available at the start of the walk. Not all participants carried these handouts with them during the walk, but with or without the handout, the presence of the marker posts was influential in determining the stimuli attended to during the walk.

DATA COLLECTION

The method selected for data collection in *Making Sense of the Land* was Environmental N (N = nature, or natural), a natural landscape adaptation of the cognitive mapping tool associated with Wood and Beck's (1976) experimental mapping language, Environmental A[1] (A more comprehensive description and discussion on the Environmental N method is presented in Steven (in press)). The basis for this decision was Environmental A's design – a multi-layered cognitive mapping instrument upon which are recorded the path walked, the stimuli encountered, and affective, cognitive and evaluative responses to those stimuli. While acknowledging the very different geographic context for which Environmental A was developed (the urban, as opposed to natural environment), the instrument was considered to have great potential for *Making Sense of the Land*. In adapting Environmental A, the map sheets were limited to a base diagram and two transparent overlays. One overlay was used for the recording of affective responses (or *feelings*, as the sheet was labelled and referred to in the instructions to users), and the other to be used for the recording of cognitive responses (or *thoughts*, as the sheet was labelled). Participants were requested to write *thoughts* and *feelings* on the respective overlays, with the only guideline being that thoughts and feelings comments should be written in such a way that it was clear to which stimulus they referred. No constraints were placed upon the terms that could be used or the length of responses. However, in response to requests (during pilot testing) for clarification of the distinction between thoughts and feelings, a schedule of affective terms, taken from Russell and Pratt's (1980) scale of the affective qualities attributed to environments was provided to participants.

In Wood and Beck's Environmental A, the task of recording the stimuli was facilitated by a 'dictionary' of pictograms or graphic symbols, each representing the generic stimuli commonly encountered in the urban environment and designed to be easily replicable by the research participant. Just which stimuli are noticed by recreationists during nature walks in natural landscapes was one of the unknown factors in *Making Sense of the Land*, so this basic problem prevented the use of an a priori selection of symbols. Instead, 6 mm coloured, adhesive dots were substituted for the use of pictograms or symbols.

Fieldwork was conducted over a period of 5 days in September, the end of the dry season when the savanna landscape is at its driest but visitor numbers are still high, prior to the onset of "the Wet". The selection of participants was done on a non-probability basis, using purposive sampling (Judd, Smith & Kidder, 1991). This approach accorded with standard sampling practice within the field of grounded theory qualitative research (Strauss & Corbin, 1990), in which sampling is guided by what is central and crucial to the research question, with participants selected on the basis of those who are most likely to represent the phenomenon under investigation. In practice, an element of accidental sampling was also involved, which Judd et al. (1991) describe as the process of reaching out and taking the cases that are at hand, as was indeed done. At all sites, visitor activity was monitored from the trailhead or car park at the departure/return point of the walk. During periods of peak track walking, participants

---

[1] While a substantial debt to Denis Wood is acknowledged for permission to adapt his Environmental A method to Making Sense of the Land, the development of Environmental N was also influenced by the work of Lynch (1960)) and Lynch and Rivkin (1959).

were approached as they completed the walk with an invitation to participate in the research. Participants had no prior knowledge of their possible selection, and thus the prospect did not colour their experience of the walk. Some 5-8 participants were recruited and completed cognitive maps on any one day of fieldwork. Thirty eight (38) participants were surveyed for 36 completed maps (in two cases, two persons completed the one map, having walked the track together).

All participants were given a kit consisting of: detailed instructions for completing their cognitive maps, an A3 sized sheet of plain, white drawing paper, 2 x A3 sheets of transparent tracing paper (110 gsm); a black fibre-tipped drawing pen, and a sheet of 6 mm, self-adhesive coloured dots. A picnic table was available for the use of participants. Apart from clarifying instructions, the researcher took no part in the process of the drawing of the maps. The primary concern of *Making Sense of the Land* was to identify the range of stimuli sensed during the walk, and the participants' cognitive and affective responses to these stimuli. Thus it was stressed to participants that geographic accuracy in the representation of the path walked was not important, nor was strict accuracy required in the ordering of the stimuli encountered. As things transpired in the field, the focus on *stimuli*, *thoughts* and *feelings* at the expense of geographic accuracy probably contributed towards faster completion of the maps. The times taken for completion of the task ranged from 20 minutes to 60 minutes, with most participants requiring 30-40 minutes to undertake the task. Examples of completed base maps are shown in Figures 1 and 2.

Participant responses to the process of completing maps was positive, with many comments indicating that the task was stimulating, absorbing and fun to do. Participants reported that the process encouraged a reflective state of mind, and several likened it to the task of completing a journal or diary entry of the event at the end of the day. Many participants commented that the reflection required for drawing the map had consolidated and enriched the experience of the walk for them.

DATA ANALYSIS: A TYPOLOGY OF RECREATIONAL EXPERIENCES

Raw data from the completed cognitive maps was transcribed into a form that would permit further analysis. A spreadsheet (Microsoft Excel) was selected as the most suitable software for data analysis, as the structured format of the data appeared to lend itself to analysis within the spreadsheet format of rows and columns. The basic spreadsheet layout adopted is represented in Table 1. The shaded cells in the table represent the basic unit of data analysis, which is conceptualised as a *moment of experience*, comprising a stimulus and its associated thought and feeling comments. Each row of the spreadsheet represents a separate moment of experience. Each moment of experience was identified by the map number from which it was taken, together with the location and the gender of the participant. Further columns were added to the basic spreadsheet layout within which codes were ascribed to the phenomena under examination. At Kakadu, 746 individual moments of experience were recorded by the 38 participants surveyed.

Contrary to expectations, not every moment of experience was 'complete' in the sense that it was comprised of a stimulus and an associated thought and feeling. Some stimuli were recorded without accompanying thoughts and feelings, and some stimuli were recorded with thoughts but no feelings, and vice versa.

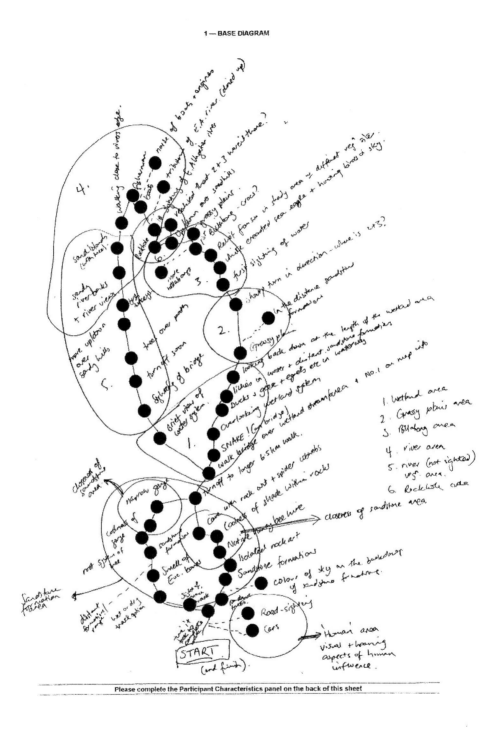

Figure 1. Environmental N Base Map, Participant K.4, Bardedjilidji Sandstone Walk, Kakadu National Park (scanned and then reduced from the original A3 sized sheet)

Sensemaking Experiences on the Bardedjilidji Sandstone Track 43

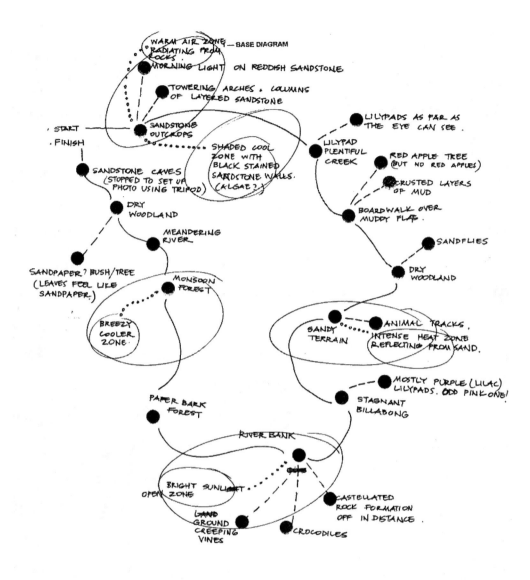

Figure 2. Environmental N Base Map, Participant K.2, Bardedjilidji Sandstone Walk, Kakadu National Park (scanned and then reduced from the original A3 sized sheet)

The principle research question[2] anticipated the construction of a theory of recreational experience, to be expressed in terms of a typology of recreation experience patterns. The process of data analysis followed was substantially that described for the discovery of grounded theory by Strauss and Corbin (1990). Central to grounded theory analysis is the process of data reduction, whereby through successive phases of re-conceptualising the data at increasing levels of abstraction, the data is reduced to a parsimonious yet rich set of concepts that account for the phenomena under investigation. Through a constant comparative process of analysis, patterns were identified within the range of stimuli and cognitive and affective responses within the data. Patterns resolved into sub-themes and then themes, to generate a typology of recreational experiences. Two versions of the typology have been developed, of which the short version, at the level of key themes and sub-themes, is presented in Table 2.

In deference to Strauss and Corbin's (1990) notion of the single 'core category', the highest-level concept to have emerged from the research is termed *'Making Sense of the Land'*. Beneath this core category are three broad patterns of recreational experiences through which sense is made of the land. While the research project was focussed primarily on the biophysical environment, many moments of experience addressed social and personal phenomena. Thus, some of the sub-themes relate to social and personal aspects of recreational experience. Collectively these three patterns are designated *recreational experience themes*. Within each recreational experience theme are nested fifteeen categories of recreational experience sub-themes.

The typology of recreational experiences outlined in Table 2 was derived from the total of all moments of experience recorded at each of the four research sites[3] – a total of 3003 moments of experience.

---

[2] Of three research questions posed, the first specifically addressed the issue of recreational experience: *"Are there patterns in people's affective and cognitive responses to landscape stimuli experienced during person-in-environment (P-i-E) transactions, and if so, can these patterns be described in terms of a typology of recreation experience?"*

[3] Data collection was conducted in Kosciuszko and Warrumbungle National Parks (New South Wales), Fraser Island (Great Sandy National Park, Queensland) and Kakadu National Park (Northern Territory).

Table 1. Indicative Example of Spreadsheet Layout for Analysis of Data.

Stimuli	Thoughts	Feelings	Location	Participant	Male / Female
forest area*	nice and cool, it is a bit too hot to walk in this heat	Pleasant, cool	Kakadu	K9	F
great old fallen tree	in parks its good to see the circle of life, not possible in cities	Sad	Kakadu	K3	M
billabong with lilies and birds	wish I could identify the birds	Beautiful, serene, interesting, lots to challenge the mind	Kakadu	K36	F

*the 3 shaded cells, representing a stimulus and its associated thought and feeling, were conceptualised as a *moment* of experience

The typology served to provide the basis for characterising recreational experiences at each research location. Table 3 presents the distribution of moments of experience for the Bardedjilidji Sandstone Track across the typology. Figure 3 shows the same data in the form of a recreational experience profile for the Bardedjilidji Sandstone Track.

ADDING AFFECTIVE TONE TO RECREATIONAL EXPERIENCES

A central issue in the analysis of the data was the complexity of many moments in experience and the problems this presented in categorisation and coding. In particular, integrating affective responses into the typology of recreational experiences proved too vexed a problem to be resolved satisfactorily, so the decision was made to 'bracket-out' the affective aspect from the typology of recreational experiences. The emotional aspect of participant responses was added to the core experience of stimulus + thought (as represented by the themes and sub-themes of the typology) in the form of an affective tone.

The terms used for the affective tone draw upon Russell and Pratt's (1980) scale of the affective quality attributed to environments[4]. The forty words that make up Russell and Pratt's (1980) scale are categorised into eight groups, which are further organised into two groups of two scales:

[4] This schema was used by participants in the field as a lexicon of feeling states for the purpose of completing their cognitive maps.

(a) Pleasant qualities — Unpleasant qualities
    Arousing qualities — Sleepy qualities

(b) Exciting qualities — Gloomy qualities
    Distressing qualities — Relaxing qualities

Table 2. Making Sense of the Land: A Short Typology of Recreational Experiences

WAYFARING
PREPARATION & ANTICIPATION
NEGOTIATING THE TERRAIN
WAYFINDING & PROGRESS
STRIVING & COPING
SUMMING UP THE WALK
SEEKING, DISCOVERING, ENCOUNTERING
IN SEARCH OF WILD THINGS
ATTENDING TO LIVING THINGS
APPRECIATING LIVING THINGS
CURIOUS ABOUT LIVING THINGS
COMING UPON BUILT OBJECTS
BEING IN THE COMPANY OF OTHERS
MAKING SENSE OF PLACE
SENSING ATMOSPHERE & AMBIENCE
ATTENDING TO ASPECTS OF PLACE
APPRECIATING ASPECTS OF PLACE

Table 3. Bardedjilidji Sandstone Track, Kakadu National Park: Incidence of Recreational Experience Sub-themes

Sub-themes	Moments of experience	%*	Cumulative %*
Appreciating aspects of place	219	29	29
Appreciating living things	126	17	46
Attending to aspects of place	70	9	56
Wondering about aspects of place	62	8	64

Attending to living things	54	7	71
Negotiating the terrain	48	6	78
Curious about living things	41	5	83
Sensing atmosphere & ambience	26	3	87
Wayfinding & progress	22	3	90
In search of wild things	20	3	92
Striving & coping	16	2	94
Being in the company of others	13	2	96
Preparation & anticipation	12	2	98
Coming upon built objects	9	1	99
Summing up the walk	8	1	100
Total moments	**746**		

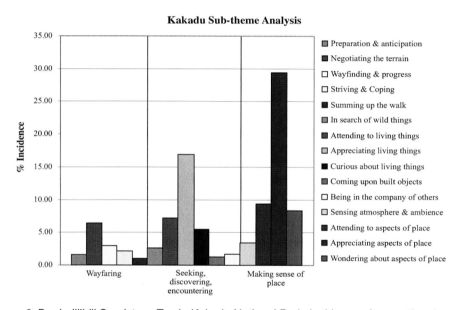

Figure 3. Bardedjilidji Sandstone Track, Kakadu National Park: Incidence of recreational experience sub-themes expressed as a percentage of total moments of experience

Wondering about aspects of place

Russell and Pratt (1980; p. 319) suggest that the Exciting–Gloomy and Distressing–Relaxing scales (Group (b), above) can be used interchangeably with the Pleasant–Unpleasant and Arousing– Sleepy scales (Group (a) above), with scores on the latter scale having been found to accurately predict scores on the former. Given the equivalence between the two scales, it was decided that rather than using one group or the other, a combination of the two groups would be appropriate, whereby equivalent words from each of Groups (a) and (b) could be matched together and used as a pair. Accordingly, the affective tone of experiences was described in terms of the following expressions:

• Feeling pleasant or relaxed

- Feeling aroused or excited
- Feeling unpleasant or distressed
- Feeling bored or unstimulated

As the data revealed, these were not necessarily the words most frequently used by participants themselves. For the first three pairs of terms, the words are an accurate enough surrogate for the purpose of summarising participant responses, and Russell and Pratt's terminology was retained. However, the exception is the pair labelled 'feeling bored or unstimulated'. In Russell and Pratt's scale, the group labelled 'gloomy' covers the words; dreary, dull, unstimulating, monotonous, and boring. The words dull, unstimulating, and boring were chosen by participants more often than gloomy and sleepy, so with due respect to Russell and Pratt's scale, the label 'sleepy or gloomy' is replaced with 'bored or unstimulated', as it gives a much more accurate picture of participants' affective responses in this part of the scale.

RECREATIONAL EXPERIENCES OF THE BARDEDJILIDJI SANDSTONE TRACK

With reference to the typology of recreational experiences, the discussion turns to the most commonly occurring experiential themes and sub-themes on the Bardedjilidji Sandstone Track, Kakadu National Park, and the affective tones that typically characterised those experiences. In keeping with a common practice of naturalistic enquiry, the discussion draws upon the participants' own words for the purpose of providing a rich picture of recreational experiences at the site, thus providing some insights into the basis of the typology of recreational experiences.

Comments are reproduced faithfully, with emphasis in the quoted comments having been transcribed exactly as in the original transcript. The quoted comments represent a composite of text taken from each of the stimuli, thoughts and feelings layers of a participant's cognitive map. As such, the quoted text did not originate in the form a single, continuous response, but rather each associated quote was collated from the three layers of a participant's cognitive map.

Throughout the discussion that follows, a convention is applied in the representation of the original text such that the text from different layers of the cognitive maps has been ordered in the sequence *Stimulus — Thought — Feeling,* with each component separated by an 'em' dash (—), as indicated in the following examples:

> rock tunnel — the miracles of nature — WOW!Interesting,exciting
> murky billabong — I thought it would be horrible if a croc killed me or one of my friends — latent fear, tense, moving on

The capitalisation and exclamation mark (!) are as was written in the original transcript. Where the original transcript contains the handwritten equivalent of an 'em' dash, a comma has been substituted to avoid confusion. In a few instances, words have been added to improve the meaning of the quoted text. Where this occurs, the added words are enclosed within square brackets and written in the non-italicised font style, i.e., [and]. At the end of each quoted moment, a pair of square brackets enclose a code identifying the participant, i.e. [K7], where K = Kakadu, and 7 represents a specific participant's map.

Contrary to expectations, not every moment of experience was 'complete' in the sense that it was comprised of a stimulus and an associated thought and feeling. In many cases stimuli were recorded with no corresponding comment on the thoughts or feeling layers of the maps, and similarly, thoughts and/or feelings comments sometimes appeared in the absence of any stimulus having been recorded: Where this occurs, the missing component of the *stimulus — thought — feeling* sequence is represented by the symbol x, as in these examples:

> beside river bed a couple of pools filled with moss and mud — x — x [response on stimulus layer only]
>
> pig family — x — felt excited to see the pigs, but also aware that they are intruders [response on stimulus and feeling layer only]
>
> x — annoying flies — x [response on thought layer only]

The incidence of sub-themes and theme is shown Table 1, ordered from most to least frequent. Relatively few of the fifteen categories of sub-theme accounted for a majority of experiences on the Bardedjilidji Sandstone Track, where experiences categorised within four sub-themes accounted for 64% of all experiences. Just two of these, *Appreciating aspects of place*, and *Appreciating living things*, accounted for 46% of all experiences. Reflecting this pattern, the discussion focuses on the experiences categorised within these four sub-themes. A summary of some of the more consequential responses to emerge from the lessor categories is also included.

APPRECIATING ASPECTS OF PLACE

The experience of the Bardedjilidji Sandstone Track in Kakadu National Park is dominated by the sub-theme, *Appreciating aspects of place*, which accounted for 29% of all recorded experiences. Given the characteristics of the physical environment of the walk, it is not surprising that moments of experience within this sub-theme frequently referred to the geology of the site. Much of the walk passes through an area of ancient, layered sandstone rock formations, sculpted by the elements into evocative forms. The predominant affective responses to these eroded forms were feelings of arousal or excitement, and pleasantness or relaxation:

> rock tunnel — the miracles of nature — WOW! Interesting, exciting [K7]
>
> natural sculptures — x — I was very impressed by the 'art' provided by nature, inspirational shapes and colours [K24]
>
> windowed rock — how can nature make these shapes? — pleasing [K35]

Beyond the sandstone rock formation the track enters an area of wetland, within which a series of billabongs[5], waterholes and dried river channels contrasts with the open sandstone rock area. The presence of water and shade from the *Melaleuca* woodland provides welcome relief from the harsh glare and radiated heat of the rocky places, instilling feelings of feeling pleasure and relaxation:

> looking back down at the length of the wetland area — how I'd love to live in a wetland system with lilies — peaceful and serene, beautiful, interesting, SPECIAL, tranquil (I love wet systems!) [K4]
>
> waterhole — nice and cool, it is a bit too hot to walk in this heat — peaceful, calm [K9]

The water features of the locality, and the dry water channels in particular, also evoked arousing or exciting responses, especially as participants contemplated the apparent contrast between wet and dry season:

> dried up creek — the wet must be incredible, how much rain! — questioning [K15]

Once through the wetland and billabongs, the track approaches, then runs adjacent to, the East Alligator River. The tree-lined banks along which the track runs afford further shade and relief from the sun:

> East Alligator River — aghhh, shade — cool, it is a nice shady spot [K14]

For a few participants, excitement was tinged with less pleasant feelings, derived from unspoken acknowledgement of the more repulsive characteristics of the sluggish, muddy river, dark waterholes and

---

[5] A *billabong* is the Australian Aboriginal term for pools, backwaters, or channels extending off a waterway. Frequently stagnant in the dry season, they are commonly covered with emergent vegetation such as waterlilies, and may contain saltwater crocodiles.

stagnant billabongs, and the imagined horrors hidden beneath their surfaces:

> stagnant billabong, insect kingdom — fascinated — half fascinated half horrified [K19]

Many participants clearly articulated their appreciation of the hazards and hidden dangers lurking within the waterways and billabongs. Indeed, for the majority of participants, the ever-present danger of crocodiles was seldom far from their minds, with consequent feelings of distress or unpleasantness:

> murky billabong — I thought it would be horrible if a croc killed me or one of my friends — latent fear, tense, moving on [K21]
>
> more dried up river — uncomfortable, tense, walking through this area I felt really uncomfortable — as I thought of the wet season and how usually this area would be full of crocs, it was like walking over someone's grave [K10]

For at least one participant, a misplaced faith in the safety afforded by the presence of a track[6] on a levee adjacent to the river was tempered by a healthy awareness of the ever-present danger lurking in the river:

> river — don't worry too much. If the track is here its because it's safe — a bit scary because of the awareness of crocodiles [K30]

While an appreciation of the biophysical characteristics of place dominated most responses, evidence of past Aboriginal occupation of the locality was to evoke an appreciation of the history of the place. Feelings of arousal and excitement coloured many responses to the several rock caves and shelters within the sandstone formations area:

> aboriginal hiding cave — unbelievable that people were hiding here — alive, intense, exciting [K32]
>
> rest at cave, rock art — I thought this place had seen a great deal of history. It must have been an important resting place for some time — very pleasant, Interesting, deciphering paintings [K21]

Some participants were drawn to the caves by the prospect of respite from the heat and glare of the sun. In the pleasure and relaxation that attended such moments, they came to appreciate some of the tangible and intangible connections between the caves and the original occupants:

> coolness of shade within rocks — how you can almost believe how spiritual and alive this place is — calm and peaceful and cool [K4]

For some participants a number of qualities and characteristics of place combined to create a sense of being somewhere special. The appreciation of such place was accompanied by feelings of pleasure and relaxation, or arousal and excitement:

> time to think, panoramic view around, hearing a lot of life — thinking how tranquil it is here, so glad it shouldn't change, I feel far better here than a city — happy, peaceful, pleased to be here [K20]
>
> picnic area — The 'human' touches in this area were reassuring, I'm not tough enough to go into the real wilderness. I admire those who do, now and in the past — relaxing, reassuring feeling from this area [K24]
>
> rocks and canyons area — x — this area of ancient rocks sculpted by the weather made me feel insignificant. I also felt elated by being in [an] area with little evidence of man's intrusion, in spite of the many visitors who came here [K24]

APPRECIATING LIVING THINGS

Moments spent *Appreciating living things* accounted for a further 17% of all experiences recorded on the Bardedjilidji Sandstone Track. The experience of feeling aroused and excited while appreciating the local fauna was common to many participants:

> skinks on rock — very cool little animals, nice to sit in shade and watch them! amazing how fast they are — active, exciting, cool critters [K12]

---

[6] A few days after these comments were recorded, the researcher observed a crocodile basking on this section of the track.

> flying cockatoos[7] — they can scream really loud, but look beautiful — active, alive [K31]
>
> pig family — x — felt excited to see the pigs[8], but also aware that they are intruders [K24]

Equally, pleasure and relaxation while appreciating wildlife attended the experience of others:

> bridge and lilies — a good spot to rest and watch the birds — beautiful, pleasing, satisfying [K18]
>
> jabiru[9] — x — beautiful, serene, interesting, lots to challenge the mind [K36]
>
> pigeon on the track — why don't you fly away? you should be scared of me — x [K30]

As well as the diversity of wildlife encountered along the track, many opportunities are afforded for appreciating plants up-close. In particular, feelings of pleasure and relaxation were associated with the appreciation of waterlilies on the billabongs:

> lilies — I love the colour of the lilies, could take a million photos — beautiful, nice to look at [K12]
>
> beautiful water lilies — wondered if I could get waterlilies to grow in a pond at home. I doubt it — pretty, beautiful [K14]

Other plants too, offered pleasurable opportunities for the appreciation of their distinctive characteristics, and not all participants were content with the visual appeal of plants up-close:

> smell of eucalyptus leaves — lovely smell if you crush them — lovely smell [K4]

The experience of appreciating plants up-close also provided for moments of arousal or excitement as participants appreciated different aspects of certain plants, particularly the fig, and its capacity to establish and grow within cracks in the sandstone:

> fig that grows through rock — how can a tree grow through a rock, wow! — beautiful [K32]
>
> the tree that grows on the rock — amazing that the tree can grow on this rock — amazing [K6]

Another species of fig attracted attention for the functional role it afforded Aboriginal toolmakers:

> the leaves from the tree — interesting that aboriginals use the leaves as sandpaper, get out of here, too many insects — interesting (the insects), interesting (aboriginals use the leaves) [K6]

Another ubiquitous plant of the locality is the pandanus, or screw pine, appreciated both for its role in Aboriginal economy, and its characteristic form. For one participant [K14], the plant was something of an obsession, an interest remarked upon by her walking companion [K13]:

> pandanus stop — thought about how cool pandanus is, was amused by how annoyed Sheryl is getting because of my love for it — interested, love the stuff [K14]
>
> pandanus info stop — boredom, heard so much about pandanus as my travel companion is in love with it — wondered if Amber's love for pandanus would fade or move to a more intense level in future, a little scared! [K13]

The appreciating of living things is by no means confined to the more conspicuous representative of the local flora and fauna, as the incidence of moments spent appreciating tiny creatures testifies. For some participants, even the smallest creatures elicited feelings of arousal and excitement:

> Joe on tree, Joe trying to climb tree, couldn't, too many ants, saw strange insect, fascinating colours — life everywhere, insect so mad to look at, so many parts and colours — apprehension turning to stimulation [K20]
>
> nest of bees up in the tree — there is a

---

[7] The *cockatoo* is an indigenous parrot.

[8] Wild, or feral pigs are not uncommon within the river flats of Kakadu National Park, where they are regarded as a pest.

[9] The *jabiru*, or black-necked stork, is a large (stands to 120 cm) indigenous bird, common around Northern Australian wetlands.

bee's nest, lets better walk off — exciting [K28]
butterflies — incredible, a grove of butterflies — wonderful, awesome, exciting, interesting [K15]

However, the context being tropical Australia, the walk also brought exposure to other less endearing bugs, including the ubiquitous Australian bush fly, the response to which was invariably displeasure or distress:

finish — irritating flies throughout the whole walk — x [K32]
x — annoying flies — x [K33]
insects — x — get on nerves (horse fly) [K34]
sandflies — must buy more insect repellent, wondered how aboriginal people coped with insects — exasperating, tense [K2]
overhead spiders webs — curious but scared — disgusted [K19]

For some participants, the focus of the interpretation program located at intervals along the track lead to boredom or lack of stimulation. However, one might speculate whether the response was due to perceived inadequacies in the interpretation rather than the stimulus itself:

termite mounds — I thought the mound not nearly as impressive as others, so thought is bizarre it was featured — dull and unimpressed. [K21]
bee? wasp? funnel formation entrance — x — unclear, anticlimax, boring [K8]

ATTENDING TO ASPECTS OF PLACE

Within the sub-theme *Attending to aspects of place* are categorised those responses to place which appeared on the stimuli sheet of participants' cognitive maps, and were unaccompanied by corresponding thoughts and feelings. Nearly 10% of all moments recorded at Kakadu fall into this category. For the most part, responses within this category were simple one or two-word responses, naming the stimulus encountered, such as "billabong", "cave", or "Alligator River". However, even without thoughts and feelings, the recording of stimuli still allowed for some richness in response, as with these examples referring to water and wetlands:

beside river bed a couple of pools filled with moss and mud — x — x [K10]
only a few pools where river usually runs — x — x [K10]

Together with water and wet places, rocks and rocky places figured prominently in participants' responses, reflecting a fascination with the geology of the place:

the 'flat pancake stack' rock formations — x — x [K21]
morning light on reddish sandstone — x — x [K2]
towering arches and columns of layered sandstone — x — x [K2]

The intimate, geologically rich nature of the place was the reason, perhaps, for the relative absence of responses to distant landmarks. Instead, general responses to characteristics of the immediate landscape were common:

dry creek bed — x — x [K36]
level open sandy area — x — x [K24]
island in the river — x — x [K3]

WONDERING ABOUT ASPECTS OF PLACE

While constituting only 8% of total responses, those that articulated a sense of wonder about place were often rich and complex. Not surprisingly, the experience of wondering about the local sandstone formations was often accompanied by feelings of arousal or excitement, as participants pondered the geological origins of the place:

sandstone outcrops — sandstone formations are a result of sedimentary deposition therefore area must have been a sea bed at some time — majestic, mysterious, ancient, forceful [K2]
rocky formations, interesting and highly visually pleasing, excellent erosion and interlaced flora — great rocks, how long has this taken, wonder what it was like in

> wet season, is the erosion water/flood or wind/rain, I think the latter? — restful, although stimulated. [K20]
>
> cave — why are just these rocks left? what has eroded them? Wish I had more time on my own to come back and do some sketching. Reminds me of some West Coast formations — interested, a little gloomy, curious, frustrated (not being able to make out paintings) [K22]

Moments spent wondering about water & wetlands evoked a diverse range of feelings as participants pondered aspects of billabongs, waterholes and wetlands. Some participants also noted the wet season/dry season differences, while others sought to imagine the environment during wet weather:

> lilypad plentiful creek — thought of sharks and rays able to swim up into creek through tidal flows — calming, still, pretty [K2]
>
> billabong — such a sudden change in habitat — unstimulating, apparently lifeless [K22]
>
> dry, sandy creek crossings — how much does the creek line change over the years? — thirsty, hot, tired [K3]
>
> water channel, water lilies — recognised lilies from yesterday, wondered if crocs would be in there, jacana[10] — slow, pleasing, alive [K8]

Wondering about the history of place lead to feelings of arousal or excitement, especially the experience of the caves showing evidence of Aboriginal occupation. These were eagerly sought out for shade and relief from the heat of the day by many of the participants, affording pleasure and relaxation, while providing an opportunity to contemplate aspects of Aboriginal history:

> cave — different pictures, one of a hand, how old are they? a hole in a stone, for what? when did people live here? — interesting [K17]
>
> cave of paintings — wondered what past inhabitants of caves lives were like, what the paintings were and meant — one of the best bits, mystified and amazed and a little in awe of history involved [K13]
>
> caves — imagined aboriginal people sheltering in the cave and looking for food in the area — relaxing, cool, pleasant, interesting [K1]

Not all comments, however, related to Aboriginal heritage, and the early European exploration of the place also caused participants to wonder:

> river — [thought] how early explorers were able to traverse country — caution (possible crocs) [K23]

OTHER EXPERIENCES

Of the remaining categories of experiences, a number of responses warrant attention for the further insight they provide on experiences characteristic of the Bardedjilidji Sandstone Track. While not significant in terms of the overall weight of reported experiences, some of these moments were nonetheless reflective of the general experiences of participants and should be recognised for the potential they have to colour the overall experience of the walk. After all, a single, momentary encounter with a crocodile has the potential to override many other more frequently occurring moments of experience!

Moments of experience within the *Wayfaring* sub-themes generally commenced at the very start of the walk, with comments related to participants' intentions, preparations, expectations and anticipated experiences:

> start — this is supposed to be a great walk, will it be? — high anticipation, exciting [K35]
>
> start — we are carrying too much stuff, flies are a nuisance — inactive (heat) [K23]
>
> start — wondered about what we were going to see — anticipation [K24]

---

[10] jacana: a bird of the wetlands and waterways of northern Australia, noted for its practice of walking on floating plants such as waterlilies.

Invariably, upon reaching the dry riverbed section of the walk, questions of the adequacy of preparations arose:
> sandy part — have we brought enough water?— hot and bothered, monotonous [K15]
> dry sandy open area —I need a drink — monotonous hard walking in sand, too hot [K36]

For some, though, the effort of walking the dry riverbeds did not preclude reflecting upon the connections between the experience of the moment and Aboriginal conceptualisations of walking the land:
> sand dunes, empty riverbeds — thinking about Bruce Chatwin's book "Songlines" and that reading it, I sort of could understand a little bit what it meant: singing your way through your county, a beautiful thought — pretty to look at but uncomfortable feeling, tiring, slowing down [K18]

Comments relating to *Wayfinding & progress* generally referred to the temporal dimensions of the overall experience, particularly the time taken to complete the walk. Those participants who commented on this aspect generally found time to be passing more quickly than they had anticipated:
> picnic area — has been a short walk up to the river — astonished about the short distance to the river, alive [K28]
> signs returning us to start — I thought the walk would be longer — happy, but slightly unsatisfied [K21]

While wayfinding itself was not an issue for most participants, the existence of a wet season route option was cause for some ambiguity on the part of at least one participant, emphasising the need for clear directions and route information:
> branch in the path (wet/dry) — which way home? — panicky, unpleasant [K3]

Several participants also commented on the absence of interpretation markers along the route:
> realised that 2+3 weren't there — x — dissatisfying for an info walk [K4]

For other participants, though, the track signage was evidently of a high standard:
> Car park junction — pleased and surprised, as always, to see how well the tracks are marked — happy to be active again [K18]

The proximity of the track to an adjacent access road was an issue for some participants, adding an unpleasant dimension to their experience:
> cars — couldn't the road be further away? — unpleasant to hear, see [K4]
> car — how cars may go everywhere even here you don't get rid of them — astonished, unpleasant [K9]

For others, the presence of boats on the East Alligator River proved an unpleasant distraction form the wilderness-like experience being enjoyed:
> cruise boat — too bad, I was enjoying feeling like an explorer — yucky, noisy, disruptive [K15]

The sub-theme, *Seeking, discovering, encountering* recognises the role that living things, whether animate or inanimate, play in the experience of natural landscapes. Moments spent searching for wildlife reveal expectations that participants would encounter wildlife, and not surprisingly the main object of attention along the Bardedjilidji Sandstone Track was the saltwater crocodile. Thoughts of crocodiles were a pervasive aspect of participant experiences, and while few participants were rewarded (or not, as the case may be) with a crocodile encounter, hopes, and fears were high:
> 'croc spotting' at river's edge — thinking I really want to see one in action, from a distance — very wary, nervous [K19]

There were some exceptions to the feelings of fear and anxiety, which characterised many responses about crocodiles, including the opportunity for some whimsy:
> boardwalk — where are the crocs? do they ever pop up with lilies on their heads? — exciting, interesting [K3]

The capacity for wildlife sightings to

transform and enrich the experience of natural landscapes, is suggested by this participant, whose response suggests the importance of the animate environment in colouring recreationists' experiences:

> Site 6 - drab forest corridor — fatigued, wished there were some wallabies — dull [K5]

Discovering and encountering living things provided many opportunities for thoughts of an interpretive nature. Thoughts about plants up-close tended to dwell on the rigours of plant establishment and growth, especially among the rocky sandstone formations:

> trees on the rocks with roots in cracks — must have taken a long time, wondering whether there's a story existing about it, there's a story to everything — sensational [K18]

> tunnel through rock — how do trees stay put? why do they persevere? do they realise where they're growing or does it just happen? — relief nice patch of shade, amazing rock formations and fig roots in rocks [K13]

Evidence of natural cycles and processes of decomposition, rarely observed within the management regimes of urban parks, were also acknowledged:

> great old fallen tree — in parks its good to see the circle of life, not possible in cities — sad [K18]

In an environment that, for many participants, was remote from their everyday experience, it was not surprising that *Curious about birds & animals* should turn to questions of identification:

> saw guilododo[11] — heard clapping sound they make, heavy looking bird — interesting, glad just read about it so we knew what it was [K12]

> billabong with lilies and birds — wish I could identify the birds — beautiful, serene, interesting, lots to challenge the mind [K36]

However, for one participant, a contrary position was suggested, whereby the possibility was entertained that the knowledge that comes with the naming of living things could diminish the sense of wonder inherent in encountering wild nature:

> x — what birds are they? wonder if knowing the answer would lessen my appreciation and interest — x [K13]

Not all thoughts about crocodiles centred upon the personal safety of the participant. Pondering the place of the crocodile in the food chain lead to variations in the concerns expressed about the crocodile as carnivore:

> crocodiles in the river — there are the crocs, I wonder if aborigines ate crocs? how many people died finding out which bush tucker is good? — exhilarating, exciting [K3]

> wetlands area — could spend hours here watching the birds, how did the pigs get here? Do crocs eat pigs? yes, bloody flies — x [K3]

Contrary to popular conceptions of the Australian outback as a malign, snake-infested void, encounters with snakes were uncommon, but certain to evoke strong responses:

> SNAKE! (on bridge) — Oh my God, how is Fiona going to react (she hates snakes) — peaceful and serene, beautiful, interesting, SPECIAL, tranquil (I love wet systems!) [K4]

We are left wondering how Fiona did indeed react, as no record of a close encounter with a snake appeared on her cognitive map – perhaps the snake had gone by the time she passed this point.

Wonder and fascination with living things was not confined to the larger and more dramatic examples of Kakadu wildlife, as this fascination with the microcosmic world of tiny creatures reveals:

> termite mound, dead meat ants, noticed the mound had been invaded by the ants, fascinating to watch — fascinated by the wildlife, wondering what goes on in there — fascination and wonder [K20]

A fascination with living things extends from a narrowly focussed fascination with

---

[11] Aboriginal name for the rock dove

the miniature worlds of the termites and ants to the scale of the wider landscape. In being curious about the savanna woodlands, participants revealed the capacity to consider the functioning of more general landscape processes implied by encounters with specific stimuli, as in the case of connecting the presence of long dry grasses with bushfires:

> very long bush grass for end of dry[12] — just wondering about burning cycles — interested, intensely arousing, disbelief [K19]

Similarly, a connection made to trees, presumably encountered elsewhere in tropical Australia, prompts thoughts on the processes operating in the distribution of plant species:

> enclosed amphitheatre of monsoon forest trees, also Qld [transcript unclear][13] — birds must travel long distances to spread forest seed — beautiful, serene, tranquil [K8]

The discussion on encounters with living things would not be complete without reference to the experience of *Being in the company of others*. Such moments may evoke very different responses, according to one's expectations. The Bardedjilidji Sandstone Track has a circuit layout rather than an 'out-and-back' configuration, and as such, the perceived absence of others afforded some opportunity for enriching experiences:

> finish — we hardly passed any people on the walk, this made it feel more adventurous than places where bus loads have turned up —x [K11]

Other encounters were deemed useful in terms of alerting participants to interesting features along the track:

> meet couple with video camera, tell me I've missed some art on rocks — frustrated at missing art, thought I'd been looking closely — x [K5]

For others, however, the social context of the walk is perhaps the pervading experiential theme.

> Joe on rock — wish I wasn't having such a bad day and was taking more notice — x [K22]

Joe was assumed to be a companion, and not a representative of the local wildlife.

Finally, this discussion must acknowledge the role that *Sensing atmosphere & ambience* played in the appreciation of the Bardedjilidji Sandstone Track. As the quotations below suggest, the rigors of walking within the challenging context of a tropical landscape of exposed rock and sand were only partially offset by the cool shade of the monsoon rainforest section of the track:

> sandy — ugh!! hot, tired — unpleasant, tiring [K15]
>
> intense heat zone reflecting from sand — x — x [K2]
>
> forest area — nice and cool, it is a bit too hot to walk in this heat — pleasant, cool [K9]

SUMMARY AND DISCUSSION

The World Heritage designation of Kakadu National Park suggests that the experience of the Bardedjilidji Sandstone Track might be dominated by place-related moments of experience, and indeed, this proved to be the case. *Making sense of place* emerged as the most common experience at Kakadu, and moments of experience within this theme were dominated by the single sub-theme, *Appreciating aspects of place*. The distinctive natural landscape of sandstone rock formations, tropical savanna woodlands, monsoon forest, billabongs and the East Alligator River are beyond the normal landscape experience of the majority of

---

[12] the 'dry' is the colloquial name for the dry season. Kakadu has a monsoonal weather pattern, with dry weather prevailing from April until October, and monsoonal rains for the remainder of the year, which is known as the wet.

[13] Qld is the conventional abbreviation for the Australian state of Queensland. It is assumed, but not known for certain, that this is a reference to the participant having encountered similar vegetation communities during travels in tropical Queensland.

visitors to the locality, and on this basis might be considered enough to warrant appreciative responses from visitors. However, when these natural landscape elements are combined with the evidence of over 40,000 years of continuous human occupation, a high incidence of affectively toned responses to place is understandable. While these were invariably of a positive nature, *feeling unpleasant or distressed* in response to *atmosphere & ambience* was not uncommon – it is a harsh environment for active recreation if one is ill prepared for the rigours of a tropical landscape.

No less than the landscape itself, the *living things* of the Bardedjilidji Sandstone Track evoked strong affective responses from participants, leading to moments within the sub-theme *Appreciating living things* being second to *Appreciating aspects of place* in occurrence. The prospect that certain of the living things – the saltwater crocodile – would not hesitate to make a meal of the unwary or careless nature walker helps focus the mind and add piquancy to the experience of the landscape. Actual crocodiles sightings, however, were scarcely noted by participants among the stimuli recorded on their maps, yet the pervading sense of anxiety and caution which accompanied the overall experience of most participants is testimony to a general awareness of their presence, even if unseen. Crocodiles aside, many other living things, including butterflies, birds and reptiles, were present in abundance, and elicited feelings of arousal or excitement, and pleasure or relaxation. Except, that is, for the ubiquitous and persistent Australian bushfly, enamoured by no nature walker at Kakadu, nor anywhere else.

Moments spent *Wayfaring* fared low on the scale of experiences, accounting for just 14% of experiences overall. Only moments spent *Negotiating the terrain*, had any appreciable impact on overall patterns of experience, yet still accounting for just 6% of all experiences at Kakadu. For the most part, the track was easily negotiated, yet a section of dry sandy riverbed elicited feelings of unpleasantness and distress. By comparison, the deep shade of the monsoon forest section of the walk was greeted by feelings of pleasure or relaxation.

As to the Environmental N method itself, the experience of data collection in the field and subsequent data analysis has indicated Environmental N to be a survey instrument with considerable potential for researching environmental experiences. In the case of *Making sense of the land*, the data was analysed for the purpose of characterising natural environment recreational experiences in sensemaking terms. Whether utilising the typology of recreational experiences, or adopting different approaches to data analysis to yield alternative schemas or typologies, Environmental N data has considerable potential to contribute towards the planning, design and management of protected natural areas. The sensemaking focus of *Making sense of the land* has revealed the method to have potential for the evaluation of interpretation programs, and for the design of walking tracks to optimise recreational experiences..

Informal discussions with park interpretation staff at Kakadu National Park raised the possibility that Environmental N may have application in field-based environmental education programs with children. Comments from adult participants that the completion of cognitive maps required a reflective mode of thought that served to consolidate the experience suggest that the same may occur with children. A recurring comment during fieldwork for *Making Sense of the Land* was the wish that participants could take their cognitive maps with them as tangible mementos of their walks. In the case of school children undertaking environmental education programs, Environmental N cognitive maps could be drawn for this very purpose and used as the basis of post-trip discussions back in the classroom. This task could be further tailored to the needs of school children through the development of a vocabulary of symbols or pictograms of common landscape stimuli, as was developed and used by Wood (Wood and Beck, 1976) in his original work with Environmental A.

LITERATURE CITED

AUSLIG (Cartographer). (1996). *Kakadu* [Topographic map].

Barker, R. G. (1968). *Ecological psychology*. Stanford, Ca.: Stanford University Press.

Beck, L. (1993). Optimal experiences in nature: Implications for interpreters. *Legacy* (January/February), 27-30.

Judd, C. M., Smith, E. R., & Kidder, L. H. (1991). *Research methods in social relations* (6 ed.). Fort Worth: Harcourt, Brace, Jovanovich Inc.

Knopf, R. C. (1983). Recreational needs and behavior in natural settings. In I. Altman & J. F. Wohwill (Eds.), *Behavior and the natural environment* (Vol. 7, pp. 205-240). New York: Plenum Press.

Lynch, K. (1960). *The image of the city*. Cambridge, MA: MIT Press.

Lynch, K., & Rivkin, M. (1959). A walk around the block. *Landscape, 8*(3), 24-34.

Markwell, K. (1996). Challenging the pedagogical basis of contemporary environmental interpretation. *Australian Journal of Environmental Education, 12*, 9-14.

Roggenbuck, J., Loomis, R., & Dagostino, J. (1990). The learning benefits of leisure. *Journal of Leisure Research, 22*(2), 112-124.

Russell, J. A., & Pratt, G. (1980). A description of the affective quality attributed to environments. *Journal of personality and Social Psychology, 38*(2), 311-322.

Starbuck, W. H., & Milliken, F. (1988). Executive's perceptual filters: What they notice and how they make sense. In D. Hambrick (Ed.), *The effective executive: Concepts and methods for studying top managers*. Greenwich, Connecticut: JAI Press Inc.

Steven, M.L. (2004). *Making Sense of the Land: A Sensemaking Approach to Environmental Knowing*. Unpublished PhD thesis. NSW, Australia: Faculty of Architecture, University of Sydney.

Steven, ML. Researching environmental experiences with Environmental N. In G.T. Moore, R. Lamb & D. Lu (Eds.), *Environment, Behaviour and Society*. Dordrecht, Netherlands: Springer (in press).

Strauss, A., & Corbin, J. (1990). *Basics of qualitative research*. Newbury Park, Ca: Sage Publications Ltd.

Thomas, J., Clark, S., & Gioia, D. (1993). Strategic sensemaking and organisational performance: Linkages among scanning, interpretation, action and outcomes. *Academy of Management Journal, 36*(2).

Uzzell, D. L. (1989). The hot interpretation of war and conflict. In D. L. Uzzell (Ed.), *Heritage interpretation: Volume 1: The natural and built environment*. London: Belhaven Press.

Weick, K. (1995). *Sensemaking in organizations*. Thousand Oaks: Sage Publications.

Weick, K. E. (1979). *The social psychology of organising* (2nd ed.). Reading, MA: Addison-Wesley.

Wicker, A. W. (1992). Making sense of environments. In W. B. Walsh, K. H. Craik & R. H. Price (Eds.), *Person-environment psychology: Models and perspectives*. Hilldale, NJ: Lawrence Erlbaum Associates.

Wood, D., & Beck, R. J. (1976). Talking with environmental a, an experimental mapping language. In G. T. Moore & R. G. Gollege (Eds.), *Environmental knowing* (pp. 251-361). Stroudsburg, Pa.: Dowden, Hutchinson & Ross.

CHAPTER 4

KEEP AN EYE ON NATURE EXPERIENCES: IMPLICATIONS FOR SIMULATION AND MANAGEMENT

Birgit H.M. Elands and Ramona van Marwijk

Abstract: Both ecologists and social scientists have focused on behavioral practices of visitors in nature areas, but less on the existential basis of 'what makes people visit nature areas?' This is especially valid in the modeling society. In this chapter, we conceptualize recreation quality by means of different experiential worlds of nature visitors. We link these experiential worlds conceptually with environmental values of nature areas that can be described in use value, perception value, narrative value and appropriation value. Two parameters related to use value (attraction) and perception value (crowding) are developed within an exploratory simulation study. Theoretical reflections and problems encountered within the simulation gave rise to develop an empirical study that focuses on the interrelation between nature experiences, environmental values and visitor behaviour. Data were collected by means of GPS and questionnaires and combined factor and cluster analyses revealed four hiker groups. Significant relationships between nature experiences and behaviour were found. However, not all inter-group differences in behaviour could be explained. A crucial factor for both nature experience and related behaviour is the level of attachment to the area. Empirical (monitoring) studies seem to offer valuable input and insight for simulation models, both at meso (area) and micro (path) scale. Still, setting standards for dimensions of environmental values and determining relative weights for model parameters is problematic. Management practices should not only be based on simulation outcomes; monitoring studies, discussions and public deliberations are equally important regarding public acceptance of strategies.

Key words: recreation, national park, simulation models, nature management, typology, nature experiences

INTRODUCTION

In densely populated countries both the protection of biodiversity and providing room for people to experience nature are important functions of nature areas. The combination of both functions is not entirely unproblematic. Large visitor numbers might threaten the protection of biodiversity and create problems for the ecological management of these areas. On the other hand, high levels of visitation may impact the quality of the visitor experience. Besides, not every nature experience is interchangeable. For example, a visitor in search of a wilderness experience wants to meet neither many other visitors nor visitors searching for fun and instant pleasure experiences. Moreover, the ecological management of a nature area might negatively influence nature experiences as well. Residents, for example, prefer in some cases managed landscapes over wilderness landscapes without active human intervention (Berg and Koole, 2006).

This chapter focuses on the protection and integration of nature experiences into management –and in particular- simulation practices. Traditionally, visitor management and simulation focus on the provision of path network and facilities for different kind of recreational activities. Visitors, however, use nature areas not only as physical setting for recreational activities, but as part of a search for meaningful nature experiences (Elands and Lengkeek, 2000; Davenport and Anderson, 2005)

The challenge for park managers is to create the appropriate conditions for nature experiences. Moreover, if the conditions are known, simulation models can integrate these conditions. Recent research shows that computer-based modelling can assist in evaluating the consequences of management measures for recreational quality (for an overview on recreational modelling: Lawson, 2006). Landscape ecology has already attained significant knowledge on spatial conditions necessary for the conservation of biodiversity and this knowledge has been integrated into models aiming at the assessment of ecological quality of landscapes (see Pouwels et al., 2007 in this volume).

Most recreation modelling activities have stayed at the level of visitor distribution and number of recreational encounters and they have not successfully evaluated experiential conditions of the landscape.

In this chapter, we want to explore the possibilities of integrating nature experiences into recreation management and simulation practices. We will conceptualize the different experiential worlds of nature visitors through characteristics of nature areas. For this purpose we will use the concept of 'environmental values'. We would like to investigate the relationship between nature experiences and subjective interpretations of the environment and spatial behaviour in order to find parameters that should be taken into account in modelling. We will then describe the methods used. Next, we will elaborate our theoretical framework empirically on the basis of two studies carried out in a Dutch National Park. Our first study concerns an exploratory simulation study. Theoretical reflections and encountered problems within this simulation study gave rise to our second empirical study, which validates our theoretical assumptions on empirical results. We will conclude with implications for visitor simulation and management.

DEFINING RECREATIONAL QUALITY THROUGH NATURE EXPERIENCES

Diversity of nature experiences

Standards of quality have appeared in recent decades as an important element of planning and managing parks and outdoor recreation. In the Netherlands, recreation quality is defined primarily in spatial terms, such as facilities for recreational activities (e.g. path network for walking or biking), attractive landscape elements (e.g. forests, waterfalls and architectural objects) or other area characteristics (e.g. accessible, quiet). However, this can be rather problematic as quality is not one-dimensional; it is also defined by place, time, scale, social and cultural conditions (Bourassa, 1990; Hull and Stewart, 1995; Daniel, 2001). Above all, individuals experience spatial quality in a different way, following from the specific goal or interest they advocate. Within the same line of thought, Jacobs (2006) explains that although people's experience of landscapes are related to the properties of the physical landscape, landscape experience is *not determined* by the physical properties of the landscape alone. Biological, cultural and personal factors influence the way people experience the landscape (Bourassa, 1990). Or as Hull and Stewart (1995: 422) state *"the quality of one's experience while viewing a landscape (represented by mood and satisfaction) seemed dependent on more than the views one encountered"*. Arler (2000) advocates that this does not mean that it is impossible for different persons to experience the atmosphere of a natural environment in a similar way. As a matter of fact, experiences that are *"altoghether private and idiosyncratic"* (p. 298), thus not in relation to environments, do not exist. Although we are aware that a certain object in a landscape can recall different emotional reactions to different persons, we assume that we can distinguish groups or typologies that show similarities in their behaviour and experience.

Consequently, we argue that recreation quality may be defined in terms of experiences (see also Elands and Lengkeek, 2000; Borrie and Birzell, 2001; Chhetri et al. 2004). Moreover, we argue that by using a typology of nature experiences, researchers, but also resource managers and planners, can understand the multiple meanings of a nature area for the visitor.

Different authors conceptualize nature experiences of people in various ways (Table 1). Lengkeek (1994) argues from a phenomenological perspective that recreation is a temporary switch from our everyday world to another imaginary world, which is not necessarily another place, but also another sense of time, another bodily feeling or a different awareness. Recreation is a search for experiences in 'out-there-ness', in which human beings distance themselves from everyday live (Lengkeek, 1994). Similarly, Urry (1990) views tourists as being in search for the 'extraordinary' or 'another reality'. Elands and Lengkeek (2000), based on a theory initially developed by Cohen (1979), argue that there is a range of experiences for which people are searching in nature areas. This differs from individual to individual and depends on the extent to which people feel attached to their everyday life world and on the extent to which people are searching for meaningful experiences in non-ordinary realities. Elands and Lengkeek (ibid.) conceptualize 'out-there-ness' into five different experiences or modes of experiences that can be viewed as motives giving substance and direction to free time behaviour. A visit to a nature area allows out-there-ness to be experienced: for having fun and pleasure in an unproblematic way *(amusement)*; for breaking loose from everyday reality and to regain energy *(change)*; for fascination,

Table 1. Experiences of nature

Modes of experience	Nature experience types	Purism scale
Based on sociological and philosophical thoughts (Elands and Lengkeek, 2002)	Based on philosophical thoughts (Abma, 2003)	Based on recreation management strategies (Stankey, 1973)
Amusement		
Change	Apolline	Urbanist
Interest	Socratic	Neutralist
Rapture	Dionysian	Purist
Dedication		

authenticity and symbolics of the landscape (*interest*); for experiencing (limitations of) the 'self' (*rapture*); and finally, for merging with the ultimate other world, which needs to be explored and appropriated (*dedication*). The 'other' that the recreationist is striving for, does not necessarily need to be another social spatial context than that of the everyday experience; the recreationist may be on a day trip to be together with the family (Elands, 2002). Moreover, the mode is context related: a person can vary in his type of experience depending on his life phase, the recreational environment and the company. However, a certain mode will be dominant during a longer period of time (Elands, 2002). In their book *Typical Tourists* Elands and Lengkeek (2000) have empirically tested their typology of experiences in seven case studies throughout the world. They found stable structures of experience, which motivate recreationist activities.

Similar work has been carried out by the philosopher Schouten (in: Abma 2003) and Abma (2003) who were inspired by Nietzsche's philosophy regarding Dionysian and Apolline drives and Socratic ratio. The Dionysian nature experience stands for unpredictability, uncertainty and immersion in natural forces; the Apolline nature experience represents peacefulness, familiarity and search for personal identity; the Socratic nature experience stands for knowing and mastering nature. Their nature experience typology is derived from the most prevailing nature images used by policy makers, namely Arcadian and wilderness nature (Keulartz 2004).

In the purism scale developed by Stankey (1973), nature as such can be conceptualized as a means or as an end in experiencing 'out-there-ness'. This typology provides a means to differentiate between various types of nature users, based on how well their definition of wilderness coincided with the definition of the US Wilderness Act. Purists are the most sensitive group concerning 'disturbance': they prefer to experience freedom and loneliness. The urbanists are the opposite group who are looking for a kind of organized experience and accept (and sometimes prefer) other users. The group of neutralists is in between (Ankre, 2005).

Environmental values[1]

These typologies all assume varying presence of human artefacts and influence in natural settings: decreasing from the mode of amusement to the mode of dedication, from Apolline to Dionysian, and from urbanist to purist. Empirical studies have proven that 'rapture' and 'dedication' modes of experience prefer wilderness nature, with pure and simple facilities and extensive nature management, while 'amusement' and 'change' modes of experience prefer cultivated nature, with good facilities and intensive nature management (Koole and Berg, 2004). Similarly, 'purists' prefer wilderness areas with few other visitors, facilities and regulations. 'Urbanists', on the other hand, prefer well-managed forests and nature areas (Shin and Jaakson, 1997); they tolerate other users – meeting others can even be a positive experience – and prefer a higher level of services and regulations (Vuorio, 2003). Dutch people in search of Dionysian nature prefer wilderness areas, but they do not, however, necessarily visit them, as in the Netherlands, such areas are either not accessible or over-managed (Abma, 2003).

We can conclude that the physical environment is relevant to its users in a variety of different ways. This person-environment interaction can be conceptualized in four different environmental values (Leng-

---

[1] Part of this section has also appeared in Marwijk et al. 2007.

Table 2. Environmental values (Lengkeek et al., 1997)

Value	Description
Use value	Instrumental or economic value, refers to functions.
Perception value	Good or bad, beautiful or ugly, this value refers to qualitative schemata or filters people have in their minds.
Narrative value	The articulation of a variety of interesting facts and stories about an area (history, symbols, etc.).
Appropriation value	The intensity of being mentally linked to the environment, to own or possess a specific place.

keek et al. 1997) (Table 2). Two of them go back as far as the book De Architectura, written by the Roman writer Vitruvius some decades B.C. He distinguished between *utilitas* (use, functionality), *venustas* (beauty) and *firmitas* (solidity). The last quality has been related to sustainability and points to the perspectives of maintaining functionality and beauty over a longer period of time. This notion represents a different order of conception (Daniel 2001); consequently, we do not include it in our discussion (Marwijk et al. 2007).

The notion of utilitas – or *use value* according to Lengkeek (1997) is usually well elaborated and acknowledged in recreation management. It is basically determined by the opportunities it offers for activities. A recreational product consists of core resources and supporting elements (Ritchie and Crouch, 2003). A National Park can be seen as a core element (it attracts visitors because of being a protected nature area), while the paths, parking places and way markers are supportive elements. A second important concept with regard to the use value is orientation (Lynch, 1960). A visitor's cognitive representation of the spatial environment influences the potential range of action as well as ideas on the preferred or possible use. Recent research shows that landmarks are distinct anchor points for tourists in nature areas (Young, 1999). Paths, signage and marked trails can also serve as anchor points for orientation and way finding in nature areas.

Within environmental psychology, the *perception value* can be explained by the arousal theory of Berlyne (1974), prospect-refuge theory of Appleton (1996) and the desire or people to understand and explore their environment (Kaplan and Kaplan, 1989). The environment has the potential to stimulate a person's level of arousal. Over-stimulation or under-stimulation creates uneasy feelings. The 'right' level of arousal creates a 'hedonic value'. Berlyne gives an evolutionary explanation for this preferred level of stimulation from the environment. Appleton (1996) suggests that evolution of homo sapiens in the savannah of East Africa created a 'hard wired' neurological preference for half-open landscapes in which people could simultaneously gain visual prospect over and refuge from hazards and other impediments to continued evolution. The Kaplans (1989) expand this theory by claiming that evolution required an ability to simultaneously understand and explore environment. For this reason, they introduced the concepts of coherence, legibility, complexity, and mystery. These psychological theories claim the existence of universal mechanisms within all human beings. The underlying assumption of the evolutionary approach is that landscape perception relates directly to physical attributes of the natural landscape. In Dutch research, this approach is theoretically and empirically elaborated by means of eight indicators, i.e. abundance of vegetation, degree of naturalness, degree of variation, abundance of water, abundance of relief, degree of landscape identity, degree of skyline disturbance, and degree of noise pollution (Buijs and Kralingen, 2003). Jacobs (2006) states that, although this theory ignores socio-cultural aspects in landscape appreciation, research suggests that these indicators are able to successfully predict the average perception value of the landscape. Theories on the social and cultural backgrounds of appreciation and preferences underline the differences between individuals according to groups to which they belong. Bourdieu (1979) pointed out that cultural preferences are passed on from generation to generation in the form of capital and embodied in 'habitus'. This implies that nature managers should not only focus on average perceptions, but also on individual differences in landscape appreciation.

Understanding beauty as the perception of aesthetics is not entirely unproblematic. Our appreciations are not only mobilized by physical appearances, but also by the cognitive dimension of 'knowing' what the object is about. MacCannell (1989) introduces the concept of 'attraction', the notion that the narratives related to objects define whether any object (landscape, building, etc.) becomes articulated as an object that is attractive to tourists. The observer who does not know the narratives of an object is able to experience beauty, which is related to general mental schemes of appreciation. The same observer, nevertheless, is not able to discern the object's touristic attractiveness. This notion made us separate – at least analytically – the perception value from narrative value. The *narrative value* refers to the construction of specific stories on an environment. It is in many ways embedded in the very concept of landscape itself Schama 1995). The physical appearances of natural environments are linked to symbols, meanings, talks and narratives, which are stored in the human mind and form the basis for understanding or even 'reading' a landscape. This reading of our natural environment is dynamic, as over time natural settings accrue new layers of symbolic representations (Corner, 1999). These layers of symbolic representations of our natural environment become especially relevant for tourism and recreation purposes (MacCannell 1989; Lengkeek et al. 1997). Interpretive facilities such as sheets and brochures, maps, roadside signs, walking trail signs and leaflets, information centres, and guided walks and talks (Ballentyne et al. 1998) are considered to be relevant in the construction and dissemination of collective stories. When the cultural stories or history of an area is recognizable for people (i.e. it can be 'read' in the environment), this is part of the narrative value (Buijs et al. 2004; Stichting Natuur en Milieu 2005). Besides those collective stories, people also have personal stories and memories upon which places are valued.

Finally, the *appropriation value* refers to the fact that people can symbolically 'own' the environment (Brouwer, 1999). This mental ownership is not intrinsic to the physical setting itself, but resides in human interpretations of the environment, which are constructed through experience with it. It is an evaluative (based on what you experience…) and responsive (…you develop a special bond with a place) form of transaction (Stokols 1978). Place attachment is a positive emotional bond that develops between people and specific places. Through these bonds, people acquire a sense of belonging and purpose that gives meaning to their lives (Relph 1976; Tuan 1977; Bricker and Kerstetter 2000). However, the plurality of the phenomenon of people's emotional bonds to places is diversely framed and studied by different researchers[2]. In this research, we build upon conceptual work by Stedman (2003) who demonstrated in his social psychological study that landscape characteristics underpin place attachment, which can be measured by asking people the importance of a specific place to them. People are more emotional in their experiences when they consider something their own.

Recreation management and simulation

Traditionally, visitor management focuses on facilitating recreational activities such as hiking, cycling and horse riding. This emphasis on the use value is logical, as it is seems to be the most direct link to visitor behaviour. By focusing on the instrumental use of the environment, the subjective interpretation on the environment might be overlooked. Perception value and narrative value are in general less considered in visitor management and simulation. Nature management in the Netherlands can be characterized as technocratic, in which experts define what nature is valuable from a natural scientific paradigm (Innovation Network Green Space and Agrocluster, 2004). However, recently nature management organizations started to show interest in visitor experiences (Association of Nature Monuments, 1999; National Forest Service, 2004). Also, they are more aware of the appropriation value, because these organisations have encountered public opposition to planned nature development issues. As a result they put more effort into communicating with and informing visitors and residents living close to protected nature areas. Nearby residents often hold stronger appropriation value related to the area. Interest in perception, narrative and appropriation values, however, is still not a strategic issue, that receives constant attention by nature managers. Moreover, the interplay between the different environmental values is not taken into consideration at all.

The environmental values help to unravel the different layers of meanings associated with nature areas. The recognition of these four environmental values allows recreation planners and managers to understand how different aspects of the natural environment give meaning to nature experiences. However, differences among individuals infuse multiple meanings on an environment. We suggest that different combinations of meanings can be clustered into different types of nature experiences. The multiplicity of experiences can be conceptualized as the symbolic landscape.

For management and simulation issues, it is interesting to know how different nature experiences as well as subjective interpretations of the environment relate to spatial behaviour. Figure 1 suggests that the search for meaningful nature experiences implies preferences for multiple combinations of environmental values. Actualization of these values results in spatial behaviour in natural settings. Behaviour often impacts the ability of an environment to produce desired levels of experiential quality and biodiversity. The diminishment of experiential quality and/or biodiversity values by visitor behaviour leads managers to take action that will protect the resource or the nature experience. As behaviour is not only an actualization of preferences for environmental values, but also a function of area design and choice of activities, facilities and information (Galloway 2002), managers are able to guide visitors. Our point is that models attempting to simulate visitor behaviour do not necessarily consider

---

[2] "Geographers have commonly taken a phenomenological approach, examining how spaces become places through personal activities and experiences. (…) Sociologists have applied a social constructionist perspective, exploring the shared values and symbols that when applied to a landscape creates common meanings.(…) Psychologists have taken a cognitive approach to sense of place." (Davenport & Anderson, 2005: 627)

underlying purposes of the behaviour related to realization of use, perception, narrative, and appropriation value in the context of the recreational experience. As such it is difficult to accurately predict implications of management actions designed to modify visitor behaviour without also considering the nature and purposes of behaviour. A next step is to find the most crucial parameters that predict where, when and what visitors will do in visiting a nature area. Summarizing our thoughts and ideas, we present a conceptual framework as a base for visitor management and simulation (Figure 1).

METHODS

We elaborate our conceptual model by means of two studies: (1) a simulation study and an (2) empirical recreation study, both carried out in the Dwingelderveld National Park in the Netherlands. Both studies serve as an example for the operationalisation of our conceptual framework and its potential benefits for recreation management and simulation. The first study conceptualizes both nature experiences as well as two environmental values, i.e. use and perception value in the simulation model MASOOR. The second study is directed towards the understanding of visitor behaviour through the perception of environmental values of nature areas. Although the background and methodology of both projects is rather different, the resemblances in theoretical approach and problem definition with respect to visitor experiences and behaviour provide valuable insights for visitor management and simulation. A final limitation of both projects is that they focus exclusively on hiking. Hiking, however, is the most popular outdoor recreational activity in the Netherlands (Statistics Netherlands, 2005).

Case study area

Dwingelderveld National Park (DNP) is a nature area in the north eastern part of the Netherlands. It was chosen as a case study area because of its recreational attractiveness and ecological quality. It contains 3,700 ha and consists of wet heath land (1550 ha) and a mixture of native deciduous and pine forests (2000 ha). The DNP belongs to the European Natura 2000 network of important nature conservation areas. The DNP receives at least 1.2 million visitors yearly (Visschedijk, 1990), although recently it is estimated that there are at least 1.6 million visitors each year (Milieufederatie Drenthe, 2003). It is a typical Dutch nature recreation area with an extensive recreational network for both short strolls (60 km marked trails that are each less than 7 km in length) and long walks, for cycling ('normal', racing, ATB) and for horse riding. Visitors can obtain information in the visitor centre or in two un-staffed information centres. They can

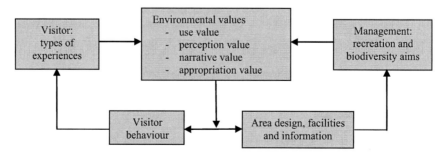

Figure 1. Conceptual framework

watch birds from two bird watch huts. Two sheep herds that contribute to the management of the heath land are very popular tourist attractions. The increased use of the area by recreationists has led to fragmentation of the heath land area and populations of animals have become isolated and threatened with extinction (Provincie Drenthe, 2000). At the same time, its National Park status implies that both recreation and ecology are important management goals.

Simulation study

The aim of our simulation study was to improve the theoretical foundation of the visitor simulation model MASOOR in order to assess the recreation quality of nature areas in the current situation as well as evaluate the recreational effects of management interventions in future scenarios. MASOOR (Multi Agent Simulation of Outdoor Recreation) is an Individual-Based Model, which means that it is capable of modeling variation among individuals as well as interaction among individuals (Jochem et al. Chapter 15). Moreover, it offers multi-agent simulations with autonomous agents, whose behaviour is specified by defined rules. The behaviour of recreationists is an outcome of the dynamic interaction of cognitive agents, capable of determining an individual hiking track according to their own nature preferences and characteristics with the social and physical configuration of the environment. MASOOR uses GIS data to represent the environment that consists of path networks. For the purpose of our study we have also included site facilities and attractions. Furthermore, MASOOR requires input in the form of visitor parameters (defined entry points, number of groups that will enter the area, length of stay, speed, and a frequency distribution of the entry time). MASOOR is able to distinguish between browsing visitors, who create their own track out of the existing path segments, and marked trail visitors, who follow a predefined track. Principles such as path type, crowding and other attractions influence the time-space behavior of browsing agents. Marked trail hikers are yet (because of pragmatic reasons) not able to leave the predefined route. A visit to a nature area consists in MASOOR of three phases, of which the final phase is split into two subphases: (1) "entry", the agent decides which attractions he is going to visit, (2) "immersion", after visits to attractions, the agent walks around until it is time to go back, (3) "exit (a) heading back towards entrance where the agent started the walk, and (b) take shortest way back to entrance (for more in depth information see Jochem et al. Chapter 15). The main visitor concepts of our theoretical framework have been operationalised into a social and a nature hiker (nature experiences), an attraction and a crowding parameter (based on use and perception value). Nature management practices have been operationalised in the current situation (for more in depth information see Elands and Marwijk, 2005.

Empirical study

The aim of our empirical study[3] is to develop a deeper understanding of visitor behaviour by means of gaining insight into the individual perception of the environmental values of nature areas by different types of visitors. Apart from the environmental values, we will also investigate the relation to background characteristics of visitors (age and group composition) and to their familiarity with the area. The results of this study will contribute to the definition and operationalisation of recreational parameters to be used in management practices and simulation models.

A survey was carried out to investigate

the symbolic landscape as defined by hikers. The survey consisted of two instruments:

- a questionnaire containing items exploring visitor motivations, environmental values, behaviour (e.g. entry point, destinations, attractions visited, marked/unmarked trails), perception of special places, and socio-demographics;
- a geographical position system (GPS) device carried by the visitors during their visit registered their spatial behaviour. The questionnaire contains some behavioural information as well (spatial goal in visit, places visited, following of marked trails, choice of starting point, place of rest during hike).

The survey population was targeted at hikers as they form the largest part of visitors to Dutch National Parks. Moreover, – contradictory to bikers – their points of origin and departure are often the same, thus making it easier to return the GPS device to the researcher. Since it is not possible to stay overnight in the park, the study automatically focuses only on day visitors. Visitors have been asked to participate in the research at five different entrances in the park: two main entrances close to a visitor or information centre, and three smaller ones. They survey was carried out during 7 days (weekend and working days) in spring and summer in 2006. The total research population consists of 461 hikers, including as many men as women, with ages ranging from 17 to 85 years. The response rate of the survey is 63%. When they arrived, visitors were asked to carry a GPS device during their visit. They completed the questionnaire when they returned from their visit and handed in the GPS device. As the number of available GPS devices was limited 400 hikers carried a GPS during their visit (for more in depth information see Marwijk, 2006; Marwijk et al. 2007).

SIMULATION STUDY

As described in the method section, this study aimed at improving the theoretical foundation of the visitor simulation model MASOOR. Behavioural rules that exist within the MASOOR model are mentioned in Table 1 of Jochem et al. (Chapter 15). In applying the concept of environmental values to the simulation of visitor behaviour at DNP, we decided to focus on use and perception value as there is hardly any knowledge with respect to norms and standards available on the narrative and appropriation value. Based on research on recreation experiences in natural landscapes, we selected the following two parameters: (1) views of and visits to specific spots, typified as 'attraction'[4] and (2) the specific experience with the presence of others, also characterized as 'crowding'[5]. Moreover, to maintain understanding of the internal dynamics of the simulation model, simplicity was another important reason to focus on only two experiential parameters. In this section, we focus on the elaboration behavioural rules related to attraction and crowding. Other behavioural rules that exist within MASOOR were kept constant for different hiker types in our study. For the description of those other behavioural rules[6], see Jochem et al. (Chapter 15).

The parameter of use value, *attraction*, can be interpreted as a pull factor: a feature of the world external to a person (Galloway 2002). According to Jansen-Verbeke (1988), the recreation function of an environment is based on the spatial concentration of a wide variety of facilities as well as on its characteristic features, i.e. the lei-

---

[3] This study belongs to the PhD project of Ramona van Marwijk that aims to theoretically and empirically ground relationships between the values of landscape characteristics and patterns of visitor use, in order to improve a management tool (simulation model) for effective ecosystem management (2005-2009).

sure setting and the activity place. This applies also to nature areas. Judging a natural site for specific activities, for example hiking, includes not only reflections on scenic quality, but also considerations on site attributes (Shelby et al. 2005). The activity place consists mainly of human-made attributes (e.g. visitor centers, accommodation facilities, catering, a network of paths, parking spaces, information signs, benches etc), whereas the leisure setting refers to more natural elements of the site (eg. wildlife, scenery). The provision of physical facilities in recreational areas often serves a dual purpose: they offer service to the visitors, but they might also include management actions designed to limit impacts on the natural environment (Vistad, 2003). The use value provided by attractions depends partly on its provision of services and facilities to afford (Gibson, 1986) realization of visitor experience expectations. From a semiotic perspective, attractions may also be 'signs' or manifestations of an underlying story about place. Fullest appreciation of attraction may require visitor understanding of the place story (MacCannell 1989). In this sense, use value is linked to narrative value.

Crowding, a parameter of perception value, is often viewed as the most direct social impact of outdoor recreation (Fredman and Hörnsten 2004). Crowding related research has been a dominant theme in the social psychology of leisure literature for several decades (Kyle et al. 2004). However, setting standards for crowding is problematic, because they depend upon visitor characteristics, number and type of encounters as well as situational variables such as configuration of trails and historic precedent (Manning et al. 1999; Fredman and Hörnsten 2004). Moreover, social norms can vary widely depending upon the measurement approach used. Manning et al. (1999) found that by using the traditional 'numerical' approach (where respondents are asked to evaluate a range of encounters with other groups along a trail) crowding norms were set higher than while using a 'visual' approach (using photographs depicting a range of use levels). However, this was only the case for visitors, as residents of near-by communities were more sensitive to the same number of people when represented in photographs than in numerical estimates. A weak point of the numeric approach is that it defines an encounter as a passing of persons, whereas the visual approach will also include visual encounters at a distance. Visitors tend to substantially underreport encounters if use levels are high (Shelby & Colvin, in: Manning et al. 1999).

Westover (1987) noticed that research on crowding in recreational settings has focused primarily on backcountry settings with relatively few visitors. Relatively few studies apply to front country settings (e.g. Fredman and Hörnsten 2004). In our MASOOR simulation, we chose the visual approach of Jacobi and Manning (1999) with respect to persons per viewscape. Because of landscape characteristics and trail configuration in the DNP, we maintained a viewscape of 50 meters.

We distinguished two visitor types based upon their search for particular nature experiences: a nature hiker (walking to discover and to get away from other persons) and a social hiker (walking to relax and eventually meet other persons). This distinction is consistent with the

---
[4] Behavioural rules that enable agents to visit attractions are 'global heading' and 'shortest distance' (see Jochem et al. Chapter 15).
[5] 'Crowding' is defined as behavioural rule (see Jochem et al. Chapter 15).
[6] Other behavioural rules are 'homing direction', 'chunking direction', and 'path segment history'. The weighting for those rules are kept similar for social and nature hikers.

research of Gimblett et al. (2000) who developed a 'landscape' and a 'social' recreationist. Within the spectrum of experiences, the social hiker resembles the amusement and change type, the urbanist and the Apolline nature experience orientation, whereas the nature hiker resembles the rapture and dedication type, the purist, and the Socratic and Dionysian nature experience orientations. This simulation study did not include the 'mid' types. Social hikers prefer comfortable and familiar areas, where they walk a relatively short trail, go to the visitor information centre, drink a cup of coffee at a cafeteria and experiencesa certain amount of crowding as pleasant. Nature hikers, on the other hand, like to visit unknown and special places, combined with a longer distance hike, take small and unpaved trails, avoid crowds, visit the bird observation hut or nature connoisseur places where they can discover the mysteries of nature.

We assumed the two hiker types to differ on average walking speed, time spend in the area, preference for certain types of attractions, and crowding (see Table 3).

In addition to this classification, we discriminate between 'marked trail' hikers, those who follow a marked trail or one that is described in a leaflet, and 'browsers', hikers who will choose their own route

Table 3. Characteristics of the defined visitor groups

	Social hiker (marked trail)	Social hiker (browser)	Nature hiker (marked trail)	Nature hiker (browser)
Part of total sample	64%	16%	4%	16%
Walking speed	3.5 km/hr	3.5 km/hr	4 km/hr	4 km/hr
Time budget	Duration of limited distance track (excl. visits to attractions)	2 hours (incl. visits to attractions)	Duration of long distance hiking track (excl. visits to attractions)	3 hours (incl. visits to attractions)
Attractions: points, areas, routes	Preference for marked walking trail. No special preference for other attractions.	Visit max. 3 attractions, one per category (visitor centre, sheep farm, catering, bird watch tower, special nature). Preference for attractions from categories 'catering and visitor centre'	Preference for described long-distance walking trail. No special preference for other attractions	Visit max. 3 attractions, one per category (visitor centre, sheep farm, catering, bird watch tower). Preference for attractions from categories 'special nature'
Attractions: path type	Not applicable (stays on route)	Preference for small (<2m) paved paths	Not applicable (strays on route)	Preference for small (<2m) unpaved paths
Crowding	Do not change direction because of crowding	Crowding moderately influences spatial behavior and experience	Do not change direction because of crowding	Crowding strongly influences spatial behavior and experience

based upon the existing paths in the nature area[7]. For the 'browsers' we developed decision-rules according to their preferences, in particular regarding available time budget, walking speed, attraction preferences (point, areas, routes, path type), and crowding norms. In the current version of MASOOR the marked trail hikers are not able to leave the predefined route. This means that when they start walking a marked trail, they have to completely finish it. They do not make shortcuts or walk extra parts. However, they are able to visit desired attractions that are situated adjacent to the marked trail upon which they hiked.

To run the simulation study, we collected quantitative data about the daily number of visitors and the number and size of all parking places. Research on visitor countings has shown that on one ordinary Sunday at least 7,000 people visit the National Park (Visschedijk, 1990). Half of the visitors are hikers and the average group size is 2. According to the National Forest Service, 80% of the hikers can be characterized as social hikers, while 20% are nature hikers. In addition, 80% of the social hikers walk marked trails, while 80% of the nature hikers construct their own trail by piecing together various parts of the path system. Both nature and social hikers were assigned to nine entry parking areas proportional to the number of parking spaces at each site. We assume no differences in entry preferences between social and nature hikers.

The spatial distribution results of the simulation run are depicted in Figure 2. This information can be used for an assessment of recreational pressure on the ecological quality of the area (Pouwels et al., Chapter 14).

The model assessed the extent to which each visitor group has been able to visit the preferred attractions and the numbers of encountered people during the visit. It is remarkable to observe that, while browsing agents were able to visit up to three attractions per agent, only 27% of them indeed visited one or more attractions. The other 73% did not visit an attraction. This might be caused by a

[7] In the Netherlands, it is compulsory for visitors to confine their hikes to only designated paths. A follower of a 'marked trail' is one whose visit takes place on a trail that is identified and designated by managers. 'Browsers' construct 'social' trails by piecing together various part of the entire path system, whether marked or not

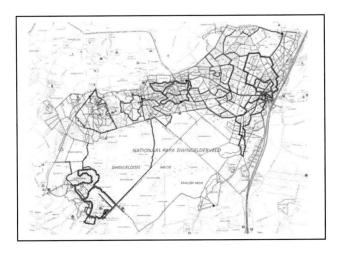

Figure 2. Visitor density map based on MASOOR simulation in DNP

scarce supply of attractions within their range of action or by high crowding densities, which make them less preferred by especially the nature hiker. Of all hikers that do visit an attraction, social hikers are more likely to visit larger attractions such as the visitor centre, the sheep farm and catering services. Nature hiker visits favour special nature areas. These outcomes are results of the preferences within the behaviour rules. However, we have to conclude that most hikers do not succeed in visiting an attraction.

With respect to crowding we found that an agent encounters an average of 51 other hiker groups on an ordinary Sunday. The browsing nature hiker meets 49 other agents compared to 48 for the browsing social hiker. This is remarkable, since nature hikers are more sensitive to crowding and thus prefer paths containing fewer agents. We are not able to explain this result. It might be that there are few quiet paths on an ordinary Sunday. Browsing agents have fewer encounters than marked trail agents. Social marked trail hikers have 53 encounters, compared to 50 of the nature marked trail hiker. This result implies that browsing agents are more able to avoid busy paths. What does the absolute number of encounters tell us? In another front country setting, Fredman and Hörnsten (2004) found that during a three kilometer hike (lasting 1.5–2 hours), less than 25% of the visitors felt crowded when encountering less than 50 others, while 60% felt some degree of crowding when the number of perceived encounters is above 50. Besides, we have to bear in mind that the actual number of encounters is higher as we did only take hikers (who account for 50% of the all visitors to DNP) into account. Encounters with bikers horse riders are not taken into account.

Lessons learned

This simulation study is a first attempt to conceptualize visitor groups based on nature experiences with different expectations regarding the environmental values programmed into MASOOR. By doing this we encountered several problems.

First, it proved to be difficult to set standards for experiential parameters for different visitor groups. A literature review on crowding norms for example made us aware of this difficulty. Not only do scientists disagree on crowding norms, the perception of crowding is culturally defined. The standards that have been set for nature areas in the USA (e.g. Manning et al., 1999) or Sweden (e.g. Fredman and Hörnsten, 2004) are likely not to apply to the less remote Dutch nature areas. Moreover, as people have different modes of experiences, crowding does not always negatively affect the quality of the recreation experience. In contradiction to research of Manning and Cole (2001) found that use density has little effect on the quality of recreation experiences. Moreover, Hammitt and Patterson (1993) found that the number of parties encountered during a hike even enhanced the recreation experience of 42% of their sample. According to Bacon et al. (2001) crowding norms appear to be relatively stable over time, while Cole and Stewart (2002) found substantial variation in the standards provided both among individuals and over time within individuals. All this makes it difficult to set an evaluative standard for crowding.

Next, the operationalisation of the parameters in the simulation model forced us to make choices on the relative importance and the mutual influence of the different experiential parameters. Determining the relative weights between crowding and other factors such as time, path types, visits to attractions proved to be difficult to do. It is also unclear how exactly people will adapt their on-site behaviour. Cavens et al. (2004:74) had the same problem: *"For instance, is it better for an agent to spend slightly longer than expected on a 3 hour hike in order to avoid a particularly steep section or visit a scenic point?"*.

Moreover, we integrated several assumptions with respect to length of hike, available time budget, distribution between marked trail and browsing hikers, etc. In the absence of literature to rationalize our decisions, our assumptions were based on managerial best judgment as opposed to empirically based information.

The outlined simulation problems as well as our managerial assumptions regarding characteristics of different hiker types, gave rise to the development of an empirical study that exclusively focuses on the interrelation between nature experiences, environmental values and visitor behaviour.

EMPIRICAL STUDY[8]

Based on the results of the MASOOR simulation study as well as a scientific objective to gain theoretical insight into experiential parameters, the empirical validation of the environmental values was further elaborated within the PhD-research of Ramona van Marwijk (Marwijk, 2006). Marwijk et al. (2007) have used the four environmental values to gain insight into the symbolic meanings of an environment and its relationship to spatial behaviour. Hikers at Dwingelderveld National Park (DNP) were asked to rate 42 statements that measure the environmental value of the area they had visit. A factor analysis (using Varimax rotation) on these statements revealed 10 underlying dimensions. Table 4 shows the 10 dimensions that were derived from analysis of the 42 statements pertaining to environmental values.

Based on a cluster analysis of the factor scores, four different groups were identified (for a detailed description of the methods used see Marwijk et al., 2007). The defined names of the four groups are based on their scores across the 10 factor dimensions.

The first group is the *happy hiker* (31%). This group evaluates both the use and perception value of DNP positively. Although happy hikers are not familiar with stories about the park, nor are they attached to it, they perceive the park as prototypical of landscapes in this part of the Netherlands, but otherwise very unique within the country. Compared to the other three visitor groups, happy hikers are especially inclined to report DNP as being attractive. They are not annoyed by management actions or other visitors in the park.

A second group is the *connoisseur* (25%). This group of visitors is highly attached to the park and is familiar with many park narratives. The environmental values relating to use and perception are evaluated positively. The connoisseurs' high familiarity with the park apparently grants them the prerogative of being critical toward management actions. They are disturbed by the logging of exotic trees and the raising of groundwater levels as the connoisseur believes the area should not be turned into a 'primeval forest'. On the other hand the area should not be too highly developed for tourism.

The *demanding hiker* (25%) is much less oriented in the park than are other visitor types, is less familiar with park facilities and services and finds DNP to be not very accessible. Although the area is perceived as being very tranquil, demanding hikers rate its attractiveness the lowest of all groups. They do not feel attached, nor do they know about park narratives. Demanding hikers desire to have additional and clearer signage in the area. They are easily irritated by issues such as a closed tourist information office in neighbouring villages on Sundays, the size of the places to eat (too small), the menu (limited offer),

---

[8] The empirical study is more extensively described in Marwijk et al. (2007).

Table 4. Ten dimensions of environmental values

Environmental value	Dimension
Use Value	orientation number of facilities accessibility
Perception value	attractiveness tranquillity naturalness non-annoyance (by both other people and management actions)
Narrative value	familiarity with cultural history/stories of park, park environment park environment is uniquely prototypical in the Netherlands
Appropriation value	personal attachment to park

bikers on walking tracks, wet paths, and areas that are closed for hikers.

The final group is called the *disturbed hiker* (19%). Like the happy hiker, this visitor group assesses the use value of the park positively and does not recognize the narrative and appropriation values. However, regarding perception value, they feel that the park is very busy and noisy, more artificial than natural, and is not very attractive. Moreover, the disturbed hiker is more annoyed by management actions and other visitors than other hiker groups. The disturbed hiker emphasizes the importance of creating only basic recreational facilities in order to keep the area as close to nature as possible and separating different user groups spatially (hikers, bikers, and dog walkers).

Summarizing, the happy hiker has the least problematic experience of nature. Most values are perceived in the landscape by this group in a positive manner. The other groups are more critical. The connoisseur is critical towards management actions, the demanding hiker towards availability of services and facilities relating to comfort, and the disturbed hiker relating to crowding, noise, and naturalness. The views of the latter two groups seem to be a result of high expectations and little knowledge/acquaintance with the area. These groups are unable to escape places perceived to be overdeveloped for tourism or to find user friendly facilities and services.

Visitor behavior

Interestingly, the behaviour of these experiential groups is diverse (Table 5). All four groups tend to begin their visits at approximately the same time of day. However, the disturbed hiker visits the park mainly during the weekends, while the other groups come both during weekdays and weekends.

Two of the five parking areas included in this study can be classified as large (>100 car places), well equipped (information centre, catering), and well advertised from the highway. The other three are relatively small with few facilities (only picnic tables). As expected, connoisseurs know their way and consequently tend to use the smaller parking facilities. Surprisingly, a large part of the demanding hikers find their way to the smaller and more poorly served parking places. Visits of the disturbed hiker start at bigger parking places. This might explain the fact that this group is most likely to be annoyed by disturbance of other people and thinks that the area is very busy and noisy. The happy and the disturbed hiker start also predominantly at the large parking places. When they start at the main visitor centre, it is not because of its accessibility but because of its proximity to a popular attraction relating to the park's sheep herds. Happy hikers choosing to start at a small parking area are motivated by its high level of accessibility.

Except for the connoisseur, most visitors walk marked trails. This is not surprising since the connoisseurs are most acquainted with the area. This reinforces the idea that familiarity increases the possibility of 'off the beaten track' behaviour (Hwang et al. 2006). Especially the demanding hiker, who feels much less orientated in the park compared to other groups, follows marked trails.

Connoisseurs spend the least amount of time during their visits (1hr 36min.), while disturbed hiker visits are longer in time (2hr 11min). Average hike length is not significantly different among the four groups, ranging from 5.6 km for the connoisseur to 6.5 km for the disturbed hiker.

Concerning places that have been visited, results show differences between the four groups. Whereas the happy hiker visits the main attractions (a local tea house and the sheep farm), the connoisseur knows where to find special nature places (to pick currants) and seems to avoid the tea house. The main visitor centre is popular amongst disturbed hikers, which might be the reason for their critical attitude towards crowding. There is no place that is particularly more likely to be visited by the demanding hiker. In fact, this group does not explicitly mention to have visited specific places. The hike itself seems to be the main activity of the day.

Interestingly, 71% of the connoisseur lives in the region of the DNP compared to less than 29% of each other group. Also, most connoisseurs (95%) are repeated visitors, whereas at least one third of the other groups visited the protected nature area for the first time. Almost half (49%) of the

Table 5. Behavioural characteristics of different experiential groups in National Park Dwingelderveld

	Happy hiker	Connoisseur	Demanding hiker	Disturbed hiker
Day of visit (%)				
• Weekday	53	56	46	29
• Weekend day	47	44	54	71
Parking place (%)				
• large and equipped	57	39	42	75
• small and simple	43	61	58	25
Follow marked trail (%)	73	44	77	69
Average time of stay (hrs)	1:56	1:36	1:44	2:11
Places visited during hike (%)				
• visitor centre	30	19	19	37
• tea house	20	10	7	18
• currant trees	3	12	3	3
• sheep farm	22	11	8	24
Living locally in the area (%)	29	71	20	22
First time visitor (%)	32	5	49	37
Visit frequency (%)				
• Seldom (once a year)	50	14	62	63
• 2-12 times per year	41	35	25	31
• Weekly/daily	9	51	13	6

All differences are significant ($p<0.01$)

demanding hiker group is a first-time visitor. This might explain why this group focuses on use values of the area.

To summarize, we can conclude that nature experiences in DNP, based upon the National Park's environmental values, affect spatial (starting point, visited places, walking marked trails or not) characteristics of visitor behaviour. However, insight into nature experiences does not give a complete insight into visitor behaviour. We have seen that small parking places attract both connoisseurs and demanding hikers (see also Figure 3), while the groups construct very different patterns of symbolic meaning of DNP.

The existence of insignificant difference between average hike lengths among the four groups for example suggests that there may be other factors besides symbolic meaning that should be considered. Therefore we examined relationships between behavioural characteristics of different demographic groups. We found several significant relationships. The majority of adults who come alone, for example, (59%) do not follow a marked trail, while over 64% of other demographic groups (adult couples and groups and families with children) do follow marked trails. However, the relationship between type of day and behavioural group is not significant, in contrast with the findings of the four experiential groups (happy hiker etc.). Most interesting finding is that we found significant differences in lengths of hikes. Adults alone walk an average of 5.4 km, while groups of adults walk 7.2 km.

Lessons learned

First, our study suggests that subsets of symbolic meanings of an environment do exist among visitors to a nature area. Furthermore, different types of visitors appear to construct varying symbolic meanings based on use, perception, narrative, and appropriation value. We found significant relationships between the groups and behavioural characteristics. For this reason we think it useful for management and simulation purposes to construct some kind of proto-typology for nature visitors in front country settings, with differing preferences for different dimensions of environmental values that should be linked to area characteristics. The path network should thus not only be characterised by, for example paved-unpaved (use value), but also by the other dimensions of use value (e.g. orientation) as well the dimensions of the other environmental values.

Second, we found that variation in construction of symbolic meaning among visi-

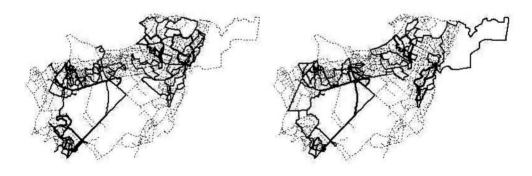

Figure 3. Spatial behaviour differences between demanding hiker (left) and connoisseur (right)

tor groups is able to explain some but not all inter-group differences in visitor behaviour. Therefore, we assume, besides nature experiences, that other visitor characteristics, such as demographics, group composition, and motivation influence visitors' spatial behaviour. Consequently, these elements should be incorporated in simulation models. Currently we are dealing with these issues and we will try to give a comprehensive picture in 2009.

Third, the outcomes of a monitoring study among nature visitors are helpful in defining the input for simulation models at different scales. At the meso scale, i.e. the level of a National Park, information is needed about spatial and temporal preferences of visitors: certain visitor types prefer larger parking places, weekend days and start halfway through the day, while others start at smaller parking places, avoid weekend days and start early in the morning. Combined with information valid for the micro scale (i.e. preferences for types of attractions, tranquillity, orientation, etc.), we expect outcomes of the simulation models to be more valid.

Fourth, the outcomes of monitoring studies among nature visitors help nature managers to improve their management. Disturbed hikers, for example, find DNP busy and noisy, and less natural than other groups. This might be related to the fact that they start at large and well equipped parking places. If nature management is able to direct these people to smaller parking places, their nature experience might improve. Other improvements can be made regarding to signage. Visitors were asked in the questionnaire if they walked a marked trail, whether they completed it. Some visitors noted that they got lost. Based on the recorded GPS tracks, nature managers are able to find out at which location visitors got lost, and improve signage. These examples show the importance of monitoring studies that, according to the researchers' opinion, should be continued even if simulation models are being used.

IMPLICATIONS FOR MANAGEMENT AND SIMULATION

Theoretical reflections

The conceptual framework we use gives valuable insight into major aspects of the interaction of visitors and nature areas. It is theoretically grounded as well as inspired by empirical recreation studies in nature areas. We claimed that by means of this model both visitor experiences and multiple environmental meanings can be understood.

Although both studies contributed to this claim, it is clear that the operationalisation of the conceptual framework in the simulation study was rather limited in its approach. Consequently, the outcomes of our simulation efforts raised several questions, which need further evaluation.

We also assumed that the conceptual framework could explain visitor behaviour. This assumption proved to be partly valid. The empirical study showed there are significant relations between multiple combinations of environmental values and behaviour, but that not all inter-group differences in visitor behaviour could be explained. Demographic factors, such as age and life phase, and familiarity demonstrated to be important predictors of behaviour as well.

We want to underline the complexity of understanding nature experiences and behaviour and its determinants. We realize, however, that due to its complexity the conceptual framework is not easy to operationalise into relevant variables for managers. Besides, we do not think that the conceptual ideas should be dealt equally within visitor management practices. Next, not every environmental value, and its subsequent operationalisation into dimensions, needs to be integrated in simulation models. As discussed earlier, the relation between experience, environmental values and time-space behaviour is not linear, but influenced by several other factors. A crucial factor determining modes of nature experiences as well as evaluation of environmental values is the extent of attachment to or the level of familiarity with an area. In our studies, this was substantiated by the nature hiker and the connoisseur. According to Farnum et al. (2005) resource managers, planners and researchers are beginning to view place attachment as a critical concept both in understanding how to provide optimal recreation experiences and in understanding the public's reaction to and proper role in management actions. We doubt whether simulation models can offer a solution for these kinds of situations. Simulation models can anticipate influences on visitor use of the area. However, public acceptance of management strategies should also be taken into account. Therefore qualitative research work, discussions and public deliberations (e.g. with participation groups) should be part of management practises as well.

From theory to simulation

Out of both studies some valuable insights for the elaboration of simulation models can be derived. First of all, area information is needed at two different spatial levels. At the meso scale, which is the level of the entire nature area, information with respect to the amount of visitors as well as with respect to the spatial and temporal distribution of visitors across the area is needed. At the micro level, information with respect to environmental values should be included in the model. For example, information regarding trail configuration and functional path characteristics (use value); aesthetic attractiveness of path (perception value) and information on distribution and attractiveness of marked trails, nature attractions (narrative value) could be more widely distributed to visitors. Appropriation value is problematic as it cannot be assigned objectively to path characteristics. Rather it is a sign of long term relationship between a visitor and a specific area, which influences the perception and interpretation of the other environmental values.

Another important source of information is related to the visitor. One should distinguish different visitor types based on the experience of multiple combinations of environmental values. Next, information about demographics and familiarity with the area is needed. Finally, insights into basic characteristics of behaviour, such as activity, spatial aspects (e.g. entry point and activity length) and temporal aspects (e.g. starting time, activity duration, type of day and season) is essential. On the basis of this information, a proto-typology of nature visitors in front country settings could be developed. This proto-typology of nature visitors includes preferences for the different dimensions of environmental values, which can be linked during the simulation process to path characteristics.

Reflections on simulation modeling practices

Simulation models like MASOOR are constantly in development. Model validation is important for a simulation model to be an acceptable research or management tool. We would like to discuss the internal validity – the logical correctness of the argumentation of the model, and the external validity – whether the outcomes of the simulation model have meaning outside the realm of the model (Nooij, 1990). The internal validity, the logical correctness of the argumentation of the model, is a main concern. As discussed in the simulation study section, both setting standards for dimensions of environmental values, and thus for recreational quality, as well as determining relative weights for the parameters in the model is rather problematic. Either theoretical studies reveal contradictory results or theoretical knowledge is simply absent. A way to validate the predictions of the simulator is to compare the simulation outcomes with the outcomes of the empirical study (see Jochem et al., Chapter 15). A more fundamental question concerning the internal validity relates to the approach of simulating. We have decomposed the landscape into discrete components (paths, attractions, other visitors), yet experiencing landscapes is a process of perceiving landscape as a whole entity.

After checking the internal validity of the simulation model, we have to assess the external validity. This means the model will be applicable in other natural settings in densely populated countries. The construction of proto-typology of nature visitor might be a helpful tool. At present, the external validity is tested in other natural settings, such as in the Fontainebleau Forest (France) and in the New Forest (UK). However, no correlations between the modelled and actual behaviour have been calculated. In the future, the model might be applicable to other recreational settings, such as amusement parks or zoos.

Next, we would like to state that the strength of a model lies in its simplicity, which is especially relevant for the model user, who prefers a simulation model that requires a minimum of data input, yet produces a maximum of recognizable and manageable output. With a small amount of parameters that predict behaviour, the cumulative chances of uncertainty are relatively small. Currently, the basic assumption of the model is ambiguous in the sense that is originally a visitor behaviour model, but with aspirations towards determining the recreational quality. For validity reasons (and the simplicity and usability wish) it might be advisable to focus on spatial behaviour exclusively, and supplement the simulation outcomes with qualitative studies to explore specific nature experiences. The strength of simulation lies in the testing of the effectiveness of alternative management practices.

Finally, we would like to discuss the legitimacy of simulation models as a source of scientific knowledge in management and decision making processes. If simulation models are developed as decision making tools, what role do scientists play? Do scientists only describe the facts or do they set the normative standards? For pragmatic reasons we, as researchers, have chosen to set the standard in this research, although we agree with Cole (2001) that science can produce helpful descriptive information, but it cannot make the decision-making process less subjective and judgmental in nature. Scientists need to be careful not to mislead managers about the utility of their empirical research, for example when using simulation models. Manning (2002) agrees that empirical relationships can be helpful in making informed decisions about management actions, but they must be supplemented with management judgments. According to Evans (2000: 223) in his paper on economic models, we should develop new ways of using models. "Rather than using models as 'truth machines', successful policy-making in the new context requires decision makers to re-define economic models as a 'discursive space' in which users and economists [experts], through a continuous dialogue, can develop shared understandings."

ACKNOWLEDGEMENTS

We would like to thank David Pitt for his extensive, helpful comments and language checking on earlier versions of this manuscript. The elaborated and useful comments of our reviewers have also helped us to improve the chapter.

LITERATURE CITED

Abma, R. 2003 De recreant als subject. Over de natuurbeleving als basis voor de beeldvorming van natuur. Thesis, Staatsbosbeheer/Wageningen, University, Driebergen/Wageningen.

Ankre, R. 2005 Visitor activities and attitudes in coastal areas: A Case Study of the Luleå archipelago, Sweden. ETOUR, Östersund.

Appleton, J. 1996 The experience of landscape. John Wiley & Sons, London.

Arler, F. 2000 Aspects of landscape or nature quality. Landscape Ecology 15, 291–302.

Association of Nature Monuments 1999 Enjoy nature: recreation policy at Natuurmonumenten (in Dutch). Vereniging Natuurmonumenten, 'sGraveland.

Bacon, J., Manning, R.E., Johnson, D. and Kamp, M.V. 2001 Norm Stability: A Longitudinal Analysis of Crowding and Related Norms in the Wilderness of Denali National Park & Preserve. Managing Recreational Use 18, 62-71.

Ballentyne, R., Packer, J. and Beckmann, E. 1998 Targeted interpretation: exploring relationships among visitors' motivations, activities, attitudes, information needs and preferences. Journal of Tourism Studies 9, 14-25.

Berg, A.E.v.d. and Koole, S.L. 2006 New wilderness in the Netherlands: An investigation of visual preferences for nature development landscapes. Landscape and Urban Planning 78, 362-372.

Berleyne, D.E. 1974 New environmental aesthetics. In Studies in the new experimental aesthetics: steps toward an objective psychology of aesthetic appreciation D.E. Berleyne (ed). Academic Press, London, New York, pp. 1-33.

Borrie, W.T. and Birzell, R.M. 2001 Approaches to measuring quality of the wilderness experience. In Visitor use density and wilderness experience: proceedings; 2000 June 1–3; Missoula, MT. Proc. RMRS-P-20. W.A. Freimund and D.N. Cole (eds). Department of Agriculture, Forest Service, Rocky Mountain Research Station, Ogden, pp. 29-38.

Bourassa, S.C. 1990 A paradigm for landscape aesthetics. Environment and Behavior 22, 787-812.

Bourdieu, P. 1979 Distinction. A Social Critique of the Judgement of Taste. Routledge and Kegan Paul, London.

Bricker, K.S. and Kerstetter, D.L. 2000 Level of specialization and place attachment: an exploratory study of whitewater recreationists. Leisure Sciences 233-257.

Brouwer, R. 1999 Tourism in the arena; a sociological reflection on the meaning of tourist processes for the social and physical environment in the Euregion Rhine-Meuse (in Dutch). Wageningen University, Dissertation, Wageningen.

Buijs, A.E., Boer, T.A.d., Gerritsen, A.L., Langers, F., Vries, S.d., Winsum-Westra, M.v. and Ruijgrok, E.C.M. 2004 Gevoelsrendement van natuurontwikkeling langs de rivieren. In Reeks Belevingsonderzoek. Alterra, Wageningen.

Buijs, A.E. and Kralingen, R.B.A.S.v. 2003 Het meten van beleving; inventarisatie van bestaande indicatoren en meetmethoden. Alterra, Wageningen.

Cavens, D., Gloor, C., Nagel, K., Lange, E. and Schmid, W.A. 2004 A Framework for Integrating Visual Quality Modelling within an Agent-Based Hiking Simulation for the Swiss Alps. In Monitoring and Management of Visitor Flows in Recreational and Protected Areas: Policies, methods and tools for visitor management, June 16–20, 2004. T. Sievänen, J. Jokimäki, J. Saarinen, S. Tuulentie and E. Virtanen (eds). Finnish Forest Research Institute, Rovaniemi, Finland, pp. 69-76.

Chhetri, P., Arrowsmith, C. and Jackson, M. 2004 Determining hiking experiences in nature-based tourist destinations. Tourism Management 25, 31-43.

Cohen, E. 1979 Rethinking the sociology of tourism. Annals of Tourism Research Jan/Mar 1979, 18-35.

Cole, D.N. 2001 Visitor use density and wilderness experiences: a historical review of research. In Visitor use density and wilderness experience: proceedings; 2000 June 1–3; Missoula, MT. Proc. RMRS-P-20. W.A. Freimund and D.N. Cole (eds). Department of Agriculture, Forest Service, Rocky Mountain Research Station, Missoula, Montana, pp. 11-20.

Cole, D.N. and Stewart, W.P. 2002 Variability of

user based evaluative standards for backcountry encounters. Leisure Sciences 24, 313-324.
Corner, J. 1999 Recovering landscape: essays in contemporary landscape architecture. Princeton Architectural Press, New York.
Daniel, T.C. 2001 Whither scenic beauty? Visual landscape quality assessment in the 21st century Landscape and Urban Planning 54, 267-281
Davenport, M.D. and Anderson, D.H. 2005 Getting from sense of place to place-based management: an interpretive investigation of place meanings and perceptions of landscape change. Society and Natural Resources 18, 625-641.
Elands, B. 2002 De toerist op het spoor. Dissertation, Wageningen University, Wageningen.
Elands, B. and Lengkeek, J. 2000 Typical Tourists: research into the theoretical and methodological foundations of a typology of tourism and recreation experiences. Mansholt Studies 21. Wageningen University, Wageningen.
Elands, B. and Marwijk, R.v. 2005 Expressing recreation quality through simulation models: useful management tool or wishful thinking? In 11th International Symposium on Society and Natural Resource Management. Presented paper. Östersund, Sweden, June 16-19.
Evans, R. 2000 Economic models and economic policy: what economic forecasters can do for government. In Empirical Models and Policy-Making: interaction and Institutions. F.A.G.d. Butter and M.S. Morgan (eds). Routledge, New York, pp. 206-228.
Farnum, J., Hall, T. and Kruger, L.E. 2005 Sense of Place In Natural Resource Recreation and Tourism: An Evaluation and Assessment of Research Findings. In U.S. Department of Agriculture, Forest Service, Pacific Northwest Research Station., Portland, OR, pp. 59.
Fredman, P. and Hörnsten, L. 2004 Social Capacity and Visitor Satisfaction in National Park Tourism. Reprint from "Proceedings 12th Nordic Symposium in Tourism and Hospitality Research, October 2nd - 5th 2003".
Galloway, G. 2002 Psychographic segmentation of park visitor markets: evidence for the utility of sensation seeking. Tourism Management 23, 581-596.
Gibson, J.J. 1986 The Ecological Approach to Visual Perception. Lawrence Earlbaum Associates, Hillsdale, NJ.
Gimblett, H.R., Daniel, T. and Meitner, M.J. 2000 An Individual-Based Modeling Approach to Simulating Recreation Use in Wilderness Settings. In Proceedings: Wilderness Science in a Time of Change. Proc. RMRS-P-000. D. Cole and S.F. McCool (eds). Department of Agriculture, Forest Science, Rocky Mountain Research Station, USDA Forest Service, Ogden, UT.
Hammitt, W.E. and Patterson, M.E. 1993 Use Patterns and Solitude Preferences of Shelter Campers in Great Smoky Mountains National, U.S.A. Journal of Environmental Management 38, 43-53.
Hull, R.B. and Stewart, W.P. 1995 The landscape encountered and experienced while hiking. Environment and behavior 27, 404-426.
Hwang, Y.-H., Gretzel, U. and Fesenmaier, D.R. 2006 Multicity trip patterns: Tourists to the United States Annals of Tourism Research 33, 1057-1078.
Innovation Network Green Space and Agrocluster 2004 Vulnerable casualities around nature: a search to innovations linked to the tension between nature experiences and nature policy (in Dutch). Innovatienetwerk Groene Ruimte en Agrocluster, The Hague.
Jacobi, C. and Manning, R.E. 1999 Crowding and conflict on the carriage roads of Acadia National Park: an application of the visitor experience resource protection framework. Park Science 19(2).
Jacobs, M. 2006 The production of mindscapes: a comprehensive theory of landscape experience. Wageningen University, PhD-thesis. Wageningen.
Jansen-Verbeke, M.C. 1988 Leisure, recreation and tourism in inner cities: explorative casestudies. Nederlandse Geografische Studies 58, Koninklijk Nederlands Aardrijkskundig Genootschap/Geografisch Instituut, Amsterdam/ Utrecht.
Jochem, R., Marwijk, R.v., Pouwels, R. and Pitt, D.G. 2008 MASOOR: modeling the transaction of people and environment on dense trail networks in natural resource settings. In

Monitoring, Simulation and Management of Visitor Landscapes. R. Gimblett and H. Skov-Petersen (eds). University of Arizona Press, Arizona.

Kaplan, R. and Kaplan, S. 1989 The experience of nature: A psychological perspective. Cambridge University Press, Cambridge.

Keulartz, J. 2004 How to create a common vocabulary for nature policy? In Symposium Political Ecology: How can different nature perceptions reflect upon policy? S. Welting, A.v. Dijk, S.d. Held, S. Moed and V. Aerts (eds). Wageningen, pp. 14-25.

Koole, S.L. and Berg, A.E.v.d. 2004 Paradise Lost and Reclaimed: A Motivational Analysis of Human-Nature Relations. In Handbook of experimental existential psychology. J. Greenberg, S.L. Koole and T. Pyszczynski (eds). The Guilford Press, New York, pp. 86-103.

Kyle, G.T., Graefe, A., Manning, R.E. and Bacon, J. 2004 Effect of Activity Involvement and Place Attachment on recreationists' Perceptions of Setting Density. Journal of Leisure Research 36, 209-231.

Lawson, S.R. 2006 Computer Simulation as a Tool for Planning and Management of Visitor Use in Protected Natural Areas. Journal of Sustainable Tourism 14, 200-217.

Lengkeek, J. 1994 Een meervoudige werkelijkheid. Een sociologisch-filosofisch essay over het collectieve belang van recreatie en toerisme. Dissertation. Landbouwuniversiteit Wageningen, Wageningen.

Lengkeek, J., Te Kloeze, J.W. and Brouwer, R. 1997 The multiple realities of the rural environment. The significance of tourist images for the countryside. In Images and realities of rural life, Wageningen perspectives on rural transformations. H. de Haan and N. Long (eds). Van Gorcum, Assen.

Lynch, K. 1960 The image of the city. M.I.T. Press, Cambridge.

MacCannell, D. 1989 The tourist: a new theory of the leisure class. Schocken Books Inc., New York.

Manning, R.E. 2002 How mucht is too much? Carrying capacity of national parks and protected areas. In Monitoring and management of visitor flows in recreational and protected areas. A. Arnberger, C. Brandenburg and A. Muhar (eds). Vienna, Austria, pp. 306-313.

Manning, R.E., Valliere, W. and Wang, B. 1999 Crowding Norms: Alternative Measurement Approaches. Leisure sciences 21, 97-115.

Marwijk, R.v. 2006 Managing the transaction of nature and recreational behaviour: Experiences in Dwingelderveld NP (working paper). Wageningen University, Wageningen.

Marwijk, R.v. and Elands, B.H.M. 2007 Experiencing nature: The recognition of the symbolic environment within research and management of visitor flows. In Forest Snow and Landscape Research 81, ½, 59-76.

Milieufederatie Drenthe 2003 Meer recreatie met meer winst voor de natuur. Gespreksagenda duurzame toeristische ontwikkeling in Drenthe. Milieufederatie Drenthe, Assen.

National Forest Service 2004 At home with Staatsbosbeheer; naturally! A vision on recreation and experiencing nature and landscape (in Dutch). Staatsbosbeheer, Driebergen.

Nooij, A. 1990 Sociale methodiek : normatieve en beschrijvende methodiek in grondvormen. Stenfert Kroese, Leiden.

Pouwels, R., Verboom, J. and Jochem, R. 2008 Linking Ecological and Recreation Models for Management and Plans. In: Monitoring, Simulation and Management of Visitor Landscapes. R. Gimblett and H. Skov-Petersen (eds). University of Arizona Press, Arizona.

Provincie Drenthe 2000 Oude Vaart: gebiedsvisie Natuur, Bos en Landschap. Ministerie van Landbouw, Natuurbeheer en Visserij, Directie Noord, Assen.

Relph, E. 1976 Place and placelessness. Pion, London.

Ritchie, J.R.B. and Crouch, G.I. 2003 The Competitive Destination: A sustainable tourism perspective. Cabi Publishing, Cambridge, MA.

Schama, S. 1995 Landscape and Memory. Alfred A. Knopf, New York.

Shelby, B., Thompson, J.R., Brunson, M. and Johnson, R. 2005 A decade of recreation ratings for six silviculture treatments in Western Oregon. Journal of Environmental Management 75, 239-246.

Shin, W.S. and Jaakson, R. 1997 Wilderness Quality and Visitors' Wilderness Attitudes: Management Implications. Environmental Management 21, 225-232.

Stankey, G.H. 1973 Visitor perception of wilderness carrying capacity. Research Paper INT-142. USDA Forest Service, Intermountain Research Station, Ogden, Utah, pp. 61.

Statistics Netherlands 2005 Tourism and recreation in figures (in Dutch). Statistics Netherlands, The Hague.

Stedman, R. 2003 Is it really just a social construction? The contribution of the physical environment to sense of place. Society and Natural Resources 16, 671-685.

Stichting Natuur en Milieu 2005 Mooi Land 2005/2006: 100 natuurgebieden getest. Veen Magazines, Diemen.

Stokols, D. 1978 Environmental Psychology. Annual Review of Psychology 29, 253-295.

Tuan, Y.F. 1977 Space and place: the perspective of experience. University of Minnesota Press, Minneapolis.

Urry, J. 1990 The Tourist Gaze: Leisure and Travel in Contemporary Societies. Sage Publications, London.

Vistad, O.I. 2003 Experience and management of recreational impact on the ground – a study among visitors and managers. Journal for Nature Conservation 11, 363–369.

Visschedijk, P. A. M. 1990 Recreatie in het Nationaal Park Dwingelderveld (No. 582). Instituut voor Bosbouw en Groenbeheer, Wageningen.

Vuorio, T. 2003 Information on recreation and tourism in spatial planning in the Swedish mountains – methods and need for knowledge. Department of Spatial Planning Blekinge Institute of Technology & European Tourism Research Institute Sweden. PhD thesis., Karlskrona.

Westover, T.N. 1987 Perceived crowding in recreational settings: an environment-behavior model. Environment and Behavior 21, 258-276.

Young, M. 1999 Cognitive maps of nature-based tourists. Annals of Tourism Research 26, 817-839.

CHAPTER 5

TECHNIQUES FOR COUNTING AND TRACKING THE SPATIAL AND TEMPORAL MOVEMENT OF VISITORS

Jianhong Xia
Colin A. Arrowsmith

Abstract: Data collection can become an overwhelming task for the simulation modeler and, indeed, can become the single most important factor that restricts the development and ultimate use of that model. This chapter serves as a resource for the simulation modeler to assist in determining what data is required and what techniques and technologies are currently available. The chapter discusses the data required for describing the spatial and-temporal movement of visitors. A review of the techniques for the tracking and counting of visitor movements at various spatial scales will be undertaken and each method will be assessed for its suitability as a mechanism for acquiring data for simulation modeling. These techniques are summarized assessing their advantages and disadvantages and suitable corresponding applications.

Key words: Monitoring, Visitor Tracking, Counting, Global Positioning Systems, Recreation Behavior.

INTRODUCTION

Collection of visitor information for monitoring visitor landscapes requires significant amounts of data. Data collection can become an overwhelming task for the simulation modeller and, indeed, can become the single most factor that restricts the development and ultimate use of that model. This chapter serves as a resource for the simulation modeller to assist in determining what data is required and what techniques and technologies are currently available. Perhaps the single-most limiting factor in monitoring and modelling visitor landscapes is the availability of data. This is almost entirely due to the time and costs associated with collecting the data, but also in part, to not fully understanding the nature of the data collected (for example the detection limits required) and not understanding the actual data sets required for effective simulation. Limited or poor data will result in a limited use simulation model. For example, to effectively simulate behaviour at a particular location for a given time period, will require "typical" actual movements and behaviours of visitors for these time periods. It is not enough to acquire data at a convenient time to the observer, for example during normal business hours, if the simulation model is to predict movements outside these hours. Likewise seasonal visitation patterns at popular tourist locations will fluctuate throughout the year. Therefore it may be necessary to monitor such landscapes throughout the year, or at a minimum collecting data a "strategic" time intervals at different seasons if the simulator is to predict behaviour throughout the year. As Manning et al. (2005) state ...*The sampling period must be appropriate to the needs of the simulation* (page 13). Arrowsmith et al. (2005) for a study in western Victoria, Australia, for example, found that the limited availability of global positioning systems (GPS) receivers used to record visitor movements, meant that diurnal changes in visitor itineraries throughout the day went largely undetected.

The chapter will commence by discussing the data required for describing the spatial and-temporal movement of visitors. A review of the techniques for the tracking and counting of visitor movements at various spatial scales will be undertaken and each method will be assessed for its suitability as a mechanism for acquiring data for simulation modelling. These techniques will be summarised in a table listing their advantages and disadvantages and suitable corresponding applications.

DATA REQUIRED FOR MODELLING

In this section, two types of data regarding visitor movements will be examined: spatial and temporal data and socio-economic data. Socio-demographic data are regarded as independent variables for modelling spatio-temporal movement of visitors. They can be used to explain or predict the spatial and temporal movement of visitors. Certain spatial and temporal behaviour can largely be predicted from socio-demographic data. Elderly citizens walking along tracks in visitor landscapes, are likely to move somewhat slower, in a more "predictable" path and be more spatially restrictive than pre or early teenage visitors. Those people of a more athletic constitution may opt for more challenging pathways when confronted with pathway choices, than the less physically able. Therefore, not only are spatial and temporal patterns required, but also knowledge relating to the individual to support decision making algorithms within the simulator

SPATIAL AND TEMPORAL DATA

Models for simulating visitor behaviour require specific spatial and temporal attributes of movement of current visitors. Simulations are generated from raw visitor data based on existing itineraries observed in the field. Essentially these data can be categorised into spatial and temporal parameters (Arrowsmith et al. 2006). These include:

Identity is essentially the identification of an individual object within the visitor landscape. That object may be a boat, an individual animal or human or a group of animals or humans. The identity of the object at any point in time and space allows continuous tracking of individual entities. Data collection techniques such as global positioning system (GPS) tracking, radio-frequency identification (RFID) or on-site observation can identify individuals. In other instance individuals cannot be tracked but can be grouped by travel mode (car, bike, horseback, foot) or by transport type for example cars, buses, sports utility vehicles (SUVs).

Position refers to a geographic location at which an activity or an event occurs. Position information can be measured in absolute terms (as an x and y, easting and northing or latitude and longitude) or as a relative location using a polar coordinate system as a given distance and bearing from a previous position. It can also be defined on a relative location using a spatial framework giving position in relation to other objects of interest.

Distance can be measured as a linear distance in plan view or travel distance along a slope. Distance can be measured in the field using range finders, GPS receivers, or by ground measure. Typically, distance is measured from maps. The accuracy of distance measures affects other metrics such as speed.

Time in monitoring of visitor landscapes needs to be considered at two levels – time point and duration. Section 1.1.5 deals with duration which is dependent on the "time point" or simply time being recorded. The "time point" or instance of an event is the time and location of where some important activity within the visitor landscape takes place. This activity may simply be the change in pace or direction of a hiker walking along a pathway. It may be the time and location when a hiker confronted with several pathway options elects to walk along a particular pathway; that is, when a conscious decision is made. Alternatively, an event may be the time a visitor enters a point of interest, be it the landscape itself, or a component of the landscape, for example a lookout, and the time that same visitor exits that component.

Duration is defined as a period of time of an activity or a trip or period of time between events (Koncz and Adams, 2002). It consists of an interval between at least two or more points in time. Each activity has a given duration that takes place at specific points or locations within the space-time framework (Huisman and Forer 1998). Duration can be calculated at several scales ranging from entry and exit to specific locations, through to the total duration of a complete itinerary.

Direction of movement can be measured if identity, position and time are known for at least two locations. Direction of movement is fundamental in the construction of trip itineraries from observed data.

Speed or velocity relates distance with time. Speed is defined as the distance covered by an object divided by the unit of time taken to cover that distance.

The sequence is the order in which an individual mobile object visits a number of attractions. The sequence provides information about how mobile objects travel along networks by following a particular order. Such sequences can exhibit decision-making processes carried out by individuals. From sequences it is possible to derive individual itineraries. The ease in which sequence can be determined varies greatly depending on how information is collected. To generate sequence directly, one must know identity, location and time at each destination. Often however we may be missing identity, as in the case of simple traffic counters. Constructing itineraries, if any one of these variables is unknown or measured with poor accuracy, can complicate the task enormously.

Flow is used to measure the intensity of traffic. Flow is measured as the number of objects passing a point per unit time. Flow can therefore be calculated if location, identity and time are known. If direction is also known then flow rates can be calculated for each direction. Flow rates are affected by the width of the network link, the speed, direction and the density of traffic. Flow will be reduced and networks will become congested where the networks are narrow or offer mobile objects a means to slow their progress (for example attractions to visitors).

SOCIO-DEMOGRAPHIC DATA

The above parameters are those required for inanimate objects such as boats on a river. Further data relating to individual characteristics of visitors are required for true representation of visitor behaviour. Research has shown that nationality might have a significant effect on visitor behaviour (Choi and Chu, 2000; Kozak, 2001). In a cross-cultural comparison, Hofstede

(2001) described four distinguishing visitor characteristics or dimensions based on cultural background. These were "Power distance", the extent to which that culture will accept an unequal distribution of power in society, "Individualism-Collectivism", the extent of grouping of visitors based on an individuals' need to belong to a visitor group, "Uncertainty avoidance", the extent to which an individual feels threatened by uncertain situations and finally "Masculinity-femininity" or how a particular culture values so-called "masculine" behaviours such as assertiveness and material wealth. "Australians" or the so-called Australian culture, for example, is characterised by a small power distance and strong individualism, low tolerance of inequality and authority, a loose social network, and a weak sense of social obligation (Hofstede, 1980). This contrasts with the Indian culture which is characterised by a large power distance (Ting-toomay, 1999) sensitive to contextual demands and is orientated to combine collectivist and individualistic behaviour (Sinha and Kumar (2004). Reisinger and Turner (2003) suggest that visitors from oriental eastern countries such as China tend to follow crowds whilst those from western cultures prefer to avoid these situations.

Arrowsmith et al. (2005) found individual characteristics revealed four distinct spatial patterns at Port Campbell National Park in Victoria, Australia. For example they found that young Australian friends who travelled together as a group, tended to stay longer and travelled more widely, than say the elderly couple or local family. International visitors also tended to travel more widely. In particular the following attributes should be collected to provide enough background and motivational data for modelling visitor simulation:

Visitor profile relates to the individual characteristics of a visitor and may include age, sex, education, travel group type (such as travelling alone or travel with the family), country of residence, country of birth, lifecycle (such as young single, young family, mature family), income and occupation.

Visiting characteristics refer to methods of mode of travel to and from the visitor site (such as car, bicycle, tour and bus etc.), type of accommodation (such as hotel or motel, caravan, apartment, self-contained cottages, and guest house), activities of the visit, motivation for the visit and frequency of visit.

METHODS USED TO ACQUIRE DATA FOR PROPAGATING SIMULATION MODELS

There are numerous approaches that can be adopted to acquire data for simulation modelling for visitor landscapes. These range from direct observation through to the more technologically complex methods that rely on sensor technology or global positioning via the Global Positioning System (GPS) a constellation of "known" satellites that can provide location. Often it can be advantageous to adopt a combination of approaches where terrain, geographic extent and complexity and visitor statistics require multiple approaches. The following section outlines a review of systematic approaches for data collection starting with the rudimentary direct observational approaches through to the sophisticated tracking technologies. This discussion is organised into two sections. The first section deals with approaches for acquiring basic count statistics of visitors. Data provided from these approaches are limited to position, time and at most direction of movement (although many counting techniques are unable to this). For the remaining identity, distance, speed, duration, sequence and itinerary, and flow data individual visitors are required to be tracked through their visit to a landscape.

Basic visitor count statistics provide policy makers with an estimate of visitors to and from particular landscapes through time, and are often used for marketing or influencing government investment strategies. However, simulation modelling requires data beyond basic visitor numbers and requires the modeller to have some understanding of the spatial and temporal behaviour of visitors to these landscapes. It is not enough for the modeller to "know" how many people visit a landscape or what the visitor patterns into and out from a landscape are, but rather how these visitors interact with these landscapes. In a study of visitor behaviour at Port Campbell National Park in western Victoria, Australia, visitor statistics indicate that capacities are often exceeded at car parks, walking tracks and lookouts for certain locations in summer and autumn (Parks Victoria, 1998). Figure 1 for example, shows one particular lookout that is capable of supporting only two or at most 3 visitors. For visitors arriving by tour bus with limited time, groups of people results in clusters of visitors that create overcrowding problems at these sites.

Therefore it is imperative that park managers are able to simulate concentrations of visitors at specific geographic locations with respect to time.

Methods for count data

Counting techniques are used to count the number of visitors and/or vehicles on roads, in buildings or at specified locations. Basic position, time and with certain technologies, direction of movement can be obtained from counting techniques. Vehicular counts can provide estimates of visitors to particular locations provided a reasonable estimate of people travelling in a vehicle can be made. Parks Victoria in Australia, have made estimates of visitors to their parks using vehicular counts. This section will discuss some of the more common methods that have been used to capture count data. Some of these techniques can be used to count vehicles or people only whilst other approaches can be used to capture count data for both vehicles and individual people at set locations.

Wang and Manning (1999) used *direct*

Figure 1. Lookout at the "Blowhole" at Loch Ard Gorge, Port Campbell National Park, Victoria, Australia. Photo: Colin Arrowsmith

observation, in combination with a variety of other approaches, for their study of Acadia National Park in Maine, USA. Observers were stationed at eight main entrance points to the park and were asked to record the number of visitors that entered the park per hour from 09:00 to 17:00. Direct observation has the limitation of being expensive (the need to employ observers over extended periods) and providing limited information (numbers into or out from a park).

Pressure pads are pads or mats buried under a path surface or mounted in the entrance or exit of doors or on the steps of a bus, which can sense a person's weight and registers a count (Infodev 2004). Pressure pad track counters have been used extensively to count the number of visitors in the national parks in Australia and New Zealand (TRAFX, 2004). Dixon (2004) used pressure pads to monitor visitor numbers in the Creag Meagaidh Nature Reserve, Scotland, to identify whether or not the provision of visitor facilities were effective. He also used the counts to compare alteration of vegetation and soil along paths with changes in the number of visitors who passed by these paths during a certain period of time to estimate the carrying capacity of the paths.

Pressure pads have a number of advantages. Firstly, they can be easily hidden. Therefore, they are less prone to vandalism and count pedestrians without intrusion. Secondly, extended battery life means that the data loggers for pressure pads can be left unattended for up to a month before the data has to be downloaded into a computer. However, pressure pads also have a number of limitations. Pressure pads are expensive to install and can only be used along tracks or at gateways. Instrumentation used with pressure pads can freeze up in winter, and incorrect counts can be made if people step over or around them. Provided two pressure pads are placed in succession, direction of flow can also be made.

Pressure road counters use rubber tubes to determine the number of axles that pass a given location. Pneumatic tubes can be stretched across a road to give precise wheel axle numbers for vehicles ranging from motor cycles through to large trucks (Metrocount, 2005). Parks Victoria use road counters for traffic counts to ascertain visitation statistics for their website (see http://www.parkweb.vic.gov.au/resources/14_1448.pdf).

Camera-based counting systems are usually comprised of two components: a counting system and image-based counting software. The major component of the counting system is a camera that is sensitive to light. The camera transfers light signals into electronic signals and records a sequence of images. These images can then be analysed using software to identify and count people or vehicles.

Camera-based counting systems have been used for various applications. The Space Syntax Group (http://www.space-syntax.com/main-nav/home.aspx) has adopted the closed circuit television to record movements of pedestrians in buildings. This technology can be used to obtain the density of people for a given area. However tracking people's movements is somewhat more difficult. Heikkila and Silven (2004) have used camera based counting systems adopting Kalman Filtering techniques to classify and count based pedestrians and cyclists. Kalman filtering is a technique that can be used to provide continuously updated information on the predicted position and velocity of a moving object by removing the effects of random error.

In a study of the Danube Floodplains National Park, Arneberger at al. (2005) found that video-monitoring was accurate to within 4% to those numbers obtained using traditional count methods.

The advantage of camera-based counting systems is that they can collect movement data with high resolution and work in diverse ways including surveillance, counting and tracking. In addition, camera based systems are non-intrusive provided that image quality is such that no individual can be identified. Significant advantages can be obtained from improved accuracy in count numbers at peak visiting times (Arnberger et al., 2005). Their main disadvantage is that these systems are sensitive to the vibrations and changes in light levels and height and temperature can also degrade their performance. Another disadvantage of camera-based counting systems is the high equipment cost. Arnberger et al. (2005) in their study also found that rapid movement activities such as jogging and bicycling were overlooked when reviewing video-images due to the speed of playback.

Infrared sensors make use of infrared electromagnetic radiation. Whilst infrared radiation is invisible to the human eye, it can be detected by heat sensitive sensors and therefore can detect heat generated by objects such as humans and animals.

Infrared sensors can be classified as active or passive. Active infrared sensors are comprised of a transmitter and receiver. The transmitter emits infrared radiation and the receiver detects the reflected or scattered radiation from a human or vehicle and converts it into electrical signals. These type of sensors are used in buildings for automated door safety where a transmitter constantly emits infrared waves into a scan zone and when the emitted light is broken by a person entering a doorway, a signal is sent back to the door controls to remain open (Bircher Reglomat, 2003; Hamamatsu, 2004). Due to cost and constant interference from biophysical parameters, their use in monitoring visitor landscapes is somewhat restricted. Their potential advantages stem from the fact that they are not influenced by external temperature and light conditions as they have their own radiation source. Additionally they are unobtrusive, easy to install and can cover a field of view.

Passive infrared sensors detect the change in luminosity or temperature in the sensor's field of view when a person or vehicle enters the target area. The object passing through the infrared beam breaks the ambient temperature eminating from the background source. Therefore, for an object to be detected it must be moving for a count to be registered (Infodev, 2004). Passive infrared sensors have been used by Hashimoto et al. (1998) to build thermal images of groups of people from multiple sensors, to detect numbers of people in a room. This has application in controlling air-conditioning remotely.

The advantage of passive infrared sensors is that they can detect objects at a greater range than sensors that use visible wavelengths. Passive infrared sensors are also non-intrusive and low cost. However, the accuracy of the detector can be degraded by heavy rain or snow (McFadden et al., 2001). In addition, passive infrared counters cannot distinguish individuals in a group, which can lead to undercounting. Another disadvantage is that changes in the background can be recorded inadvertently as a count. Passive infrared systems are sometimes used to trigger external cameras that record pictures of objects passing their field of view to support count data (Gasvoda, 1999).

Magnetic detectors can detect objects such as vehicles by recording a bias in the earth's magnetic field as automobiles pass over them. They are usually buried in the road or mounted on the side of the road. The primary use of magnetic detectors is to supplement or enhance data for other types of traffic detectors, although they are occasionally used in stand-alone applications (Klein, 1995). The advantages of magnetic detectors are they are relatively low cost, unobtrusive and unaffected by weather. However, their main disadvantages are the low resolution and difficulty in discriminating longitudinal separation between closely spaced vehicles.

Inductive loop detectors are composed of one or more turns of insulated wire, a lead-in cable, and an electronic unit. This type of detector can sense a vehicle by detecting the decrease of its inductance as the vehicle passes by. This is converted into a digital signal and a count is registered (Klein, 1995).

Inductive loop detectors are a mature technology. They can be easily installed, have excellent counting accuracy and are not affected by extreme environmental conditions. For these reasons inductive loop detectors have become the most widely used traffic detector technology today. However, these detectors can be damaged by heavy vehicles and through road reconstruction or repair. Inductive loops can only be installed in well paved roads.

Two types of *microwave detectors* are used in vehicle counting. The first type of microwave detector emits electromagnetic energy at a constant frequency and analyses the differences in frequency between the transmitted and received signals using the Doppler principle. A count is registered as the difference is detected. Microwave detectors can also be used to measure the speed of vehicles. The second type of microwave radar emits a "frequency-modulated continuous wave" (FMCW) that changes in frequency with time. Therefore this type of microwave detectors can sense the presence of both a moving and a stationary vehicle (McFadden et al., 2001). Microwave detectors are low-cost, small, lightweight and easy to install and can perform well in inclement weather. Microwave detectors have been used extensively in security applications (Klein, 1995).

Ultrasonic detectors transmit sound waves at a selected frequency between 20 and 65 KHZ, and measure the time for signal to return to the detectors. Ultrasonic detectors can sense the presence of a vehicle and register a count. Ultrasonic detectors are non-intrusive, small in size, easy to install and perform well in inclement weather, but they can be sensitive to changes of temperature, air turbulence and humidity.

Methods for tracking data

Tracking allows for more detailed information to be provided to the simulation modeller. Rather than providing simple statistical visitation data, tracking enables the modeller to observe movements through a network of pathways. This can be extremely useful to the modeller through providing probability statistics on decision making behaviour that can be used within the simulation model.

There are a number of tracking techniques that are available for the collection of spatial and temporal visitor movement data. This section reviews several tracking techniques used including, observation and interviews, self-administered questionnaires, Global Positioning System (GPS) receivers, timing systems, Personal Digital Assistants (PDA), mobile phone tracking, and closed circuit television..

Direct observation is an easy yet flexible tracking technique. This procedure relies on direct observation where the researcher follows or observes individual visitors and records their route, arrival time and length of stay at each attraction (Dumont et al., 2004). Batty (2003) modelled pedestrian movement throughout the Tate Gallery in London using an agent-based simulation based on movement data obtained by observation. Similarly, the visitor behaviours at the Melbourne Zoo were recorded by Broad and Smith (2004) using this technique. In a study of boat traffic along the Yarra and Maribyrnong rivers in Melbourne, Australia, Arrowsmith used direct observation techniques to count the movements of boats and ships at seven locations along the lower reaches of these rivers over two weekends in January and February 2007. An earlier survey reported in Arrowsmith et al. (2005) provided movement data for the development of a simulation model of boat movements for Parks Victoria taken from direct observation at thirteen locations along the river.

Whilst direct observation techniques provide an unobtrusive and easy method for acquiring movement patterns amongst mobile objects such as visitors to visitor destinations, it lacks the detail necessary for understanding and classifying behaviour patterns exhibited by different user groups. For this level of detail, it becomes necessary to *interrogate the individual users of the pathways in question*. This technique adopts tracking via direct observation but in addition the observer intercepts visitors and asks a number of questions aimed at gathering socio-demographic, perception and wayfinding data. Whilst this technique provides both spatial and non-spatial data for modelling purposes, it is both expensive and obtrusive and can in fact modify the behaviour of the individual.

Self-administered questionnaires are a traditional method used to track visitor movement. As a part of a questionnaire, participants are asked to retrace their spatial movements through a designated area on a cartographic map (Fennell, 1996; Wang and Manning, 1999). This requires participants to recall from memory, and draw their trip route and write down any visited attractions, their approximate arrival time and length of stay on a map. Survey accuracy is dependent on when the survey was conducted, the participant's spatial ability, spatial knowledge and levels of familiarity with the environment (Li, 2004). The accuracy of the data also depends on the memory of participants; the less time between visiting a location and drawing it on the map, the more accurate the data will be. The ease at which a visitor is able to recall their travel pathways will increase if the participant has a higher spatial ability and knowledge. Also, the more familiar with an environment the participant is, the easier it is for them to draw their travel routes on a map.

The advantages and disadvantages of the self-administered questionnaire technique are similar to those of observation and interview techniques. However, more sample data can be collected using the self-administered questionnaire technique than with observation and interviews. One significant disadvantage with self administered questionnaires over observation and interview techniques, are the expected response rates which can vary from 97% as experienced by Day (1999) in her survey of the Grampians region of Scotland conducted on an intercity train, to typically less than 50% (Leeworthy et al. 2001). In addition, any ambiguity or uncertainty in the questionnaire cannot be clarified unless the survey distributor/author is close at hand to answer unforeseen questions. Another problem is that because the questionnaire has been constructed by the observer, any qualitative additional information to support decisions made in the field that have not been adequately handled within the questionnaire, will be lost.

The *Global Positioning System (GPS)* is a worldwide radio-satellite navigation system that relies on a constellation of 24 satellites and their ground stations providing precise position, velocity and time data to users around the world, 24 hours a day (Chadha and Osthimer, 2004). A GPS receiver receives signals from the satellites and determines its geographic location. GPS receivers can work in any weather condition and there are no subscription fees or setup charges to use GPS. Compared to manual tracking techniques, GPS tracking has the advantage of spatial accuracy, but also can provide details relating to speed and direction of visitor movements.

Recent GPS research has been undertaken to track both pedestrian and vehicle movements. Ashbrook (2002) clustered the location data of people collected by GPS and incorporated these locations into a Markov model to predict their next movement. Arrowsmith et al. (2005) used GPS receivers to track the spatial behaviour of visitors to Loch Ard Gorge in western Victoria, Australia.

There are two fundamental limitations of GPS. Firstly it is necessary to have at least four satellites for three-dimensional locations to be determined. Canopy cover can restrict the number of satellites required for high positional accuracy. Arrowsmith et al. (2005) found spatial accuracies within plus or minus 5 to 8 metres. However on a number of occasions there was evidence of "drift" as satellite contact was lost. Secondly, GPS recorded locations alone are not capable of interpreting the nature of object behaviour.

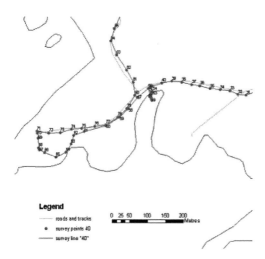

Figure 2. A GPS tracking example for Port Campbell National Park.

In fact visitor behaviour can be modified if the participant knows that he or she is being monitored for spatial behaviour.

Timing systems are widely used in sports tracking. These systems are able to record the time and location of movements for individual competitors. Most timing systems are comprised of two components: 1) transmitters or transponders that are attached to an individual's clothing or shoelaces and, 2) data loggers and sensors (receivers) that are installed at necessary locations. O'Connor et al. (2005) used the Alge timing system (ALGE-TIMING, 2005) to track visitors at the Twelve Apostles. Battery life and the data logger memory capacity were found to be a limitation in this study. In addition, the high cost of equipment limits the density of the receiver network. The key advantages of timing systems are their capability to log a unique identifier for individuals, as well as being able to track their spatial and temporal movement at high resolution.

Personal Digital Assistant (PDA) tracking is an integration technology of GPS and Geographic Information Systems (GIS). The PDA was originally conceived as a small device with the basic functions of a personal organiser such as maintaining contact lists, calendars, diaries and calculators (Casademont et al., 2004). Current PDA's are equipped with up to 112 Mbytes of memory, processing capabilities of up to 400 MHz plus addition functions (Casademont et al., 2004). In PDA tracking applications, the PDA is used as a platform to communicate with mobile phone networks or a GPS and download spatial and temporal data from the Internet. The GIS is integrated into the PDA so that spatial and temporal data can be displayed in real-time on a screen map and users can access information for a particular time or location using the map. Through the PDA, users are also able to access Location-

Based Services (LBS) including detail related to a user-defined location.

Hadley et al. (2003) combined GPS with a PDA fitted with GIS software that enabled location based information relevant to the location of the user to be obtained for points of interest. Research by Loiterton and Bishop (2005) involved the use of PDA and mobile phone technology to track visitor wayfinding decision-making processes in urban parks and gardens. Location-based questionnaires were integrated into the PDA so that as the visitor moved around the park with the PDA questions were relayed to the user for choice selection when making decision regarding the pathways selected.

Because the main component of PDA tracking is GPS, PDA tracking has the same disadvantages as GPS. However, PDA tracking can be designed to enable observers to communicate with visitors in real-time to track not only the physical movement of visitors but also the decision-making process they make as they walk around particular locations.

PDA tracking is still very much theoretical. Very little research by industry has taken place with respect to the application of PDA tracking. Even so, there is great potential for its integration into LBS applications for research purposes.

Mobile phone tracking is a technique used to track visitors by obtaining visitor mobile phone locations. Permission is required from visitors and their phone company. VeriLocation is a web-based system that enables registered users to track the locations of consenting mobile phone users. The mobile phone signals of users from the United Kingdom's major network operators (such as T-Mobile, Orange, O2, and Vodafone) are detected, recorded and displayed on corresponding street maps in a web browser. The price of tracking through VeriLocation is £30 (around AUD$70) to locate 100 mobile phones, or £125 (around AUD$292) to locate 500 mobile phones (VeriLocation, 2004).

Mobile phone tracking is a simple and low cost tracking technology. However, the big issue for this technology is security and many visitors are reluctant to provide their mobile phone number. In addition, this technology is still developing and the current accuracy of location information ranges from 500 metres to 5000 metres, depending on its distance to the nearest mobile phone network ((Mateos, 2005; VeriLocation, 2004).

Closed circuit television (CCTV) monitoring has been extensively used for surveillance or security. For example, Bogaert (1996) designed PASSWORDS, an intelligent video image analysis system used to detect dangerous situations (such as a person with agitated behaviour) and inform the appropriate authorities. Similarly, W4 (Who? When? Where? What?), constructed by Haritaoglu (1998), is a real time system used to detect and track people in monochromatic imagery in order to acquire information like what are people doing, where and when they act. KaewTrakulPong and Bowden (2003) developed a variety of probabilistic models for tracking multiple independent targets in outdoor visual surveillance scenes. Closed circuit television monitoring is a technique that is limited to small areas.

There are many factors that can affect the accuracy of CCTV monitoring. This includes light variation and vibrations, the quality of the optical components and the image analysis software. Light variation, especially a sudden intensive light in a monitored area, can cause the camera to record images incorrectly. The relocation of a camera lens caused by vibrations can also lead to erroneous image recordings and inaccurate counts. Therefore, camera-based counting systems are better suited for interior environments.

Different image analysis software can

also influence the accuracy of counting. Recognising objects, and tracking their movements in a complex real scene using a sequence of images are the most difficult tasks in computer vision (Bogaert et al., 1996). Early image analysis software had a low accuracy rate. However, Heikkila (2004) has utilised a Kalman filtering algorithm, a technique that can be used to provide continuously updated information on the predicted position and velocity of a moving object by removing the effects of random error, to improve the quality of tracked images of people. Using a developed Learning Vector Quantization algorithm for classifying the observations into pedestrian and cyclist, accuracies of counting and classification was around 80-90%.

EVALUATION AND COMPARISON OF COUNTING AND TRACKING TECHNIQUES

This section will compare the technologies discussed in the previous section for data acquisition for agent modelling. In table 1, the list of data discussed in section 1 required for modelling is cross-tabulated with technologies discussed in section 2. Each technology is stated as either having or not having the ability to measure a specified data item, and the associated costs, levels of intrusion and reliability are rated as being low, moderate or high.

Table 2 summarises the advantages and disadvantages of the various counting and tracking techniques discussed in section 2. The advantages and disadvantages are allocated to each technique based upon the following significant factors: resolution, cost, intrusive, non-intrusive and sample size. Table 3 provides a list of applications and the counting and tracking techniques that are best suited for a given application.

Tables 1, 2 and 3 summarise the capabilities, advantages and disadvantages and applications of both counting and tracking techniques. All the counting techniques mentioned above can count vehicles. However, only camera-based systems, active and passive infrared sensors, pressure pads and direct observing can be used to count visitors. Furthermore, camera-based systems, active infrared sensors, inductive loop detectors and microwave detectors can count visitors or vehicles at higher resolution compared to the other counting techniques such as observation, passive infrared sensors, pressure pads and magnetic detectors.

Of the eight tracking techniques reviewed in section 2, all techniques can be used to track individual visitor movement. GPS, timing systems, and mobile phone tracking techniques can also be used to track the movement of vehicles with high or moderately high resolution. Traditional tracking techniques such as direct observation, interview and self-administered questionnaires can be used to obtain spatial and temporal information related to the movement of people in low resolution but could obtain a great deal of detailed information. However, modern tracking techniques such as GPS, timing systems, PDA tracking and mobile phone tracking can track not only the spatial and temporal motion information of visitors, but also their speed and direction, which could be very useful information for some applications such as movement prediction. One important disadvantage of these modern tracking techniques is that the equipment is usually expensive and has a high risk of being lost.

To sum up, each technique has its own advantage and disadvantage, and it is important to apply them in appropriate situations (see table.3).

Movements of visitors can be ascertained at two spatial and temporal scales. At the macro-level, spatial scales of tens or hundreds of kilometres, and temporal scales of hours, days or even weeks are modelled.

Table 1. Summary of capabilities of various counting and tracking technologies

Attributes	Methods for acquiring count only data							Methods for tracking								
	Direction observation	Pressure pads	Camera-based	Infrared sensors	Magnetic detectors	Inductive loops	Microwave detectors	Direct observation	Obsn and interviews	Questionnaires	GPS	Timing systems	PDA and LBS	Mobile phone	Virtual questionnaires	CCTV
Identity	YES	NO	YES	NO	NO	NO	NO	YES	YES	YES	YES	YES	YES	YES	YES	YES
Position	YES	YES	YES	YES	YES	YES	YES	YES	YES	YES	YES	YES	YES	YES	YES	YES
Distance	YES	NO	NO	NO	NO	NO	NO	YES	YES	YES	YES	YES	YES	YES	YES	YES
Time	YES	NO	NO	NO	NO	NO	NO	YES	YES	YES	YES	YES	YES	YES	YES	YES
Direction of movement	YES	YES	YES	NO	NO	NO	NO	YES	YES	YES	YES	YES	YES	YES	YES	YES
Speed	YES	NO	NO	NO	NO	NO	NO	YES	YES	YES	YES	YES	YES	YES	YES	YES
Duration	YES	NO	NO	NO	NO	NO	NO	YES	YES	YES	YES	YES	YES	YES	YES	YES
Sequence or itinerary	YES	NO	NO	NO	NO	NO	NO	YES	YES	YES	YES	YES	YES	YES	YES	YES
Flow	YES	NO	YES	NO	NO	NO	NO	YES	YES	YES	YES	YES	YES	YES	YES	YES
Socio-demo data	NO	NO	NO	NO	NO	NO	NO	NO	YES	YES	NO	NO	NO	NO	YES	NO
Cost of equipment	LOW	LOW	HIGH	HIGH	HIGH	HIGH	HIGH	LOW	LOW	LOW	MOD	HIGH	HIGH	LOW	HIGH	HIGH
Cost of implement	HIGH	LOW	MOD	LOW	LOW	LOW	LOW	HIGH	HIGH	HIGH	MOD	MOD	MOD	LOW	MOD	HIGH
Level of intrusion	HIGH	LOW	HIGH	LOW	LOW	LOW	LOW	HIGH	HIGH	HIGH	HIGH	HIGH	HIGH	HIGH	HIGH	HIGH
Reliability	HIGH	LOW	MOD	LOW	LOW	LOW	LOW	HIGH	HIGH	MOD	MOD	HIGH	HIGH	MOD	MOD	MOD

Table 2. Summary of advantages and disadvantages of various counting and tracking technologies (based on Klein, 1995; Skszek, 2001; Infodev, 2004)

Technique	Advantage	Disadvantage
Pressure pads or mats (People/bicycle counting)	• Low cost • Non-intrusive • Variety of size suitable for different purpose • Can work with small long-life batteries up to one month	• Low resolution • Inclement weather or too low temperature could degrade performance. • Can take more time to install depending on ground condition
Camera based systems (People/vehicle counting)	• Non-intrusive • High resolution • Can work as surveillance, counting, and tracking tool	• Vibrations and changes in light, height, and temperature could degrade performance. • Expensive
Active infrared (People/vehicle counting)	• Non-intrusive • Uninfluenced by sudden temperature and light change • Ease of install covering wide entrance	• Affected by obscurants in the atmosphere and weather • Expensive
Passive infrared sensors (People/vehicle counting)	• Non-intrusive • Low cost • Uninfluenced by sudden temperature and light change	• Affected by heavy rain or snow • Affected by chance of background
Microwave detectors (Vehicle counting)	• Non-intrusive • High resolution • Mature technology • Ease of install • Good performance in inclement weather	• Overhead installation requires the presence of existing structure for mounting the device
Inductive loop detectors (Traffic counting)	• High accuracy • Mature technology • Good performance in inclement weather	• Intrusive • Could be damaged by heavy vehicles and road repair
Magnetic detectors (Vehicle counting)	• Non-intrusive • Low cost • Unaffected by weather	• Low resolution • Difficulty in discriminating longitudinal separation between closely spaced vehicles
Ultrasonic detectors	• Non-intrusive • High resolution • Mature technology • Ease of install • Good performance in inclement weather	• Affected by temperature and air turbulence
GPS tracking	• High resolution • Mature technology • Ease to use	• Intrusive • Low sample size • GPS signals blockage from buildings and foliage • High risk of loss of GPS equipment • Expensive

Timing systems	• Non-Intrusive • High resolution • High sample size	• Expensive • Limited by battery life and data logger memory capacity
Mobile phone tracking	• Non-intrusive • High resolution	• Low-cost • Privacy security issue • signals blockage from buildings and foliage
PDA tracking	• High resolution • Can communicate with visitors in real-time • Can track their decision-making process	• Intrusive • Low sample size • GPS signals blockage from buildings and foliage • High risk of loss of PDA equipment • Expensive
Manual observation and interview	• Can communicate with visitors deeply • Can classify vehicles more accurately • Can count or track visitors more flexibly	• Intrusive • Time-consuming • Low resolution • Low sample size
Self-administered questionnaire	• Can acquire non-spatial and approximate spatio-temporal information of visitor • Can count or track visitors more flexibly	• Intrusive • Time-consuming • Low resolution • Low sample size

Table 3. Counting and tracking technology applications to tourism management

Applications	Objectives	Tracking Technique	Comment
Route planning (Slater 2002; Loiterton and Bishop 2005)	To explore the decision-making process of visitors movement to identify optimum route for them	• PDA tracking • GPS+self-administered questionnaire	There needs to be route recording as well as wayfinding decision tracking when estimating visitors' movement and their wayfinding decisions.
Wildlife protection (Chen and Morley 2005)	To monitor visitors' movement patterns in parks or reserves in order to protect wildlife	• GPS tracking • Mobile phone tracking • Self-administered questionnaire • Observing	There needs to be route recording as well as simple counting when estimating visitors' movement and their number.
Visitor behavior research (Chhetri, Arrowsmith et al. 2004)	To identify and understand visitor experiences of natural landscapes	• GPS Tracking • Pressure pads and mads • PDA Tracking • Self-administrative questionnaire • Observing • Timing system	There needs to be route recording, visitor experience tracking as well as simple counting when estimating visitors' movement, their experiences of natural landscapes and number of them.

Visitor behavior research (Fennell 1996)	To examine the relationship between tourism group movement patterns and their motivation for visiting parks	• GPS Tracking • Timing system • Self-administered questionnaire	There needs to be route as well as visiting characteristics recording when deciding the visitor group based on movement information
Visitor behavior research (Arrowsmith and Chhetri 2003; O'Connor, Zerger et al. 2005)	To validate model of visitor behavior prediction	• GPS Tracking • Video image processor • Timing systems • PDA Tracking • Self-administered questionnaire	There needs to be route recording when estimating visitors' movement behavior.
Traffic congestion (O'Connor and Zerger 2002)	To assess the volume or density of pedestrians at attractions or road intersection in order to improve traffic conditions in parks	• Timing system • Pressure Pad or mad • Passive infrared • Light barriers • Video image processor	There needs to be time recording as well as simple counting when estimating traffic congestion.
Traffic congestion (Itami 2003)	To assess the volume or density of vehicle at car park or road intersection or attractions in order to improve traffic conditions in parks	• Passive infrared • Laser radar • True-presence microwave radar • Doppler microwave detector • Ultrasound • Video image processor	There needs to be time recording as well as simple counting cars when estimating traffic congestion.

At this level movement information including location visited, corresponding arrival or departure time, duration of stay, visiting characteristics, and visitor demographic information needs to be collected. Generally low resolutions of time and space are required and therefore interviews, self-administered questionnaires, and virtual questionnaires can be used to track the movement of visitors.

At the micro level, spatial movements are measured in metres and temporal scales are in terms of minutes. At this level higher resolution movement data is required. Movement information such as location, time, direction and speed needs to be collected. Closed circuit television monitoring (camera based systems), GPS and timing systems can be used to track visitors at this level. PDA tracking, self-administered questionnaires and interviews are able to acquire visitor wayfinding decision-making processes at the micro-level.

DATA ANALYSIS FOR SIMULATION MODELLING

In our discussion so far, we have concentrated on investigating the data requirements and the various approaches and technologies to acquiring this data for simulation modelling. Depending upon the technology used managing the data can be quite simple say from direct observation methods, through to complex when using automated systems that capture movements and behaviours at fine temporal and spatial scales. This section will introduce some simple methodologies for aggregating and simplifying collected data and converting that into useful simulation information.

In two separate yet contemporaneous studies conducted at Port Campbell National Park in western Victoria, Australia, Arrowsmith et al. (2005) and O'Connor et al. (2005) utilised two different technologies yet similar methods for extracting information for simulation modelling. Arrowsmith et al. (2005) used GPS supported by on-site socio-demographic surveys. Basic socio-demographic data were collected at the entrance to the park and participants were asked to carry a GPS receiver throughout their visit. Spatial movements were handled within a geographic information system, via two distinct spatial analytical methods. The first method required "regions" or polygons to be drawn around distinct pathway segments that would encompass all GPS recorded "legs" (line drawn between each successive recorded GPS location). For each GPS leg, speed and direction were calculated. For each region the average walking speed and duration within that region could be derived. For simulation modelling average velocities and duration times within different geographic regions (including point locations such as lookouts and car parks) could then be calculated. The second method enabled typical movement sequences to be derived. Using network analysis, key nodal points on the pathways were labelled alphabetically. From observed sequences of movements, classification of movements for each survey participant were derived as a succession of nodal movements depending on the order of the sequence and the nodes that were passed through. Using a dendritic system (see figure 3), total number of similar sequence patterns and visitation numbers could be determined. From this information it was then possible for the simulation modeller to determine probabilities for alternate pathways in complex networks.

Simple statistical cluster analysis was used to classify visitors according to completed questionnaires based on age, visitor group and place of residence. Each participant was then allocated to one of four clusters which were then mapped to ascertain spatial patterns.

A similar approach was adopted by O'Connor et al. (2005) where the Alge timing system (http://www.alge-timing.com/alge-e.htm) was used to track visitors wearing transmitters as they passed strategically located receivers. Using "typology detection techniques" using methods of ordination, spatial data could be simplified through classification. Using a series of axes in two dimensions (where each axis represented a single variable), points representing individuals were plotted between these axes. Points that fell close together were considered more similar than those further apart. Variables that were plotted against each other included time spent on site, time spent on each path segment, total time spent on the main viewing platform and time spent at a second viewing platform. Whereas in the first study only just over 100 participants were mapped due to equipment costs, the second study enabled 900 individuals to be mapped.

The approach developed by Arrowsmith et al. (2005) was used in a second study of boat movements along the Yarra River in Melbourne, Australia (Arrowsmith et al., 2006). This paper gives a good overview of a generalised methodology for analysing mobile objects that move along linear pathways or networks from inventory development, detection, observation and recording of movements, analysis, and identification through to simulation of mobile objects.

As an extension of the dendritic analysis shown in figure 3, Xia et al. (in preparation) used Markov Chains to analyse the spatial and temporal movements of visitors to Phillip Island, a well-visited island within one days drive south east of Melbourne, Australia. A Markov Chain is a discrete stochastic process where the future state of the process is independent of any past state. From a total of 800 questionnaires on movement patterns, 464 valid responses were used in the analysis. Nine attraction nodes and one "absorbing state" (outside the island) were used in the analysis. Conditional probabilities of moving from one attraction to the next enabled transitional probabilities within the Markov Chain to be determined. The study found that Markov Chains were appropriate for shorter length itineraries (for less than five visited point attractions or nodes).

CONCLUSIONS

This chapter has concentrated on the practical aspects relating to acquiring data required for building simulation models for visitor landscapes. Before reviewing methods that can be used to acquire these data, the chapter commences by defining what data is actually required. Essentially there are two fundamental data types required for simulation models – spatial and temporal, and these are discussed in section 1. People behave in various and often unpredictable ways, yet managers of visitor landscapes are required to implement measures to control this behaviour. Yet most behaviour can be predicted, and it is this predictability that enables modellers to build rules into simulation models that can be then used by managers. To understand and model these behaviours, additional socio-demographic data are required. These data relate to the visitor "profile", including age, sex, socio-demographic background, and "visiting characteristics", such as travel mode, visitor group make-up and frequency of visitation.

There are a range of options for capturing such data, ranging from manual, labour intensive techniques through to remote automated systems. Section 3 summarised and compared these techniques and this should provide a useful resource to the simulation modeller of visitor landscapes. Each technique has been evaluated based on the capability of the technique and the type of information each technique can provide. Each technique has also been evaluated based on the level of resolution that can be achieved, whether it is seen as an intrusive or non-intrusive approach, the size of sample that can be recorded and its suitability for particular applications. Section 4 provided some interesting methods for extracting relevant and useful information from raw data. This section should also assist the modeller when confronted by what might at first appear an overwhelming set of data.

LITERATURE CITED

ALGE-TIMING (2005), ALGE-TIMING (Tdc 8001), http://www.alge-timing.com/alge-e.htm, accessed on 8/8/05.

Arnberger A, Haider W and Brandenburg C (2005). "Evaluating Visitor-Monitoring Techniques: A Comparison of Counting and Video Observation Data", Environmental Management Vol 36 No 2, pp317-327.

Arrowsmith C, Zanon D and Chhetri P (2005), "Monitoring Visitor Patterns of Use in Natural Tourist Destinations" in Chris Ryan, Stephen Page and Michelle Aicken (eds) Taking Tourism to the Limits: Issues, Concepts and Managerial Perspectives, Elsevier, The Netherlands, ISBN 0-08-044644-2, pp33-52.

Arrowsmith C, Itami R and Kim S (2006) "The application of information technology in mobile object simulation" Networks and Communication Studies NETCOM, vol 20, no 1-2.

Arrowsmith, C. and Chhetri, P. (2003), Port Campbell National Park: Patterns of Use. RMIT University.

Ashbrook, D. (2002), Learning Significant Locations and Predicting User Movement with GPS, paper presented to Proceedings of the 6th IEEE International Symposium on Wearable Computers, Seattle, Washington, October 07 - 10, 2002.

Batty, M. (2003), Agent-based Pedestrian Modelling, in P. Longley and M. Batty (eds),

Advanced spatial analysis : the CASA book of GIS, ESRI Press, Redlands, Calif, pp. 81-106.

Bircher Reglomat (2003). http://www.bircheramerica.com/automatic-door/act-infrared-sensors.htm accessed on 17 May 2007.

Bogaert, M., Chleq, N., Cornez, P., Regazzoni, C. S., Teschioni, A. and Thonnat, M. (1996), Passwords project, IEEE International Conference on Image Processing, vol. 3, pp. 675-678.

Broad, S. and Smith, L. (2004), Who Educates the Public about Conservation Issues? Examining the role of zoos and the media, paper presented to the International Tourism and Media Conference, Tourism Research Unit, Melbourne.

Carvell, J.., Balke, K., Ullman, J., Fitzpatrick, K., Nowlin, L. and Brehmer, C. (1997), Freeway Management Handbook, http://ntl.bts.gov/lib/6000/6400/6416/fmh.pdf 27/09/04, viewed 27/09/04.

Casademont, J., Lopez-Aguilera, E., Paradells, J., Rojas, A., Calveras, A., Barcelo, F. and Cotrina, J. (2004), Wireless technology applied to GIS, Computers & Geosciences, vol. 30, no. 6, pp. 671-682.

Chadha, K. and Osthimer, S. (2004), GPS Technology White Paper, http://www.sirf.com/gps_tech_white_paper.htm accessed on 20/09/04.

Chen, S. and Morley, R. S. (2005), Observed herd size and animal association, Ecological Modelling, vol. 189 no. 3-4, pp. 425-435.

Chhetri, P., Arrowsmith, C. and Jackson, M. (2004), Determining hiking experiences in nature-based tourist destinations, Tourism Management, vol. 25, no. 1, pp. 31-43.

Choi T and Chu R (2000). "Levels of satisfaction among Asian and Western travellers", The International Journal of Quality and Reliability Management, Vol 17 No 2, pp116.

Day G. (1999). "Trainspotting: Data collection from a captive audience", Notes in Tourism Management, Vol 20, pp153-155.

Dixon, T. (2004), People in the Scottish Countryside and Automatic People Counters (APC), Countryside Recreation, vol. 12, no. 2, pp. 19-23.

Dumont, B., Roovers, P. and Gulinck, H. (2004), Estimation of off-track visits in a nature reserve: a case study in central Belgium, Landscape and Urban Planning, vol. 71 no. 2-4, pp. 311-321.

Fennell, D. A. (1996), A tourist space-time budget in the Shetland Islands, Annals of Tourism Research, vol. 23, no. 4, pp. 811-829.

Gasvoda, D. (1999), Trail Traffic-Counter, 9E92A46, Missoula Technology and Development Center (MTDC).

Hadley, D., Grenfell, R. and Arrowsmith, C. (2003), Deploying Location-Based Services for nature-based tourism in non-urban environments, paper presented to Spatial Sciences Coalition Conference, Canberra, September 2003.

Hamamatsu (2004). "Characteristics and use of infrared detectors: Technical Information SD-12". Available online at: http://sales.hamamatsu.com/assets/applications/SSD/Characteristics_and_use_of_infrared_detectors.pdf accessed 17/05/07.

Haritaoglu, I., Harwood, D. and Davis, L. (1998), W4: Who, When, Where, What: A Real Time System for Detecting and Tracking People, paper presented to Third Face and Gesture Recognition Conference.

Hashimoto, K., Morinaka, K., Yoshiike, N., Kawaguchi, C. and Matsueda, S. (1998), People-counting system using multisensing application, Sensors and Actuators A: Physical, vol. 66, no. 1-3, pp. 50-55.

Heikkila, J. and Silven, O. (2004), A real-time system for monitoring of cyclists and pedestrians, Image and Vision Computing, vol. 22, no. 7, pp. 563-570.

Hofstede G (1980). Culture's consequences: International Differences in Work-Related Values. SAGE Publications London.

Hofstede, G. (2001), Culture's consequences : comparing values, behaviours, institutions, and organizations across nations, SAGE Publications, 2001.

Huisman, O. and Forer, P. (1998), Towards a Geometric Framework for Modelling Space-Time Opportunities and Interaction Potential, paper presented to International Geographical Union, Lisbon, Portugal.

Infodev (2004), Comparative analysis of counting technologies, http://www.infodev.ca/

AN/article/tecnoAN.shtml, viewed 19/08/04 2004.

Itami, R. M. (2003), RBSim3: Agent-based simulations of human behaviour in GIS environments using hierarchical spatial reasoning, paper presented to ModSim 2003 International Congress on Modelling and Simulation, Townsille, Australia.

KaewTrakulPong, P. and Bowden, R. (2003), A real time adaptive visual surveillance system for tracking low-resolution colour targets in dynamically changing scenes, Image and Vision Computing, vol. 21, no. 10, pp. 913-929.

Klein, L. A. (1995), Detection Technology: For IVHS-Volume 1: Final Report Addendum, FHWA-RD-96-100, U.S. Department of Transportation Federal Highway Administration.

Koncz, N. and Adams, T. (2002), Temporal Data Constructs for Multidimensional Transportation GIS Applications, Transportation Research Record, Journal of the Transportation Research Board, no. 1894, pp. 196-204.

Kozak, M. (2001), Repeaters' behavior at two distinct destinations, Annals of Tourism Research, vol. 28, no. 3, pp. 784-807.

Leeworthy V., Wiley, P, English, D and Kriesel W (2001). "Correcting response bias in Tourist Spending Surveys", Annals of Tourism Research Vol 28 No 1, pp83-97.

Li, C. L. (2004), Spatial Ability, Urban Wayfinding and Location-Based Services: a review and first results, in CASA Centre for Advanced Spatial Analysis http://www.casa.ucl.ac.uk/working_papers/paper77.pdf.

Loiterton, D. and Bishop, I. D. (2005), Virtual Environments and Location-Based Questioning for Understanding Visitor Movement in Urban Parks and Gardens, paper presented to Real-time Visualisation and Participation, Dessau, Germany, May 26-28.

Manning Robert E, Itami, Robert M, Cole David N and Gimblett Randy (2005). "Chapter 3: Overview of Computer Simulation Modeling Approaches and Methods", USDA Forest Service General Technical Report RMRS-GTR-143, 2005. Available online at http://www.fs.fed.us/rm/pubs/rmrs_gtr143/rmrs_gtr143_011_016.pdf accessed on 18/05/07.

Mateos, P. (2005), Mapping the space of flows: mobile phone location as a method to track the mobile society, paper presented to Proceedings of the 9 th International Conference in Computers in Urban Planning and Urban Management (CUPUM), University College London , London, 29-30 June.

McFadden, J., Graettinger, A., Hill, S. and Tucker, B. (2001), Potential Applications of Video Technology for Traffic Management and Safety in Alabama, The University of Alabama, Tuscaloosa, Alabama.

Metrocount (2005). "MetroCount 5710 Configurable Roadside Unit", March 2005, MetroCount, available online at: http://www.metrocount.com/downloads/files/MC5710.pdf accessed on 22/04/07.

O'Connor, A., Zerger, A. and Itami, B. (2005), Geo-temporal tracking and analysis of tourist movement, Mathematics and Computers in Simulation, vol. 69, no. 1-2, pp. 135-150.

O'Connor, A. N. and Zerger, A. (2002), Building Better Agents -Statistical and Spatial Analysis of Tourist Tracking Data, [On line], Available http://www.sli.unimelb.edu.au/students/ugrad/projects/2002/aliceoc/index.html [2004, June 14].

Parks Victoria (1998). "Port Campbell National Park and Bay of Islands Coastal Park: Management Plan" available online at http://www.parkweb.vic.gov.au/resources07/07_0186.pdf, Parks Victoria September 1998.

Reisinger, Y. and Turner L. (2003), Cross-cultural behaviour in tourism: concepts and analysis, Butterworth Heinemann, Oxford.

Sinha J and Kumar R (2004). "Methodology for Understanding Indian Culture", The Journal of Asian Studies 19. Copenhagen available online http://rauli.cbs.dk/index.php/cjas/article/viewFile/27/26 accessed 17/05/07.

Skszek, S. L. (2001), State-of-the-art report on non-traditional traffic counting methods, Arizona Department of Transportation Final Report 503, available online http://ntl.bts.gov/DOCS/arizona_report.html accessed on 22/04/07.

Slater, A. (2002), Specification for a dynamic vehicle routing and scheduling system,

International Journal of Transport Management, vol. 1, no. 1, pp. 29-40.

TRAFX (2004), TRAFx Pressure Pad Trail Counter, http://www.trafx.net/TRAFx_Pressure_Pad_Trail_Counter.pdf, 22/08/04.

Ting-toomay, S (1999). Communicating across culture. Guilford Press New York.

VeriLocation (2004), Now providing mobile location services to over 15,000 customers from 192.com, Phonetrack, Locate Mobiles, Easy Reach and Trace A Mobile., available online: http://www.verilocation.com/default.aspx accessed on 17/05/07.

Wang, B. and Manning, R. E. (1999), Computer simulation modelling for recreation management: A study on carriage road use in Acadia National Park, Maine, USA, Environmental Management, vol. 23, no. 2, pp. 193-203.

Xia J, Zeephongsekul P and Arrowsmith C (in preparation). "Modelling spatio-temporal movements using finite Markov Chains".

CHAPTER 6

SIMULATION, CALIBRATION AND VALIDATION OF RECREATIONAL AGENTS IN AN URBAN PARK ENVIRONMENT

Daniel Loiterton
Ian Bishop

Abstract: Autonomous agent models are becoming increasingly common in studies of human movement patterns and choice-behavior in a number of real-world situations. Simulating recreational agents in this way - essentially creating and observing virtual people in a virtual environment - can provide insight into the use of real-life environments and the impact of management decisions. This study employs iRAS (Intelligent Recreational Agent Simulator), a multi-agent system developed at The University of Melbourne to examine complex spatial patterns of visitor behavior in the Royal Botanic Gardens (RBG) in Melbourne, Australia. This study describes a multi-agent model coupled with a mixture of visitor surveys, experiments, and on-site monitoring to create acceptable behavioural algorithms, environmental attributes, and agent variables. To validate the results of the agent model, the simulation output was compared with observation, allowing differences and anomalies to be isolated, investigated and removed by adjusting system parameters accordingly. As a result, the generated movement patterns are highly correlated with reality and suggest that the system as it stands is quite valid for this particular environment. This study illustrates that through rigorous field investigation, monitoring, and experimentation along with an iterative approach to system analysis and improvement, statistically valid visitor movements can be simulated and aid planners in reducing congestion and enhancing visitor experience.

Key words: Human Behavior, Choice Behavior, Simulation, Visitor Experience, Vitural Environments

INTRODUCTION

Autonomous agent models are becoming increasingly common in studies of human movement patterns and choice-behaviour in a number of real-world situations. These include studies of pedestrian movement in urban environments and shopping centres (Dijkstra and Timmermans 2002; Zachariadis 2005), simulations of evacuations at large events or after major disasters (Masuda and Arai 2005; Murakami et al. 2002), and the generation of probable routes for tourists in recreational areas (Cavens et al. 2003; Gimblett et al. 2001; Itami et al. 2000). Simulating recreational agents in this way - essentially creating and observing virtual people in a virtual environment - can grant us great insight into the use of real-life environments and the impact of possible management decisions (Gimblett 2005). However, to ensure the validity of such simulations, pains must be taken to carefully parameterize and calibrate the models via thorough studies of the location's populace (Skov-Petersen 2005).

In the Royal Botanic Gardens (RBG) in Melbourne, Australia, complex visitor behaviour is being simulated with iRAS (Intelligent Recreational Agent Simulator), a multi-agent system developed at The University of Melbourne. iRAS agents are capable of a series of realistic behaviours ranging from deliberate, shortest-route movement towards a goal, to virtually aimless wandering with path choices based on spontaneous, aesthetically-driven decisions. Visitors may enter the gardens with preconceived goals to visit certain points-of-interest and may also generate goals on the fly in order to eat, rest, or use a toilet at nearby garden facilities. Movement decisions are based on physical and visual path attributes, the presence of other visitors, and also influences specific to each visitor. These include personal preferences for particular path features, repetition avoidance, directional motivations (eg. towards or away from exit gates, towards priority goals, or along a similar bearing) as well as dynamic feelings of exhaustion, boredom, and hunger among others.

To generate statistically valid movement patterns, a mixture of visitor surveys, experiments, and on-site monitoring were used to create acceptable behavioural algorithms, environmental attributes, and agent variables. These ranged from informal interviews with Botanic Gardens visitors and staff to carefully planned experiments using high-level technology. To validate the results of the agent model, the simulation output was compared with observation, allowing differences and anomalies to be isolated, investigated and removed by adjusting system parameters accordingly. As a result, the generated movement patterns are highly correlated with reality and suggest that the system as it stands is quite valid for this particular environment.

After ensuring the robustness of the system via sensitivity analysis, the simulation can be used to test hypothetical management scenarios. Configurations of paths, garden beds, toilets and kiosk facilities can be manipulated quickly and easily in the virtual environment with the resulting agent movement patterns compared to those observed in the original landscape.

CREATING THE AGENT SIMULATION

Agent Oriented Software's JACK™, and agent development kit built on the Java programming language, was chosen to provide the basic agent modelling constructs. JACK™ is commercially available worldwide and offers a robust set of tools for creating autonomous, goal-directed and reactive agents. The iRAS program itself is broken into a number Java objects which include several JACK™ agent-oriented features. The different modules within the software generally fall into four categories:

- Agent-based objects
- Environment objects
- Interface objects
- Miscellaneous tools

AGENT-BASED OBJECTS

Agent-based objects include the agents themselves, as well as the various events, plans, goals, and feelings, which control their behaviour. Only one type of agent is required for the Royal Botanic Gardens simulation; the *Visitor* agent, which can represent a single person or a decision-making group made up of any number of members. While it would be possible to include *Staff* agents, this was not deemed necessary as their effect on visitor movement would be minimal and ultimately it is only visitor movement that we are interested in. Wildlife agents could present an interesting addition to the simulation as some visitors indicated the presence of birds and bats as an influence in their decisions. It was decided however, that the added complexity involved in simulating the behaviour of both animals and humans, and the interactions between them, would far outweigh any benefits. The effect of bats was greatly diminished around the time the project began thanks to a relocation project which drastically reduced their numbers. We can also assume that the effect of waterbirds, fish, and eels (which are always described in a positive light) is included in the appeal of water features in general. Finally, other birds around the gardens are spread fairly randomly, and are hardly mentioned compared with the water-dwelling swans, ducks and geese; thus their impact on visitor movement patterns can be assumed to be minimal.

While there is only one type of agent, the numerous visitors populating the simulation are far from identical. Each has a large set of parameters, which are either defined in an input file or calculated at the beginning of the simulation and updated throughout. While these variables essentially allow for infinite variations in behaviour, all visitors nevertheless follow the same set of basic rules. Computationally, each agent will experience the same events and will utilise the same set of plans and methods to handle these events.

Only two event types are used in the simulation, and these are implemented very simply using the JACK™ BDIGoal-Event construct. Basically, the software uses these to indicate the next type of action the agent is to take; namely, whether it is to Stop or Go. When one of these events is triggered, the agent will assess each of the plans able to handle the event until it comes across one that is relevant. This plan will inturn carry out a number of tasks until certain criteria are satisfied, at which point another event is triggered and the process repeats itself until the agent exits the environment.

For instance, if the agent experiences a Stop event, it will consider each of the plans at its disposal which involve pausing at the current location. Depending on the agent's position, its set of goals, and how it is feeling at the time, it can choose to sit and rest, sit and eat, use the toilet, or simply admire the scenery. If a Go event is posted, the agent can either; take the shortest route to one of its goal destinations, take its favourite immediate path based on physical and visual properties, or simply return along the path it came if a dead end is encountered.

AGENT PARAMETERS

Throughout the agent's journey, its decisions are based not only on environmental factors, but on numerous internal influences controlled by various parameters. These range from simple integer values to complex programmable objects, themselves made up of numerous variables. Agent parameters generally fall into five categories:

- Static input details
- Updatable details
- Path preferences
- Feelings
- Goals

Static Input Details are assigned values from input data files at the beginning of the simulation, and remain constant throughout. These include variables such as the agent's age, gender, group size, start time, desired entrance & exit nodes, mobility status, and whether the agent has brought a picnic.

Updatable Details mainly describe the agent's progress within the environment as its visit progresses. These include the agent's current state (ie. whether wandering, admiring, eating, resting, etc), position, speed, slope, bearing, distance walked, time elapsed, path, node and destination.

Path Preferences describe the agent's individual attraction to or repulsion from certain physical and aesthetic path qualities. These values incorporate both the actual preference (whether positive or negative) and a level of importance (where -1 & 1 are extreme and 0 denotes no preference). As opposed to Feelings (described below), preferences are not influenced by any other factors; individual preference values are obtained from an input file and remain static over the course of the simulation. While in reality a visitor will possibly alter their preferences depending on the scenery they have already encountered, these particular agent variables are not designed to simulate such real-time impulses but rather general inclinations one might have for certain features. In particular, an agent can have preferences relating to path width, surface, shape, slope, enclosure, views, water visibility, traffic noise, and crowds.

Feelings are designed to replicate personal physical or emotional influences one encounters when making path choices. For instance, while an agent may have a general preference for or against path slope, the agent's level of exhaustion will also affect the appeal of this attribute. Furthermore, a path that leads directly to a possible rest point (eg. bench, grassed area, or rest house) will become more attractive as exhaustion increases. Finally, when an agent's level of exhaustion reaches a critical point, the agent will generate a Goal to rest, which in turn will cause the agent to move directly to the nearest rest point in order to recover for a short period of time. All feelings are dynamic and range in intensity from zero to one. They are always dependent on a specific corresponding 'feeling-factor', which generally affects the feeling's variation over time or distance. Exhaustion, for example, is a function of exertion (distance multiplied by slope) and agent fitness (the exhaustion-specific feeling-factor). Table 1 describes each of the four feelings in more detail.

Goals are spatially enabled objects which essentially simulate the desire to visit a particular location, and influence the appeal of paths leading in their direction. An agent can have any number of goals and each of these will have an importance value (between 0 and 1), which allows the agent to prioritise its goals. While the direction of the highest priority goal will always be attractive, if the importance is extreme or if a goal is within close proximity, the agent can choose to complete it immediately by moving directly to that goal's destination via the shortest route. An agent may have a number of 'visit' goals at the beginning of its journey and it can generate 'rest', 'eat', 'use toilet' and 'exit' goals on the fly once a corresponding feeling reaches its critical point (see Table 1).

AGENT METHODS

While each individual visitor agent is defined by the parameters described above, all visitors follow standard procedures defined by the various methods contained in the Visitor agent class. Some of these methods, which will be discussed briefly, include:

- prioritiseGoals
- evaluatePath
- startPath
- walkPath
- leavePath
- wait

Every time an agent comes to a decision

point, its first task is to prioritise its goals using the *prioritiseGoals* method. This involves adding any new goals (if a feeling has reached its critical point) and then sorting all incomplete goals based on importance & distance. If the agent is within close proximity to a certain goal, this goal may become a higher priority even though other goals may have higher importance values. If a goal is close enough and/or important enough, the agent will choose to move directly to that location. If not, it may choose to evaluate all path choices, before deciding on its favourite.

The *evaluatePath* method calculates an appeal value (a) for any given path, based on the agent's current position, preferences, goals and feelings. When an agent is deciding on a preferred path, the path with the highest appeal value will be chosen. The appeal of any given path is the sum of the appeal values for all path features, each of which is calculated using a specific algorithm. These algorithms will

Table 1. Agent feeling algorithms and descriptions

Exhaustion (f_E) $\Delta f_E = (1+b) \cdot \Delta d / f$	Exhaustion increases over distance (d) with respect to individual agent fitness (f), which is defined as 'distance to exhaustion over flat terrain'. This rate will hence increase with (uphill) slope (b). In turn, exhaustion will influence the appeal of paths with upward slope and also paths leading to a rest point. If f_E reaches 1, a **Rest** goal is generated. Exhaustion decreases with rest.
Boredom (f_B) $f_B = t / t_d$	Boredom is simply a function of total visit duration (t) as a fraction of desired visit duration (t_d). Boredom always begins at 0 and uniformly increases to 1 as the visitor's elapsed time approaches its desired exit time. In reality, time spent would be a function of boredom and not vice-versa, however this distinction is not required since visit durations (and hence boredom) are directly based on statistical data. Boredom influences the appeal of path direction relative to the visitor's desired exit; as boredom approaches 1, the appeal of the exit direction increases. If f_B reaches 1, an **Exit** goal is generated. Boredom cannot decrease during the visit.
Hunger (f_H) $\Delta f_H = \Delta t / t_H$	Hunger increases over time (t) with respect to the agent's individual hunger duration factor (t_H). Hunger influences the appeal of paths leading to an eating point. If f_H reaches 1, an **Eat** goal is generated. Hunger decreases while the agent is eating.
Toilet Need (f_T) $\Delta f_T = \Delta t / t_T$	Toilet Need increases over time (t) with respect to the agent's individual toilet duration factor (t_T). This feeling influences the appeal of paths leading to a toilet. If f_T reaches 1, a **Use Toilet** goal is generated. The feeling decreases when the agent visits a toilet.

always involve at least one path-dependent variable and at least one agent-dependent variable.

Most path attribute values range between 0 and 1, but are often transformed within the algorithm so that the possible range is actually between -1 and 1. Similarly, feelings and goal importance values are usually modified within the formula, while preferences naturally range between -1 and 1. The agent variable (preference, feeling, goal importance, or a combination of these) is then usually multiplied with the appropriate path attribute so that the resulting appeal value is also between -1 and 1. In this way, the majority of path features either add to or detract from the overall appeal of a path by a fraction of one (Table 2 lists some examples of path feature appeal algorithms and resulting ranges). Path feature weights (w) are then used to uniformly increase or decrease the effect of any given path attribute in order to increase or decrease its relative importance. The resulting equation for final path appeal is hence:

$$a_{Total} = (a_1 \cdot w_1) + (a_2 \cdot w_2) + \ldots + (a_n \cdot w_n)$$

Once an agent has chosen a path or route, it will next call the *startPath* method to update a number of path-specific variables. Then, for each time step, the *walkPath* method is used to move the agent a certain distance along its chosen path. This method first verifies that the agent is on the correct path, determines the direction based on the agent's previous node, then calculates the distance to walk based on the length of the time step and the agent's speed (which is in turn based on the slope of the current path). The new position is then calculated by moving the agent along a number of path segments (from vertex to vertex) until the appropriate distance is covered. The agent's current slope, bearing, and speed are then re-calculated based on this new position as well as the agent's feelings and other time and distance-related variables. If the end of the path is reached within the time step, the *leavePath* method is called, which simply removes the agent from the path.

When an agent chooses to pause at a specific node in order to admire the scenery, sit and rest, sit and eat, or use the toilet; the *wait* method is invoked. As with *walk-*

Table 2. Path appeal algorithm examples

Path Feature	Appeal Algorithm	Range
Direction (relative to priority goal)	$a_{q2} = g_1 \cdot \left(1 - \dfrac{2f}{p}\right)$, $f = \left|\Delta q\right|_{p \to g_1}$	$-1 \leq a \leq 1$
Direction (relative to exit)	$a_{q3} = (2f_B - 1) \cdot \left(1 - \dfrac{2f}{p}\right)$, $f = \left|\Delta q\right|_{p \to x}$	$-1 \leq a \leq 1$
Slope	$a_b = b \cdot (r_b - f_E)$	$-1 \leq a \leq 1$
Horizontal Enclosure	$a_h = (2 \cdot h - 1) \cdot r_h$	$-1 \leq a \leq 1$
Leads to Rest Point	$a_{\ell R} = \ell R \cdot (3 \cdot f_E - 2)$	$0 \leq a \leq 1$

Path, this method is called once per time step for the duration of the pause, and is used to update all time-related variables and feelings. The manner in which these are updated depends on the status of the agent; that is, what it is doing while paused. For example, if the agent is sitting and eating, its feelings of exhaustion and hunger will both decrease, while its need to use the toilet will continue to increase.

THE AGENT ENVIRONMENT

The agent environment consists of a basic network of paths and intersections (nodes). Each path contains a node at either end, as well as a series of vertices which define its shape. All nodes and vertices contain x, y, and z coordinates, making the environment three-dimensional in some respects. Just as each path contains two nodes, each node subsequently contains a number of paths that are attached to that intersection. Paths and nodes also have several attributes which define their visual and physical properties, as well as the activities that an agent may perform whilst visiting them.

Node objects contain a set of boolean variables indicating whether or not the node is an exit point, toilet point, food point, rest point, grassed area, or point of interest. *Path* attributes include the path's distance and width, as well as 'factor' values (usually ranging between 0 and 1) which describe the path's surface, shape, horizontal enclosure, vertical enclosure, traffic noise level (proximity), and water feature visibility (proximity). Two slope factors also define the amount of upward slope in each direction, while two other variables record the number of visitors currently on the path and the day's total usage.

A *Network* object *stores* all path and node objects, but also *creates* them at the beginning of the simulation from a series of input files (one each for path, node, and vertex data). Positional data is read and stored allowing the given three-dimensional positions to be compared so that each node can be linked to its connecting paths and vice versa. A number of search methods are also contained within the *Network* class, allowing a node to be identified by position, ID, proximity or feature.

Some discussion took place as to whether the environment should in fact be continuous, allowing agents to wander anywhere in three-dimensional space. It was argued however, that a constrained discrete network such as this is certainly able to replicate the most common movement patterns observed in the Royal Botanic Gardens (that is, movement along the many paths), and that the addition of important grass routes goes some way towards simulating unconstrained movement. Furthermore, the end result of a given simulation is generally focussed on the compilation and comparison of *path usage*, and so exact positioning away from the path network becomes somewhat superfluous.

THE USER INTERFACE

Autonomous agent models are, by nature, very difficult to debug. Due to the complex, sometimes random-like behaviour of the agents, errors are often impossible to replicate and thus much effort is required to locate the cause within actual code. Fortunately, with a model based in real space, it is possible to visualize the simulation as it is running. This enables far greater transparency while designing, building, testing, and presenting the system.

The iRAS user interface (figure 1) consists of a control panel, an agent status panel, and a main map. The control panel uses simple icons to begin, pause, and reset the simulation, as well as to display the path network and/or areal photograph in the map panel. The agent status panel displays the details, preferences, feelings and goals of any agent that is selected (by clicking on a given agent within the map panel). Finally, the map panel itself offers a two-dimensional, overhead view of all agents currently within the path network including text relating to the current status of the selected agent. The agents' progress, along with all textual information in both the map panel and the status panel is updated in real-time as the simulation runs.

MISCELLANEOUS TOOLS

Apart from the agent, environment, and interface objects, the remaining classes generally contain numerous objects, tools, and functions for generic, frequently-used tasks. The Data class, for example, primarily contains a number of global variables – objects and values that need to be accessed from various methods within the program. The Mapping class on the other hand, contains methods which perform all manner of calculations in three-dimensional space from simple point-to-point distance calculations, to more complex computations which take multiple paths into account, or depend on the shape of the network at a given location.

Of particular interest is the *RouteCalculator* class, which allows agents to compute the quickest route between any two nodes. Due to the complexity of the network, and the need for relatively short calculations, it is not reasonable to assess every possible path combination and simply choose the shortest. Instead, a specific process was devised that somewhat resembles the way in which actual people will travel from one point to another when they don't have a map to guide them. That is, the route is chosen in steps from start to finish based on each section's ability to progress the

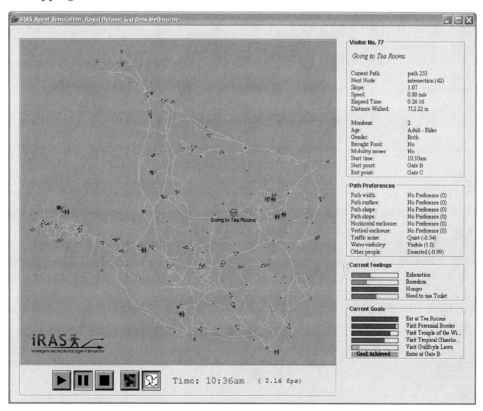

Figure 1. iRAS user interface

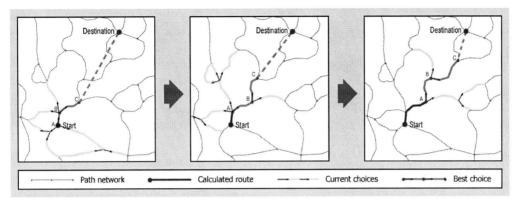

Figure 2. Calculating the shortest route

visitor closer to its destination. The actual process is discussed in detail below.

First the function assesses each path connected to the start node, along with each path connected to each of their end nodes (ie. two levels are analysed). The best two-path route is determined based on its distance and the distance from its end node to the final destination. That is:

$$path.A_{(node.A \to node.B)} + path.B_{(node.B \to node.C)} + remainder_{(node.C \to destination)}$$

Of the best two-path route, the first path (*path. A*) is added to the final route, while the function then moves to the middle node (*node. B*) and repeats the process to find the next shortest two-path route. Once the destination node is reached, the process terminates and returns the final path set. Figure 2 further illustrates the steps involved in the route-calculation process.

PARAMETERISATION & CALIBRATION

Any computer simulation is only ever as good as the data on which it is based. For a complicated model attempting to replicate human behaviour, this means that a great deal of quality data is required. Firstly, every single path and node within the network needs to be defined with high spatial accuracy, connecting one to another at exact positions. Similarly, visual and physical attribute data must be accurately assigned to each path and node within the network. The agents themselves require even more data than the environment – some thirty or more parameters per agent, with unique initial values for every one simulated. Furthermore, all behavioural algorithms must produce realistic, yet varied outcomes in order to accurately simulate visitor movement.

Techniques for collecting and generating this data were many and varied. No absolute rules were applied regarding which techniques should be used for certain input requirements and discussing the method for every parameter used would yield a lengthy paper in itself. The major experiments and surveys undertaken are discussed below however.

GIS DATA GENERATION

Quality map data was thankfully supplied by the Royal Botanic Gardens, who have embraced spatial technologies such as GIS (Geographic Information Systems), GPS (Global Positioning Systems), and areal photography for some time in order to create and maintain a useful geodatabase. This data provided the basis for the vast majority of agent network parameterization.

First, paths and nodes were carefully

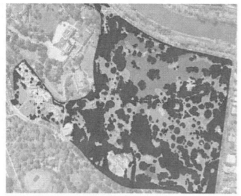
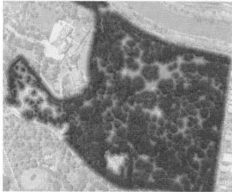
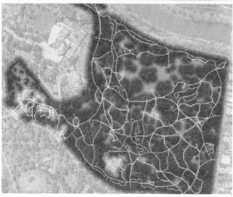

Figure 3. Generating 'vertical exposure' attributes for each path in the agent network

figure 3 illustrates the process of assigning 'vertical enclosure' values (a surrogate for shading) to each path within the network.

The 'canopy' layer is first used to create a straight-line distance raster by calculating a distance to the nearest edge of the canopy from each pixel in the raster grid. Each path is then assigned the average value of all cells through which it passes. Finally this data is standardised to yield a relative vertical enclosure factor, which all agents are able to 'sense' when choosing a path within the simulation. Figure 3 shows the canopy layer (top), straight-line distance raster (middle), and overlayed path network (bottom) displaying high (green) and low (white) levels of vertical enclosure.

LOCATION-SENSITIVE QUESTIONNAIRES

Using location-based technologies, a method for collecting information about the movement habits and influences of garden visitors was envisaged. Hand-held PDA (personal digital assistant) devices coupled with Bluetooth GPS receivers can enable participants to be tracked as they move around within the gardens, while also being asked location-specific questions throughout their journey.

The trial was broken up into two separate experiments involving two survey designs. This allowed different types of information to be gathered and, given the limited previous research in this field, allowed different approaches to be tested. Both surveys use small hand-held Pocket-PCs to display the questionnaire, and incorporate Bluetooth GPS positioning technology to locate the participant as they traverse the gardens.

digitised from areal photographs and existing path layers, and assigned height data based on a DEM (Digital Elevation Model) of the terrain. Further attributes were then assigned to these layers by actual field evaluation or by analysing further GIS layers (such as external roads, lake beds, and canopy area). For example,

The first design tracks the participant as they roam freely within the gardens. At specific time intervals, as the visitor approaches an intersection, a set of generic questions pops up, allowing the user to indicate the physical and perceived properties of the environment which are affecting their path decisions as they progress. Each time these questions are answered, the participant also gives an indication of their current feelings (ie. levels of boredom, fatigue, hunger, etc) using a simple scroll-bar design. Theoretically, this system can be used to examine spatial and temporal effects on visitor feelings, as well as how these feelings affect the importance of various path properties.

The second questionnaire design uses ESRI ArcPad™ to display a map of the RBG paths on the PDA screen. It then asks participants to follow a pre-defined route, displayed over the RBG map, while also indicating the user's location in relation to the route. At specific intersections of interest, the user is asked to select the path that they would most desire to travel, followed by the features that most affected their decision. In this way, information relating to the importance of specific features can be gathered using common intersections, while the stimuli experienced prior to each decision point remains controlled.

VIRTUAL ENVIRONMENT EXPERIMENTS

Virtual environments provide an excellent means of testing human responses to the environment. While visiting a landscape first-hand, or viewing photographs or video footage can yield a more accurate and realistic human experience, these methods only allow us to experiment with the landscape as it currently exists. At the opposite end of the scale, maps, charts, and plans are able to convey a great deal more information about both real and potential environments, yet the higher levels of abstraction greatly detract from the sense of realism. Computer visualization provides the happy medium – offering the ability to view both current and hypothetical landscapes using the same natural perspective with which we view the real world. As computer graphics technology rapidly progresses towards photo-realism, more realistic and immersive experiences can be produced, eliminating the expense and logistical burden of actually visiting a site of interest.

Visual Nature Studio™ 2 (VNS) was chosen to model the virtual environment as it is specifically designed to create realistic natural landscapes. It is also able to use GIS data to build up its scenes, and can automatically populate areas with pre-made 3D or 2D objects such as trees and plants. Vegetation was photographed and extracted to make 2D billboard textures of common trees, shrubs, and grasses. Larger trees were placed in the environment with exact positions using GIS point data, while smaller trees, shrubs and ferns were procedurally positioned within certain garden beds defined by polygonal data. Using VNS 'ecosystems', the distribution, density and size of such vegetation can be controlled, replicating the look of an area without the need for precise measurement data for every single plant.

Two intersections within the Royal Botanic Gardens were modelled in this way, and a series of short walk-through animations rendered, each leading along one of the paths up to the decision point. At the intersection, the camera pans across the possible path choices, allowing the audience to evaluate their options. A series of these animations were created, each with certain alterations to the features in the virtual model. The animations were then screened to several groups of survey participants who indicated their desired direction along with the importance of various visual features. In this way, the effect of very specific visual stimuli on path choice can be examined by ensuring that everything other than the feature in focus remains unchanged. Only by using a controlled virtual environment can we achieve this. Figure 4 provides an example

Figure 4. RBG virtual intersection – original (left) and lake views (right)

of one of the original intersection models compared with its slightly modified counterpart. By removing vegetation on the right-hand side of the path, views to the nearby lake are opened up. The differences in path choice between the two simulations can hence be attributed to this factor.

FIELD SURVEYS

Finally, a number of simple interview and observation style surveys were carried out at the Botanic Gardens site. Visitors were approached and asked simple questions relating to their visit and prompted to indicate which visual factors they thought influenced their choices at any given time. A number of specific intersections were also monitored, allowing actual visitor decisions to be recorded, along with various demographic data. Stretches of path were marked out and passers-by timed with a stop-watch in order to calculate average visitor speeds. Electronic people-counters located at each entrance to the gardens enabled the monitoring of visitor numbers per hour, per entrance. Botanic Gardens staff were also interviewed and gave invaluable information relating to the perceived popularity of certain areas within the park.

CALIBRATING THE SYSTEM

While all of this information greatly aided the creation of simulation algorithms, parameters and input data, many aspects of the software were still designed simply using sensibility and educated judgement. A good amount of freedom for further adjustments was also purposefully built in, as it was acknowledged that creating such a complex system requires an iterative approach, involving rigorous testing and analysis of multiple passes to obtain acceptable results. Even so, the initial comparison between simulated and actual visitor movement was extremely encouraging, with only small adjustments to the simulation required to facilitate the necessary improvements.

The first noticeable anomaly was a general over-usage of paths relative to the total number of agents within the environment when compared with observed path usage (during intersection observation surveys) and actual visitor numbers (based on RBG counter data). It was discovered that this was primarily due to agent speeds being generally higher than reality and with less frequent pauses and stops compared to the typical wandering visitor. Furthermore, it was hypothesised that visit durations were over-estimated due to inherent biases in the visitor surveys. While many people were interviewed within the gardens, many others declined and it is probable that visitors making shorter visits to the gardens (especially those simply passing through) are less likely to spend five minutes undertaking a survey than those who are casually wandering around with no immediate time pressure.

The other main issue with the simulated movement was a noticeable over-usage of one particular path in one particular direction. Upon further investigation it was revealed that a very high proportion of agents were congregating at the Tea Rooms node throughout the day and then moving directly to an exit as their boredom levels reached critical point whilst eating. Connecting the tea rooms to a number of highly-used exits was the path in question, explaining the one-way congestion. At this point it was wrongly assumed that any hungry agent would automatically make use of the kiosk facilities within the gardens, ignoring the fact that many people bring their own food and drinks to the gardens, or leave the gardens to find external sources of food.

By correcting these and other smaller issues with the system's logic, all major anomalies were eliminated so that the simulation output correlates extremely closely with the field observations. While it is important to note that these results are based on finite observational data, it nonetheless supports a strong argument for the validity of agent models for simulating visitor movement in this particular environment.

VALIDATION & ANALYSIS

Figure 5 compares the path usage of simulated agents with that of observed visitors at four intersections within the Melbourne Royal Botanic Gardens between 2pm and 4pm on a sunny weekday afternoon. Each total path usage column is also separated into directional components; 'in' denotes the number of visitors who approached the intersections via each path, while 'out' displays the path choices made upon reaching each intersection. While the values represented do not match perfectly, the proportions for each intersection are very similar. Out path choices are particularly accurate, demonstrating the similarity of agent and real visitor path decisions.

Figure 6 further demonstrates the high correlation observed between simulated and observed data with directional usage totals compared for each path. The correlation coefficient between the two data sets (simulated and observed) was calculated to be 0.85 for directional usage with a correlation of 0.92 when total path usage was compared. Considering the fact that the observed correlation between two

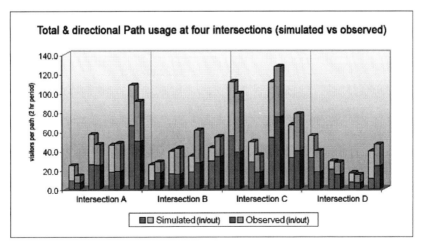

Figure 5. Comparing simulated and observed movement at test intersections. At each intersection four edge shoices were avaiblable.

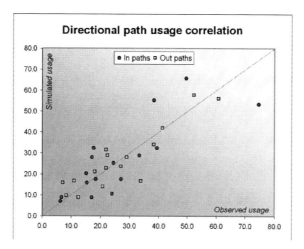

Figure 6. Correlation between simulated and observed path usage

similar afternoons in the real environment hardly improves upon these values (yielding correlations of 0.86 and 0.89 for directional and total usage respectively), this data in itself certainly suggests that the simulation as it stands is quite valid under current conditions. However, it also begs the question; why are the results so good? What factors are contributing most to these seemingly accurate movement patterns? The simulation was designed from the outset to be very complex, replicating many different styles of human behaviour in order to achieve realistic results. Now that such results have been achieved, it is worth investigating whether only certain aspects of the simulation are driving the results, or whether it really is the combination of these many 'human' traits that is causing the agents to behave successfully.

SENSITIVITY ANALYSIS

In order to study the effect of various parameters and algorithms on the resulting agent movement, a series of experiments was designed, which allowed the simulation to run without the inclusion of certain functionalities. Goal, feeling and preference parameters were each eliminated from the system in turn, with the resulting output of each test compared with actual observation data. Table 3 compares the correlation coefficients of the resulting simulated path usage data for each of the tests. While the differences are not extreme, the fully functional simulation still exceeds the trials in terms of accuracy, having the highest correlation for both directional and total path usage. This suggests that all aspects of the simulation are having some positive effect on behaviour, and so the inclusion of agent preference, feeling, and goal based influences, is indeed a valid choice.

While the features tested clearly do have an influence on the simulation results, it is still necessary to acknowledge the simulation's directional algorithms. These are somewhat difficult to eliminate from the system without causing gross errors in agent movement and hence could not be tested. We can assume however, that these parameters certainly have the greatest impact on agent movement patterns. An agent's current direction is always considered in path decisions, allowing each agent to move in a realistic 'general' direction without erratic directional changes or strange repeated loops. Similarly, the direction of the agent's exit, which is ini-

Table 3. Path usage correlation comparisons (simulation vs observation)

Dataset	Description	Directional	Total
Observation	Intersection observation data from two separate afternoons under similar conditions. (Correlation is between path usage data for the two days.)	0.86	0.89
Simulation	Fully functional simulation.	0.85	0.90
Test 1	No initial goals. Eliminates the effect of specific points of interest on visitor decisions.	0.69	0.83
Test 2	No feelings. Rest, food, and toilet stops are eliminated, as is the effect of feelings on path decisions.	0.71	0.90
Test 3	No preferences. All visual path properties are seen as identical and therefore have no effect.	0.74	0.86
Test 4	Combination of tests 1 to 3. All feelings, goals, preferences eliminated. Decisions primarily driven by direction (compared with current and exit direction).	0.46	0.54

tially a negative factor but becomes gradually more attractive as time passes, encourages exploration of large portions of the landscape and eventually influences the agent to move towards its correct exit as its desired exit time approaches. Eliminating these aspects can cause extremely long journeys as the agents can theoretically wander aimlessly ad infinitum. Hence we can assume that the correlation between such a test and reality would be zero.

Apart from analysing the simulation with variations to its functionality, the results of multiple simulation runs were also compared to determine the system's sensitivity to random variations. These variations can arise from inconsistencies in computation speed (because of hardware limitations), however they are mostly due to purposely included random variables, primarily relating to pause times. That is, it was deemed unnecessary to calculate such durations based on agent-specific variables and so universal resting, eating, and toilet durations were introduced with random variations included to simulate diversity. Such random variation was found to make little difference to the simulation output, with an average directional correlation coefficient of 0.94 calculated between three separate runs using identical input data; a much greater correlation than that of observed data from separate days under similar conditions, which revealed a coefficient of 0.86.

CONCLUSION

While the validity of multi-agent systems for replicating human movement is gradually being demonstrated for a range of applications including that of recreational behaviour in urban park environments, the future for projects such as this requires the confirmation of their value in spatial planning and management situations. Using the iRAS agent model, simulations of visitors to the Melbourne Royal Botanic Gardens in Australia have successfully replicated movement patterns under current conditions and could plausibly be used to establish likely movement patterns in hypothetical variations of the site. This application could provide professionals with an invaluable tool, able to deliver a quick and simple vision of the possible effects of a proposal without the time and expense of making actual physical modifications to the real environment.

In order to achieve this objective, a model such as this must first be tested under a variety of conditions, which can be replicated in the real environment in order to compare simulated output with observed behaviour. Undoubtedly, iterative modifications would again be necessary to achieve acceptable and consistent results; however, the open nature of software such as iRAS, with its ability to simulate a broad range of human behaviours, makes its use under numerous diverse conditions entirely feasible.

Through rigorous field investigation, monitoring, and experimentation along with an iterative approach to system analysis and improvement, statistically valid visitor movements *can* be simulated. It has become clear that the many mysteries of the human psyche do not need to be understood in full in order to successfully generate some forms of human behaviour. By isolating and simplifying those influences – both environmental and internal – which primarily control such behaviour, a logical set of rules can be derived and applied to autonomous agents in order to produce valid and useful spatial/temporal information regarding collective visitor movement patterns.

LITERATURE CITED

Cavens, D., Lange, E., and Schmid, W. 2003. "Virtual Alpine Landscapes and Autonomous Agents." *Trends in Landscape Modelling*, Heidelberg.

Dijkstra, J., and Timmermans, H. 2002. "Towards a multi-agent model for visualizing simulated user behavior to support the assessment of design performance." *Automation in Construction*, 11(2), 135-145.

Gimblett, H.R. 2005. "Human-Landscape Interactions in Spatially Complex Settings: Where are we and where are we going?" *MODSIM 2005 International Congress on Modelling and Simulation*, Melbourne, Australia, 11-20.

Gimblett, H.R., Daniel, T., and Meitner, M. 2001. "The Simulation and Visualization of Complex Human-Environment Interactions." *Landscape and Urban Planning*, 54(1-4), 63-79.

Itami, R.M., MacLaren, G.S., Hirst, K.M., Raulings, R.J., and Gimblett, H.R. 2000. "RBSIM 2: Simulating human behavior in National Parks in Australia: Integrating GIS and Intelligent Agents to predict recreation conflicts in high use natural environments." *4th International Conference on Integrating GIS and Environmental Modeling (GIS/EM4):Problems, Prospects and Research Needs*, Alberta, Canada.

Masuda, H., and Arai, T. 2005. "An agent-based simulation model of evacuation in a subway station." *Computers in Urban Planning and Urban Management*, London.

Murakami, Y., Minami, K., Kawasoe, T., and Ishida, T. 2002. "Multi-agent simulation for crisis management." *IEEE Workshop on Knowledge Media Networking*, 135-139.

Skov-Petersen, H. 2005. "Feeding the Agents: Collecting parameters for agent-based models." *Computers in Urban Planning and Urban Management*, London.

Zachariadis, V. 2005. "An agent-based approach to the simulation of pedestrian movement and factors that control it." *Computers in Urban Planning and Urban Management*, London.

CHAPTER 7

APPLYING DATA FROM AUTOMATIC VISITOR COUNTERS TO AGENT-BASED MODELS

Hans Skov-Petersen
Henrik Meilby
Frank Søndergaard Jensen

Abstract: Automatic visitor counters are used routinely for monitoring e.g. traffic loads and pedestrian activities. Data loggers attached to counters aggregate counts into time intervals. In the case of counting visitors in nature, counts will vary according to e.g. time of day, time of week, time of year, and weather conditions. When applying counter data to Agent-based Models (ABM's) it needs to be standardized into ideal time tables, representing the number of entries or the probability that a given number of visitors will enter during a given interval of time over an overall period (e.g. every hour over a day). Construction of timetables can consider different daily patterns (e.g. between days of the week), scalar courses (such as weather), time of year, and the duration of the time intervals used. Since many ABM's operate with stochastic events, which are simulated using central estimates and a random component, and since counts are characterized by considerable random variation the simulation of the random component appears to be of core importance to trustworthy simulation of timetables. In spite of this, other approaches to residuals than simple standard deviation – and the underlying assumption of a normal distribution – are rarely considered. Based on a case study in a Danish nature area methods for analysis and application of data from automatic counters are demonstrated and discussed.

Keywords: Visitor counters, time tables, agent-based models, recreation, Denmark

INTRODUCTION AND BACKGROUND

Knowledge about visitor use levels is a base requirement for management and planning in nature areas where recreation is in focus. Information about the type of visitors activities, duration of stay, and itineraries is often needed – but first and foremost the number of visitors entering the area is regarded essential (Hornback and Eagles, 1999; Kajala et al., 2007; or Cessford et al., 2002). The number of visitors is generally a good indicator of use levels because it is specific, objective (for individual areas or when area characteristics are taken into account), repeatable and manageable (Manning, 2007).

Two main classes of sources of visitor load information can be distinguished (Muhar et al., 2002):

- Indirect observations where the information about entry loads, temporal distribution and dependency of external conditions are revealed from e.g. information from commercial fishing and harbor master records (Gimblett, et al., 2005). In another example provided by Gimblett et al. (2002) the ABM launch schedule is obtained from the number of permissions granted to rafters on Colorado River in Grand Canyon.

- Direct observation where the entries are measured e.g. by means of counters which is the main focus of the present chapter or where visitors are performing self-registration (Fredman, 2006).

Registration of cars at parking lots can be seen as an intermediate technique – it involves direct registration of a parameter (cars) that indirectly indicates the number of visitors. Se for instance Koch (1980) and Jensen and Skov-Petersen (2007).

Recreational use of areas like the Danish forests and nature is mainly an individual, non-commercial, non-organized endeavor. Consequently visitor loads cannot be obtained from e.g. registration of the number of licenses granted or the number of boats hired out. In such areas direct observation – including automatic counting – is of special interest.

Besides counting the number of visits it is often of interest to further investigate:

- The distribution of different types of users – walkers, mountain bikers etc.
- The temporal profile over the day, week and/or year which can also include information about entry loads at special events, on holidays, and other special days of the year (Manning et al., 2005).
- The dependence on external factors, e.g. the weather (Ploner and Brandenburg, 2003).

A variety of technologies has been applied in automatic counters, ranging from pneumatic pressure pads, over seismic, magnetic and optical sensors, to cameras and video recorders (e.g. Muhar et al. 2002, Cessford et al., 2002 and Kajala et al., 2007). Recordings made by counting devices (excluding photo and video equipment) can be transmitted to data loggers which can aggregate counts over specified periods of time (e.g. hourly) or store the time of each individual event. 'Translation' of video-tapes and images into visitor counts has to be performed later by a human interpreter (Arnberger and Eder, 2007). The counters applied in the context of the present chapter are equipped with active optical, infra-red sensors. Individual counts registered by automatic counters do not always correspond 1:1 to an individual visitor passing. Depending on visitor and equipment type over- and underestimation of actual visitor loads will take place. Accordingly it is necessary to calibrate automatic counts using visual registrations obtained in field (Hornback and Eagles, 1999 and Kajala et al., 2007).

In agent-based models (ABM's) and simulation of recreational behavior the main purpose of counting visitors is to determine the number of visitors entering the target area at given points in space and time. That is, specific entry profiles or time tables for different entry points. These profiles provide the expected number of entries over a given period of time, typically a day. Profiles may vary between different types of periods, e.g. different days of the week or times of the year. Automatic counters can also be applied at strategic locations inside the investigated area to enable model verification.

A major concern when applying counter data to ABM's is the construction of statistically trustworthy time tables. For a given period of time, e.g. an hour, a time table may specify the expected number of entries including parameters describing the random variation around it, which later in the chapter is referred to as the stochastic approach. Alternatively a time table can list the probabilities of given numbers of entries (e.g. 0-5, 5-10, 10-20… >100) during a certain period of time, the probabilistic approach. Central points of interest include:

- The duration of the time periods used in the time table. Longer periods of time provide simpler time tables which are easier to handle and communicate. Shorter time periods may better express the variation of visitor activity during a day.
- The number of alternative time tables to apply for different period types. How many different time tables are necessary and can be supported statistically by the counter data for different days of the week? Are the activity patterns of Saturday and Sunday similar? Does the pattern observed on Fridays differ from those of other weekdays?
- The use of scalars to adjust activity levels for different period types or different entry points. For instance visitation may display the same pattern over the day in summer and winter, but at different levels. In that case a single time table can be defined but issued to parameters accommodating the differences between periods.
- In a similar way application of scalar coefficients for externalities e.g. effects of weather can be considered.

When applying visitor loads to an ABM it can be of interest to investigate the similarity of temporal profiles between stations within the area investigated. Scalars accounting for differences between stations can be estimated and applied. Likewise in cases where the consequences of hypothetical changes in visitor loads are of interest changes can be applied by adjustment of the scales.

Randomness is a premise of many ABM 'engines'. Traveling speed, duration of stay, and – not the least – number of entries at a specific point in time and space are characterized by given random distributions around central estimates. Consideration of the type of random distribution is of core interest to the design of recreational ABMs.

Data from automatic counters will often reveal that visitor loads are not evenly or symmetrically distributed for a given time of day or at a given time of year. Especially at fairly low visit frequencies, the visitor loads will not be normally distributed; sudden, high levels of visitation will appear at unpredictable times. Accordingly, simulating visitor loads simply by plain average or normally distributed visitor loads for given time periods can be problematic because of the lacking ability to cope with infrequent, high visitor load situations. In many inferential statistical models residuals play a rather passive role. As long as they are independent and normally distributed, everything is fine. In the context of stochastic simulation models things are different. Often the way that ABMs handle situations that differ from expectations is what makes them special and of particular interest. Accordingly, the behavior of the random events and, hence, the model describing the residuals requires special attention.

For the study presented in this chapter data from nine counters in three nature areas of Denmark was analyzed. Data from two stations was used to construct standard timetables. A special issue realized during the study is the non-normal distribution of the data. Often sudden spikes in visitor numbers occur, requiring attention to be drawn to problems related to the use of the average visitor load during time periods as good representations of visitation. A statistical solution to the special problems related to this phenomenon will be provided.

COUNTER OVERVIEW, LOCATIONS AND TECHNICAL DETAILS

Automatic counting of forest visitors has a relatively long tradition in Denmark. The first (car) counters were established in 1975 – and are still in operation at four forest car parking lots. This continuous car counting is based on inductive-loop detection in 15 minutes intervals (Koch 1984).

For the present study nine automatic counters were installed in three different Danish nature areas. The locations of the tree areas, Rude Skov, Mols Bjerge, and Hestehave Skov are shown in Figure 1. The counter survey is part of the project 'The impact of outdoor recration on the nature' funded by the Danish Outdoor Council (www.friluftseffekter.dk). The overall aim of the project is to investigate and document the possible and actual effects of recreation – both in social and biological terms. The three areas were selected to represent different types of recreational usage patterns and different locations in relation to urbanized areas and potential users: Rude Skov is situated in a suburban area with approximately 33,700 inhabitants, less than 20 km north of the center of Greater Copenhagen (approximately 1.2 Million inhabitants). It is very popular and heavily used for recreation. A major proportion of the visitors are local, everyday users. Hestehave Skov is also popular for recreational purposes, including both everyday and weekend activities. It is situated just next to the small town of Rønde (2,300 inhabitants), 33 km to the northeast of Denmark's second-largest city, Aarhus (227,000 inhabitants). The third area, Mols Bjerge, is – relative to the other two – more remotely located. Only approximately 270 people dwell in the nearby townships. By virtue of its hilly terrain, spectacular views and scenic beauty, it is assigned to be a core part of one of the planned Danish national parks. Most of the visitors come from tourist areas along the coast (10-15 km away)

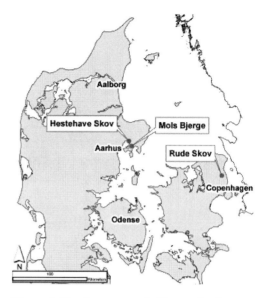

Figure 1. The three areas in Denmark where the counters were installed: Rude skov, Mols Bjerge and Hestehave skov. Background map: © Danisk National Surey and Cadarstre (GX-04)

or from other parts of the country. The distance to the city of Aarhus is 45 km.

The counters were manufactured on demand by a Danish company (Info-Scan Instruments, 2007). They were equipped with an infra-red optical sensor with separate transmitter and receiver. Transmitter and receiver were mounted in natural wooden poles. When the infra-red beam is broken, e.g. when a visitor is passing, a registration is made.

In the present configuration, registrations are aggregated to 30 minutes laps and stored by a data logger. Data can be transferred to a computer by a serial cable for further processing. Counter electronics, data logger and power supply are mounted in a watertight plastic barrel, partially buried in the ground and thus hidden to avoid vandalism.

ADDITIONAL DATA BACKGROUND

In addition to the automatic counter recordings a questionnaire survey was conducted in the three areas. The interviews included questions related to visitor behavior, preferences, and motivations. In addition, information related to spatial behavior and reasoning was collected. Counter maintenance and interview campaigns continued through 2005, 2006 and 2007. During the same period the actual number and types of entries were recorded to enable calibration of the data collected by the counters. In accordance with the settings of the data loggers of the counters, the number of entries was aggregated to 30-minute periods. For each 30 minute period in- and out-going traffic was divided into the categories: walkers, bikers, horse riders, and others. In the present context we will utilize the calibration data on number and types of entries and data on group size which is needed for grouping the visitors in simulations. Counter and calibration data for 2005 was used.

Daily meteorological data, mean temperature, sunshine hours, precipitation and mean wind, was purchased from the Danish Meteorological Office. The data were recorded at two stations, one located 10 km from both Mols Bjerge and Hestehave Skov (Tirstrup airport, sun hours were recorded at Oedum 25 km to the west), and one 5 km from Rude Skov (Sjaelsmark). Only weather data for Hestehaven was applied in the present study.

ANALYTICAL METHOD

Calibration

Raw data from the automatic counters have to be calibrated with data from field surveys to reflect the actual number of visitors. In this section, the applied methods are described. Calibration data recorded during the 2005 field campaign from the three stations where manual counts took place will be used.

Assuming that the measurement error characterizing manual counts is small compared to that of automatic counts, the overall relationship between manual and automatic counts was examined taking the following model as the point of departure:

$$N_{it,automatic} = \beta_0 + \beta_1 N_{it,manual} + \varepsilon_{it} \quad (1)$$

where $N_{it,automatic}$ is the total number of counts at station i within time interval t. Similarly, $N_{it,manual}$ is the total number of visitor passages counted manually. β_0 and β_1 are parameters to be estimated and ε_{it} are random error terms. To make $N_{it,automatic} = N_{it,manual}$, the parameters should be: $\beta_0 = 0$ and $\beta_1 = 1$ but in practice these ideals are unlikely to be met. It will therefore be necessary to use the resulting model for correcting predicted counts.

The capability of the automatic counters to reliably count various categories of visitors was examined by estimating the parameters of the model in eq. 2.

Time tables

In this section, the analytical methods applied for compilation of time tables will be described. First the duration of time periods used in the tables will be considered. Next the potential of using different timetables for different days of the week and scalars for time of year and weather conditions will be discussed. Only data from Hestehaven will be applied (cf. Figure 2). There are three entrances to the forest; the two main entrances are covered by

$$N_{it,automatic} = \alpha_0 + \alpha_{walk} N_{it,walk} + \alpha_{bicycle} N_{it,bicycle} + \alpha_{dog} N_{it,dog} + \alpha_{rider} N_{it,rider} + \alpha_{other} N_{it,other} + \varepsilon_{it} \quad (2)$$

Where $N_{it,walk}$ for instance is the number of passing walkers counted within time interval t at station i and α_0.... α_{other} are parameters to be estimated.

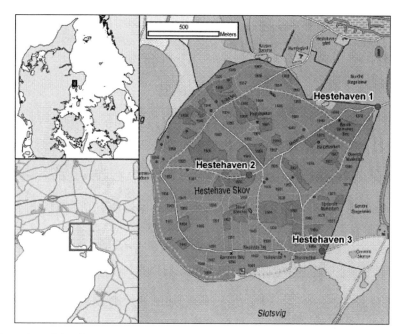

Figure 2. Location of the three counters in Hestehave Skov. Background maps: © Danish national Surey and Cadarstre (GX-04) and Danish Forest and Nature agency.

two of the counting stations applied (Stations 1 and 3). To enable verification of the simulation model Station 2 is installed in the centre of the area. Accordingly, visitor loads at Station 2 are expected to be much lower than at Stations 1 and 3.

The automatic counting devices provided counts per half an hour. Periods where the instruments were known to be out of order were omitted. Further, it emerged that artificially high counts occurred in some periods. After discussion with local managers, it was decided that no counts over 40 visitors per half an hour could be expected. Accordingly, periods with counts exceeding 40 were also excluded from the data set.

The variation of the counts per 30 minutes during each day of the year is illustrated in Figure 3 for Stations 1 and 3 in Hestehaven. Counting was started on the 18th of June 2005 and the last day included in the dataset is 2 May 2006. In Figure 3 periods for which no data is available are indicated by gray bands. As will appear from the graphs the greatest passage frequencies are observed for Hestehaven 3.

The half-hourly counts were in most cases low. In fact, the two stations applied here, Hestehaven 1 and 3, yielded counts ≤ 5 in 92 and 89 per cent of the cases, respectively. Moreover, the percentages of zero counts were as high as 60-61. Consequently, based on half-hourly counts it would be necessary to model the probabilities of given counts rather than modeling the expected counts and the distribution of the residual variation. As will appear from the graphs in Figure 3, counts are generally low from midnight till around 6 am. During the morning hours they gradually increase, reaching their daily maximum between noon and 3 pm. After 3 pm the counts decrease again and after 9 pm they end up close to zero, except during the peak of the summer. A similar variation in the visitation pattern is found in the pioneer work of Koch (1984) for four other forest areas, based on car counting from 1976-1981.

Based on the observed patterns it appeared reasonable to consider aggregating counts within periods of 1-4 hours, thereby also smoothing the dataset somewhat, leading to a broader distribution of outcomes and a lower percentage of time intervals with zero counts, and enabling preparation of models predicting expected counts.

Figure 4 shows that the correlation between successive half-hourly counts was a bit higher at Station 3 than at Station 1 but in both cases the autocorrelation decayed almost linearly with increasing lag number, at least from lag 3 onwards. Hence, the serial structure of the data did not provide a clear basis for a decision regarding the duration of periods within which to aggregate. It was therefore decided to seek a compromise that did not smooth the data too much and yet allowed the use of modeling approaches focusing on expected counts rather than probabilities of certain outcomes. Based on numerous trials it emerged that operating with two-hour intervals constituted such a compromise. Moreover, the strong variation of the mean counts suggested that factors such as station, day of week and month of year interacted in a multiplicative fashion. Based on the considerable length of the right-hand tails of the distributions of counts and the assumption that the magnitude of errors was likely to be at least proportional to the mean, operating with logarithmic counts seemed the most promising approach. In practice, due to the common-

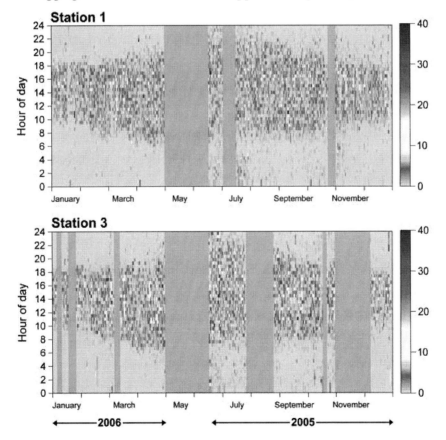

Figure 3. Plots showing observed number of counts per 30 minutes at Stations 1 and 3 in Hestehaven. Gray bands indicate periods for which there are no observations. Plot design based on Koch (1984).

ness of zero counts, the transformation $\ln(N_{i,t} + 1)$ was applied.

One of the ideals of the modeling effort was to prepare models that allowed calculation of time table entries in a simple fashion. Due to the different patterns characterizing the variation of counts during a day at each of the two stations it was decided to test effects of each combination of station, day (working days, Saturdays and Sundays), weather type and time interval (two hour intervals). In addition, modifier coefficients were considered for time of the year and weather. With regard to time of the year various alternatives were tested, including periods of 1, 2 and 3 months. With regard to the weather three classes were defined, 'sunny', 'rainy', and 'neither-nor', sunny days being defined as days with more than 300 minutes (5 hours) of sun, rainy days being days which were not 'sunny' and where the precipitation was at least 1 millimeter, and all other days ending up in the 'neither-nor' category. Using these definitions the percentages of days categorized as sunny, rainy and neither-nor were 43, 24 and 33, respectively.

In line with one of the approaches applied by Ploner and Brandenburg (2003), the first model tested was an analysis of variance model, including effects of station, time of the year, hour of the day and weather (eq. 3):

It appeared that sunny weather was rare during the winter. Since in the winter period (December-February) there were no sunny Saturdays and it was considered important to have separate time table entries for working days, Saturdays and Sundays it was necessary to omit weather effects from the final models. These included both additive and multiplicative models:

$$\ln(N_{i,t} + 1) = M_t \times S_{i,t} + \varepsilon_{i,t} \qquad (4)$$

$$\ln(N_{i,t} + 1) = M_t + S_{i,t} + \varepsilon_{i,t} \qquad (5)$$

where $N_{i,t}$ is the count at Station i, time t, $S_{i,t}$ is a parameter estimated for a particular station, day of the week (three categories) and hour of the day (two-hour periods), M_t is a coefficient estimated for the relevant three-month period and the $\varepsilon_{i,t}$s are random disturbances. Model (5) assumes additive interaction in logarithmic space and, hence, multiplicative interaction with regard to visitor counts, but errors are assumed log-normal. Model (4) is more extreme and is based on the assumption of multiplicative effects in logarithmic space. Initially, the Models (4)-(5) were estimated by ordinary least squares non-linear regression using the procedure NLIN in the statistical software package SAS™ (v. 8.02). However, to correct for the fact that the residual terms, $\varepsilon_{i,t'}$ were not likely to be truly independent an autoregressive process was assumed:

$$\ln(N_{i,t} + 1) = \mu + S_i + M_t + D_t + H_t + W_t + \\ + S_i \times M_t + S_i \times D_t + S_i \times H_t + S_i \times W_t + M_t \times D_t + \\ + M_t \times H_t + M_t \times W_t + D_t \times H_t + D_t \times W_t + H_t \times W_t + \varepsilon_{i,t} \qquad (3)$$

Where:
- $N_{i,t}$ is the count at Station i, time t, μ is an intercept,
- S_i is the station effect,
- M_t is the seasonal effect (three-month categories),
- D_t is a parameter specific to the relevant day (category) of the week,
- H_t is the effect of the hour of the day (two-hour interval),
- W_t is the effect of the weather (sunny, rainy, neither-nor), and
- $\varepsilon_{i,t}$ are random errors which are, mistakenly it turned out, assumed to be normally and independently distributed.

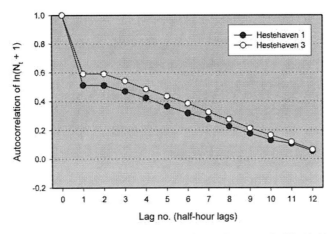

Figure 4 : Autocorrelation between half-hourly, transformed counts, ln(N_t+1). Note that lag 12 corresponds to 6 hours.

$$\varepsilon_{i,t} = \phi_1 \varepsilon_{i,t-1} + \phi_2 \varepsilon_{i,t-2} + \ldots + \phi_L \varepsilon_{i,t-L} + \upsilon_{i,t} \quad (6)$$

where $\phi_1 \ldots \phi_L$ are coefficients of the autoregressive terms for lag $1 \ldots L$ and $\upsilon_{i,t} \sim N(0,\sigma^2)$ is white noise. Models including autoregressive terms were estimated using the procedure MODEL of the SAS™ package (SAS Institute, 1993).

RESULTS

Calibration

The overall relationship between manual and automatic counts was tested using Model 1. The parameter estimates are reported in Table 1. It emerges that the intercept, β_0, is positive and significantly different from zero, indicating that the automatic counters tend to record passages even in time intervals when no passages occur. It should be noted however that this result is based on the assumption of independent observations, an assumption that is presumably not fully in agreement with reality. With the present dataset, though, a Durbin-Watson test did not lead to rejection of the independence assumption. Assuming that manual counts are approximately correct, the true number of passages can be estimated as

$$\hat{N}_{true} \approx \hat{N}_{manual} = (N_{automatic} - 1.33)/1.02.$$

However, for predicted values less than 1.33 this is obviously not a viable approach and, therefore, we also estimated the parameters of a reduced model, not including an intercept (Table 1). Using the reduced model, the true number of passages is estimated as

$$\hat{N}_{true} \approx N_{automatic}/1.13 = 0.88 N_{automatic}.$$

Intuitively, there should be no intercept but sometimes counters may respond to waving branches, butterflies and falling leaves and presumably this is the reason for the significant value of the intercept.

The parameter estimates of Model 2 are reported in Table 2 and it is noted that generally the estimates are not statistically distinguishable from 1, thus corroborating that the counting devices are equally capable of detecting passages of all categories of visitors. Nevertheless, the coefficient estimated for the category 'Other' appears low. This category included visitor types that could not immediately be categorized by the field staff, e.g. cars, baby trolleys, and Nordic walkers, and was rarely applied.

To make sure that the counters

Table 1: Models – derived from Model 1 – describing the relationship between automatic and manual counts. Parameters were estimated using the procedure REG in the SAS™ (v. 8.02) software package. Standard errors are given in square brackets.

Model	RMSE	Parameter $_0$	Parameter $_1$
$N_{it,automatic} = \beta_0 + \beta_1 N_{it,manual}$	4.00	1.3255 [0.270]	1.0238 [0.032]
$N_{it,automatic} = \beta_1 N_{it,manual}$	4.10	-	1.1348 [0.023]

responded identically to inward and outward passages a model similar to Model 2 was set up including separate variables expressing inward and outward passages. Next the equality of coefficients for inward and outward passages of a given category of visitor was tested (F tests). It appeared that there was no significant difference between the effects of in- and outgoing visitors on the counts.

In general it appears that, except for the fact that the counters tend to record a few 'ghost' counts now and then, the number of visitors registered by the counters corresponds well to those observed by the field staff. Thus, if simulated counts are distributed to visitor types the use of a calibration factor of 1 for all visitor types is not in conflict with the nature of the present dataset. However, at least in part the reason for this is that the present dataset is too small, particularly for dog walkers, riders and the miscellaneous class 'other'. If no distinction is made between visitor types it should be considered to correct the counts using one of the models in Table 1.

Time tables

As a first step Model 3 was tested. Since the counts were not independent the tests of the five main effects and ten interaction effects are not reliable. Nevertheless, they provide hints as to the relative importance of each of the effects specified in the model. Among the main effects, station, day type, time of the year, time of the day and weather, the most important effect turned out to be the time of day, followed by the time of the year. The least important main effect was the weather category. With regard to time of the year it emerged that operating with a winter season defined as

Table 2: Left: frequencies and percentages of passage at the three stations. Right: parameter estimates and tests of unity in Model 2. The analysis was done in the REG procedure of the SAS™ v. 8.02 software package.

	Observed frequencies of passage			Model 2: Counter reliability	
	Hestehaven 3[†] (n = 1090)	Mols Bjerge 1[†] (n = 1094)	Rude 1[†] (n = 798)	Parameter $_j$[‡]	Pr>F for $H_0: _j = 1$
Intercept				1.39 [0.28]	Irrelevant
Walker	774 (71.0%)	1063 (97.2%)	375 (47.0%)	1.04 [0.03]	0.2378
Bicyclist	145 (13.3%)	5 (0.5%)	286 (35.8%)	1.02 [0.11]	0.8164
Dogwalker	141 (12.9%)	24 (2.2%)	23 (2.9%)	0.84 [0.16]	0.3251
Rider	0 (0%)	0 (0%)	101 (12.7%)	1.05 [0.37]	0.8901
Other	30 (2.8%)	2 (0.2%)	13 (1.6%)	0.26 [0.47]	0.1166

[†] Percentage of all passages in rounded brackets
[‡] Standard error in square brackets

December-February, spring as March-May, summer as June-August and, autumn as September-November yielded slightly better results than three-month periods matching the calendar year. The coefficient of determination of the model was 0.68.

Among the ten interaction effects, those that accounted for the greatest part of the sum of squares were the interactions between (a) time of day and time of year and (b) type of day and time of day. The former (a) is clearly related to the fact that because of the great difference between the number of daylight hours during winter and summer, the daily schedule of recreational activities varies considerably over the year. Similarly the latter (b) is related to the difference between daily schedules on working days, Saturdays and Sundays. Not surprisingly, the least important of the interaction effects turned out to be the ones between (a) type of day and weather, (b) station and weather, and (c) type of day and station. Among the remaining effects of weather, the interactions between (a) weather and time of year and (b) weather and time of day were considerably more important, presumably because (a) the available number of hours of sun varies over the year and because (b) some activities are less weather dependent (e.g. walking the dog) than others.

Including the intercept the total number of parameters in Model 3 is 131 so although aggregated counts from 6293 two-hour intervals were available it was clearly necessary to simplify the model. First it was decided to leave out effects of the weather. In part this was due to its limited contribution to the sum of squares in Model 3, in part it was a consequence of the fact that the dataset included only 319 days, implying that certain combinations of weather, season and day type were rare. For example, it emerged that the dataset lacked Saturdays with sunny weather during the winter. Next, based on the observed relative importance of the effects tested in Model 3 and considerations regarding the practical use of the time table it was decided to operate with seasonal effects (M_t in Models 4-5) and combined effects of station, type of day and time of the day ($S_{i,t}$ in Models 4-5). Thus, for each station and each type of day, separate parameters are estimated for each two-hour interval so the total number of structural parameters to be estimated in Models 4-5 is 75 (2 stations × 3 day types × 12 two-hour intervals + 4 seasons − 1), roughly half the number of parameters in Model 3.

First Models 4-5 were estimated using ordinary least squares (OLS). It emerged that Model 5 yielded a slightly better fit and more well-behaved residuals than Model 4 so subsequent analyses focused on Model 5. The left-hand part of Figure 5 includes a scatter plot of the variation of the studentized OLS residuals (top), a normal probability plot (top) and a histogram showing the autocorrelation (bottom) of the residuals. In the scatter plot groups of studentized residuals numerically greater than 2 indicate that the tails of the distribution are longer than expected for a normal distribution. This is also obvious from the probability plot. In addition, dense horizontal bands of residuals occurring within the range from -1 to 0 indicate that the distribution is characterized by a marked spike close to the median. Finally, the correlogram shows that within 8 lags (16 hours) the autocorrelation is positive and decreases geometrically, at least for the first 3-4 lags.

To account for the autocorrelation of the residuals, the autoregressive process in Model 6 was assumed and in agreement with the autocorrelation pattern observed for the OLS residuals in the left-hand part of Figure 5 it emerged that only the first-order term was significant. However, the tails of the distribution of the residuals were still too long to be in agreement with the assumption of a normal distribution and, therefore, it was decided to use Iterative Reweighted Least Squares (IRLS) estimation to reduce the weight of observations resulting in large residuals and thereby limit their impact on the parameter estimates (Beaton and Tukey, 1974). A scatter plot, a probability plot and a correlogram of the IRLS residuals are shown in the right-hand part of Figure 5. Compared to the OLS residuals the IRLS residuals are better-behaved, with virtually no autocorrelation, no long tails and less pronounced 'horizontal banding'.

For IRLS the normal probability plot shows that the distribution of the residuals resembles that of a normal distribution quite well. Nevertheless, the assumption of normality was rejected at the 1 per cent level by all diagnostic tests, the main reason presumably being that a bit too much of the probability mass is concentrated close to zero. There are two reasons for this spike: (a) the total number of different values predicted by Model 5 is only 288 (2 stations × 12 time intervals per day × 3 day types × 4 seasons); (b) although counts were aggregated to two-hour intervals the percentage of intervals with zero counts was still rather high (38-39%). As further aggregation would lead to time tables which would be of little practical interest, it seems that had the frequencies of passage been a bit lower than observed at the two stations examined here it would have been necessary to operate with models predicting the probability of a given number of passages during a time interval instead of regression models.

The final parameter estimates of Model 5 are reported in Table 3. For practical purposes the parameter estimates need to

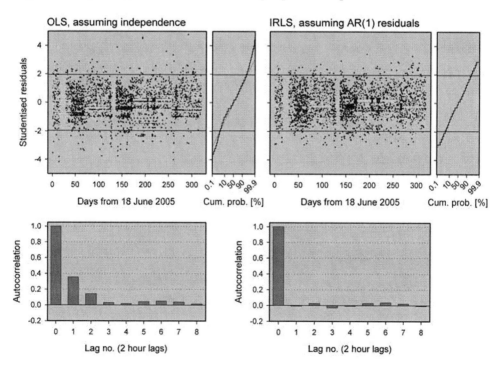

Figure 5. Distributions and autocorrelation of residuals in Model 5. Left: using ordinary least squares estimation. Right: using Iteratively Reweighted Least Squares estimation and modeling the residuals using an order 1 autoregressive process, AR(1). In the cumulative probability plots the residuals are compared to normal distributions. Blue dots (dark): Station 1, green dots (light): Station 3.

be transformed using the exponential function but to allow us to specify simple, symmetric measures of uncertainty (standard errors) Table 3 reports the original parameter estimates. It should be noticed, however, that since the estimation method is not ordinary least squares the standard error estimates are likely to be inaccurate. Since it may be difficult to imagine the consequences of these estimates, in part because there are many, in part because they are on a logarithmic scale, 95% confidence intervals for the predicted counts are presented in Figure 6 for the winter season (December-February). First it may be noticed that Monday-Friday the level of the predicted counts are similar for the two stations but Saturday and, particularly, Sunday the levels are considerably higher for Station 3 than for Station 1. Moreover, Monday-Friday it emerges that the activity at Station 3 peaks between 2 and 4 pm while at Station 1 it peaks between 4 and 6 pm but generally the number of counts remains within the range 5-8 from 10 am to 6 pm. This could be an indication of more every-day, after working hour use of entrance 1 which is expected because of its location closer to town. On Saturdays and Sundays the activity at both stations is characterized by a peak from 2 to 4 pm. Sunday the decrease from the maximum level happens gradually but on Saturdays the activity remains almost the same from 2 to 6 pm and then drops to a much lower level. These variations clearly show the necessity of distinguishing stations, time of day and day type in the time table. The seasonal effects also turn out to be important.

APPLICATION

Based on the results in the previous section time tables can be prepared in two ways; deterministically or stochastically, both providing the number of entries within a given period of time based on a random draw. A core difference between the two approaches is the type of data that constitutes the time table at model runtime: For each time period a *deterministic time table* stores the exact number of expected entries or the *probabilities* of given numbers of entries; a probabilistic/deterministic approach. A *stochastic time table* stores, still for each time period, a central

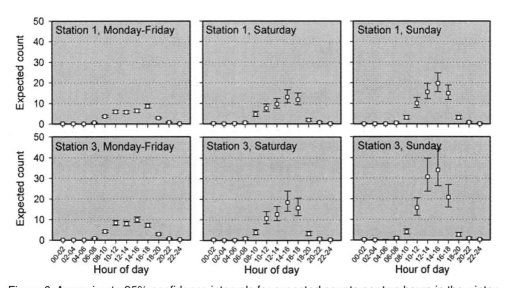

Figure 6. Approximate 95% confidence intervals for expected counts per two hours in the winter season, December – February, using Model 5.

estimate of the expected number of entries and parameters describing the distribution of the random variation around it.

In simple stochastic time tables where residuals around a central estimate are assumed to be normally distributed and not influenced by autocorrelation the expected number of counts and a standard deviation are supplied, either for each time period or, more likely, for the entire table. The actual number of counts at a given point in time is decided by drawing random terms from a normal distribution, $e \sim N(0,\sigma^2)$ where σ is the standard deviation. In the present case where both (approximately) log-normal distribution of residuals and autocorrelation have been detected additional measures have to be implemented. Using Model 5 and the AR(1) version of Model 6 stochastic simulation of counts is performed by drawing random terms from a normal distribution, $\upsilon_{i,t} \sim N(0,\hat{\sigma}^2)$ where σ is the estimated standard deviation of the residuals (RMSE), and using the model expression in eq. 7 & 8.

By repeated application of the model a stochastic realization of the counting process is generated, yielding a sequence of two-hour counts, distributed in accordance with the observations. It should be noted, however, that since the actual distribution of the data was characterized by even longer tails than implied by the RMSE applied here the realizations produced by (7) and (8) will produce less extreme outcomes than might actually be observed in reality. However, as will appear below even this imperfect model produces a lot of variation.

The first logical step therefore is to transform the parameters listed in Table 3 using the exponential function, leading to a table of coefficients for the expression in (7). Next, random terms are drawn from a normal distribution, $\upsilon_{i,t} \sim N(0,\sigma^2)$, and used to calculate the random coefficients, $\exp(\varepsilon_{i,t})$. Initially, no value of $\varepsilon_{i,t-1}$ is available and it is therefore necessary to initialize the sequence with a random value. A burn-in sequence of, e.g., 50-100 realizations of (8) will practically eliminate traces of the initial value.

The resulting predicted count can be corrected using one of the models in Table 1 and the passages distributed to categories of user types using the percentages in Table 2.

To illustrate the stochastic simulation approach the generation of a twelve-hour simulated sequence is exemplified in Table 4. Here it is assumed that the simulation has already been running for some time so that an initial value of is available from the 22:00-00:00 period of the previous day.

Applying stochastic time tables in a simulation requires:

- a basic time table of the expected number of counts which might be different for different types of days ($S_{i,t}$ in (7)),
- a set of scalars for e.g. time of year (which is the only one applied in the present case, M_t in (7)), weather, 'handles' for construction of scenarios, etc.,
- an initialized random coefficient ($\varepsilon_{i,t-1}$ in (8)),
- an auto-regression parameter (ϕ_1 in (8)) in the range of -1 to 1, and
- a normally distributed random number ($\upsilon_{i,t}$ in (8)).

$$\ln(N_{i,t}+1) = S_{i,t} + M_t + \varepsilon_{i,t}$$
$$\Updownarrow \qquad (7)$$
$$N_{i,t} = \exp(S_{i,t} + M_t + \varepsilon_{i,t}) - 1 = \exp(S_{i,t}) \times \exp(M_t) \times \exp(\varepsilon_{i,t}) - 1,$$

where $\varepsilon_{i,t} = \phi_1 \varepsilon_{i,t-1} + \upsilon_{i,t}$ \qquad (8)

Table 3. Parameters in Model 5 describing logarithmic counts, $\ln(N_{i,t} + 1)$, within two-hour intervals. The parameters were estimated using the procedure MODEL of the SAS™ (v. 8.02) software package. Correction for autocorrelation was made using an AR(1) version of (6). N = 6293, RMSE = 0.6668, R^2 = 0.7040

	Dec-Feb	Mar-May	Jun-Aug	Sep-Nov		AR(1)
M_t	1 (-)	1.425 (0.04)	1.734 (0.04)	1.426 (0.03)		
AR(1), ϕ_1	-	-	-	-		0.351 (0.01)

	Station 1			Station 3		
$S_{i,t}$	Mon-Fri	Saturday	Sunday	Mon-Fri	Saturday	Sunday
00:00 – 01:59	-1.077 (0.05)	-1.028 (0.11)	-1.103 (0.11)	-1.134 (0.06)	-0.978 (0.12)	-0.990 (0.12)
02:00 – 03:59	-1.125 (0.05)	-1.161 (0.11)	-1.178 (0.11)	-1.144 (0.06)	-1.136 (0.13)	-1.162 (0.12)
04:00 – 05:59	-1.193 (0.05)	-1.069 (0.11)	-1.221 (0.11)	-1.167 (0.06)	-1.127 (0.13)	-1.269 (0.12)
06:00 – 07:59	-0.732 (0.05)	-0.787 (0.11)	-0.850 (0.11)	-0.624 (0.06)	-0.594 (0.13)	-0.539 (0.13)
07:00 – 09:59	0.312 (0.05)	0.505 (0.11)	0.168 (0.11)	0.421 (0.06)	0.348 (0.13)	0.377 (0.13)
10:00 – 11:59	0.704 (0.05)	0.926 (0.11)	1.180 (0.11)	1.011 (0.06)	1.231 (0.13)	1.591 (0.13)
12:00 – 13:59	0.672 (0.05)	1.143 (0.11)	1.582 (0.11)	0.968 (0.06)	1.386 (0.13)	2.233 (0.13)
14:00 – 15:59	0.768 (0.05)	1.419 (0.11)	1.807 (0.11)	1.159 (0.06)	1.741 (0.13)	2.333 (0.12)
16:00 – 17:59	1.042 (0.05)	1.324 (0.11)	1.545 (0.11)	0.877 (0.06)	1.594 (0.13)	1.857 (0.12)
18:00 – 19:59	0.118 (0.05)	-0.176 (0.11)	0.173 (0.11)	0.132 (0.06)	0.208 (0.13)	0.062 (0.13)
20:00 – 21:59	-0.659 (0.05)	-0.707 (0.11)	-0.679 (0.11)	-0.661 (0.06)	-0.677 (0.13)	-0.601 (0.12)
22:00 – 23:59	-1.098 (0.05)	-1.103 (0.11)	-1.083 (0.10)	-0.991 (0.06)	-0.966 (0.13)	-0.970 (0.12)

In the present study an auto-regression coefficient of 0.351 was estimated. If a higher degree of autocorrelation is expected in a particular case or assumed in a given scenario, greater coefficients would have to be applied. Negative coefficients would result in a situation where, e.g., the likelihood of a low number of counts following higher numbers would be greater.

A deterministic application predicting the expected count may be of interest. In this case it will be necessary to correct for logarithmic bias (eq. 9).

The estimated standard deviation of the residuals is only σ = 0.6668, perhaps yielding the impression that the difference between simulating counts using the deterministic and stochastic approaches is lim-

$$\hat{N}_{i,t} = \exp\left(S_{i,t} + M_t + \frac{\hat{\sigma}^2}{2}\right) - 1 = \exp(S_{i,t}) \times \exp(M_t) \times \exp(\hat{\sigma}^2/2) - 1 \quad (9)$$

Where

σ is the estimated standard deviation of the residuals (RMSE). In this case, all that is needed is a table of coefficients based on the estimates in Table 4.

Table 4. Counts per two hours simulated for Station 3 on a Saturday in March-May using (7) and (8) with parameters ϕ_1 = 0.351, σ = 0.6668, $S_{i,t}$ and M_t from Table 3. Equation (8) is initiated with $\varepsilon_{i,t-1}$ = 0.3182.

Time of day	$S_{i,t}$	M_t	Exp($S_{i,t}$)	exp(M_t)	$\varepsilon_{i,t}$	$\upsilon_{i,t}$	exp($\varepsilon_{i,t}$)	Count, $N_{i,t}$
22:00-00:00					0.3182			
00:00-02:00	-0.978	1.425	0.3761	4.1579	-0.5267	-0.6383	0.5906	0.9234
02:00-04:00	-1.136	1.425	0.3211	4.1579	-0.0880	0.0968	0.9157	1.2226
04:00-06:00	-1.127	1.425	0.3240	4.1579	-0.2125	-0.1816	0.8085	1.0892
06:00-08:00	-0.594	1.425	0.5521	4.1579	-0.2717	-0.1971	0.7621	1.7495
08:00-10:00	0.348	1.425	1.4162	4.1579	-0.3282	-0.2329	0.7202	4.2408
10:00-12:00	1.231	1.425	3.4247	4.1579	-0.0475	0.0677	0.9536	13.5782

ited. However, due to the multiplicative structure of the model and the serial correlation of random events this is far from true. To illustrate this 100,000 sequences were simulated for Station 3, Monday-Friday and Sunday during the months of June-August. For each two-hour interval of a day the number of outcomes was recorded for classes with a width of 1. Subsequently, the relative frequency was calculated for each class and in Figure 7 the resulting distributions of the outcomes are shown (color coded) for each two-hour interval. Since the number of realizations was 100,000 the lowest relative frequency that can be shown is 0.00001 (light color). For each two-hour period the simulated relative frequencies are shown (color coded) in Figure 7. For comparison the graphs include expected counts obtained using (9). The graphs clearly show that stochastic sequences are characterized by very large variation, even for low expected values. Moreover, due to the long tails upwards the expected value is generally located above the modes of the distributions.

Application of probabilistic/deterministic time tables in a simulation model would require a pre-runtime simulation of probabilities of all possible counts (or ranges of counts) for all time periods. In the form presented the simulation requires:

- a basic time table of the expected number of counts which might be different for different types of days ($S_{i,t}$ in (9))
- a set of scalars for e.g. time of year (which is the only one applied in the present case, M_t in (9)), weather, 'handles' for construction of scenario, etc., and
- a standard deviation (σ in (9))

Once simulated counts have been obtained, the problem of distributing them over time within each two-hour interval remains. This can be done in a number of ways, e.g.:

1. By spacing the events uniformly within each two-hour period according to the (corrected) predicted count, e.g. if $N_{i,t}$ = 3.124, the time between events will be 120 / 3.124 = 38.41 minutes (38 minutes, 25 seconds).

2. By rounding off the (corrected) predicted count and launching each event at random points in time within the two-hour interval.

3. By defining time periods shorter than two hours, formulating a Poisson process in time, using the (corrected and rescaled) predicted count as the intensity of this process and either ignoring that the actual number of events occurring within each two-hour period may

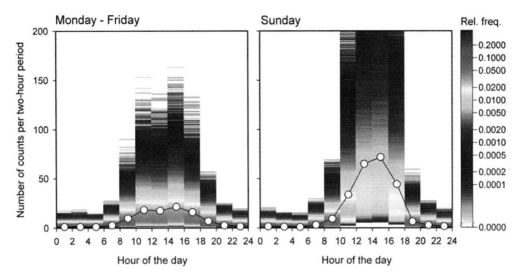

Figure 7. Graphical representation of a probabilistic/deterministic. Relative frequency of counts per two hours, simulated using Model 5 for Station 3 on the days Monday-Friday and Sunday in June-August. Relative frequencies are based on 100,000 replications. The expected count (deterministic prediction using (9)) is shown with open circles.

not be equal to the predicted number or constraining the number of events within a certain range around the predicted value.

4. By distributing the predicted count to groups of various sizes, thereby forming agents which are launched at random points in time.

The first of these options will produce patterns that are homogeneous within each two-hour period, which is clearly not in agreement with reality. The second and third options are based on the assumption that events are essentially independent. Since this is unlikely to be a good description of reality, at least for most categories of forest visitors, the fourth alternative appears most promising. To use the fourth approach data on the distribution of forest visitors to group sizes must be available, ideally for all categories of visitors. An example of such a distribution is shown in Table 7. Since only approx. 28 % of the respondents visited the forest alone, approaches 1-3 are clearly poor descriptions of reality. A possible application of the table would be to determine the group size based on a random draw. In each two-hour interval the number of such draws would depend on the predicted count and the mean group size observed. In the case shown in Table 5 the mean group size is 2.23 so if, e.g., in a given two-hour time interval the predicted and corrected count is 50, then the number of random draws would be 50/2.23 ≈ 22. In some instances the total number of persons in groups launched during a time period would go over or under the expected, but on average it would match the expectation.

Table 5. Percentage of respondents arriving in groups of 1, 2, ... Based on interviews in Hestehaven, 2006. All visitor categories.

Group size	1	2	3	4	5	>5
% of respondents	27.93	47.49	8.38	8.94	4.47	2.79

DISCUSSION AND PERSPECTIVES

Automatic visitor counters can conveniently be applied to visitor monitoring in nature areas. Simple assessments of visitor loads at different locations and at different times – in terms of aggregated measures - are routinely applied. The use of data from automatic counters as input parameters to ABM's seems obvious and promising. For the counting stations used in the present study it appears that use of simple, independent averages, and the assumption of normally distributed random variation around them is not a very good description of the temporal structure of the visitor loads. Two main deviations from the normally and independently distributed case emerged:

- Residuals have very 'long tails'; meaning that sudden, rare cases of unexpected high counts appear, in spite of extremely high counts being removed during the pre-processing of data. A prominent feature of agent-based models is their ability to simulate situations beyond the average. Omitting the capability to produce sudden, rare events would undermine this important feature.

- Counts appear to be positively autocorrelated – if a high number of visitors appear in a given time interval, the probability of a high number of visitors in the following period will be high. If for instance an ABM was aiming to assess the frequency of visitor encounters and an even distribution of entries without recognition of autocorrelation was applied, the periods during which agents were encountering each other and the frequencies of such encounters would be underestimated. A partial reason for the observed autocorrelation may be that the effect of weather was removed from the model. Any effect of nice weather at one point in time will also apply in the next couple of periods. Likewise the effect of special days, e.g. public holidays, which was not considered in the model, might also explain part of the autocorrelation. Negative autocorrelation might be observed in cases where a high number of cars at a parking lot would have a repelling effect. Effects of this sort has been included in ABM's (see e.g. Bishop and Gimblett, 2000) but not supported by statistical inference.

In this chapter a distinction between stochastic vs. deterministic (and probabilistic/deterministic) representations of time tables has been addressed. Stochastic representation appears to have a number of advantages:

- They take up less memory and are easier to handle at run time,

- they require less pre-processing, and

- they are easier to define and parameterize 'from scratch' e.g. in cases where no counter data is available.

Deterministic and probabilistic/deterministic time tables on the other hand require less advanced computer code at run time. A look-up in a pre-simulated table is a very simple operation. If for some reason there is a predefined need for deterministic time tables, the statistical assessment of 'raw' data from automatic counters should have taken its point of departure in the probability of a certain count occurring within a certain time interval, rather than looking for patterns in actual counts. This probably would give rise to problems with regard to the number of degrees of freedom required by such models, essentially because all possible counts (1, 2, 3 etc.) have to be dealt with in individual equations (although simplifications can be made). The problems would be further evident if autoregressive processes was assessed as well. To some extent this could be handled, by operating with probabilities of counts falling within broader classes (1-5, 5-10, 10-15 counts, etc.). For the reasons mentioned above, stochastic time tables are preferred in the present context.

As it stands, it appears that in the present context the number of visitors observed at a location on average, or normally distributed around the average, does not provide an adequate description of reality. Assuming that this is generally true it appears that application of averages on the outcomes side of an ABM is also problematic. Accordingly focus has to be put on alternatives to using statistics such as, e.g., the mean period of time that agents feel disturbed or that the nature area is congested. For instance percentile approaches could be applied.

It is expected that a more comprehensive data set, covering a longer period of time and with fewer periods of missing registration would provide a better background for statistical treatment. The automatic counters addressed here are still in operation. After a period of three years (June 2008) data from all nine counters will be reassessed. It is expected that

- a better description of the residual variation will be possible,
- weather effects will be more clearly expressed,
- effects of different day-types and different times of the year (short days in winter, longer in summer etc.) can be taken into account, and
- finally, the regional differences between the three sites, in terms of visitor load and distribution over the day, week and year will be a major concern. The data from the four permanent car counters (Koch, 1984) will also be considered in this context. This additional dataset will in principle include visitation information from 1975 to the present in 15 minutes intervals.

LITERATURE CITED

Arnberger, A. and Eder, R. 2007. Monitoring recreational activities in urban forests using long-term video observation. Forestry, 80(1), 1-15.

Beaton, A.E. and Tukey, J.W. 1974. The fitting of power series, meaning polynomials, illustrated on band-spectroscopic data. Technometrics 16, 147-185.

Bishop, I.D. and Gimblett, H.R.. 2000. Management of recreational areas: GIS, autonomous agents, and virtual reality. Environment and Planning B: Planning and Design 27(3) 423-435.

Cessford, G., Cockburn, S. and Douglas, M. 2002. Developing new visitor counters and their application for management. In Arnberger, A., Brandenburg, C, and Muhar, A. 2006. Monitoring and management of visitor flows in recreational and protected areas. Conference proceedings. Vienna, Austria. January 2002.

Fredman, P. 2006. Modeling visitor expenditures at Fulefjället National Park, Sweden. In Siegrist, D., Clivaz, C., Hinziker, M. and Iten, S. 2006. Exploring the nature of management. Proceedings of the third international conference on monitoring and management of visitor flows in recreational and protected areas. University of Applied Sciences Rapperswil, Switzerland. 13-17 September 2006. 513 pages.

Gimblett, R., Itami, R. and Cable, S. 2005. Recreation Visitation in Misty Fjords National Monument in the Tongass National Forest. In Cole, D. (compiler). Computer simulation modelling of recreation use: Current status, case studies and futures directions. USDA/FS, Rocky Mountain Research Station. General technical report, RMRS-GTR-143. Pp. 75 pages.

Gimblett, R., Roberts, C., Daniel, T., Ratcliff, M., Meitner, M., Cherry, S., Stallman, D., Bogle, R., Allred, R., Kilbourne, D., and Bieri, J. 2002. An intelligent agent-based model for simulating and evaluating river trips scenarios along the Colorado River in Grand Canyon National Park. In Gimblett, R (ed). Integrating Geographic Information Systems and Agent-based Modeling Techniques for simulating social and ecological processes. Oxford university press. ISBN0-19-514336-1. 327 pages.

Hornback, K.E. and Eagles, P.F.J. 1999. Guidelines for public use measurement and reporting at parks and protected areas. IUCN, Gland, Switzerland and Cambridge, UK. 90 pages. ISBN: 2-8317-0476-6. Available at http://www.ahs.uwaterloo.ca/~eagles/parks.pdf (last accessed June 2007)

Jensen, F. Søndergaard and Skov-Petersen, H. 2006. Visitor monitoring in Denmark. In: Kajala, L. (eds.), 2006: Monitoring outdoor recreation in the Nordic and Baltic countries. TemaNord 2006:530, Nordic Council of Ministers, Copenhagen 2006. p. 27-33.

Kajala, L., Almik, A., Dahl, R., Dikšaitė, L., Erkkonen, J., Fredman, P., Jensen, F. Søndergaard, Karoles, K., Sievänen, T., Skov-Petersen, H., Vistad, O. I. and Wallsten, P. 2007. Visitor monitoring in nature areas – a manual based on experiences from the Nordic and Baltic countries. Nordic Council of Ministers, Naturvårdsverket and Metsähallitus. ISBN 91-620-1258-4. 205 pages.

Koch, N.E. 1980. Skovenes friluftsfunktion i Danmark. II. del. Anvendelsen af skovene, regionalt betragtet. (Forest Recreation in Denmark. Part II: The Use of the Forests Considered Regionally). Forstl. Forsøgsv. Danm., København, 37(1980): 73-383.

Koch, N.E. 1984. Skovenes friluftsfunktion i Danmark. III. del. Anvendelsen af skovene, lokalt betragtet. (Forest Recreation in Denmark. Part III: The Use of the Forests Considered Locally). Forstl. Forsøgsv. Danm., København, 39(1984): 121-362.

Manning, R.E. 2007. Parks and carrying capacity. Commons without tragedy. Island press. ISBN 978-1-55963-105-1. 314 pages.

Manning, R., Itami, R., Cole, D. and Gimblett, R. 2005. Overview of computer simulation modelling approaches and methods. In Cole, D. (compiler). Computer simulation modelling of recreation use: Current status, case studies and futures directions. USDA/FS, Rocky Mountain Research Station. General technical report, RMRS-GTR-143. 75 pages.

Muhar, A., Arnberger, A. and Brandenburg, C. 2002. Methods for visitor monitoring in recreational and protected areas: An overview. In Arnberger, A., Brandenburg, C. and Muhar, A. 2006 Monitoring and management of visitor flows in recreational and protected areas. Conference proceedings. Vienna, Austria. January 2002. 481 pages.

Ploner, A., Brandenburg, C. 2003: Modeling Visitor Attendance Levels Subject to Day of the Week and Weather: A Comparison between Linear Regression Models and Regression Trees. Journal for Nature Conservation, 11, 4, 297-308.

Poe, A., Gimblett, R.H., Goldstein, M. and Guertin, P. 2007. Evaluating Spatiotemporal Interactions between Winter Recreation and Wildlife Using Agent-Based Simulation Modeling on the Kenai Peninsula, Alaska. In the present volume.

SAS Institute, 1993. SAS/ETS® User's Guide, Version 6, Second Edition, Cary NC: SAS Institute Inc. 1022 pp.

Web-pages and company references:

Info Scan Instruments. 2007. Poul Erik Madsen, Selmersvej 40. 8260 Viby J. Denmark.

www.friluftseffekter.dk. Home page for the project 'The impact of outdoor recreations on the nature' (only in Danish). Last accessed May 2007.

CHAPTER 8

PROBABILITY OF VISUAL ENCOUNTERS

Hans Skov-Petersen
Bernhard Snizek

Abstract: In various contemporary Geographical Information Systems (GIS) and Computer Aided Design (CAD) systems visibility has been assessed as a Boolean phenomenon: things either can or cannot be seen. Probabilistic Visibility (PV) is presented as an advanced alternative concept. PV describes the probability that one location of object can be seen from another. Results of PV analysis, the probability of visual contact, are stored in Probabilistic Visibility Graphs (PVG) for all relevant pairs of locations (e.g. raster cells).. Based on field experiments and literature surveys, visibility decay - as a function of terrain, distance, viewing-angle and vegetation type - a visibility graph will be calculated. This chapter starts with an analysis of visibility in forested environments and describes the field experiment and how the graph is calculated, visualized and implemented in agent-based models. Furthermore, perspectives of the further development and application of probabilistic visibility graphs will be given.

Keywords: Visibility, viewshed, probabilistic visibility, Visibility graph

INTRODUCTION

In Geographical Information Systems (GIS) and Computer Aided Design (CAD) systems visibility is normally treated as Boolean; either one location can be seen or it cannot be seen from another. A location of origin (a vantage point) is most often represented as a point. In some cases the visibility from lines (e.g. roads) or areas (e.g. parks) can be assessed. An extension to this Boolean approach is necessary when the visibility of more than one object or location is in question; any target objects, which can be seen from a given location is recorded. The entire area that can be seen from a given object is termed view shed (Kim et al. 2004) or isovist (Benedict, 1979 and Snizek, 2003). The term 'isovist' itself indicates an area that is evenly visible. Often the outer boundary of view sheds can be set as a maximum range of the analysis. But again, it is Boolean. Being outside the maximum range, is interpreted as completely invisible. Being just slightly inside, is completely visible.

But is visibility really only Boolean? Would one expect that it only takes a single step on the ground – across the top of a hill or beyond the maximum range of analysis - to make you invisible? As so often in Geography the first idea that comes into mind is that scale matters: Analyzing visibility in large landscapes the Boolean approach can make good sense, but working in finer scales, more detailed objects like trees and other on-ground objects that might interrupt visual contact must be taken into account. Then more facets have to be dealt with. The ruggedness of the terrain plays a major role. In mountainous regions the terrain itself will overrule possible effects of transparent on-ground objects. In rather flat environments - like Denmark – the terrain has less influence on visibility than in more rugged ones. Accordingly, assessment of probabilistic visibility will be more significant in such landscapes and environments than elsewhere.

A number of parameters can be taken into consideration when considering visibility decay. Including:

- *Increasing distance*. Even along an uninterrupted line of sight, the probability of paying attention to near things will be higher than to those at greater distances. This corresponds well to Tobler's 'First law of Geography': "Everything is related to everything else, but near things are more related than distant things." (Tobler 1970).

- *Transparency*. On-ground objects – or compounds of objects – provide different degrees of transparency. For instance different types of vegetation which will interrupt visual contact more or less. The same is true for weather- and light conditions, such as mist and twilight. The transparency decay of vegetation is the main scope of the present paper.

- *Viewing angle*. The orientation of the viewer can influence the probability of visual contact. You are more likely to pay attention to objects that happen right in front of you than thing that are behind your back. The effect of viewing angle is of course affected by a range of additional stimuli, including: other perceived information (e.g. noises and smells), is the viewer looking for something (scanning) or is he/she moving straight on towards the target or is the viewer standing still or moving. The angle decay makes most sense when applied to human or animal perception. It the visibility analysis is used for assessment of telemobile transmitter coverage, the probability of contact will be even all 360 degrees.

The formulation of visibility as graphs of mutual visibility of viewer and target locations or cells is introduced by O'Sullivan and Turner (2001). The work presented extends these Boolean visibility graphs to include probabilities – accordingly termed Probabilistic Visibility Graphs (PVGs).

Agent-base models (ABMs) and computer games are potential applications of PVGs. When evaluating whether one agent can see another, the probability can be revealed from a PVG. For each agent the probability is juxtaposed with a random number (between 0 and 1), to determine if visual contact is actually made. This suggests that two agents standing next to each other might not visually perceive the same thing. The relevance of applying probabilistic visibility to ABMs is amplified if more detailed phenomena's, in spatial or temporal terms, are modeled. The finer the details of the landscape anticipated, the shorter the model time step, the more heterogeneous the landscape elements (e.g. vegetation), the denser the path- and road network, the higher the number of visitors; the greater the need for finer detail in the perceptive abilities of the model agents. In order to capture the finer details in the perceptive abilities of model agents the classical Boolean realization of visibility needs to be broadened to include probabilistic visibility.

The remainder of the paper will first provide some background on how probabilistic visibility has been assessed by other authors. Following is a description of the field experiment applied. Then the applied analytical method will be described, accompanied by examples of results. Finally remarks on the future application and further development will be given.

BACKGROUND

Handling visibility in natural environments involves both aspects related to monitoring (how real world phenomena can be assessed, measured and parameterized) and modeling (how measures and indicators can be brought into action as part of a simulation).

Monitoring visibility in natural environments (especially though forest vegetation) has been approached in terms of

- Quantitative assessment of the physical/geometric configuration of the vegetation (measured in field or calculated theoretically),
- Qualitative evaluation in the field or by means of photographs or
- A combination of quantitative and qualitative approaches.

If the representation of forest vegetation is reduced to rather primitive, idealized geometric figures, the probability of visual contact between two locations can be calculated. Rasmussen (1992) acknowledges the need for probabilistic visibility when evaluating visibility of construction in forested land (e.g. buildings or roads). He proposes two purely geometric models of probabilistic visibility in forests. The models are based on the distribution, number and diameter of tree stems. One model calculates the average probability of any point on an object, whereas the other estimates the probability of having a clear view of the entire object. The models are not able to cope with screening by forest under storage or branches.

Another example of a purely physically based approach is provided by Kumsap et al (2005). To adjust the Level of Detail (LOD) in forest 3D visualization, distance-decay parameters are applied depending on different types of forest. Decay parameters obtained from Preston (2003) are related exponentially to increasing distance. This way the required LOD of a given object (e.g. a tree), given the location of the viewer, can be estimated from the mutual distance and the forest types penetrated. The parameters applied were not based on configuration of individual trees (as in Rasmussen, 1992, mentioned above), but on type of vegetation.

In a project conducted by Ruddell et al. (1989) the psychological and the physical approaches are combined to study visual penetration to explain parameters for perceived aesthetic value for various forest types. The basic idea was to compare (physical) measures of visual penetration to (psychological) evaluation by respondents. The measurement of penetration was devised by a 'screenometer' which optically measures the 'depth' of the forest up to 15.2 meters from the viewer (Rudis, 1985). Respondents were asked to rank images (color slides) according to perceived scenic beauty. Comparing measured penetration with preference ranks revealed that measured penetration was a very good indicator of scenic beauty.

Preston (2003) provides a very rare case of a planning application of probabilistic visibility: In Ipswich Council situated in Queensland (Australia), the visual exposure of the landscape was assessed by a combination of locations where people would go (roads, viewing points etc.) and the transparency of different land cover types. For each land cover type, a set of decay parameters was suggested. Decay parameters for a selection of land cover types are shown in table 1 below. The parameters were intuitively based and not tested empirically in field (Preston,

Table 1. Selected transparency distance decay factors in different land cover types (Source: Preston, 2002). The decay parameter for the class 'open' was adjusted to 0.975. In the original table the class was termed 'water' and was set to 1.0 indicating that objects would be evenly seen no matter the distance.

Land cover type	Decay factor, per 25 m	Distance from viewer (m)					
		0	25	50	75	100	500
1: Open	0.975	100%	98%	95%	93%	90%	60%
3: Pasture	0.975	100%	98%	95%	93%	90%	60%
6: Low density trees	0.900	100%	90%	81%	73%	66%	12%
7: Dense trees	0.750	100%	75%	56%	42%	32%	0%
8: Very dense trees	0.500	100%	50%	25%	13%	6%	0%

pers. com. 2005). The visual exposure is combined with maps of scenic amenity to assess the landscape's visual vulnerability. This way it is expected that the visual impact of proposed development can be minimized or avoided.

The present study is based on a combined psychological/physical approach to modeling probabilistic visibility. A field study was conducted in which the respondents' ability to see a target person at different distances in a forest environment is recorded. The measures are used to justify the parameters provided by Preston (2002).

Many disaggregated ABM's – where individual human and/or animal agents are the smallest modeled objects – has potential visual contact as a basic perceptive component. Often visibility is the only perceptive mechanism involved in models. In its most straight forward form, (visible) contacts and herby encounters are assumed to take place at a given distance. If an agent and a target – whether being another agent or an amenity that is closer than a given distance, the agent will see or encounter the target. An example is provided by Bishop and Gimblett (2000) where encounters are assumed when more than one agent is present within the same 10 meter raster-cell. In a more advanced form, viewshed analysis – based on a digital terrain model – is applied. This can be performed by means of standard GIS procedures or by in-application calculation. Itami (2002) demonstrates how inter-visibility algorithms can be applied to the recreational ABM-package RBSim. RBSim contains a visibility class which is responsible for inter-agent visibility. According to Itami (2002) it is a modification of standard GIS line of sight analysis. It takes both elevation and vegetation cover into consideration. Apparently vegetation cover is treated as opaque.

Likewise MASOOR, which is another contemporary agent-based model of recreative behavior, is not explicitly applying visibility. Encounters are entirely determined by a function based on the sheer distance between agents (pers. com Johem, 2007).

The screening characteristics of the terrain can be extended by on-terrain objects such as buildings, or stands of trees. Jiang and Gimblett (2002) demonstrate how view sheds in urban environments can be assessed in ABM-applications.

Only in rare cases the direction of movement of the agent is acknowledged as significant for obtaining visual contact. This way it is assumed that objects behind the viewer are just as visually significant as those in front – or put another way, that two agents standing back to back can see each other. Manning et al. (2005) provides an infrequent – but indirect example of this. Encounters are aggregated as sums of agents present along 100 m sections of the path networks in Yosemite National Park. This means that agents in front or behind are realized where as agents next to the path (or at a parallel path) are not perceived.

None of these examples acknowledge the need for, or applications of the gradual decline of visual impact of effect, given increasing distance or various types of screening objects. However, there are good examples found in advanced computer games. In the game 'Splinter Cell Double Agent' (Ubicon, 2007) agents are gradually seen and recognized according to distance and light conditions. This sort of probabilistic visibility in relation to ABMs in general and recreational ABMs specifically, will be pursued in the remainder of this chapter. It is believed that development of more accurate and realistic simulation models requires application where visibility decay, based on distance, obscuring objects and viewing direction is taken into consideration.

SETTING TRANSPARENCY PARAMETERS

Very few documented attempts have been made to set actual decay parameters for different land cover types. A rare example is provided by Preston (2002). In Ipswich Council situated in Queensland (Australia) the visual exposure of the landscape was assessed by a combination of locations where people would go (roads, viewing points etc.) and the transparency of different land cover types. For each land cover type, a set of decay parameters was suggested. Decay parameters for a selection of land cover types are shown in table 1 below. The parameters were not tested empirically in the field (Preston, Pers. com. 2005).

FIELDWORK

To empirically justify the parameters found in table 1 (Preston, 2002) a pilot field experiment was set up in a fairly open, old (approximately 100 years, 339 stems per ha), fairly homogeneous beech forest. The aim was to investigate to what extent a person could see another person given different distances and angles away from a central line of sight.

Three persons were involved:

- A controller, taking care of time and records.
- A respondent, seeing or not seeing the target.
- A target, roaming the forest.

Five sessions at five different locations in the beech stand were conducted. Before each session, the location of a base station was recorded. Likewise the angle of a central line of sight was decided and recorded. During a session the controller and the respondent were standing at the base station. The outlook of the respondent in the direction of the central line was blocked. The target was moving around in the forest stand. At uneven intervals the target would stand still and the respondent was given either one or five seconds to realize whether the target could or could not be seen. Every time the target would record its location by GPS. At each session ten recordings based on one second's exposure and ten based on five seconds exposure were obtained. This way a total of 100 recordings were made. Figure 1 shows the location of the controller/respondent and the target of one such session.

STATISTICAL TREATMENT

Based on the GPS recordings of the distances and angles between base stations and target locations the probability of seeing the target could be calculated and assessed as explaining variables. Logistic regression analysis was carried out in SAS, using the LOGISTIC procedure.

Table 2 shows that only the distance and the intercept are significant for the explanation of visibility. The estimate for the effect

Table 2. Parameter estimates and P values revealed from the logistic regression for the full model.

Parameter	Standard Estimate	Pr > ChiSq
Intercept	2.8644	0.0051
Angle	-0.0136	0.3611
Distance	-0.0236	0.0112
Session	-0.0287	0.9022
Exposure time	-0.0433	0.8724

of deviating angle from the central viewing line has the expected sign (the further away from the central line the less probability of visual contact). The P value for the angle deviation indicates that a larger number of observations could be required to reveal higher significance. The relation between visibility, angle and distances is shown in a 3D graph in figure 2.

The interpretation of the parameter estimate for distance (table 3) is that the probability of visual contact is reduced by a factor of 0.0223 per meter:

$$df = (1 - 0.0223)^n \qquad \text{Formula 1}$$

Where:
df: Decay factor for the given distance (n)
n: is the distance in meters.

In table 4 the estimate is applied to the distances used in table 1.

As it can be seen, the decay parameters revealed from the field experiment are in the magnitude as the ones in table 1,

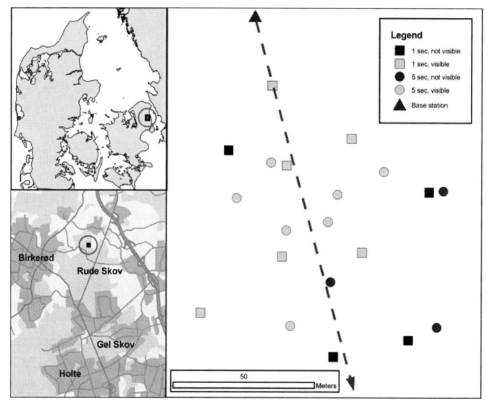

Figure 1. Location of fieldwork site and one out of five sessions. The controller and the respondent were located at the yellow triangle. The targets location and the respondent's ability to see the target is indicated according to the legend. The 'central viewing angle' is indicated by the yellow dashed line.

somewhere between 'Dense trees' and 'Very dense trees'. Optimally it would be expected that the measures should apply to 'Low density trees', but many courses can influence the measures. Including:

- Moving targets will be more visible (in the present experiment, the target was immobile)
- Different color outfits will be more or less visible (in the present case all targets were wearing black jackets).
- Definition and conception of what is realized as 'Low density forest' or 'Very dense forest' obviously differ in different environmental settings.

The results of the field experiment cannot be used as a formal inductive proof of the

Parameter	Standard Estimate	Pr > ChiSq
Intercept	2.3235	0.0013
Distance	-0.0223	0.0100

Table 3. Parameter estimates and P values revealed from the logistic regression for a model including only the distance and intercept.

figures found in table 1, but it points in the right direction. In the present context the parameters of figure 1 will be used in the methodological development of the remainder of the presentation. Angular deviation from the central line will not be taken into further account.

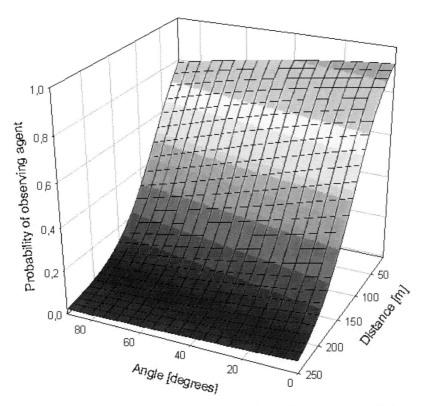

Figure 2: Illustration of the transparency decay as a function of distance and angle, based on logistic regression. Please bear in mind that the angle was shown not to be significant and cannot be taken as authoritative. The illustration and statistical treatment was done by Associate Professor Henrik Meilby, Danish Forest & Landscape.

CALCULATION METHOD

A Viewshed Calculation Application and a Visibility Datastore were developed. In this section we will look at the technical background for the development, the testdata, the actual methods used for calculate line-of-sights and probabilistic viewsheds and storage of the resulting PVG's. Finally an introduction to the data acquisition methods for accessing the data store will be given. The Visual Calculation Application is a standalone JAVA application. The Visibility Data store is based on Postgresql, an open source relational database (Postgresql). In addition a module for export of viewsheds from the data store was developed. The exporter extracts viewsheds for given points as an ASCII raster grid which can be imported in a range of GIS systems. Basically it adds a probabilistic component to well established viewshed calculation methods. All developed software will be available as open source.

The Viewshed Calculation Module consists of two ASCII raster files, a Digital Elevation Model (DEM) representing the terrain as well as a digital representation of different forest types, classified according to transparency (table 1). A maximum line-of-sight range can be set to bring down calculation time. While developing the examples for the present text, a 25 x 25 m DEM (resampled to 5x5 m), provided by the Danish Survey and Cadastre was used. Data representing the vegetation was provided as polygons by the Danish Forest and Nature Agency. Land cover type - tree species and –age - were reclassified in accordance with Preston (2002). Finally, visibility classes were converted to a 5x5 m raster. The reason for the resampling to 5x5 m, was to ensure that the integrity of narrow forest roads were maintained.

A simple line-of-sight calculation algorithm was implemented. First the overall slope between the vantage point and the target point is calculated and stored. The algorithm is then used to sample points along a line from vantage point to target point using the Bresenham Algorithm (Bresenham, 1965). In every sampled cell the z-coordinate is retrieved from the DEM and used to calculate the partial slope from the vantage to the sampled point. Then the overall slope is compared to the partial slope. If the partial slope is bigger than the overall slope, the line of sight is interrupted, and the sampling is aborted. The visibility between vantage and target points is set to 0. If the line-of-sight is not interrupted, the land use class of the current cell is extracted and added to a list. If the target point is reached without being interrupted by the terrain, i.e. the partial slope never gets bigger than the overall slope, the target point can be seen from the vantage point and we can proceed to the next step: the calculation of the probabilistic visibility. Given a maximum distance (e.g. 200 meters), the Line Of Sight algorithm will terminate when reaching the specified range. The method can be seen in pseudo code in Figure 3.

The list of land use classes from the line-of-sight algorithm is used for calculation of the Probabilistic Visibility. At first it performs a lookup of the visual decay according to the landuse classes. Here the cell size of the model has to match the dis-

Table 4. Distance decay based on regression estimates of decay parameter.

		Distance from viewer (m)					
	Decay factor, per 25 m	0	25	50	75	100	500
Based on estimate	0.57	100%	57%	32%	18%	10%	0%

```
function line_of_sight(Vantage Point v, Target Point t, Double maxlength):
    initialize list l
    compute the slope sc from v to t
    repeat
        sample a point p between v and t
        calculate slope sp between v and p
        if sp > sc then
            return None
        else
            append landuseclass(p) to l
    until p == t or distance (v,p) >= maxlength
    return l
```

Figure 3. The line of sight calculation algorithm in pseudo code. The function returns a list of cells and their land use/transparency classes.

tance of the visual decay table. Preston (2002) state the visual decay factors in percent per 25 meters, the algorithm has to adapt the visual decay factor to the present cell size (5 m in the present examples):

$$df\_cs = df^{(cellsize/cw)} \qquad \text{Formula 2}$$

Where:
df_cs: Decay factor, for a cell size
df: Decay factor as per table 1
cw: Cellwith (in the present case 25 m)

The visual decay factors every cell in the list of visible cells are multiplied. The method returns a double precision number. The method can be seen in pseudo code in Figure 4.

The calculation of an entire probabilistic visibility graph is done by double looping through the list of cells of the input grid and applying the line-of-sight and probabilistic visibility method described above. The result for a single pair of cells is stored as a triplet of cell1, cell2 and probabilistic visibility in a relational database. We assume that probabilistic visibility from point A to B is identical to the one from B to A, so we avoid calculating each relation twice by storing relation data in an index and checking whether we have already calculated a relation before we start the line-of-sight calculation.

The current version of the module implements only very simple procedures to make the code more efficient. Definitely, there is a great potential for further improvement. The only optimization parameter applied, is the maximum range. Using the hardly optimized, present methods a complete calculation of a grid sized 624 x 519 cells took about 24 hours on a MacBook Pro 2Ghz Intel Core Duo with 2GB RAM.

Postgresql (Postgresql) was selected for storage of the resulting probabilistic visibility graphs. Postgresql is an open source relational database with possibility to be extended by geospatial functionality (PostGIS). Postgresql is free and fast, and has virtually no limits regarding size as it can be distributed over more than one computer. Functionalities like views, triggers etc. brings it very close to commercial databases like an Oracle Database (Oracle), visual administration GUIs make it very

```
function calc_prob_viz(List l):
    p_viz = 1
    for element in l:
        decay = lookup_decay(element)
        p_viz = p_viz * decay
    return p_viz
```

Figure 4. Calculation of the probabilistic visibility in pseudo code. The function returns a visibility probability between 0 and 1.

productive to work with. Implementation in the latest ArcGIS suite (ESRI) is an indication of maturity and usability within geographical information systems.

The database consists of two tables: viewgraphs and line-of-sight. Viewgraphs contains one record per calculated visibility graph. It constitutes data about grid location and extent as well as other metadata. Los contains one record per visual relation and a backlink to viewgraphs.

The Visibility Datastore is responsible for connecting the Agent Based System to the viewgraph database and deliver probabilistic visibility given two locations or two grid cells. Implementation of the Visibility Datastore is programmed as a JAVA module. Basic functionalities can be seen in Figure 5.

Free roaming agents spatially referred to grid cells (e.g. animals), as well as agents locked to vectors, which typically could be humans following routes can access the datastore. The grid cell access approach might be slightly faster, as conversion from geographic coordinates to grid cell references is avoided. The present model applies both types of agents, as it is constituted by animals (freely roaming), humans on round trips (vector locked) and hybrid agents (e.g. humans following a round trip vector and getting unlocked as roaming agents at given points of time).

RESULTS

Based on the parameters of table 1 and the application described above, resulting probabilistic viewsheds and PVG's will be shown. All examples are based on a viewing height of 1 m and height of the target object of 1 m. Relevant land cover classes of the case study area, situated in the northern part of Rude Skov (Denmark) are shown in figure 6

In figure 7 (B, C, and D) probabilistic viewsheds around three viewing positions are shown. All three points are along the paved road leading through Rude skov from the North to the South. As expected the visibility is in general relative good along the road when compared to the surrounding forest.

In figure 7B (viewpoint 2), the forest on both sides of the road is quite lightly vegetated. The effect of the little path leading to the Northwest can be seen; this is also the case of the effect of the path network to the Southeast of the road. In 7C (viewpoint 3), the viewing point has been moved to a location, where the dense forest cover to the Southeast and the lighter cover to the Northeast are divided by a path. The effect on visibility is easily recognized as an immediate blockage in the dense forest as compared to the lighter forest. 7D (viewpoint 4) shows a probabilistic viewshed from a location with dense forest to the

```
// First a VisibilityDataStore object is instantiated by passing database connection
// parameters to the constructor.

from org.kvintus.visualgraph import VisibilityDataStore
VisibitityDataStore vds = new VisibilityDataStore(<pgsql connection string>,
                                                  <viewshed_id>);

// Then the class VisibilityDataStore can use two public methods for data retrieval:

// returns the Probabilistic Visibility given two points
double getProbVisibility(Point p1, Point p2) :

//returns the Probabilistic Visibility given two grid cells
double getProbVisibility(Integer x1, Integer y1, Integer x2, Integer y2) :
```

Figure 5. Implementation of the Visibility Datastore in JAVA.

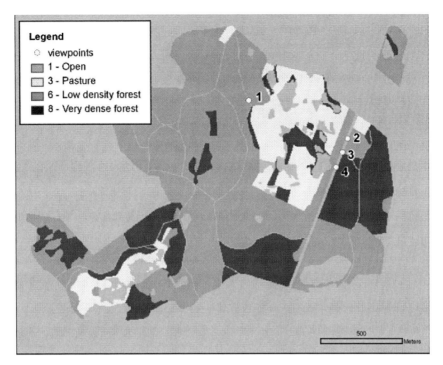

Figure 6. Northern part of Rude Skov, Zealand (Location in Denmark can be seen in figure 1). Land cover/Transparency classes are applied in accordance with table 1. Viewpoints 1 are used in the examples in figure 8. Viewpoints 2, 3, and 4 applies to figure 7.

East and an open patch with thin strip of light forest in front to the West. The effect is as expected a high probability of visible contact raying out to the West and an almost full blockage to the East.

I figure 8 a 3D representation of a viewshed can be seen. The wiewpoint is just in the pass enabling vision to the back and some way a head in the valley in front. The valley immediately in front is invisible due to the terrain. Figure 9 shows an extract from a probabilistic visibility graph (PVG). For each combination of cells the probability is stored. In the context of simulation modeling the advantage of application of a precalculated PVG is that assessment of visibility between two locations is a simple look up in the graph, rather than calculation of visibility as the simulation model executes.

PERSPECTIVES AND FURTHER DEVELOPMENT

This is only the beginning. No doubt probabilistic visibility enhances our ability to analyze – and maybe understand – our surrounding environment. But there is still more to investigate, develop and apply. Both field monitoring techniques and quantitative analytical methods can be further developed.

Topics that can be further assessed include:

- The present application does not take the effects of viewing angles into account.
- The findings from the field study indicate that correlation of viewing angle and visual probability exists. Further it is intuitively expected that there is a relationship between viewing angle and speed of the viewer: A mountain

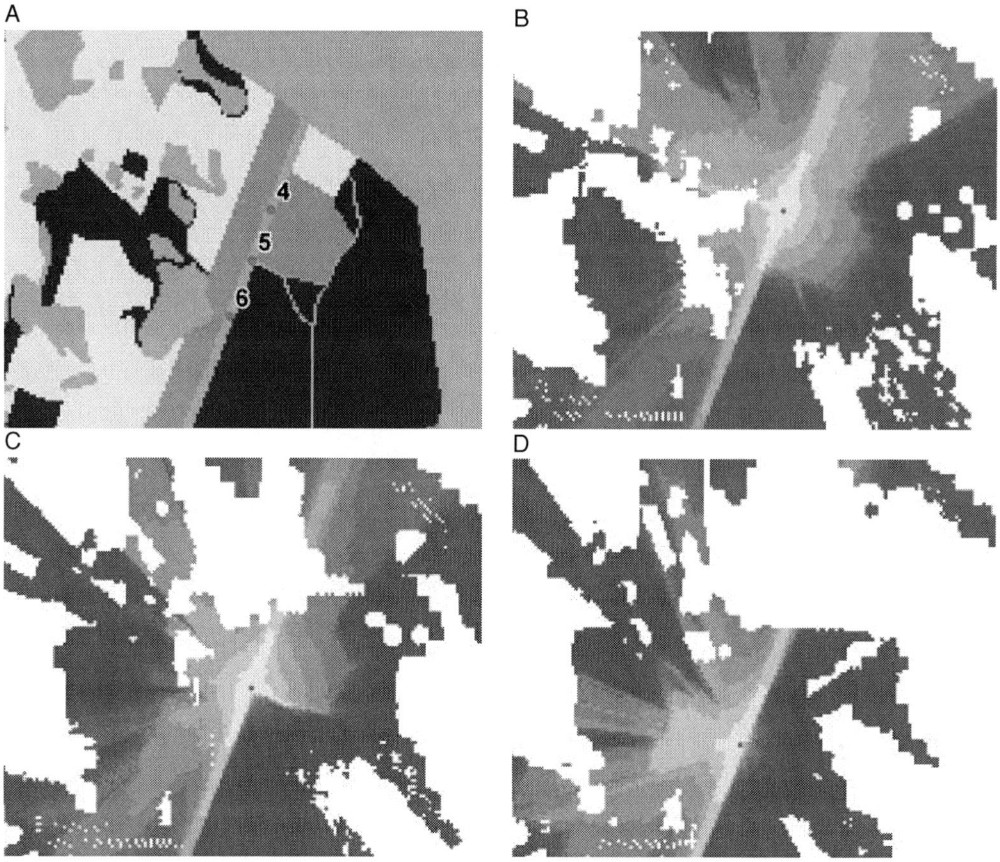

Figure 7. Examples of probabilistic viewsheds, based on viewpoints 4, 5, and 6 of figure 2. The three locations of the viewer are situated along the paved road leading though the forest. The yellow to red color scale displays probabilities in equal intervals of 0.1. Yellow (light gray in grayscale) represents a visibility probability of 1.0-0.9. Red (dark gray in grayscale) represents 0.1-0.0. White areas are invisible due to the terrain.

biker sprinting through the landscape will tend to keep a very close eye on the track, not paying much attention to the surroundings. On the other hand a group sitting still at a picnic table will be looking around having a much broader angular perspective. Walkers and runners will be somewhere in between. To further assess the effects of angular decay, more data has to be retrieved from literature and/or empirically through experiments.

- Visibility analysis could take into account the screening effect of obstacles that might change during the time span of a model run. For instance smoke, weather- and light conditions could be applied as scalars depending on global conditions of the model. Involvement of the number and type of other agents as screening objects between the viewer and the target is a more complex matter. It has to be assessed dynamically by ray tracing techniques at every step of the model. Accordingly it will demand high computer capacity. This sort of dynamic assessment of visual probability could

Figure 8. Details from probabilistic viewshed in 3D. Point 1can be identified in figure 6. Black/white legend: The view shed is indicated by the lightest gray colors among which the light shades have a high visibility probability. The darker have lower probabilities. The invisible, background terrain is shown in darker gray. Color legend: Yellow color indicates high probability of visual contact. Red indicates low probability. The invisible background terrain is green and gray.

be of interest in relation to pedestrian- and escape models with multiple moving agents.

- Currently, probabilistic visibility is treated evenly disregarding the direction, i.e. a given relation from agent a to agent b and from agent b to agent a, is the same. In reality there might be differences looking from a bush into the open and vice versa. An agent within the bush might clearly see another agent in the open, while it might be visually covered by the bush. Along the notions of the general graph theory, bidirectional visibility graphs can be developed. Directional dependency could be handled by the order of the screening cells of the raster. At present – applied as a multiplicative effect – the order of the screening cells is insignificant.

- It can be of interest to include different decay factors in different heights of the screening objects (e.g. trees). Existing literature deals only with one distance decay factor, assuming a viewer with a fixed eye height of e.g. 1.7 meters. Imagining that a valley of 20 meters depth is overlooked, objects higher that 20 m in the valleys center will be screening as well. More over, the decay factor applied to the top of the tree is likely to differ from the factors applied closer to the terrain. This calls for a need to establish a vertical decay profile for different forest types. For instance spruce will definitely have a different decay profile than maple. Gaining vertical decay profile might be a quite timely and costly affair and need a lot of equipment. Modern airborne radar equipment

Col_1	Row_1	Col_2	Row_2	Probability
1	1	1	2	0.9852
1	2	1	3	0.9912
...				
234	412	389	256	0.0012
234	413	389	257	0.0021
...				
624	519	624	517	0.8898
624	519	624	518	0.8791

Figure 9. Selected records from the resulting PVG, based on a 624x519 cell grid. E.g. is cell 389, 257 all most invisible from cell 1, 2, where as 1,2 is 99% visible from 1,3.

might be a future option, as it is able to measure ground elevation, canopy elevation and canopy density/structure.

- While the probability of seeing a moving object is higher than seeing a stationary, monitoring and application of different effects of stabile (as in the present case) vs. moving objects could be of interest. This calls for further development of the present empirical and modeling approach.

- Different types of agents and the visual appearance of agents will influence the probability of visual contact. Different types of agents –horse riders, cars, walkers, mountain bikers, and various animals – will visually be perceived differently. If this was to be included in future models further development of the present empirical and modeling approach has to take place.

CONCLUSION AND DISCUSSION

In the present chapter it has been demonstrated how probabilistic visibility (PV) can be measured, modeled, and applied to simulation models. Theories and techniques have been presented and discussed. Technically, empirically and analytically, assessment PV can be performed. It can be further developed if even more detail is required for future applications. One big question is why it has not been done before and under which circumstances there is a reason to do so now.

First of all PV requires relatively high performing computers. Even at present run-time application of probabilistic line-of-sight analysis is not regarded as an operational option. One of the reasons for applying probabilistic visibility graphs (PVG's) is to perform the rather resource demanding analytical processes as a pre-process. Requests about actual visibility at run-time are simple lookups in the PVG. As computer speed and memory capacity increases application of PV in simulation models will be a more appropriate option.

But having an operational option is not necessarily the same as using it. As indicated earlier the degree of detail of a simulation model – or any other model for that matter – has to do with the degree of detail of the phenomena in question. In cases where a high degree of detail is needed for the perceptive processes involved, application of PV appears to be an obvious requirement. These cases potentially includes models where different vegetation patterns (e.g in parks and urban forests) and evacuation models (where visibility might be obscured by smoke are examined. On the other hand if the objects of a model – landscape, transport network, number of agents etc – are well represented in less detail, application of PV is less required.

Another question stemming from behavioral psychology addresses the differences between 'seeing', 'encountering', 'paying attention', and 'making an impact'? How is registration of visual contact applied to transitions (changes in state-of-mind) of the agents? An agent might see another agent without actually reacting or changing state-of-mind. Explicit transition rules have to be considered for different agent types in direct relation to the visibility model applied.

Over and above, a core feature of a model is that it is a representative simplification of a real word phenomenon. The simplified representation provided by models enables us to comprehend relations and functions. If the models built become too complex; if it is too close to what occurs in the real world that it is representing, it will be harder to operate and understand. It will be difficult to see the relation between courses (including the 'handles' turned by the operator) and the effects (the model results). Accordingly models should not be made more complex than required. If there is no need for PV – considering the phenomenon as discussed above – it should not be applied. Or applying Orcham's razor: 'If things can be described simply; complex description should be avoided'.

If – on the other hand –visibility at a finer scale is required to solve the problem being modeled then PV appears to be a necessary addition to the existing stock of methods that can be applied to agent-based models.

ACKNOWLEDGEMENT

The work presented is part of the development of kvintus.org, an agent-based model of recreational behavior. The development of the model is supported by the Danish Forest and Natures agency and the Danish Outdoor Council. Further, we would also like to thank associate professor Henrik Meilby of Danish centre of Forest, Landscape and planning for the statistical treatment of the field experiment data.'

REFERENCES

Benedikt, M L. 1979. To take hold of space: isovist and isovist fields. Environment and Planning B, vol 6. 47-65.

Bishop, I. and Gimblett, R. 2000. Management of recreational areas: GIS, autonomous agents, and virtual reality. Env. and Planning B: Planning and design. Vol 27. Pp. 423-345.

Itami, B. 2002. Mobile agents with spatial intelligence. In Gimblett, R. (ed.). Integrating geographic information systems and agent-based modeling techniques for simuating social and ecological processes. Pp. 191-210. Oxford University Press.

Jiang, B. and Gimblett, R. 2002. An agent-based approach to environmental and urban systems within geographic information systems. In Gimblett, R. (ed.). Integrating geographic information systems and agent-based modeling techniques for simuating social and ecological processes. Pp. 171-190. Oxford University Press.

Jochem, R. 2007. Pers. com. Rene.Jochem@wur.nl. University of Wageningen, the Netherlands.

Kim, Y. and Rana, S. and Wise, S. 2004. Exploring multiple viewshed analysis using terrain features and optimisation techniques. Computers and Geosciences, 30 (9-10). pp. 1019-1032.

Kumsap, C., Borne, F. and Moss, D. 2005. The technique of distance decayed visibility for forest landscape visualization. International Journal of Geographical Information Science, Vol. 19, Number 6.

Manning, R., Valliere, W., Wang, B., Lawson, S. and Newman. 2005. Frontcountry trails and attraction sites in Yosemite national park: Estimating the maximum use that can be accommodated without violating standards of quality. In Cole, D. (ed). Computer simulation modeling of recreation use: Current status, case studies, and future directions. Pp. 36-38. USDA/FS. Report RMRS-GTR-143.

O'Sullivan, D. and Turner, A. 2001. Visibility graphs and landscape visibility analysis. International Journal of Geographical Information Science. Volume 15, Number 3 / April 1, 2001

Preston, R. 2002. Visual Exposure of Landscapes in the Bremer River Catchment and the Middle Brisbane River Catchment.

Preston, R. 2005. Pers. com. robert.preston@forestimages.com.au. Forest Images Ltd.

Ruddell, E., Gramann, J., Rudis, V. and Westphal, J. 1989. The Psychological Utility of Visual Penetration in near-view Forest Scenic-Beauty Models. Environment and Behavior 1989 21: 393-412.

Rudis, V. 1985. Screenometer: a device for sampling vegetative screening in forested areas. Canadian Journal of Forest Research. Vol 15: 996-999.

Snizek, B. 2003. Quantitative spatial analysis of landscapes. M.Sc. thesis. Royal Veterinary and Agriculture University. Copenhagen, Denmark.

Tobler, W. R. (1970). A computer model simulation of urban growth in the Detroit region. Economic Geography, 46(2): 234-240.

Ubisoft. 2007. Splinter cell double agent. URL: http://splintercell.uk.ubi.com. Last accessed 8 February 2007.

Bresenham J.E. (1965) - Algorithm for computer control of a digital plotter, IBM System Journal vol 4 no 1

URL's:

Oracle-http://www.oracle.com. Accessed February 2007.

PostGIS-http://postgis.refractions.net/. Accessed February 2007.

Postgresql-http://www.postgresql.org. Accessed February 2007.

CHAPTER 9

EXPLORING SPATIAL BEHAVIOR OF INDIVIDUAL VISITORS AS A BASIS FOR AGENT-BASED SIMULATION

Karolina Taczanowska
Arne Arnberger
Andreas Muhar

Abstract: Generic assumptions and theories concerning human recreational behavior require validation against real phenomena. Monitoring campaigns provide comprehensive data about visitors and their activities in natural settings. In particular, high-resolution, spatio-temporal data at an individual level are highly relevant as input for agent-based models and simulations. This chapter deals with questions of what types of analysis techniques are important to access monitoring data for simulation and how they contribute to model parameterization. The considerations are illustrated by examples from an empirical study (Danube Floodplains National Park, Austria), where the main emphasis was placed on exploring human-environment interactions as well as on the characteristics of individual routes. GIS, relational database management and statistics software packages were used for data storage and analysis. The results were applied for setting parameters at the MASOOR simulation platform.

Keywords: spatial behavior, human-environment interaction, individual-level data, visitor flows, agent-based simulation, GIS, recreation, MASOOR

INTRODUCTION

Successful use of computational modeling and simulation for supporting planning and management decisions in natural recreation areas depends on the reliability of their outputs. It is believed that the design of individual-based models must be supported with actual data (Gimblett et al. 1996). Generic assumptions and theories concerning human spatial behavior and recreational experience require validation against real phenomena (Cole 2005b; Elands and van Marvijk 2007; O'Connor et al. 2005; Skov-Petersen 2005). Monitoring campaigns may provide comprehensive information about visitors and their activities in natural settings (Cessford and Muhar 2003). Collecting demographic, psychographic, behavioral, spatial and temporal data contributes to better understanding and more effective management of visitor flows in recreational sites (Daniel 2002). Reliable knowledge concerning spatial behavior of people outdoors as well as human-environment interactions is particularly important in establishing inputs for agent-based simulations of visitor landscapes (O'Connor et al. 2005).

Figure 1 illustrates steps leading from the actual recreational behavior in a leisure setting towards the simulation of this phenomenon. Monitoring is an important step, but it is already a selective observation of recreation use in an outdoor environment. Further steps leading to the simulation comprise: data storage, data analyses, and defining behavioral rules of agents and their formalization. All those stages additionally interpret the genuine phenomena and consequently have significant effect on the simulation results.

STUDY OBJECTIVES

The aim of this study was to explore spatial behavior of visitors in a leisure setting by characterizing their individual routes. In this chapter we focus on those aspects of collecting, storing and analysis of spatial behavior data, involved in the preparation of input for agent-based simulations. All the above-mentioned steps are illustrated by practical examples from the case study in a suburban recreational area in Austria. Particular attention is paid to the structure of the physical environment and measurements describing routes of individual visitors. Discussion of the utility of the results as input to agent-based simulations is supported by the application of the study outcomes within the MASOOR simulation platform.

The term 'spatial behavior' is understood as 'spatially manifested and overt acts of people performing a range of daily or other episodic activities' (Golledge 2001). These features are represented and analyzed as occurrences in space (Golledge and Stimson 1997). A different term, namely 'behavior in space', involves investigating choices underlying spatially manifested acts (Golledge 2001). This aspect, however, is not the subject of this chapter and can be found in other sections of this book.

MONITORING METHODS FOR RECREATION USE

Defining the monitoring and modeling goals determines the type of information to be collected (Cessford and Muhar 2003). Monitoring campaigns can serve exploratory purposes, where general information

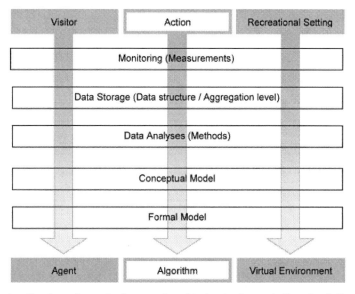

Figure 1. Steps leading from the actual recreational behavior towards the simulation of this phenomenon.

about recreation use at an area is collected (Arnberger and Eder 2007; Brandenburg 2001). It may also address specific management questions, such as identifying bottlenecks of use (Itami et al. 2002), zones where recreation and nature protection goals contradict, or social conflicts (Arnberger and Haider 2005). In order to gather information about the behavior of people in natural leisure settings, several methods have been developed to systematically monitor visitors. A general distinction can be made between aggregate and individual level data, both of which might be applied for simulation purposes.

Bottom-up simulation approaches are often based on information about the activities of individual recreationists. This requires collecting trip itinerary data including: routes, visited points of interest, speed and direction of movements. This data can be collected using traditional survey methods (Gimblett et al. 1996), or modern tracking technologies such as sport timing equipment (O'Connor et al. 2005), or GPS devices (Elands and van Marvijk 2008).

For simulation purposes, the modeler makes use not only of specific small resolution spatio-temporal data, but also needs general figures describing visitor load at particular locations, shares of activity types, as well as seasonal, weekly and daily visitation patterns (Cole et al. 2005). In order to capture aggregate level data, a systematic long-term observation is required. Several methods have so far been applied for indirect observation of visitors worldwide e.g. photoelectric counters, pressure-sensitive devices, and automatic cameras (Cessford and Muhar 2003; Kajala et al. 2007). The use of cameras is reported to be an effective and reliable method to make long-term observations of the recreation use (Arnberger and Eder 2007). Besides a record of visitation levels, video recordings provide additional information, such as types of performed activities, group size, direction of movement, visitor behavior, and interaction between users (Brandenburg 2001).

Documenting recreation use exclusively as an 'occurrence' in space might be insufficient for modeling visitor flows. Incorporating additional information, including social and environmental components, supports understanding of the phenomenon (Skov-Petersen 2005).

REPRESENTATION OF RECREATIONAL SETTING – DEFINITION AND STRUCTURE OF SPATIAL ENTITIES

In order to analyze and model the spatial dimension of recreational behavior it is necessary to define the environmental components of a leisure setting. As the physical environment is extremely complex, it is essential to make choices about 'what to represent, at what level of detail, over what time period' (Longley et al. 2001). Several questions, such as abstraction of the real world features into virtual ones, the level of space granularity, and the meaning of certain spatial configurations must be addressed in order to create a useful computational model (Benenson and Torrens 2004).

Geographic Information Systems (GIS) are frequently used to represent recreational space digitally (Gimblett 2002). The main reason for this is the widespread use of GIS in the fields of spatial planning and natural resources management, and therefore, the existence of ready-to-use structured spatial data for simulations (Gimblett 2002). There are several ways to structure spatial information, but the following three approaches are most common in present recreation behavior simulations: vector representation of environment, raster data structure, and mixed approaches. Figure 2 illustrates environmental components of a recreational setting and presents different approaches to structuring this information.

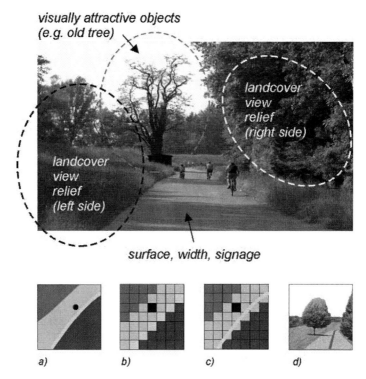

Figure 2. An example of environmental components of a recreational setting and different methods of structuring this information: a) vector data model, b) raster data model, c) mixed approach: raster and vector data, d) mixed approach: 3D landscape visualization and vector data

Several recreation behavior models make use of vector representation of the environment (Itami and Gimblett 2001; Jochem et al. 2008). This coordinate-based data model represents geographic features as points, lines, and polygons, characterized by associated attributes (Longley et al. 2001). In agent-based simulations the environmental components are usually represented in the form of points and lines (Itami and Gimblett 2001; Jochem et al. 2008). Line features carry a central meaning and are typically used to represent trail networks and waterways in recreational areas (Cole 2005a). Line features can easily be transformed into network datasets, consisting of links and nodes. Such structure enables the storage of geometric and topological information about a network. Point features often represent tourist attractions. They might be either a part of a network, using nodes to define places of special interest ('locales' in RBSim (Itami et al. 2002)) or may be classified as separate point features ('attraction points' in MASOOR (Jochem et al. 2008)). Points are also used to define entrances/exits in the area (Itami et al. 2002). Agents use the network to move within a recreational setting. Such a structure enables storage of the overall environmental characteristics, for instance type of surface, width of trails, and land cover along the paths, as attributes of trail network. In this case the modeler must be aware of certain data aggregation biases required to reduce complex information about the environment to path segment characteristics. In this case, artificial individuals may acquire knowledge about the surroundings directly from the network's path segments and nodes.

A raster data model represents landscape as a surface divided into a regular grid of cells (Longley et al. 2001). Each cell contains attribute value defining the environmental components, for instance landscape type, trail network, terrain elevation, etc. This type of environment representation was used in the early recreation behavior simulations (Deadman and Gimblett 1994). Artificial agents move through a grid of cells and acquire landscape information stored in the neighboring raster cells. The raster resolution determines the level of detail used in the simulation (Gimblett 2002) and depends on the scale of a case study area, behavioral phenomena being modeled, temporal dimension, and capacity of computer used, etc. The cell size selected for modeling purposes in the case studies reported by Gimblett (2002) varied from 10 meters in the Broken Arrow Canyon, Arizona up to 90 meters in the Grand Canyon simulation model.

There are also attempts to use mixed approaches to represent recreational settings. They combine the advantages of vector and raster data models used within GIS (Gimblett et al. 2001), or integrate advanced 3D landscape visualization techniques with vector representation of linear features (Gloor et al. 2004). Raster data model is frequently used to represent terrain relief, which is particularly valuable for mountainous sites. In addition, a raster-based digital elevation model may support visual analyses, which are used by several RBSim models to calculate the number of encounters with other visitors during a given trip (Itami et al. 2002). Modeling visual appeal in recreational settings makes use of advanced 3D landscape visualization techniques (Cavens et al. 2004); there are also successful examples of using GIS data for visual quality analyses and analyses at effects at vegetation transparency (Skov-Petersen and Snizek 2008). The above-mentioned agent-based models, such as AlpSIM (Cavens et al. 2004) or RBSim2 (Itami and Gimblett 2001), in addition to raster or 3-D landscape representation, use linear vector features to represent trails. Although agents move along the trail network, they acquire environmental information, which is stored separately from the path segments and nodes for instance as raster information.

CASE STUDY AREA

The case study area, Lobau, is situated east of Vienna, Austria, and is part of the Danube Floodplains National Park. It lies within the capital city boundaries and is a traditional local recreation site (Brandenburg 2001). Long-term monitoring of visitor flows in the Lobau permitted identification of spatio-temporal patterns of recreation use as well as characterizations of the visitors (Arnberger et al. 2001). This relatively small area (approx. 10 km long and on average 2 km wide) attracted 600,000 visits in 1999. The dominating leisure activities here are biking and hiking (Arnberger et al. 2000). According to the National Park regulations, recreation use is allowed at designated sites and along marked trails (Donau-Auen Nationalpark 2006). The entire area comprises a complex mixture of riverine forests, meadows, and old branches of the Danube River. There are no distinct attraction points in the Lobau. Therefore, the setting is suitable for analyzing spatial behaviors of visitors, which are mostly independent from particular destination choices.

METHODS

In order to investigate the spatial distribution of visitors and to determine factors affecting human spatial behavior a quantitative approach was applied. The following components were used for the investigation (see figure 3):

- Demographic and behavioral data about park visitors;

- Records of visitors' routes;
- Descriptions of spatial structure of the study area.

Data concerning visitors and their recreational behavior, as well as route information, were collected via on-site interviews combined with trip reports. Additionally, environmental data was gathered to describe the spatial structure of the Lobau area. The spatial context of visitor flows in this recreational setting was the core subject of the research. Measurements describing the morphology of space as well as spatial attributes of routes were developed and analyzed. The outcomes of the empirical study were used as the basis for setting parameters within MASOOR. Figure 3 presents the methodological steps of the case study, leading from the data collection towards setting the parameters of the simulation.

DATA COLLECTION

On-site visitors were interviewed at main entrance or intersection points (12 interview locations) on 4 sample days. The interviews took place in spring and in summer, on weekdays and weekends. The visitors were interviewed about their outdoor activities, visitation motives, length of stay, local knowledge, etc. As part of the interviews, the respondents were asked to mark on a map (1:25,000) the route, which they took or planned to take on that day. Additionally, the interviewers collected data on the visitor's performed activity, group size, and the company of children and dogs (Arnberger et al. 2000). In total, 780 questionnaires and 627 routes (532 valid routes) were collected, which resulted in 511 complete datasets comprising visitor characteristics coupled with a valid trip report (Arnberger et al. 2000).

The spatial data was derived from existing cartographic material and complemented during fieldwork in the Lobau recreational area. The path network played a central role for data collection. In the first step, all the paths in the area were digitized, based on the Austrian topographic map 1:50,000 and then checked in the field (Hinterberger 2000). The path segments were used as the basic unit for the collection of environmental data. Structured data collection forms were used in the field in order to characterize the path segments and intersection points. Detailed data covering environmental features of the area such as type of surface, width of trail, land cover along paths, views, accompanying tourist infrastructure, location of on-site information boards and signs was collected.

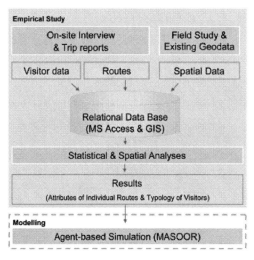

Figure 3. Methodological steps leading from data collection towards setting the parameters of the simulation.

DATA STORAGE

In a first step, all spatially referenced information was stored in a GIS, and the non-spatial in MS Access. Next, all the data was integrated in the MS Access database, due to the ease of establishing relations between different entities and extensive possibilities of performing SQL queries. Table 1 illustrates the general structure of the gathered data. The final

Table 1. The general structure of the gathered data

Input data	Source	Type	Input data storage framework
Visitor characteristics	Survey	Non-spatial	MS Access
Routes	Survey (trip report)	Spatially referenced	ArcGIS > MS Access
Environmental characteristics	Map & field study	Spatially referenced	ArcGIS

database comprised three interrelated data blocks: visitor characteristics, routes, and environmental characteristics.

The route information was the main subject of this investigation; therefore its data structure was carefully considered in order to conform to further analyses. The reported routes were spatially referenced and stored into a database (Hinterberger et al. 2002). The data structure was based on the trail network (GIS). A route was defined as a sequence of visited path segments. This information was stored in a relational database (MS Access) where all the records were checked for topological consistency, e.g. contiguity of route segments (Hinterberger et al. 2002). Figure 4 illustrates an example of the route data structure.

For the purpose of this study, environmental data were stored either as attributes of path segments or as attributes of network nodes. Figure 5 presents an example of a thematic map illustrating surfaces and widths of paths in the Lobau area.

ANALYSES

The raw input data described above did not provide route attributes directly. These were additionally generated with the help of SQL queries within the database. Table 2 summarizes route attributes covering the route geometry and topology as well as the physical appearance of trails.

Exploratory methods were applied to investigate systematic patterns in the spatial behavior of individual recreationists.

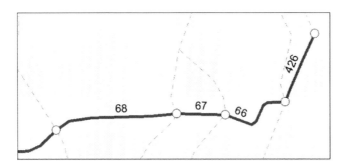

Figure 4. Example of a route information. Route as a sequence of visited path segments in a trail network.

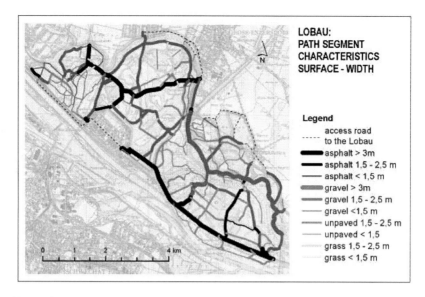

Figure 5. Thematic map illustrating the surface and the width of paths in the Lobau area

Basic statistical methods, such as distributions of the variables, correlation matrices and crosstabulations were used in this phase. In the next step, a hierarchical cluster analysis - Ward method (Ward 1963) was applied to identify groups of visitors with similar spatial behavior. The analyses of route attributes were performed with the help of the SPSS statistical software.

RESULTS

The results of the on-site survey demonstrate a great diversity of routes, which the respondents took or planned to take during their visit to the Lobau. The reported trips ranged from long-distance loops, following marked and well-paved trails, up to destination-oriented shortcuts leading from a car park to the nearest picnic or swimming spot. Selected attributes of routes are presented below (Table 3). The mean value of the route length was 7.0 km. The shortest distance reported in the Lobau was 163 meters, the longest one 25.7 km. Generally, two types of route shapes were identified: loop and traverse. The large majority of the

Table 2. Summary of route attributes

Attribute class	Variables / Measurement
Length	Total sum of used path segments (m)
Location of entrance/exit	Loop (entrance = exit) or traverse (entrance ≠ exit) (category)
Level of retracing paths	Share of path segments used more than once during a given trip (%)
Signage	Marked trail; marked multi-use trail (m); share of marked and non-marked trails (%)
Surface	Asphalt; gravel; unpaved track; grass; other (m); share of different surface types (%)
Width	Width classes: >4 m; 3-4 m; 2-3 m; 1,5m; <1m (m); share of different width classes (%)
Landscape type	Riverine forest; pine forest; bushes; meadows; agriculture; water, industry, building (m); share of different landscapes types (%)
Infrastructure	Number of benches per trail km (count)
Attractions	Picnic spot; restaurant; museum; (category)

respondents finished their trips in the start locations by making loops (80%). Twenty percent of interviewed visitors traversed the area. More than half of the recreationists (52%) did not retrace their paths. However, a considerably large share of visitors partly (32%) or completely (16%) repeated the trail on their way back. Figure 6 illustrates examples of route shapes reported by visitors to the Lobau recreational area.

Only 52% of the respondents staged completely on the marked trails, as prescribed by the park regulations. A large share of visitors (41%) only partly followed a signage (marked trail covered 50-90% of the entire trail system. About seven percent of the interviewees were predominantly wild-paths users. The visitors tended to use prepared and well maintained paths. Gravel and asphalt surfaces were most willingly used among the visitors while performing different types of recreational activities. Unpaved and narrow paths in the forest as well as grass trails in open landscapes were less frequently used. The analyzed routes differ considerably in terms of length, type of surface, width, shape and signage. However, those large spatial differences could not have been easily explained by visitor characteristics.

The typological segments of visitors were

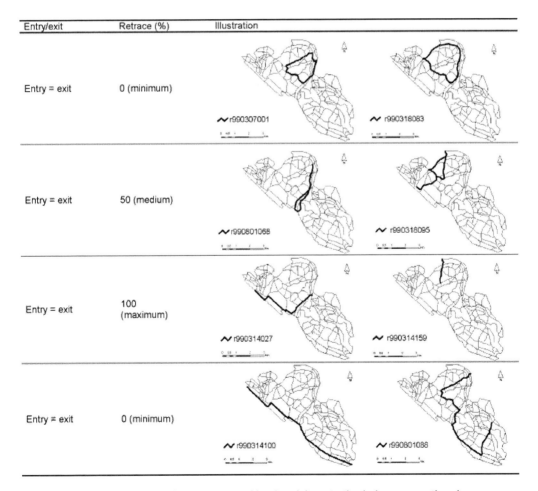

Figure 6. Examples of route shapes reported by the visitors to the Lobau recreational area.

Table 3. Typology based on the revealed trail preference of the Lobau visitors

Selected attributes of routes	Classical visitor	Infrastructure-guided visitor	Wild-paths visitor
Share of N=532 (%)	59	31	10
Mean values:			
Asphalt (%) *	12.8	57.2	9.4
Gravel (%)	59.7	27.7	20.9
Unpaved (%)	18.5	11.5	64.2
Wide paths > 2.5 m (%)	17.0	53.2	7.2
Narrow paths < 1.5 m (%)	16.8	8.6	65.7
Signage (%)	85.1	89.9	53.8
Length of route (km)	6.6	9.1	3.1

*) % of surfaces, widths and signage categories describe the mean share of a given category (e.g. percentage of asphalt surface) within an entire route

developed for the simulation purposes. As we found no correlation between the spatial behavior of recreationists and their social attributes in this study, the typology was based exclusively on the revealed trail preferences of the respondents. The three general types were identified among the Lobau visitors: *classical visitor, infrastructure-guided visitor, and wild-paths visitor*. Table 3 presents selected characteristics of the defined types. Classical visitors prevail among the recreationists, which spend their leisure time in the Lobau area (59%). They typically use marked trails; choose gravel and paths of medium-width, and on average cover distances of 6.6 km. Infrastructure-guided visitors use mostly paved trails, often asphalt and wide ones, and they follow almost exclusively marked trails. Infrastructure-guided visitors belong to the active part of the recreationists; traveling on average 9.1 km. Wild-paths users concentrate their recreational activities mostly on unpaved and narrow paths. Signage seems to have little meaning for this user group - on average only 54% of their route runs along marked trails. Wild-paths users tend to travel short distances ranging around 3 km.

APPLICATION WITHIN THE MASOOR SIMULATION PLATFORM

The study results were applied to set the parameters at MASOOR (Jochem et al. 2008). In order to design an artificial population of agents, the visitors were grouped into preference-activity types and distributed among the main entrances to the recreational area. Table 4 illustrates the types of agents used in the case study.

Three behavioral strategies, corresponding to the trail preference types, were applied in this model: *classical, infrastructure-guided* and *wild-path agent*. Each of these groups was additionally divided into three activity types, like hiking, jogging and biking. In this simulation, the type of activity influenced exclusively the speed of agents' movement.

Table 4. The types of agents used for simulating visitor flows in the Lobau recreational area.

Preference - Activity Types (%)	Classical agent	Infrastructure-guided agent	Wild-path agent
Hiker	36	18	8
Jogger	3	1	1
Biker	18	13	2

N = 455 (100%)

Activating and weighting parameters define the behavioral strategies of agents within the MASOOR simulation framework. A detailed description of its functionality can be found in Jochem et al. (2008). Three hierarchically nested sub-modules: Planner, Navigator and Pilot, decide how agents navigate through the path network. The results of individual route analyses were particularly useful when defining input data for the 'Planner' and the 'Pilot' levels, for instance, information about the route geometry and topology (route length, topological relationship of the entrance and exit point, level of retracing a path). As there are no distinct highlights in the Lobau, the agents did not use attraction points to navigate in the outdoor environment and started their movement straight in the 'immersion' phase. The individuals were sensitive to the environmental features, such as the type of surface, width of paths and trail signage in the area. Different types of agents reacted to the environmental features mentioned above in a type specific way, for instance infrastructure-guided agents preferred well-paved, wide and marked trails, whereas the wild-paths agents regarded such trails as unattractive. Table 5 summarizes the MASOOR 'Pilot' parameters and presents which of them were used to simulate visitor flows in the Lobau recreational area. As the relation between the physical appearance of trails and agents' movement was well supported by emperical data, the sensitivity to the path type characteristics was a central parameter in our case study. It was decided to minimize the role of other parameters in the simulation, due to the lack of sufficient empirical information supporting their use in the Lobau case study.

The simulation delivered satisfying results. There is a high correlation between the observed and the simulated spatial distribution of visitors, expressed as visitor load per network path segment. The simulated figures are mean values of 30 simulation runs. Figure 7 presents the comparison between the observed and simulated distributions of visitors in the Lobau area.

DISCUSSION

Spatial behavior of individual visitors

So far, investigation on human spatial behavior at individual scale belongs to the underreported fields in outdoor recreation research; although the first European research on this subject seems to appear in the 1970s (Koch 1984). This study provids practical description of how individuals use a recreational setting. Studies of this

Table 5. Usage of the parametres within the MASOOR platform

Motivation behavior	Phase of movement			
	Entry	Immersion	Exit	Direct Exit
Global heading	-	-	-	-
Path type (Surface & width)	-	++	+	(+)
Path type (Signage)	-	++	(+)	-
Homing direction	-	-	-	-
'Chunking' direction	-	-	-	-
Path segment history	-	-	-	-
Shortest distance	-	-	+	++
Crowding	-	-	-	-

Importance of the MASOOR parameters at the 'Pilot' level in the Lobau case study:
++ very important; + important; (+) less important; - not used

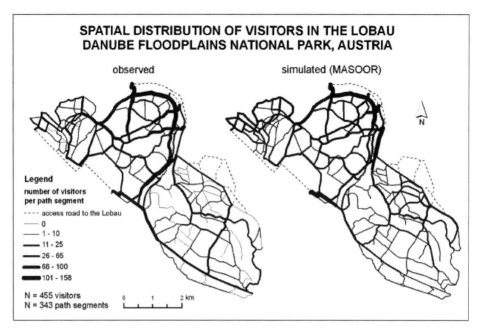

Figure 7. A comparison between the observed and the simulated distribution of visitors in the Lobau area. The correlation coefficient between the observed and the simulated visitor load per path segment is 0.709 (R) for p < .01

kind may be very helpful for verifying a priori assumptions concerning distribution of individual visitors in the outdoors.

The analyzed routes differ considerably in terms of length, type of surface, width, shape and signage. However, those large spatial differences cannot easily be explained by visitor characteristics. It implies that spatial behavior should be considered an additional feature for defining visitor profiles and building recreationists typologies. Nevertheless, in the ongoing studies at public use in natural settings we extend social characteristics of visitors and intend to repeat the inquiry concerning the relationship between the social and spatial aspects of recreational behavior. Such a comprehensive 'picture' of a visitor, covering demographic data, recreational behavior characteristics and spatial dimension of his/her acts, would be of great value for visitor flow simulation and management purposes.

This investigation demonstrates several spatio-temporal similarities of recreation use with other natural areas in Europe. The analogies refer particularly to the quantities of route lengths and duration of stay (Elands and van Marvijk 2008; Zundel and Völksen 2002). The layout of the trail system, the design of recreational infrastructure and provision of information affect the distribution and experiences of visitors (Barth 1982). Feeling secure without the danger of getting lost, is necessary when spending time in outdoor leisure areas (Findlay and Southwell 2004). The results of presented empirical study also confirm these findings. Visitors to the Lobau generally use well-defined paths and follow marked trails. Nevertheless, our investigation revealed a group of visitors who do not follow the site regulations. Cyclists, particularly, tend to use all possible trail combinations, not considering any limitations for this type of activity. The number of users of non-marked paths is an important measure from the nature

Route characteristics as input for agent-based simulation

Findings of the investigation concerning individual visitor routes proved to be useful for defining the input for agent-based simulation. MASOOR is a flexible simulation platform allowing manipulation of number and weights of model parameters (Jochem et al. 2008). Information about visitors' preferences for certain types of surfaces, widths and signage of paths were particularly valuable for defining the behavioral rules of agents. Additional quantitative information concerning the geometry and topology of routes was also helpful while preparing the input for simulation. The results concerning the length of routes combined with the information about the speed of movement were very practical when defining the distribution of effective travel time values within the MASOOR framework. Knowledge about re-tracing the paths on the way back also proved to be constructive information for the model parameterization. Most of the measurements of individual spatial behavior developed in our study were compatible with the MASOOR input requirements.

Potential application of simulation as a tool supporting management decisions in leisure settigs

As soon as a simulation generates reliable results, it might be considered as a tool supporting management decisions in natural leisure settings (Cole 2005a; Gimblett 2002). There are several issues in the Lobau area where the application of an agent-based simulation could be a promising approach to aid the discussion on future management initiatives in the area. One of the biggest challenges in the Vieneese part of the Danube Floodplains National Park (Lobau) is a rapid urban development in the north-eastern part of the City of Vienna. The population in residential areas neighbouring the Lobau, is estimated to increase by more than 20% within the next fifteen years (Institut für Demographie der Österreichischen Akademie des Wissenschaften 2002). The growth of population in the Lobau's catchment area will most likely contribute to the increase of visitation level in the leisure setting (Arnberger et al. 2000). By this time, several management alternatives concerning strategic guidance of visitors, provision and allocation of tourist infrastructure and information should be developed in order to mitigate effects of a gradually increasing public use. Experiences from other protected areas in the world show that the evaluation of alternative management scenarios can be well supported by the application of agent-based simulations of recreation use (Gimblett et al. 2005; Itami et al. 2002).

CONCLUSIONS & OUTLOOK

From monitoring to simulation

There is no single monitoring method to address all the simulation input requirements, therefore parallel application of various techniques to gather data is necessary. Such combinations of monitoring methods should enable us to gather both individual as well as aggregate level data. The use of automatic cameras for long-term observations is reported to be a reliable and effective method for collecting information on visitor load of various user groups and its temporal differentiation (Arnberger and Eder 2007). Additionally, high-resolution, geo-temporal data combined with the social characteristics of visitors would be desirable. Presented case study emphasizes the role of individual-level visitor data as a necessary input for bottom-up simulations. The application of new technologies, such as GPS tracking to record actual visitor behavior seems to be a promising solution (Elands and van Marvijk 2008). There is, however, a need to evaluate the method against traditional data collection techniques (e.g. trip diaries and sketch maps).

As the recreation activities take place in an outdoor environment and it is believed that there is a relationship between the environment and human behavior, it is necessary to include environmental characteristics into the data collection procedure. The quantitative dimension of a computational simulation requires fracturing the complex physical environment to spatial entities that reflect structures meaningful for recreation behavior. That is why questions as to which environmental components are relevant for spatial decision-making and what is the meaning of spatial configuration for recreation experience of visitors need to be answered. In addition to studies concerning the representation of the physical environment, more attention should be paid to the research concerning social and perceived environment issues. In parallel to the discussion about the environmental determinants of human behavior, investigating the effects of applying different space resolutions and data structures in recreation behavior models would be a desirable direction of future research.

The investigation concerning the accuracy of different data collection methods, ways of describing human behavior and the representation of landscape as well as the effects of different ways of storing and analyzing the same raw information would be of great value for future studies.

Implications for management actions

Precise questions regarding the management of recreational sites need to be formulated in order to specify the objectives of monitoring, analyses and simulation. Integrated approaches could enhance the effectiveness of data collection and data processing by tailoring them from the very beginning to the simulation requirements. Only in this way can simulation become a reliable tool in supporting management decisions, and monitoring campaigns might be focused exclusively on the contextually relevant aspects of recreation use.

ACKNOWLEDGEMENTS

The Forest Department of the City of Vienna commissioned the Institute of Landscape Development, Recreation and Conservation Planning to collect data on public use. Mrs. Christiane Brandenburg coordinated the project, where part of the data comes from. The authors would like to express gratitude to Rene Jochem, Ecological Modeling and Monitoring Team, Landscape Centre at the Wageningen University and Research in the Netherlands, for the possibility to apply presented case study outcomes to the MASOOR simulation platform.

LITERATURE CITED

Arnberger, A., Brandenburg, C., Cermak, P., and Hinterberger, B. 2000. „Besucherstromanalyse für den Wiener Anteil am Nationalpark Donau-Auen, Bereich Lobau." Im Auftrag des Magistrats der Stadt Wien, MA 49, Forstverwaltung Lobau, Institut für Freiraumgestaltung und Landschaftspflege, Universität für Bodenkultur, Vienna.

Arnberger, A., Brandenburg, C., and Muhar, A. 2001. „An Integrative Concept for Visitor Monitoring in Heavily used Conservation Area in the Vicinity of a Large City: The Danube Floodplains National Park, Vienna." Northeastern Recreation Research Symposium, Bolton Landing, New York, 195-201.

Arnberger, A., and Eder, R. 2007. „Monitoring recreational activities in urban forests using long-term video observation." Forestry, 80 (1), 1-15.

Arnberger, A., and Haider, W. 2005. „Social effects on crowding preferences of urban for-

est visitors." Urban Forestry & Urban Greening, 3, 125-136.
Barth, W., E. 1982. „Tourismus in Waldgebieten. Erfahrungen über Steuerungsmöglichkeiten von Touristenströmen." Neues Arch. F. Niedersachsen, Göttingen, 31 (3), 270-289.
Benenson, I., and Torrens, P. M. 2004. „Geosimulation: automata-based modeling of urban phenomena", Wiley, Chichester
Brandenburg, C. 2001. "Erfassung und Modellierung von Besuchsfrequenzen in Erholungs- und Schutzgebieten - Anwendungsbeispiel Nationalpark Donau-Auen, Teilgebiet Lobau," Dissertation, Universität für Bodenkultur, Wien.
Cavens, D., Gloor, C., Nagel, K., Lange, E., and Schmid, W. A. 2004. "A Framework for Integrating Visual Quality Modelling within an Agent-Based Hiking Simulation for the Swiss Alps." Working Papers of the Finnish Forest Research Institute 2.
Cessford, G., and Muhar, A. 2003. "Monitoring options for visitor numbers in national parks and natural areas." Journal for Nature Conservation, 11 (4), 240-250.
Cole, D. 2005a. "Computer simulation modeling of recreation use: current status, case studies, and future directions." Gen. Tech. Rep. RMRS-GTR-143, Department of Agriculture, Forest Service, Rocky Mountain Research Station, Fort Collins, CO, U.S.
Cole, D. 2005b. "Future Directions for Simulation of Recreation Use." Computer simulation modeling of recreation use: current status, case studies, and future directions. Gen. Tech. Rep. RMRS-GTR-143, D. Cole, ed., Department of Agriculture, Forest Service, Rocky Mountain Research Station, Fort Collins, CO, U.S., 3-9.
Cole, D. N., Gimblett, H. R., Cable, S., Itami, R. M., Lawson, S. R., Manning, R. E., Valliere, W., Wang, B., Newman, P., Mayo, K., Hallo, J. C., and Vande Kamp, M. 2005. "Case Studies of Simulation Models of Recreation Use." Computer simulation modeling of recreation use: current status, case studies, and future directions. Gen. Tech. Rep. RMRS-GTR-143, D. Cole, ed., Department of Agriculture, Forest Service, Rocky Mountain Research Station, Fort Collins, CO, U.S., 17-64.
Daniel, T. 2002. "Modeling Visitor Flow from the Visitor Perspective: The Psychology of Landscape Navigation." Monitoring and Management of Visitor Flows in Recreational and Protected Areas, Vienna, Austria, 159-165.
Deadman, P., and Gimblett, H. R. 1994. "The Role of Goal-Oriented Autonomous Agents in Modeling People-Environment Interactions in Forest Recreation." Mathematical and Computer Modelling, 20 (8), 121-133.
Donau-Auen Nationalpark. 2006 [http://www.donauauen.at/].
Elands, B. H. M., and van Marvijk, R. 2008. Keep an eye on nature experiences; implications for management and simulation. In this Volume.
Findlay, C., and Southwell, K. 2004. "'I just followed my nose': understanding visitor wayfinding and information needs at forest recreation sites." Journal of Managing Leisure, 9 (Oct), 227-240.
Gimblett, H. R., Durnota, B., and Itami, B. 1996. "Spatially-Explicit Autonomous Agents for Modelling Recreation Use in Complex Wilderness Landscapes." Complexity International, 3.
Gimblett, H. R., Itami, R. M., and Cable, S. 2005. "Recreation Visitation in Misty Fjords National Monument in the Tongass National Forest." Computer simulation modeling of recreation use: current status, case studies, and future directions. Gen. Tech. Rep. RMRS-GTR-143, D. Cole, ed., Department of Agriculture, Forest Service, Rocky Mountain Research Station, Fort Collins, CO, U.S., 22-27.
Gimblett, R. 2002. "Integrating Geographic Information Systems and Agent-Based Modeling Techniques for Simulating Social and Ecological Proccesses", Oxford University Press, New York
Gimblett, R., Daniel, T., Cherry, S., and Meitner, M. J. 2001. "The simulation and visualisation of complex human-environment interactions." Landscape and Urban Planning, 54, 63-78.
Gloor, C., Stucki, P., and Nagel, K. 2004. "Hybrid techniques for pedestrian simulations." ACRI 2004: 6th International Conference on Cellular Automata for Research and Industry Amsterdam, The Netherlands.
Golledge, R. G. 2001. "Behavioral Geography." International Encyclopedia of the Social &

Behavioral Sciences, Elsevier Science Ltd, 1105-1111.

Golledge, R. G., and Stimson, R. J. 1997. "Spatial Behavior: a Geographic Perspective", Guilford Press, New York, NY

Hinterberger, B. 2000. "Besucherstromanalyse im Wiener Anteil des Nationalpark Donau-Auen, der Lobau: Routenanalyse mit GIS," Master Thesis, Universität für Bodenkultur in Wien, Wien.

Hinterberger, B., Arnberger, A., and Muhar, A. 2002. „GIS-Supported Network Analysis of Visitor Flows in Recreational Areas." Monitoring and Management of Visitor Flows in Recreational and Protected Areas, Vienna, Austria, 28-32.

Institut für Demographie der Österreichischen Akademie des Wissenschaften. 2002. "Bevölkerungsvorausschätzung 2000 bis 2030 nach Teilgebieten der Wiener Stadtregion", Stadtentwicklung Wien, Magistratsabteilung 18, Wien. Werkstattberichte / Stadtentwicklung 49.

Itami, R., Raulings, R., MacLaren, G., Hirst, K., Gimblett, R., Zanon, D., and Chladek, P. 2002. „RBSim2: Simulating the Complex Interactions between Human Movement and the Outdoor Recreation Environment." Monitoring and Management of Visitor Flows in Recreational and Protected Areas, Vienna, Austria, 191-198.

Itami, R. M., and Gimblett, H. R. 2001. „Intelligent recreation agents in a virtual GIS world." Complexity International, 08.

Job, H. 1991. "Freizeit und Erholung mit oder ohne Naturschutz?: Umweltauswirkungen der Erholungsnutzung und Möglichkeiten ressourcenschonender Erholungsformen, erörtert insbesondere am Beispiel Naturpark Pfälzerwald", Bad Dürkheim: Pfalzmuseum für Naturkunde

Jochem, R., van Marvijk, R., Pouwels, R., and Pitt, D. G. 2008. MASOOR: modeling the transaction of people and environment on dense trail networks in natural resource settings. In this Volume.

Kajala, L., Almik, A., Dahl, R., Dikšaitė, L., Erkkonen, J., Fredman, P., Jensen, F. S., Karoles, K., Sievänen, T., Skov-Petersen, H., Vistad, O. I., and Wallsten, P. 2007. "Visitor monitoring in nature areas - a manual based on experiences from the Nordic and Baltic countries", Swedish Environmental Protection Agency, Nordic Council of Ministers. TemaNord. 534.

Koch, N. E. 1984. "Skovenes friluftsfunktion i Danmark. III. del. Anvendelsen af skovene, lokalt betragtet. (Forest Recreation in Denmark. Part III: The Use of the Forests Considered Locally)." Forstl. Forsøgsv. Danm, København, 39 (1984), 121-362.

Longley, P. A., Goodchild, M. F., Maguire, D. J., and Rhind, D. W. 2001. „Geographic Information Systems and Science", John Wiley & Sons, Ltd, Chichester, New York, Weinheim, Brisbaine, Singapore, Toronto

O'Connor, A., Zerger, A., and Itami, B. 2005. „Geo-temporal tracking and analysis of tourist movemnet." Mathematics and Computers in Simulation, 69, 135-160.

Skov-Petersen, H. 2005. „Feeding the agents - collecting parameters for agent-based models." Computers in Urban Planning and Urban Management, London, paper nr 60.

Skov-Petersen, H., and Snizek, B. 2008. Probability of Visual Encounters. In this Volume.

Ward, J. H. 1963. „Hierarchical grouping to optimize an objective function." Journal of the American Statistical Association (58), 236.

Zundel, R., and Völksen, G. 2002. "Ergebnisse der Walderholungsforschung : Eine vergleichende Darstellung deutschsprachiger Untersuchungen", Dr. Kessel, Oberwinter.

CHAPTER 10

MEASURING, MONITORING AND MANAGING VISITOR USE IN PARKS AND PROTECTED AREAS USING COMPUTER-BASED SIMULATION MODELING

Steven R. Lawson
Jeffrey C. Hallo
Robert E. Manning

Abstract: Recent research has identified at least five ways in which computer simulation can facilitate more informed planning and management of protected natural areas, including 1) describing existing visitor use conditions; 2) monitoring "hard to measure" indicator variables; 3) "proactively" managing carrying capacity; 4) testing the effectiveness of alternative visitor use management practices; and 5) guiding the design of research on public attitudes. The purpose of this chapter is to demonstrate, using empirical data from studies conducted in several U.S. protected natural areas, each of these five potential contributions of computer simulation to sustainable management and planning of visitor landscapes.

Keywords: Indicators of quality, standards of quality, wilderness, alternative transportation

INTRODUCTION

The popularity of outdoor recreation in parks and protected areas has grown substantially in recent decades. For example, the U.S. national park system now accommodates over 280 million visits per year, while the U.S. national forests receive over 200 million annual visits. This represents both an opportunity and a challenge. The opportunity is to provide visitors with high quality experiences that enhance appreciation and public support for parks and related areas. The challenge is to protect the natural and cultural resources of these areas and the quality of the visitor experience in the face of increasing recreational use. For example, high levels of visitor use can lead to trail erosion, loss of ground cover vegetation at and around campsites, extensive networks of social trails, and disturbance of wildlife (Hammitt & Cole, 1998). In addition, increased recreational use of these areas can cause crowding and conflict, degrading the quality of the visitor experience (Manning, 1999).

To effectively manage parks and related areas, information is needed about visitor use. For example, measures of the spatio-temporal distribution of visitor use can help managers identify potential recreation-related threats to the natural and cultural resources of an area and the quality of the visitor experience. While in some cases it may be possible to measure and monitor visitor use through "on-the-ground" observation, this becomes increasingly difficult in larger areas where recreation tends to be highly dispersed. Consequently, efficient measurement and monitoring of visitor use in national parks, wilderness and related areas is important but challenging.

Recent research suggests that computer-based simulation modeling can be an effective tool for facilitating planning and management of visitor use in outdoor recreation areas, such as national parks, wilderness, and related areas (Daniel & Gimblett, 2000; Gimblett et al., 2001; Lawson et al., 2006; Lawson & Manning, 2003a; Lawson et al., 2003a; Wang & Manning, 1999). This research has identified five ways in which simulation modeling of recreation use can facilitate more informed management.

First, simulation modeling can be used to measure existing visitor use levels and patterns. That is, given current management practices and existing levels of visitor use, where and when is visitor use occurring in a park or protected area? By providing managers with detailed information about how visitors are currently using an area, this baseline information can assist managers in identifying "hotspots" that warrant management attention. For example, does visitor use tend to concentrate in certain locations or at certain times within a recreation area, which may lead to crowding or conflicts among different types of visitors? Is visitor use occurring within zones that contain fragile ecological and/or cultural resources that are highly sensitive to recreation use?

Second, simulation modeling can be used to monitor the condition of indicator variables that are inherently difficult to measure through direct observation (Lawson et al., 2006; Wang & Manning, 1999). For example, how many encounters do backpacking parties have with other groups per day while hiking? How does the number of people at a popular attraction site change throughout the course of a day, a visitor use season, or with increasing or decreasing levels of visitor use?

Third, contemporary park and outdoor recreation carrying capacity and related management frameworks rely on monitoring indicators to ensure that minimum acceptable conditions (i.e., standards) are maintained (Manning, 2001; Manning, 2007). Computer simulation modeling provides a tool to "proactively" manage carrying capacity by providing estimates of the number of people that can be allowed to visit an outdoor recreation area without violating standards for crowding-related indicators (Lawson et al., 2003a; Hallo et al., 2005a). Such estimates of carrying capacity can be used to design management strategies, such as a trailhead quota or permit system, that ensure crowding-related standards are maintained.

Fourth, simulation modeling can be used to test the effectiveness of alternative management practices in a manner that is more comprehensive, less costly, and less politically risky than "on-the-ground" trial and error (Lawson & Manning, 2003a). For example, what effect does a permit quota have on the number of encounters visitors have with other groups while hiking? How many new campsites would need to be built in order to ensure that visitors do not have to share campsites with other people not in their group? How do alternative transportation systems affect the density of visitor use along trails and at attraction sites? How does the addition of a new trail or road affect the level and pattern of visitor use?

Fifth, simulation modeling data can be used to guide the design of more realistic research on public attitudes concerning the management of visitor use in parks and recreation areas (Lawson & Manning, 2003b; Lawson et al., 2003b). For example, rather than asking the public to evaluate purely hypothetical management scenarios, questions can be designed to measure public attitudes about management alternatives that simulation modeling suggests are feasible, effective and, therefore, realistic.

The purpose of this chapter is to demonstrate, using data from several studies, the above five broad uses of computer simulation modeling in parks and protected areas. The next section of the chapter briefly describes the computer simulation modeling approach used in these studies. This is followed by five applications of computer simulation modeling designed to illustrate the five uses outlined above.

SIMULATION MODELING

Modeling approaches that are dynamic, discrete, and stochastic are best suited for simulating visitor use in parks and related areas, since most outdoor recreation systems (e.g., trail networks, campgrounds, scenic roads) possess these characteristics. Dynamic models are those that represent the internal interactions of a system as they change over time (Banks et al., 2001). For example, a simulation model designed to measure use levels on a trail throughout a day would be considered a dynamic model. Dynamic models in which the values of variables change at separated points of time, when an event occurs, are referred to as discrete-event simulation models (Banks et al., 2001). Simulation of hikers on a trail system is a good example of a discrete-event model, since the number of hikers on the trail system changes only when an event occurs (i.e., a hiker enters or leaves the trail system). In contrast, the values of variables in a continuous model change continuously over time. A model of stream flow is a good example of a continuous simulation model, in that stream flow changes continuously over time (Wang & Manning, 1999). In a stochastic simulation model, some of the components of the system being modeled are based on probability distributions in order to account for variation within the system (Banks et al., 2001). For example, the number of visitors that pass through an entrance kiosk and into a park varies throughout a day, from day to day, and throughout the visitor use season. To account for this variability, a stochastic simulation model generates simulated visitors based on an empirical or theoretical probability distribution.

The simulation models described in this chapter use Extend (Version 6.0), a software package developed by Imagine That, Incorporated (Diamond et al., 2002), to develop probabilistic, discrete-event simulations (Law & Kelton, 2000). Extend is a commercially-available, generalized simulation modeling software package. Extend provides modelers with a library of "blocks" that each perform a defined function. These blocks may be used to represent "parts" of a park or protected area. For example, a "generator block" represents visitors arriving at a park entrance. This block can be modified to vary how often visitors arrive, at what time they arrive, or the stochasticity of arrivals. A "generator block" may be linked in Extend to other types of blocks (with other functions) that together simulate more complex park or protected area processes.

The models built using Extend have relied primarily on empirical data about visitor use to determine the movement of visitors in a park or protected area. For example, the number and frequency of visitors entering an area is simulated based on "headway counts" – field-based counts of the number of visitors entering into a park or protected area. For example, headway counts for a trail system could be recorded by placing a counter at a trailhead and recording the number of visitors entering at specific time intervals throughout the day. Where visitors go in a simulation model and how fast they travel (i.e., their travel route) is also collected empirically. These data may be gathered by asking visitors to record their travel routes on paper map diaries or by having visitors carry a Global Positioning System (GPS) unit with them as they travel through a park (Hallo et al., 2005b). Rules (which could be based on expert opinions, physical attributes, field-based data, theories, hypotheses, or policies) may be built into the model that will also guide where simulated visitors go or can go and their travel behavior in the model.

Simulation models of visitor use in parks and related areas are often designed to provide outputs related to visitor numbers. For example, models have been designed to estimate the number of visitors at one time at attraction sites and along roads and trails, and the number of encounters between groups of visitors. However, simulation models of parks and related areas may be built to provide many types of outputs that can be expressed in numerical and graphic formats. The types of outputs are ultimately based on the objectives of the simulation and the interests of managers and researchers.

CASE STUDIES OF SIMULATION MODELING

This section of the chapter provides a series of case studies designed to demonstrate the five uses of computer simulation modeling in park and protected area planning and management outlined above.

Measuring Visitor Use Levels and Patterns

The John Muir Wilderness covers 584,000 acres in the Sierra and Inyo National Forests, in the Sierra Nevada Mountains of California, USA. The area can be characterized as an alpine environment, with backcountry camping opportunities alongside high elevation lakes and in alpine meadows. Backpacking and horseback riding trips constitute the majority of visitor use of the area, however, day use occurs in the area as well. A computer-based simulation model of visitor use was developed for a portion of the Humphrey's Basin area of the John Muir Wilderness Area (Lawson et al., 2006).

A series of simulations was conducted with the model of visitor use in Humphrey's Basin, including a "Baseline Simulation". The "Baseline Simulation" was designed to generate spatially explicit estimates of hiking and camping use in the area under existing visitor use levels and management. The specific outputs generated by the "Baseline Simulation" included: 1) Average hiking use per day, by trail segment; and 2) Average camping use per night, by camping location. For the purposes of this chapter, only the outputs related to hiking use will be presented. Refer to Lawson et al. (2006) for additional information about the simulation outputs.

The results of the "Baseline Simulation" presented in Table 1 suggest that under existing conditions, hiking densities are low throughout most of the study area, with moderate levels of visitor use along several trail segments. These results illustrate how computer simulation can be used to provide managers with spatially explicitly outputs that describe existing visitor use levels and patterns. Contemporary outdoor recreation planning frameworks including the Limits of Acceptable Change (LAC) (Stankey et al. 1985) and Visitor Experience and Resource Protection (VERP) (Manning, 2001; National Park Service, 1997) rely on a zoning approach that involves prescribing alternative visitor use, resource, and management conditions for different sections of a protected area (Manning,

Table 1. Average hiking use, by trail segment – Humphrey's Basin

Trail ID	Mean Use a	Trail ID (cont.)	Mean Use a (cont.)
2	3.51	23	0.09
3	0.08	24	0.13
4	3.51	25	2.31
5	3.43	26	0.15
6	0.58	27	1.08
7	0.14	28	0.15
8	0.04	29	0.45
9	3.35	30	1.29
10	3.28	31	0.68
11	3.20	32	0.63
12	0.12	33	0.04
13	0.20	34	1.87
14	0.80	35	0.07
15	2.95	36	1.43
16	1.10	37	0.29
17	2.47	38	0.88
18	2.41	39	1.29
19	0.15	40	0.22
20	0.99	41	1.25
21	0.90	132	0.06
22	0.77		

1999). The spatially explicit nature of computer simulation outputs like those generated in the "Baseline Simulation" helps managers to assess the extent to which existing visitor use is consistent with management zoning prescriptions.

The map in Figure 1 portrays the spatial distribution of hiking use within the study area for the "Baseline Simulation". While the data in Table 1 suggest that use throughout the study area is low, the map helps to illustrate the *relative* density of hiking use. Specifically, thicker lines on the map correspond to higher use trail segments, while thinner lines correspond to lower use segments. To generate the map of baseline visitor use, tabular computer simulation outputs were exported to a GIS database. In this way, the geo-referenced estimates of visitor use generated by the simulation model allow managers to integrate information about visitor use patterns with other resource data to perform land suitability and overlay analyses.

Monitoring Hard to Measure Variables

Zion National Park, Utah is known for its red rock cliffs, deep canyons, and desert landscape. Approximately 2.5 million visitors come to the park annually, and many of these visitors take advantage of hiking opportunities within the park. Weeping Rock Trail is one of several day hiking trails that is highly accessible to visitors.

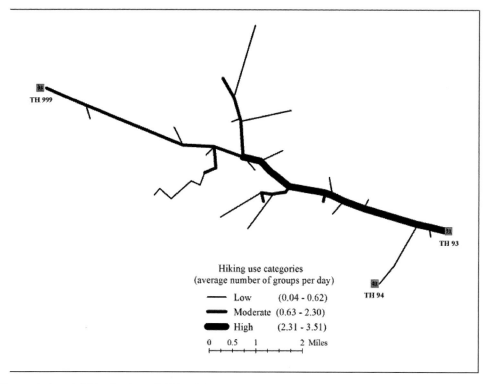

Figure 1. Spatial Distribution of Hiking Use – Humphrey's Basin

A computer simulation model of visitor use on the Weeping Rock Trail was developed to measure the number of encounters among hikers. Trail encounters are often difficult to measure since they occur at unknown intervals and at unknown locations. This makes field-based measurements of encounters difficult. However, a simulation model can use input data (e.g., headway counts and travel routes) to simulate trail use and estimate the number of encounters among simulated hikers. The purpose of the simulation model of Weeping Rock Trail was to measure trail encounters at current use levels and to understand the effect of increased use levels on encounters. The trail was modeled as four sections, labeled A through D.

Table 2 shows results from the simulation model for the number of encounters per hiker on different trail segments. Encounter numbers are shown for 0.5, 1, 2, and 4 times the current average daily use level (128 hikers per day). At current use levels, the average number of encounters visitors have while hiking on the Weeping Rock trail ranges from 5.6 on Trail Section B to 8.6 on Trail Section A. Results indicate that hikers who use parts of the trail closer to the trailhead (Trail Sections A and B) will have more encounters than hikers

Table 2. Average number of encounters per visitor on the four segments of Weeping Rock Trail at four use levels.

Proportion of Current Mean Daily Use Level (128 people per day)				
Trail Section	0.5	1	2	4
Trail Section A	4.3	8.6	17.8	34.3
Trail Section B	4.2	8.3	16.7	32.3
Trail Section C	3.1	6.3	12.5	24.7
Trail Section D	2.7	5.6	11.1	21.2

who take advantage of more distant portions of the trail (Trail Sections C and D). However, hikers who use the more distant portions of the trail still encounter a substantial number of other hikers. Also, simulation model results indicate that encounters will approximately double every time use levels double. For example, if use were to increase from the current average of 128 hikers per day to 256 hikers per day, then average hiking encounters on Trail Section C would increase from 6.3 to 12.5. This information provides managers with an empirical basis for understanding the relationship between visitor use levels and inter-group encounters. In this case, the relationship is more or less linear. However, it is likely in other cases to be a non-linear relationship that can be described relatively precisely with computer simulation model outputs like those presented in this case study.

"Proactively" Estimating Carrying Capacity

Muir Woods National Monument is a part of Golden Gate National Recreation Area in the San Francisco Bay area, California. Its establishment in 1908, as a tribute to famed preservationist John Muir, protects a cathedral-like grove of virgin redwood trees. Today, approximately three-quarters of a million visitors come to Muir Woods each year to see these trees and experience the park. The majority of visitor use is concentrated on just a few miles of paved and unpaved trails.

Simulation modeling can be used to estimate the maximum amount of visitor use that can be accommodated in a park without violating crowding-related standards (Manning, 2007). A simulation model of visitor use at Muir Woods was developed for this purpose. Crowding-related standards used in the model were formulated from a survey of visitors to Muir Woods. Two series of computer-edited photographs – one for the more accessible paved trails and one for the more remote unpaved trails – were incorporated into the visitor survey. Each series of study photographs showed a range of Persons per Viewscape (PPV) along a section of the trail (Manning et al., 2002/2003). Respondents were asked to rate the acceptability of each photograph on a scale from +4 ("very acceptable") to -4 ("very unacceptable") based on the number of visitors shown in the trail viewscape. Average acceptability ratings for the sample as a whole for each set of photographs are graphed in Figure 2. These graphs show that respondent ratings fall out of the acceptable range and into the unacceptable range (i.e., cross the neutral point of the acceptability scale) at 16.2 PPV for the paved trails and 6.8 PPV for the unpaved trails. The findings concerning the relative sensitivity of respondents to PPV on paved trails versus unpaved trails are consistent with intuition, given that the unpaved trails are representative of a backcountry setting while the paved trails convey to respondents a frontcountry environment. These values represent potential crowding-related standards.

The Muir Woods simulation model was run for a total of 20 simulated days at several use levels: at the current use level, 2 times the current use level, and 4 times the current use level. The output from these runs was the percentage of time that the

Table 3. Simulation model estimates of the percentage of time visitor use on Muir Woods' trails violate standards for crowding.

	Percent of time crowding-related standard is violated	
Multiple of Current Use	Paved Trails	Unpaved Trails
1	0.1%	0.0%
2	4.8%	0.4%
4	35.7%	3.4%

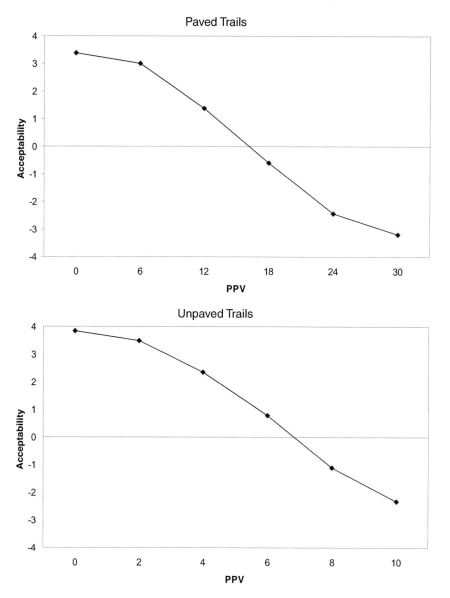

Figure 2. Average acceptability ratings for a range of persons-per-viewscape (PPV) conditions on paved and unpaved trail at Muir Woods National Monument.

standards of quality noted above were violated. Model results (Table 3) indicate that at current average use levels (3,313 visitors/day) crowding-related standards are almost never violated on either paved or unpaved trails. However, if use level doubles, then the model estimates that crowding-related standards would be violated 4.8% of the time on paved trails but only 0.4% of the time on unpaved trails. If use level were to double again, the crowding-related standards would be violated over a third of the time on paved trails. These simulation model results were used to calculate the number of visitors that could enter the park per day without violating crowding-related standards more than 10% of the time (the 10% value was used to allow for occasional peaking of use during weekend and holiday periods). Linear interpolation was applied to the results in Table 3 to calculate the factor of current use level at which the crowding-related standards would be violated 10% of the time. Finally, this factor of current use level was multiplied by the average current daily use level (3,313 visitors/day) to estimate a maximum number of visitors that could be accommodated in the park each day. Results (rounded to the nearest hundred) indicate that paved trails are the limiting trail type in managing visitor use for crowding (Table 4). The model suggests that approximately 7,700 visitors could use Muir Wood's paved trails per day without violating the "acceptability"-based crowding-related standards more than 10% of the time.

Testing Alternative Management Practices

Unusual rock formations and sandstone arches draw over three quarters of a million visitors to Arches National Park, Utah each year. One of the primary features that visitors come to see is Delicate Arch. Research conducted to examine the experience of visiting the park found that People at One Time (PAOT) in the viewing area at Delicate Arch was an important indicator of quality for the visitor experience (Manning, et al., 1995). Further research, using the visual techniques described above, found that an appropriate standard for

Table 4. Number of visitors that may enter Muir Woods per day without violating crowding-related standards more than 10% of the time.

	Number of visitors accommodated without violating crowding-related standards
Paved Trails	7,700
Unpaved trails	>13,200

Table 5. Estimated number of visitors that could hike to Delicate Arch per day without violating a crowding-related standard under alternative shuttle bus delivery systems.

Shuttle Bus Arrival Interval (minutes)	Percent Increase in Use at Delicate Arch without Violating Crowding Standards (30 PAOT)
60	29%
30	47%
15	68%

PAOT at Delicate Arch was a maximum of 30 (Manning, et. al, 1995).

How might visitor use be managed to help ensure that there are no more than 30 PAOT at Delicate Arch? One approach would be to require that visitors use a shuttle bus system to access the Delicate Arch trailhead, and then design this huttle bus system to deliver the "right" number of visitors at the "right" times. To help design this type of shuttle bus system, a simulation model of visitors at Arches National Park and Delicate Arch was constructed (Lawson et al., 2003a). The model used input data on the number of visitors entering the park and walking to Delicate Arch each day and travel routes of park visitors and hikers. The primary output of the model was PAOT at Delicate Arch.

If visitors continue to drive to the Delicate Arch trailhead, the model estimates that a maximum of 315 visitors could hike to Delicate Arch each day without violating the standard of 30 PAOT. However, the model also estimates that the number of visitors to Delicate Arch could be increased by 29 to 68 percent if a shuttle bus system were implemented (Table 5). For example, a shuttle bus system that arrived at the trailhead every 30 minutes could increase the number of visitors who could hike to Delicate Arch from 315 to 462. Model results illustrate the progressive increases

Table 6. Simulation model results showing the number of minutes out of an hour that a visitor would see a range of PPV conditions.

Range of PPV Conditions	Carriage Road Use Level					
	375	750	1,500	3,000	6,000	12,000
0 PPV	57	55	48	40	28	17
1 to 5 PPV	3	5	11	18	26	28
6 to 10 PPV	0	0	1	2	5	10
11 to 15 PPV	0	0	0	0	1	3
16 to 20 PPV	0	0	0	0	0	1
21 to 30 PPV	0	0	0	0	0	1

in visitor capacity that are possible with more frequent delivery schedules.

Guiding Research on Public Attitudes Toward Management

Acadia National Park, Maine was created largely due to efforts and donations from wealthy philanthropists. John D. Rockefeller, Jr., one key benefactor, donated land, buildings, and carriage roads to be included in the park. The carriage roads he donated are now one of the primary attractions that draw over 2 million visitors to the park each year. Preliminary research about visitor use on the carriage roads found that use levels and related perceptions of crowding were an important indicator for the quality of the visitor experience (Jacobi & Manning, 1999). A standard for the minimum acceptable condition for use levels/crowding was needed to protect the experience of visitors on the carriage roads.

Research on crowding-related standards for the carriage roads was guided by development of a computer simulation of visitor use. The model uses data on visitor use levels and patterns, including headway counts and travel routes, to estimate the range of PPV that will be found along the carriage roads. For example, the model estimates that when 1,500 visitors use the carriage roads per day (the approximate average current use level), a visitor on the carriage roads would typically see no other visitors (0 PPV) for 48 minutes out of an hour, one to five PPV for 11 minutes out of an hour, and six to ten PPV for one minute out of an hour. The model was run to obtain this type of output for a range of five levels of daily carriage road use: 750; 1,500; 3,000; 6,000; and 12,000 visitors.

Output data from these model runs are presented in Table 6. These data were organized into five scenarios that described the PPV conditions that would be experienced along the carriage roads as a result of a range of carriage road use levels. The scenarios were then incorporated into a survey of carriage road visitors whereby respondents were asked to rate the acceptability of each scenario. Study findings are shown in Figure 3. From this graph, it is clear that acceptability falls as PPV levels increase. Average visitor ratings fall out of the acceptable range and into the unacceptable range at about scenario 4, or approximately 6,000 visitors per day. This estimate is based on visitor evaluations of a range of realistic PPV conditions as derived from the simulation model.

DISCUSSION

Simulation modeling has become substantially more accessible through advances in

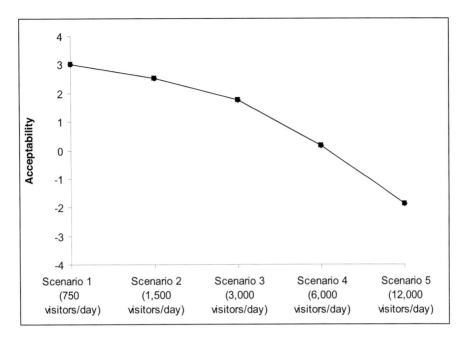

Figure 3. Average acceptability ratings for a range of persons-per-viewscape (PPV) conditions on the carriage roads of Acadia National Park.

computer hardware and software. Simulation modeling can now be performed on desktop computers with relatively inexpensive software programs. This accessibility has led to wide application in a variety of applied fields, including management of parks and related areas. These applications have begun to demonstrate the multiple ways in which simulation modeling can be useful in planning and managing parks and protected areas.

The case studies outlined in this chapter suggest five ways in which simulation modeling has been found to be useful in measuring, monitoring, and managing visitor use in the U.S. national parks and protected areas. These ways are (1) measuring visitor use levels and patterns, both spatially and temporally; (2) monitoring variables that are difficult to measure "on the ground"; (3) proactively carrying capacities for an area; (4) testing the effectiveness of alternative management practices, and (5) informing research on public attitudes toward park and protected area management. Additional uses – such as helping to improve communication of ideas between managers and the public – have been suggested by others (Cole, 2005). Furthermore, computer simulation outputs can be integrated into Geographic Information System (GIS) databases and used in overlay analyses to examine relationships between natural resource characteristics (e.g., resource fragility, resource impacts) and visitor use patterns (Lawson et al., 2006). Uses for simulation modeling and its overall utility are likely to grow as simulation modeling is applied to other locations, contexts, and issues in parks and protected areas.

While the case studies presented in this chapter demonstrate the potential utility of computer simulation modeling for planning and managing parks and protected areas, there are several challenges that need to be addressed through further research. A primary challenge has to do with collecting reliable, representative data with which to construct computer simulation models. Due to the dispersed nature of outdoor recreation in many protected natural areas contexts, it is often difficult to capture travel itineraries and other data needed to construct a computer simulation model using on-site survey techniques alone. Further research and coordination with land managers is needed to identify methods for collecting trip and visitor characteristics data in a reliable and consistent manner. Additional research to develop alternative data collection methods, including GPS, automatic timing systems (i.e., race technology), mechanical counters, and related technologies is warranted. Furthermore, in areas where permits are issued, researchers should work with managers to develop standardized permit applications with questions designed to collect the information needed to construct valid computer simulation models.

Validation of computer simulation model outputs presents another significant challenge to be addressed in future research. Existing applications of computer simulation to recreation management and planning, including the examples presented in this paper, have generally lacked or been limited in the use of quantitative validation techniques (Law & Kelton, 2000). Furthermore, while there is a relatively extensive body of literature describing validation techniques for simulation models of manufacturing systems, there is a lack of recent research concerning the appropriateness of alternative statistical techniques for validation of computer simulation models of visitor use in parks and protected areas. Consequently, more research is warranted to develop standardized methods and procedures to assess the validity of computer simulation models designed for outdoor recreation management.

Similarly, previous applications of computer simulation modeling to park and protected area planning and management have generally done little to assess the reliability of model estimates. Yet the issue of reliability is particularly important because computer simulation modeling uses random numbers to generate input variables (e.g., visitor arrival times, durations at destinations) and therefore the estimates from a model vary across replications of the model. Consequently, conclusions should not be drawn from a single replication of a model (Law & Kelton, 2000). Thus, a significant issue that should be addressed within future applications of computer simulation modeling to parks and protected areas is identifying the number of simulation replications needed to estimate model outputs at a level of precision that is useful for management purposes.

Finally, an underlying assumption of the computer simulation models presented in this chapter is that visitor headways and travel route distributions will remain the same, even as use levels rise in the study areas modeled. Since people's recreation behavior can be driven by the amount of time they have and the location of key attractions (e.g., water bodies, vistas, mountain tops), these assumption are probably reasonable in many contexts. However, these assumption may become more problematic for dramatic increases in overall use levels in which case use densities may override the attractiveness of specific landscape features and, to a lesser extent, time constraints in determining people's recreation behavior. In cases where the scenarios to be simulated deviate dramatically from existing conditions, it may be more appropriate to use a rule-based simulation approach. With this approach, simulated visitor behavior is driven by a set of rules that take into account such factors as the number of other visitors in selected locations, time constraints, and the attractiveness of landscape features. While it is possible to develop rule-based simulation models in the range of software packages the authors are familiar with, including Extend, a primary challenge of rule-based simulation lies in defining rules that accurately represent human behavior. If rules of human behavior are not specified correctly, results of rule-based simulations will not be valid. Consequently, more research is needed to develop empirical bases for defining context-specific and generalizable rules of recreation behavior within models of outdoor recreation in parks and protected areas.

CONCLUSION

Visitor use constitutes one of the primary "agents of change" in parks and protected areas. Thus, park and protected area management and planning requires information about visitor use of these areas. In many cases, managers of protected natural areas have had to rely on intuition and best guesses about visitor use and related social and resource conditions within the areas they manage. As the examples presented in this chapter illustrate, computer simulation has significant potential to provide a more informed basis for planning and management of visitor use in parks and protected areas.

To ensure that computer simulation models provide valid and reliable information in a cost-effective manner, it is important that state-of-the-art knowledge, procedures, and methods be adopted for simulation of visitor use in parks and protected areas. A primary product of future research should be a set of technical standards and procedures that guide future applications of computer simulation to visitor use in parks and protected areas. These goals can be achieved, in part, though collaboration with computer simulation scientists from other disciplines and by drawing upon existing literature that addresses computer simulation modeling and analysis.

ACKNOWLEDGMENTS

The authors would like to thank the following people for their contributions to the work presented in this chapter: William Valliere and Benjamin Wang, University of Vermont; Bob Itami, GeoDimensions Inc.; Randy Gimblett, University of Arizona; David Cole, Aldo Leopold Wilderness Research Institute; Charlie Jacobi, Jeff Bradybaugh, Mike Savidge, Mia Monroe, Nancy Hornor, and Karen McKinley-Jones of the National Park Service.

LITERATURE CITED

Banks, J., Carson, J., Nelson, B., & Nicol, D. (2001). *Discrete-event system simulation*. Upper Saddle River, NJ: Prentice Hall.

Cole, D. N. (2005). *Computer simulation modeling of recreation use: Current status, case studies, and future direction*. USDA Forest Service General Technical Report RMRS-GTR-143.

Daniel, T., & Gimblett, R. (2000). Autonomous agents in the park: An introduction to the Grand Canyon river trip simulation model. *International Journal of Wilderness*, 6(3), 39-43.

Diamond, B, Lamperti, S., Krahl, D., & Nastasi, A. (2002). Extend (Version 6.0) [Computer Software]. San Jose, CA: Imagine That Inc.

Gimblett, R., Richards, M., & Itami, R. (2001). RBSim: Geographic simulation of wilderness recreation behavior. *Journal of Forestry*, 99(4), 36-42.

Hallo, J., Manning, R., & Valliere, W. (2005a). Acadia National Park scenic roads: Estimating the relationship between increasing use and potential standards of quality. In D. N. Cole, *Computer simulation modeling of recreation use: Current status, case studies, and future direction* (pp. 55-57). USDA Forest Service General Technical Report RMRS-GTR-143.

Hallo, J., Manning, R., Valliere, W., & Budruk, M. (2005b). A case study comparison of visitor self-reported and GPS recorded travel routes. *Proceedings of the 2004 Northeastern Recreation Research Symposium*. USDA Forest

Service General Technical Report NE-326, 172-177.

Hammitt, W. & Cole, D. (1998). *Wildland recreation: Ecology and management*. New York: John Wiley & Sons, Inc.

Jacobi, C., & Manning, R. (1999). Crowding and conflict on the carriage roads of Acadia National Park: An application of the Visitor Experience and Resource Protection Framework. *Park Science*, 19(2):22-26.

Law, A. & Kelton, W. (2000). *Modeling simulation and analysis* (3rd ed.). Boston: McGraw Hill.

Lawson, S., Itami, B., Gimblett, R. & Manning, R. (2006). Benefits and challenges of computer simulation modeling of backcountry recreation use in the Inyo National Forest. *Journal of Leisure Research*, 38(2), 187-207.

Lawson, S. & Manning, R. (2003a). Research to inform management of wilderness camping at Isle Royale National Park: Part I – descriptive research. *Journal of Park and Recreation Administration*, 21(3), 22-42.

Lawson, S. & Manning, R. (2003b). Research to inform management of wilderness camping at Isle Royale National Park: Part II – prescriptive research. *Journal of Park and Recreation Administration*, 21(3), 43-56.

Lawson, S., Manning, R., Valliere, W., & Wang, B. (2003a). Proactive monitoring and adaptive management of social carrying capacity in Arches National Park: An application of computer simulation modeling. *Journal of Environmental Management*, 68, 305-313.

Lawson, S., Mayo-Kiely, A., & Manning, R. (2003b). Integrating social science into park and wilderness management at Isle Royale National Park. *George Wright Forum*, 20(3), 72 – 82.

Manning, R. (1999). *Studies in outdoor recreation: Search and research for satisfaction*. Corvallis: Oregon State University Press.

Manning, R. (2001). Visitor experience and resource protection: A framework for managing the carrying capacity of national parks. *Journal of Park and Recreation Administration*, 19(1), 93-108.

Manning, R. (2007). *Parks and carrying capacity: Commons without tragedy*. Washington, D.C.: Island Press.

Manning, R., Lime, D., Hof, M., & Freimund, W. (1995). The Visitor Experience and Resource Protection (VERP) process: The application of carrying capacity to Arches National Park. *The George Wright Forum*, 12(3): 41-55.

Manning, R., Valliere, W., Wang, B., Lawson, S., & Newman, P. (2002/2003). Estimating day use social carrying capacity in Yosemite National Park. *Leisure/Loisir*, 27(1-2), 77-102.

National Park Service (1997). *VERP: The visitor experience and resource protection (VERP) framework: A handbook for planners and managers*. USDI National Park Service Technical Report.

Stankey, G., Cole, D., Lucas, R., Peterson, G., Frissell, S. & Washburne, R. (1985). *The limits of acceptable change (LAC) system for wilderness planning*. USDA Forest Service General Technical Report INT-176.

Wang, B., & Manning, R. (1999). Computer simulation modeling for recreation management: A study on carriage road use in Acadia National Park, Maine, USA. *Environmental Management*, 23, 193-203.

CHAPTER 11

CHOICE MODELS: ESTIMATION, EVOLUTION, LIMITATIONS AND POTENTIAL APPLICATION FOR SIMULATION MODELING

Len M. Hunt

Abstract: Researchers have used choice models for over twenty-five years to study the behaviors of outdoor recreationists. Early applications of choice models represented a crude and rigid method to investigate these behaviors. Constraints of early models include limited ways to account for: varying preferences among decision-makers; substitution effects among choice alternatives (e.g., recreational sites); joint models of participation and site choice; awareness of alternatives; and dynamic-like decision-making. Advancements in methods and computers now permit researchers to develop flexible choice models that shed many of their past limitations. This chapter provides a critical review of choice models and their advancements. A number of examples are provided that aid in illustrating necessary theories and data requirements for choice model development and the intuition behind model estimation. Innovations are discussed to demonstrate the increasing flexibility of these models. Discussion will also focus on deficiencies of choice models at predicting behaviors and emerging research that examines these deficiencies. Finally, the potential of using choice models as a simulation tool for simulating and predicting outdoor recreation behaviors is discussed.

Key Words: Choice Models, Decision-Makers, Outdoor Recreation, Random Utility, Behavior

INTRODUCTION

Simulation modeling of outdoor recreation behaviors can benefit from the tremendous amount of work that has developed and applied theories of decision-making on outdoor recreation topics. Choice modeling is one development that researchers use extensively to examine outdoor recreation behaviors. Researchers use choice models to forecast the amount, timing, location, and economic value of outdoor recreation under a wide array of management scenarios. In fact, hundreds of peer-reviewed articles exist that apply choice models to study the behaviors of individuals who pursue outdoor recreation activities.

Many researchers who study outdoor recreation have little familiarity with the methods and application of choice models. This unfamiliarity arises since resource economists conduct most outdoor recreation applications and communicate their findings through economics journals. Interesting methodological developments to choice model methods are also found in transportation and marketing literature.

This chapter attempts to familiarize readers with choice models. While the chapter is specifically tailored to the study of outdoor recreational behaviors, much of the text provides relevant considerations for conducting any choice model application.

The chapter begins by defining choice models and the necessary theoretical assumptions. The following section describes data requirements and approaches to consider when estimating models with many different alternatives. The strengths and weaknesses of estimating models with observed and hypothetical data are discussed in the third section. Section four discusses ways of transforming the theoretical considerations into statistical models. Estimation of choice models is discussed in section five. Section six details approaches to address issues that arise when studying outdoor recreation behavior. Section seven discusses past attempts to marry choice and simulation modeling approaches and discusses new opportunities to strengthen this marriage. Concluding remarks are presented in the final section.

WHAT IS CHOICE MODELING?

Before answering the question 'what is choice modeling?', let us address the issue of choosing. Choosing is a common task that we all do on a daily basis. In fact, our choices can be overwhelming. What should we have for breakfast? What clothes (if any) should we wear? How should we travel to a destination? Many of these choices are of trivial importance to us while other choices may significantly affect our life (e.g., buying a home, deciding on a career). Choice models focus on the task of choosing, that is selecting one distinct alternative from a set of alternatives (e.g., one park from a set of parks).

What do we know about the process of choosing? Well, depending upon one's disciplinary perspective, one makes different assumptions about this process. For example, some social psychologists believe that choosing, or behavior more generally, arises from values that shape attitudes and norms (i.e., informal rules of behaviors). These attitudes and norms in turn influence behavioral intentions, which influence behaviors such as choices (Fishbein and Ajzen, 1975)[1].

A common economic perspective assumes that people seek to maximize utility when choosing among alternatives. Utility is an unobserved (i.e., latent) measure of well-being for an individual. In most instances, utility arises from the attributes that describe an alternative and not simply the alternative itself. For example, utility for an automobile is associated with more than the functionality of the vehicle. Attributes such as price, dependability, style, mileage, and brand all affect the overall utility for a specific vehicle. Consequently, the purchase of an automobile represents the purchase of a bundle of attributes (Lancaster, 1966). Utility, therefore, arises from attributes of alternatives and preferences for these attributes. With choice models, the researcher's goal is to estimate people's preferences for attributes from empirical data based on their chosen alternatives.

Many readers will note that people are not optimizers. People are satisficers who do not have the cognitive abilities or the interest to integrate all information to

---

[1] This proposed relationship between behavior and attitudes and norms is the Theory of Reasoned Action. More recently, Ajzen (1985) has expanded this theory into a Theory of Planned Behavior by incorporating perceived behavioral control as another element that affects behavioral intentions.

make optimal decisions. Instead, these individuals have a bounded rationality (Simon, 1955) that leads them to choose alternatives that are sufficiently good rather than optimal. Most choice model researchers agree that decision-makers have a bounded rationality. McFadden (2000) notes that assuming choices are made from utility maximization is often justifiable. This justification is based on the ability of choice models to provide reasonably accurate forecasts of actual choice behaviors. Some other researchers (e.g., Gilbride and Allenby (2004)) account for bounded rationality by modeling decision-making as a two-stage process. Individuals are assumed to have a bounded rationality when identifying a set of alternatives from all possible alternatives that they will consider choosing. The second stage involves using utility maximization to choose one alternative from the set of considered alternatives.

Random utility (Thurstone, 1927) is a second important theory that governs most choice models. Thurstone developed random utility theory to account for variations in an individual's response under identical situations. He assumed that individuals take a random draw of utility on each choice occasion, and this draw may lead to different outcomes. Manski (1977) and Ben-Akiva and Lerman (1985) provide a contemporary interpretation of random utility theory for choice modellers. These authors suggest that individuals behave deterministically (i.e., under an identical scenario, the same alternative would always be chosen) and that the randomness of utility arises from the limited abilities of researchers to model choice behaviors and not from the random nature of people's choices[2].

---
[2] Train (2003) leaves open the possibility of interpreting randomness from satisficing behaviors of decision-makers.

The inability of choice modellers to model choice behaviors with 100% accuracy relates to three limitations (Manski, 1973). First, researchers may not understand all of the possible attributes that are important to people when making decisions. For example, during a study of campsite choice, an interviewee once relayed that he only travelled to camping sites within a one-hour drive of a hospital. His rationale was that this was the maximum time threshold for getting his son to the hospital in the case of an allergic reaction to a bee sting. It is almost unimaginable that any researcher could account for this level of detail within a campsite choice model.

Second, although researchers may know that certain attributes are important to decision-makers, they may not be able to measure these attributes accurately. For example, many recreational fishing choice models use a wide array of indicators for fishing quality (e.g., species presence, reported catch rate, reported fish size, models of expected catch). The resulting measurement errors from these indicators add noise into the estimate of utility.

Third, researchers often cannot account for the differing preferences that individuals have for attributes (i.e., preference heterogeneity). For example, not all cross-country skiers have an identical preference for trail length. If a researcher assumes that all skiers have the same preference for trail length, some error will be introduced when estimating utility. These concerns about varying preferences (i.e., preference heterogeneity) have lessened with advances to choice models (see the section 'Preference Heterogeneity' on page 199).

Choice models embrace the theories of utility maximization and random utility. The randomness associated with the researcher's limitations restricts researchers to forecast only the probability of an individual selecting one alternative from a set of alternatives. Despite this restriction, researchers and others can use these choice models to evaluate the likely consequences of choosing different alternatives in specific hypothetical scenarios. For example, one can forecast expected changes in visitation at a park associated with a hypothetical fee increase. This ability to predict the consequences of scenarios such as park fees on choice behaviors is a strength of the choice modeling approach.

WHAT ARE THE DATA REQUIREMENTS FOR CHOICE MODELS?

Three data sources are required to develop and estimate choice models. Researchers need information on chosen alternatives, the attributes that describe the alternatives, and the set of alternatives available to decision-makers. While this section focuses on choices of real alternatives, it is possible to estimate these models with data from choices of hypothetical alternatives (see the following section).

Data on choices of alternatives come primarily from survey-based sources. Some researchers obtain choice information from existing data from general surveys that profile a specific group of individuals (e.g., hunters). These general surveys include questions about where and sometimes when survey respondents took past trips (i.e., choices). Other researchers develop their own surveys or diaries to collect information about chosen alternatives. Researchers can tailor these other surveys and diaries to understand information that describes the context of trips and other information relevant to explain certain choices. For example, a researcher can ask individuals who are fishing to state whether their current trip is primarily for fishing or some other activity (e.g., camping). Dillman (2000) is a necessary read for any researcher who is contemplating collecting choice or any other data through a survey instrument.

Measures of attributes for the alternatives are also required to estimate a choice model. The attributes help to describe the alternative and are often the objects of utility. Relevant attributes are typically identified from reviews of past literature and formal (e.g., focus groups) or informal discussions with decision-makers and managers. Using outdoor recreation activities as an example, one could broadly identify relevant categories of attributes as costs, environmental quality, development, congestion and regulations. The use of costs, either direct (e.g., fees or taxes) or indirect (e.g., travel), within a choice model enables one to estimate economic welfare by identifying the expected maximum utility between two scenarios. For example, one can determine the money necessary to take from (or give to) canoeists to leave them equally well off as before once a change to water flow levels occurs on a river. This money represents the compensating variation form of economic valuation.

Researchers must also know details of the set of available alternatives. These data are required since preferences for the attributes are revealed by statistically comparing the characteristics of the attributes associated with frequently chosen alternatives with the characteristics of attributes associated with infrequently chosen alternatives. For example, if anglers choose many trips to waters with bass and few trips to waters without bass, the statistical preference for the attribute of bass presence will be strong and positive.

Determining the set of available alternatives is not always straightforward since the number of possible alternatives can be almost limitless. Consider the task of identifying available alternatives for hiking trails. The set of trails available to a hiker includes all trails on the planet! Even if we narrow our focus to day hiking, the set will include all trails that a hiker could feasibly visit, hike and return home from

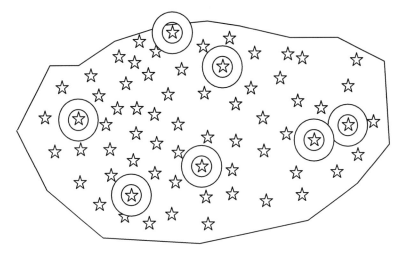

Figure 1. Use of random sampling (circled sites) to reduce the universal choice set

in one day. These millions of alternatives are part of a universal choice set.

To cope with a large universal choice set, researchers adopt at least one of three approaches: random sampling, narrowing, and aggregating. McFadden (1978) noted that one could estimate some choice models[3] from a random sample of alternatives (Figure 1) rather than from the universal choice set. While sampling error will introduce some problems, McFadden proved that a random sample will not bias the estimation of a choice model.

Many researchers also use narrowing to reduce the size of the universal choice set. Typical narrowing approaches include geographical scale (e.g. distance) (Figure 2), accessibility (e.g., road accessible recreational sites), resource quality and choice frequency for the alternatives. Narrowing will always introduce bias into choice

---
[3] These choice models must contain the Independent from Irrelevant Alternatives (IIA) property (e.g., conditional logit), which is discussed in the following section 'Substitution among alternatives'

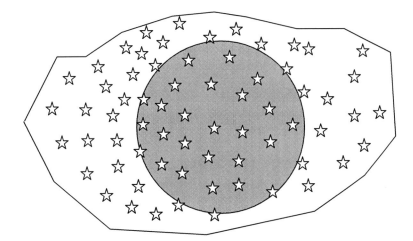

Figure 2. Use of a distance threshold (shaded circle) to narrow universal choice set.

model estimation unless the narrowing criterion is not related to reasons for choice (i.e., the narrowing criterion is the same as a random sample). While bias in estimation is introduced, several researchers (Parsons and Hauber, 1998; Hicks and Strand, 2000) suggest that it is an empirical question whether this bias results in managerially significant differences in model forecasts.

Aggregating choice alternatives into groups such as regions is a final approach to reduce the number of available alternatives. Ben-Akiva and Lerman (1985) demonstrate that aggregation will not bias choice model estimates if the researcher accounts for: the mean value of attributes, the number of actual alternatives within a region, and the variability of attribute measures among alternatives within the region. Few researchers who employ aggregated alternatives have incorporated all three of these aspects when estimating choice models (for an exception see Ferguson and Kanaroglou (1997)).

SHOULD ONE USE REAL OR HYPOTHETICAL DATA?

Researchers initially estimated choice models from data on actual (observed or reported) choices of alternatives made by individuals (i.e., revealed preference). While these revealed preference choice models (random utility models) are based on actual behavior, several problems exist with estimation, forecasting and interpretation (Louviere, Hensher, and Swait, 2000). First, the attribute measures among alternatives may be highly correlated. For example, angling catch rates may be greater as one travels further from urbanized areas. This relationship arises from the impacts that anglers may have on fish stocks. The negative association of travel distance and catch rate will confound any attempt to understand the "true" preference that anglers have for catch rate. Second, there is often little variability in measures among important attributes. For example, all parks may offer visitors facilities such as washrooms, showers, and electrical "hook ups". Since they do not vary, it is impossible to establish preferences for these attributes. Third, researchers and managers are often interested in understanding the average choice probabilities (i.e., market shares) for new or hypothetical alternatives (e.g., the choice of hybrid vehicles in the 1990s). Since these alternatives do not exist, a revealed preference choice model cannot provide any information about the expected market share for these hypothetical alternatives.

One way to address the problems of revealed preference data is to estimate models from hypothetical choices (i.e., stated preference). Louviere and Woodworth (1983) first illustrated these stated preference choice models[4]. The authors constructed hypothetical alternatives with attribute information that was based from experimental design plans[5]. The use of an experimental design plan permitted the authors to estimate the effects of different attributes on choice in isolation of other attribute measures. Despite this ability, one limitation of a stated preference choice model is that the choices are hypothetical and an individual "may tell you one thing and do another".

Ben-Akiva and Morikawa (1990) attempted to benefit from both data types by estimating choice models from revealed and stated preference choice data. With a transportation model of commuter choices of transport modes (alternatives) in Tokyo, Japan, the authors estimated both stated and revealed preference choice models. After accounting for larger noise in the

---

[4] These models are also referred to as discrete choice experiments and stated choice experiments.
[5] See Louviere, Hensher and Swait (2000) or Hensher, Rose and Greene (2005) for more information about conducting stated preference choice models.

stated preference data, the authors found no reason to conclude that the preferences for attributes differed between the two models. This joint revealed and stated preference choice model benefited from the validity of using actual choices and from the experimental control for the hypothetical choices. Adamowicz, Louviere and Williams (1994) used a similar joint revealed and stated preference choice model for the case of water-based recreational site choices.

Another way to address the limitations of revealed preference data is to employ field experimentation. Field experimentation, which is popular among ecologists and other natural scientists, helps to isolate the effects of treatments from other attributes that are of less importance to researchers and managers. For example, let us assume that parks managers are interested in knowing the implications of increasing park fees. Managers could take a sample of parks and set them into two groups. Next, one can collect information about park choices in the first year without any changes to fees. In the second year, more information about park choices could be collected at parks with an increase in fees (i.e., a treatment group) and at parks without any increase (i.e., a control group). From this design it is possible to assess the implications of fees on actual park choices without much concern of collinearity among attributes. I am not aware of any choice model study that has employed a field experiment to estimate a revealed preference choice model.

HOW DO WE TRANSFORM THEORY INTO STATISTICAL MODELS?

Since one cannot understand nor model all aspects of utility, researchers view the utility for an alternative as having deterministic and unobserved parts. Formally, one can present the utility (U):

$$U_{in} = V_{in} + \varepsilon_{in} \quad (1)$$

The utility for alternative i and individual n consists of the sum of the deterministic V_{in} and unobserved (ε_{in}) parts. The probability that an individual will choose alternative i from a set of alternatives (C_n) equals:

$$P_{in} = P(V_{in} + \varepsilon_{in} \geq V_{jn} + \varepsilon_{jn}), \forall j \in C_n, i \neq j \quad (2)$$

Simply put, the probability (P_{in}) of selecting alternative i equals the probability (P) that the utility for alternative i is greater than or equal to the utility for all other (j) alternatives that belong to set C_n. This equation comes to us directly from our utility maximization assumption (see previously in this chapter). We can rearrange equation 2 to focus on the unobserved utilities for the other (j) alternatives:

$$P_{in} = P(\varepsilon_{jn} < V_{in} - V_{jn} + \varepsilon_{in}), \forall j \in C_n, i \neq j \quad (3)$$

To solve equation 3, all we need is information about the deterministic and unobserved utilities for all alternatives and individuals. While it is assumed that the deterministic utilities are recoverable from individual choices, we don't know anything about the unobserved utilities! To develop a statistical model, we need to make an assumption about the unobserved utilities. One assumption is to treat the unobserved utilities as random variables. If we have three alternatives (J) with the first alternative being the alternative of interest (i), we can rewrite equation 3 as eq. 4.

Equation 4 includes the three random variables ($\varepsilon_{1n}, \varepsilon_{2n}, \varepsilon_{3n}$) with integrals ($\int$) that account for the values that these variables

$$P_{in} = \int_{\varepsilon_{1n}=-\infty}^{\infty} \int_{\varepsilon_{2n}=-\infty}^{V_{1n}-V_{2n}+\varepsilon_{1n}} \int_{\varepsilon_{3n}=-\infty}^{V_{1n}-V_{3n}+\varepsilon_{1n}} f(\varepsilon_{1n}, \varepsilon_{2n}, \varepsilon_{3n}) \partial \varepsilon_{1n} \partial \varepsilon_{2n} \partial \varepsilon_{3n} \quad (4)$$

can take along with weights of the likelihood that particular values will arise[6] (i.e., the joint probability density function). Since only situations when both ε_{2n} and ε_{3n} are less than V_{1n}-V_{jn}+ ε_{1n} satisfy equation 3, the integrals for ε_{2n} and ε_{3n} are truncated at a maximum of V_{1n}-V_{jn}+ ε_{1n} as shown in equation 4. If one assumes that the random variables follow a normal distribution, equation 4 becomes the multinomial probit model. If equation 4 seems intimidating, you are not alone. Until recently there were few ways for researchers to estimate multiple integrals for choice models.

McFadden (1974) developed from equation 4 a conditional logit model[7], which does not require estimation of integrals (i.e., a closed form model), from a distributional assumption for the random variables. He assumed that the random variables were distributed according to a type I extreme value distribution[8]. He also assumed that there were no relationships among the random variables (i.e., independence) and that the expected variance for the random variables was the same (identical). These assumptions along with properties of the type I extreme value distribution give rise to the conditional logit model (equation 5) (see Ben-Akiva and Lerman (1985, p. 101-106) for a well-explained proof).

$$P_{in} = \frac{e^{\mu V_{in}}}{\sum_{j=1}^{J} e^{\mu V_{jn}}}, \quad \forall j \in C_n \quad (5)$$

---

6 Hausman and Wise (1978) show a trick to remove one of the integrals from equation 4.

7 The conditional logit is the multinomial logit model. Some individuals split hairs over the differences between these models. Technically, a conditional logit allows attribute measures to vary over alternatives while a multinomial logit does not.

8 This probability density distribution is $\mu\exp(-\mu(\varepsilon_i-\eta))\exp(-\exp(-\mu(\varepsilon_i-\eta)))$, where μ is a scale parameter and η is the mode. The resulting distribution is similar in shape to a normal distribution.

Equation 5 states that the probability (P) of individual n selecting alternative i depends upon the deterministic utility V and μ, which is a scale parameter that is inversely related to the variance of the unobserved utilities. The deterministic utilities for all J alternatives in a choice set C_n are required to calculate the denominator in equation 5.

In most instances, V equals the addition of attribute measures (X) and preferences for these attributes β. Rather than writing $X_{in1}\beta_1 + X_{in2}\beta_2 + \ldots X_{ink}\beta_k$, researchers use the matrix algebra expression $X_{in}\beta$ (see equation 6). This particular form for the deterministic utility embraces compensatory decision-making since we sum each attribute and preference to estimate utility. In compensatory decision-making, individuals can offset those attributes with low desirability such as congestion with those attributes with desirable preferences such as environmental quality. This tradeoff ability is one of the greatest strengths of any choice modeling approach.

$$P_{in} = \frac{e^{\mu X_{in}\beta}}{\sum_{j=1}^{J} e^{\mu X_{jn}\beta}}, \quad \forall j \in C_n \quad (6)$$

While the attribute measures (X) are taken directly from the data, one must estimate the preference parameters (β). Positive values for β indicate that greater amounts of that attribute are preferable. For example, a positive parameter for environmental quality suggests that people are more likely to select alternatives with higher than lower levels of environmental quality. Since μ is not separable from $X_{in}\beta$, researchers restrict (i.e., normalize) the value of μ to one. When comparing different data sets such as revealed and stated preference data, it is possible to account for differing μ values, which represent different variance levels in unobserved utilities.

Train (2003) notes two important properties that govern all choice models. First, the scale of utility is arbitrary (i.e., an alternative with maximum utility still has maximum utility regardless of any change to the scale of utility). Second, only differences in utilities matter. These properties arise since utility is unobservable to researchers as choice models are estimated from choice indicator variables (i.e., the alternative with the maximum utility). These properties are important to remember as it is possible to over parameterize choice models when estimating parameters associated with unobserved utilities (see Walker (2002) for more details).

The following example should help to clarify the reasons why the above two properties exist. Imagine you divide 200 islands into 50 sets of four. An assistant provides you with information about the islands (e.g., latitude, distance to mainland, etc.) and for each set, she tells you which island has the highest average temperature. From this information, can you determine the scale used to measure temperatures (e.g., Celsius or Fahrenheit) and the actual temperatures on the islands? The answers are no since you don't have enough information about the islands. This same inability to identify actual values of utility or the scale used to measure utility affects all choice model estimates.

HOW ARE THESE MODELS ESTIMATED?

The data collected on choices, alternatives and attribute measures are necessary to estimate statistical parameters associated with choice models. For the conditional logit model, there is a need to estimate the β values that represent the preferences that individuals have for attributes. Maximum likelihood is used to estimate these parameters. The example below will show that maximum likelihood estimation is identical to model calibration.

The objective of maximum likelihood estimation is to find values (estimates) for a set of parameters that yield the greatest likelihood of reproducing the observed data. Imagine a simple example of predicting likely support for two candidates (Jane Maverick and John Stable) in an upcoming election. We decide to use equation 6 to estimate the support for Jane Maverick. The only attribute that we consider is campaign donations (X) with $2 million for Jane and $1 million for John. Through a daringly small sample of four individuals, three of whom stated they will vote for Jane, we estimate our choice model.

Our goal is to estimate the parameter (i.e., preference) for the millions of dollars in donations attribute (β_D) from the choices of candidates by our sample of four individuals. Let us arbitrarily set β_D equal to two and calculate the likelihood that this parameter estimate would reproduce our sample of choices. Normalizing $\mu=1$, equation 6 estimates a probability of 0.88 for Jane (i.e., exp(2*2) / (exp(2*2) + exp(2*1)). We calculate the likelihood statistic for this parameter estimate by multiplying these choice probabilities by the observations. Since we have three Jane votes (probability of 0.88) and one John vote (probability of 0.12) the likelihood of observing our sample with β_D =2 is 0.88 * 0.88 * 0.88 * 0.12, which equals 0.082. This value of 0.082 is our likelihood statistic.

Next, we estimate the likelihood statistic with β_D =1. With choice probabilities of 0.73 for Jane and 0.27 for John, we obtain a likelihood statistic of 0.105. Since this likelihood statistic is greater than the statistic for β_D =2, we conclude that β_D =1 has a greater chance of being the correct parameter estimate given our sample. We keep trying different values of β_D until we cannot improve the likelihood statistic (i.e., we have reached the maximum likelihood). A value of 1.04931 for β_D is this maximum. In essence, this parameter estimate is calibrated from the data on choices, alternatives and attributes.

Practical estimation of maximum likelihood differs in two ways from the example. First, one must account for the fact that multiplying many probabilities sends the likelihood statistic towards zero. Most computer programs retain only eight or 16 decimal places on numbers. This truncation may cause problems with estimating likelihood values that are close to zero. For example if we had a sample of 32 observations with 24 votes for Jane and eight for John, the maximum likelihood statistic would be 0.0000000153. One way to prevent this truncation problem is to take the natural logarithm of the likelihood function. This transformation does not affect the estimates of parameters that produce the maximum likelihood. Instead of multiplying probabilities, the log likelihood function involves adding logarithms of probabilities. These logarithms each take on values between zero and negative infinity resulting in a maximum log likelihood that is negative.

Second, researchers do not use trial and error to determine the parameter estimates that lead to a maximum for the (log) likelihood function. Instead, non-linear optimization techniques are employed (see Train (2003) for details). These techniques work by intelligently selecting values for the parameters and estimating the log likelihood function. At each step (iteration), a new set of parameter estimates is selected and the log likelihood function is calculated. This process is repeated until there is almost no detectable improvement to the (log) likelihood function. Standard errors for the parameter estimates typically come from the inverse of the matrix of second derivatives with respect to the parameters (i.e., the Hessian).

HOW ARE OTHER ISSUES ADDRESSED WITHIN CHOICE MODELS?

The conditional logit model of equation 6 provides an efficient way to estimate choice models. This efficiency, however, comes at a cost. The model imposes assumptions on choice behaviors that one may not be willing to accept. Two of the most troubling aspects of the conditional logit relate to varying preferences and substitution among alternatives. The conditional logit assumes that researchers can account for all heterogeneity in preferences through observed characteristics of individuals (e.g., market segments). This assumption limits the ways that one can account for preference heterogeneity among and even within (e.g., dynamically changing preferences) individuals. The conditional logit also imposes a potentially restrictive assumption on the ways that alternatives may act as substitutes for each other. Consequently, researchers using the conditional logit model are forced to accept potentially rigid forecasts of changes in choice behavior when assessing different management scenarios.

Researchers are quite aware of the limitations associated with the conditional logit model. Developments to choice models are now sufficiently mature to enable researchers to abandon potentially unrealistic assumptions with the conditional logit and other choice models. As Keane (1997) notes, it is time for researchers to develop statistical models to fit behavioral processes rather than assuming that behavioral processes fit the assumptions of statistical models. While this section describes some advances, readers may benefit from consulting more thorough discussions and reviews of these choice model advances (Ben-Akiva, Bolduc and Walker, 2001; Bhat, 2002, 2003; Hunt, Boots, and Kanaroglou, 2004; Train, 2003; Walker, 2002).

This section explores five important issues that one should consider when conducting a choice model on outdoor recreation behaviors. With each issue, a menu

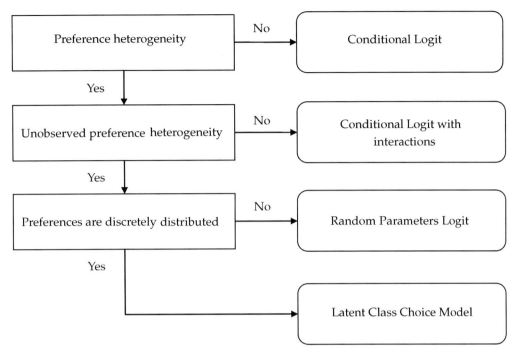

Figure 3:. Options for addressing heterogeneity in preferences for attributes

of available approaches is offered and the necessary behavioral assumptions of each approach are discussed. The issues discussed below include preference heterogeneity, substitution, choice sets, participation and site choices and dynamic decision making.

Preference Heterogeneity

Estimating preferences for attributes is a key requirement for choice modeling. Different choice models provide various ways to account for preferences that individuals have for attributes. Figure 3 provides a flowchart that describes some of the implicit assumptions associated with different ways to model preferences.

At one extreme, one may assume that all individuals have identical preferences for all attributes. While this assumption permits researchers to use a conditional logit model, this assumption is unlikely to hold in most situations since preferences for attributes are likely to vary among people.

The conditional logit model remains a viable option even when one assumes that preferences for attributes vary over individuals. To employ this conditional logit model, a researcher assumes that any variations in preferences are observable. This assumption permits one to account for varying preferences within the deterministic utility by interacting characteristics of individuals with attributes to estimate different preferences for different individuals. For example, consider the choice of a park for a recreational trip. Assume that only travel distance and park development are important to decision-makers. Finally, let us assume that we suspect that experienced and inexperienced park visitors will have different preferences for these attributes. We can estimate preferences associated with the travel cost and development attributes along with a second set of parameter estimates that interact experience level (coded one for experienced and zero for inexperienced) with the travel cost and park development attributes[9]. The first set of parameter estimates reveal the preferences of inexperienced visitors for the attributes while the sum of the first and second sets of parameter estimates reveal these same preferences for the experienced visitors. If the second set of parameter estimates is jointly equal to zero, there is no reason to conclude that preferences for these attributes differ by experience level.

Researchers may not assume that they can observe all heterogeneity in preferences through the characteristics of individuals. For example, researchers may assume that while discrete groups (i.e., market segments or classes) of preferences exist, they do not know the membership of individuals to these classes. This assumption (see Figure 3) leads to the latent class choice model (Swait, 1994). Given a number of classes specified by a researcher, the latent class choice model estimates different sets of parameters for the classes and the probability that individuals belong to each class. Researchers can also determine whether the characteristics of individuals are associated with the probabilities of class membership. In this way, one can learn about the individuals who comprise each class.

Consider the example provided earlier about the choices of parks by individuals. If we assume that two classes exist, the latent class choice model will estimate for each class a set of parameter estimates for the travel cost and park development attributes. The model also will estimate the probability that an individual will belong to a class (i.e., the class membership is latent (unobserved) to the researcher). We can include experience level and other characteristics of the individuals as independent variables that explain class membership probabilities. If experience level is significantly associated with the class membership probabilities, we can conclude that the class membership is partially explained by differing experience levels. Note that this approach is much different than the conditional logit model where we assumed that preference heterogeneity is entirely explainable by experience level. Boxall and Adamowicz (2002) provide an empirical example of a latent class choice model on park choices.

Researchers may also assume that the heterogeneous preferences follow a continuous distribution rather than discrete classes (see Figure 3). This assumption leads one to a random parameters logit model, which is a type of mixed logit model. Given the increasing importance of mixed logit models in choice modeling, the random parameters logit model is explained in some detail (see Ben-Akiva et al. (2001) for a general exposition of the mixed logit model).

To simplify the exposition of the random parameters logit, assume that the random variables are independent. The utility (see Equation 7) for the random parameters logit consists of a deterministic part ($\mathbf{X}_{in}\boldsymbol{\beta}$) and an unobserved part ($\Sigma(\mathbf{X}_{ink}\boldsymbol{\beta}_{Sk}\xi_{nk})+\varepsilon_{in}$). The ε_{in} is a random variable that is distributed to yield a conditional logit (i.e., type I extreme value with independence and equal expected variance). The new term $\Sigma(\mathbf{X}_{ink}\boldsymbol{\beta}_{Sk}\xi_{nk})$ consists of K random variables ξ_{nk}, the attribute measures ($\mathbf{X}_{ink}$), and parameters ($\boldsymbol{\beta}_{Sk}$) that account for the standard deviations of the preferences for the attributes. Researchers a priori select distributions for the random variables that they assume variations in preferences will follow (e.g., normal).

$$U^*_{in} = \mathbf{X}_{in}\boldsymbol{\beta} + \sum_{k=1}^{K} X_{ink}\beta_{Sk}\xi_{nk} + \varepsilon_{in} \qquad (7)$$

If the researcher knew the values for the

---

[9] Researchers can use other contrast schemes than the dummy variable employed here. For example, many researchers use effects based codes whereby the average score for a coded variable is zero.

random variables, the $\Sigma(X_{ink}\beta_{Sk}\xi_{nk})$ term would simply be part of the deterministic utility and one could estimate the β_{Sk} with a conditional logit model. However, since the values for the random variables are unknown, one needs to account for all possible values of these random variables and the likelihood that they would be observed (i.e., integration). This integration over all (K) random variables ξ_{nk} is shown in equation 8.

Let us consider the example of the park visitors provided earlier. While we assume that travel cost and park development are important attributes and preferences likely vary, there is no reason to believe that we can explain this variability. Instead, we simply assume that the preferences will vary according to a normal distribution. By estimating equation 8 with maximum simulated likelihood or Hierachical Bayesian approaches (see Train, 2003), the estimates for β and β_{Sk} are revealed. The β parameters tell us the median preference while the β_{Sk} tell us the standard deviations of preferences for the travel distance and park development attributes. Assume that the median preference for park development is 0.9 and the standard deviation preference is 0.6. It is possible to estimate the percentage of the population that have positive and negative preferences for park development from these estimates and the normal distribution. For example, the percentage of individuals with negative preferences equals the cumulative density function associated with a z score of -1.5 (i.e., (0-0.9)/0.6). This z-score is associated with a probability of 0.0668, which means that the model estimates that 6.68% of the population have negative preferences for park development. See Train (1999) for a good example of the random parameters logit model applied to recreational fishing site choices.

Substitution among alternatives

An important use of choice models is to forecast how changes in attributes at some sites are likely to affect choices at all sites. While researcher have long recognized that some alternatives should act as better substitutes than others (e.g., a wilderness park is a better substitute for another wilderness park than is an historical park), different choice models force different

Table 1. Predicted relative change in fishing trips from closure of Bass#1 lake (actual predicted trips in parentheses)

	Original Percentage	Conditional Logit	Nested Logit	Cross-Nested Logit	Generalized Nested Logit
Bass #1	50.00	-100.00%	-100.00%	-100.00%	-100.00%
Bass #2	10.00	100.00%	146.30%	109.17%	128.21%
Bass #3	5.00	100.00%	146.30%	118.22%	161.53%
Bass #4	5.00	100.00%	146.30%	109.17%	117.69%
Trout #1	10.00	100.00%	69.14%	89.24%	72.12%
Trout #2	10.00	100.00%	69.14%	96.79%	88.59%
Trout #3	5.00	100.00%	69.14%	92.95%	74.22%
Trout #4	5.00	100.00%	69.14%	89.24%	68.71%

$$P_{in} = \int_{\xi_1}\ldots\int_{\xi_K} \frac{e^{\mu(X_{in}\beta+\sum_{k=1}^{K}X_{ink}\beta_{Sk}\xi_{nk})}}{\sum_{j=1}^{J} e^{\mu(X_{jn}\beta+\sum_{k=1}^{K}X_{jnk}\beta_{Sk}\xi_{nk})}} f(\xi_1)\ldots f(\xi_K)\partial\xi_1\ldots\partial\xi_K, \forall j \in C_n \quad (8)$$

assumptions about the ways that alternatives act as substitutes.

The conditional logit model assumes that the unobserved utilities for alternatives are independent. This assumption invokes an independence from irrelevant alternatives (IIA) property that forces a potentially restrictive pattern of substitution among alternatives. For example, consider eight fishing lakes of which four contain bass and the other four contain trout fish species. Assume that the bass lakes receive 50, 10, 5 and 5% of fishing trips and the trout lakes receive 10, 10, 5 and 5% of fishing trips. If the bass lake with 50% of trips was closed to fishing, we would expect that the other bass lakes would act as better substitutes for this closed bass lake than would the trout lakes. However, a conditional logit would predict that the bass lakes would receive 20, 10 and 10% of trips while the trout lakes would receive 20, 20, 10, and 10% of trips. For every lake, the increase in fishing trips is exactly twice the original percentage (see Table 1). This identical relative increase for fishing trips results directly from the IIA property within the conditional logit model.

Several researchers (Daly and Zachary, 1978; McFadden, 1978; Williams, 1977) developed a nested logit model that bypassed the IIA property by assuming that unobserved utilities among alternatives consist of two parts: a unique part and a part shared among alternatives within a group (i.e., nest). McFadden (1978) proved the existence of a family of generalized extreme value (GEV) models including the conditional, nested, cross-nested, and generalized nested logit models. These GEV models provide researchers with flexible ways to accommodate substitution among choice alternatives.

Equation 9 represents a general utility form for alternative i and individual n that captures most GEV models. The equation consists of M number of nests, an allocation (α) of alternative i to nest m, deterministic utility at the nest (V_{mn}) and alternative (V_{in}) levels, and unobserved utilities at the nest (ε_{mn}) and alternative levels (ε_{in}). Assuming that the allocation parameters (α) sum to unity over the M nests, the exponential and logarithmic transformations provide a way to weight the utility by the probability that an alternative belongs to a particular nest.

$$U_{in} = \ln\left(\sum_{m=1}^{M} \alpha_{im} e^{V_{in}+\varepsilon_{in}+V_{mn}+\varepsilon_{mn}}\right) \quad (9)$$

Through various assumptions of the distributions of the unobserved utilities involving the type I extreme value distribution (see Ben-Akiva and Bierlaire, 1999), and assuming that the deterministic utility ($V_{mn} + V_{in}$) equals $X_{in}\beta$, the generalized nested logit model (Wen and Koppelman, 2001) arises (see Equation 10).

The μ_m term is a dissimilarity parameter that relates to scale parameters at the nest and alternative levels. One minus this dissimilarity parameter provides an estimate of the degree of correlation among the unobserved utilities within one nest. If all $\mu_m = 1$, there is no correlation among unobserved utilities and equation 10 collapses into the conditional logit model of equation 6.

The nested logit model arises by assuming that alternatives are only allocated to

$$P_{in} = \sum_{m=1}^{M} \left[\frac{(\alpha_{im} e^{X_{in}\beta})^{\frac{1}{\mu_m}} \left(\sum_{j=1}^{J_m}(\alpha_{jm} e^{X_{jn}\beta})^{\frac{1}{\mu_m}}\right)^{\mu_m - 1}}{\sum_{l=1}^{M}\left(\sum_{j=1}^{J_l}(\alpha_{jl} e^{X_{jn}\beta})^{\frac{1}{\mu_m}}\right)^{\mu_m}} \right] \quad (10)$$

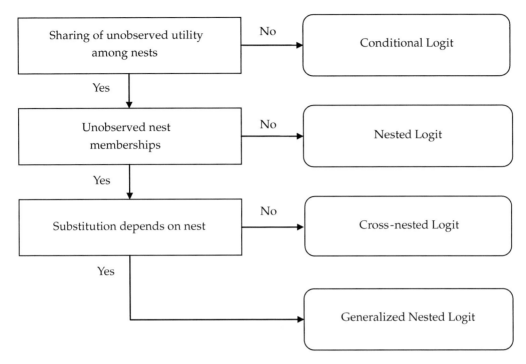

Figure 4. Options for accounting for substitution among alternatives with Generalized Extreme Value Models

one nest (α_{im} = 1 or 0) and the researcher knows the membership of these alternatives to all the M nests (see Figure 4). Consider the fishing example above with all trout and all bass alternatives in separate nests and all μ_m = 0.70. This nested logit model predicts a relatively greater percentage of fishing trips would go from the closed Bass #1 lake to all other bass lakes than to the trout lakes (see Table 1). Within a nest, the IIA property remains as the relative increase within the other bass lakes (146.30%) and within the trout lakes (69.14%) are identical. Morey (1999) provides a thorough overview of this nested logit model.

Another GEV model arises when a researcher allows alternatives to belong to more than one nest (i.e., cross-nesting). If researchers assume that all μ_m are identical, equation 10 becomes the cross-nested logit model. This model would allow the researcher to estimate a dissimilarity parameter along with the allocations of alternatives among the different nests (see Figure 4). Since the alternatives are cross-nested, the model is not affected by the IIA property.

Assume for the fishing example that the lakes are cross-nested with allocation parameters of 0.8, 0.7, 0.9, 0.7, 0.2, 0.4, 0.3, and 0.2 for the first nest and one minus these allocations for the second nest. These partial allocations of alternatives to nests reflect the fact that trout and bass fish species are not the only criteria that leads to similarity among sites (e.g., other fish species, accessibility). Again, assume that all μ_m = 0.70. Two differences become apparent in Table 1. First, the IIA property no longer affects the predicted changes to probabilities from closure of Bass #1 lake. Second, the relative impact between bass and trout lakes is lessened since these alternatives are cross-nested to some degree. Hunt, Boots and Boxall (2007) provide an example of a cross-nested logit model applied to recreational fishing.

The most general form of equation 10 allows the researcher to estimate the allocation parameters and separate dissimilarity parameters for each nest (see Figure 4). This generalized nested logit relaxes the assumption of the cross-nested logit model that the correlation of unobserved utilities for each nest is identical (i.e., the dissimilarity parameters can vary). For the fishing example, assume that all conditions are identical as the cross-nested logit example with the exception that the dissimilarity parameters equal 0.5 and 0.8 for the nests rather than 0.70 for both. The predictions in Table 1 show an even more complex pattern of substitution among the fishing sites that reflects the allocation of alternatives to nests and the dissimilarity parameters for each nest. Wen and Koppelman (2001) provide an application of the generalized nested logit on choices of alternative transportation modes.

A few cautions are provided to readers who are interested in applying GEV models. First, there is a possibility that maximum likelihood estimation procedures may not identify the true maximum likelihood and consequently, the parameter estimates may be incorrect. To avoid this possibility, one should estimate the model with several different sets of starting values for the parameters. Second, the dissimilarity parameters from equation 10 must lie between zero and one to be consistent with random utility theory. Finally, it is necessary to fix the values of some parameters from equation 10 to identify the other parameters. For example, if the allocation parameters (α) for an alternative sum to unity over the nests, it is only possible to identify M-1 allocation parameters.

Choice set limitations

Many researchers do not believe that all individuals select an alternative from a universal choice set. Instead, individuals are assumed to have different choice sets from which they make decisions. When abandoning the universal choice set concept, researchers consider several different approaches each with strengths and weaknesses.

When collecting information about choices, some researchers ask individuals to identify their own choice sets (e.g., Peters, Adamowicz, and Boxall, 1995). While appearing simple, this approach has several limitations. First, there is no clear consensus on the definition of a choice set. Some researchers believe that an individual's choice set is based on availability (i.e., alternatives that are feasible) while others believe that they are based on consideration (i.e., alternatives that pass some pre-screening phase). Limiting choice sets by awareness or consideration criteria can result in different parameter estimates and model forecasts. Even when one makes a stance about defining a choice set through availability or consideration, it is difficult to word questions appropriately (e.g., Parsons, Massey and Tomasi, 1999). A final problem is that researchers estimate choice models from a sample of individuals. Without knowing the actual choice sets for the population of individuals, researchers can have a difficult time estimating choice models that require one to know the choice sets of the individuals.

Other researchers use modeling approaches to account for differing choice sets. Many of these models follow Manski's (1977) formulation of choice as a two-stage process. In the first stage, the researcher estimates the probability of an individual selecting a choice set only from information about chosen alternatives. The second stage involves the estimation of a probability of selecting an alternative given a particular choice set. While several researchers have used this framework (e.g., Haab and Hicks (1997)), the approach is seriously compromised when there are even a moderate number of alternatives. For example, with 13 alternatives, there are 8191 unique choice sets (i.e., unique combinations of alternatives into sets). This combinatorial problem has led other researchers to implicitly model choice set formation (Cascetta and Papola, 2001). These researchers assume that while the universal choice set is relevant, limited awareness and constraints affect the likelihood that individuals will select some alternatives. The implicit choice set model works by penalizing the deterministic utility of an alternative by these awareness and constraint considerations. Finally, Gilbride and Allenby (2004) use thresholds based on attribute measures to define choice sets. Through a Hierarchical Bayesian estimation, these authors demonstrate how their model can accommodate consideration sets developed from satisficing rather than optimizing behavioral processes.

Accounting for participation and site choice

Outdoor recreation involves a decision to participate in an activity and a decision of where to conduct that activity. Choice models that first examined recreational behavior focused solely on the latter choice decision (i.e., where to recreate). However, researchers should provide managers and others with predictions about how often recreational trips will occur besides where trips will occur.

Researchers have devised two approaches to model participation and site choice decisions[10]. Both models link site choices to participation by including the expected maximum utility from the sites as an independent variable when estimating participation decisions. This linkage can account for the expected impacts to recreational participation that may occur with negative or positive changes to site attributes (e.g., better fishing quality should lead to more fishing trips, all else considered equal).

One approach uses a count regression model such as Poisson to explain and predict the number of trips that individuals take for a recreational activity. The dependent variable is the number of reported trips provided by the individuals. Independent variables include the expected maximum utility and individual characteristics. The model allows one to estimate the likely effects of a change to site attributes on the participation level for that outdoor recreation activity. Parsons, Jakus, and Tomasi (1999) describe this approach and some variations.

The repeated nested logit model (Morey, Rowe, and Watson, 1993) is the second approach to account for participation and site choices. The repeated nested logit treats the site choice decision as a nest under the participation decision. On each potential choice occasion (e.g., day), participation and non-participation decisions are modelled. When the individual chooses to participate, the site chosen for the trip is also modelled. One can express these probabilities in a generalized nested logit (GNL) form (Equation 10).

The dissimilarity parameter μ_m in equation 10 provides the link between site choice quality and participation selection. If μ_m is less than one, the likelihood of participation is positively related to the expected quality at the recreational sites. Besides expected site quality, researchers predict participation decisions through interactions between individual characteristics (e.g., age) and an intercept for the participation alternative.

---

[10] Researchers are also increasingly interested in Kuhn-Tucker models to estimate site and participation decisions since these models account for a seasonal budget for trip taking (see Phaneuf, Kling, and Herriges (1998) for an example).

Accounting for dynamic effects

Most choice model applications assume that each choice decision is independent of past decisions by individuals. However, there are many instances where people become habituated to choose sites or they strategically make choices to account for expected changes to attributes or choices by others. For example, some people may delay recreational trips to times when they expect the number of other recreationists to be low (i.e., avoid congestion).

The state dependence concept (Heckman, 1981) captures habituation as a special case. State dependence results when past choices of an alternative affect the current utility for that alternative. For example, if a person becomes attached to a recreational site through frequent visitation, this attachment would represent a form of state dependence.

Some researchers estimate state dependence by including past choices as part of the deterministic utility. If the parameter estimate for past choices is positive and significant, individuals are more likely to select an alternative than one would expect solely from the attributes of that alternative. Since researchers are not omniscient (even those who believe otherwise), the presence of state dependence may come from a substantive process (e.g., attachment to place) or from a mis-specified model. A mis-specified model may arise from making incorrect assumptions about the preferences that individuals have for attributes (e.g., assuming all individuals share identical preferences), omission of key attributes and/or the use of indicator variables for important attributes. Without observing the first choice (i.e., the initial conditions), it is difficult to conclude whether substantive state dependence is present.

To illustrate the implications of a mis-specified model, let us consider the case of modeling choices of paddlers among two canoe routes: one with and one without portages. Next, assume that most paddlers (75%) strongly dislike portages while some paddlers (25%) actually prefer portaging. If we estimated the choice model with only one parameter estimate (i.e., preference) for portage, we would observe a significant and negative preference for this attribute. Obviously, this average preference for portaging would lead to errors (over predicting the selection of a non portage site by individuals who like portaging and under estimating use of this site by paddlers who dislike portages). What would happen if we included the past site chosen by a paddler as an explanatory factor in the choice model? We would find a positive association between the past and currently chosen alternatives. This association does not represent true state dependence since it arises from our wrong assumption that all paddlers have identical preferences for portages. For paddlers who like portages, we would observe more of these individuals selecting the portage alternative during the past and current trip than we would expect by chance.

Several researchers have devised ways to incorporate state dependence or other dynamic aspects into recreational choice decision-making contexts (Provencher, and Bishop, 1997; Train, 1999; Swait, Adamowicz, and van Bueren, 2004; Hunt, forthcoming). Dubin (1995) developed a model that allows the past choices of neighbouring landowners to influence current choices made by landowners (i.e., a model of spatial diffusion). Other researchers have accounted for risks (e.g., Akabua, Adamowicz, Phillips, and Trelawny, 1999) and expected encounters with other recreationists (e.g., Jakus and Shaw, 1997; Schumann and Schwabe, 2004) within choice models. These examples demonstrate the interest that researchers place on accounting for dynamic aspects of decision-making within choice models.

WHAT ARE THE POSSIBILITIES FOR USING CHOICE MODELS TO GUIDE AGENT BEHAVIORS?

Researchers often view simulation modeling and statistical models including choice models as mutually exclusive and even adversarial approaches. With an open mind, however, one can benefit from the strengths of these two approaches. Researchers who employ choice and agent based models seek many similar objectives. These objectives include understanding spatial and temporal patterns of behaviors and forecasting how these patterns may change with hypothetical scenarios. Skov-Petersen (2005) notes some of these similarities between agent-based, choice and other behavioral models.

Hunt, Kushneriuk, and Lester (2007) describe many of the strengths and weaknesses of choice and agent based models. These authors suggest that choice models may be ideally suited to help instruct the behaviors of individual agents. Choice models provide a theoretically and statistically sound approach that links changes in the environment to the likelihood that individuals will conduct various behaviors.

The application described by Hunt et al. (under review) focuses on an agent-based model (the Landscape Fisheries Model) of recreational fishing in Ontario, Canada. This agent-based model uses results from a repeated nested logit model of revealed preference data that predicts the amount, timing, and location of recreational fishing trips by resident anglers (see Hunt et al. (2007) for choice model estimates). This choice model governs the behaviors of angling agents that exist in a simulated world of recreational fishing opportunities. The agents decide whether or not to take day or multi-day trips of different contexts (e.g., publicly accessible sites, private cottages, etc.) based on characteristics of the day (calendar events, weather), characteristics of the individual and the expected maximum utility from the available fishing sites on that day. The fishing site choice decision of an agent is influenced by angler characteristics and attributes on travel costs, fishing quality, environmental quality, facility development and regulations.

The Landscape Fisheries Model represents an initial attempt to marry the benefits of choice and agent based modeling approaches. The model is still naïve as the agents do not impact each others choices of fishing trips nor do they impact the resource (e.g., declines to fish abundance). Future research will shed some of these naiveties through the introduction of fish agents and different decision-making processes for angling agents.

There are several advantages of using choice models to guide behaviors of agents. First, choice models provide a theoretical rationale for the behavioral rules used to guide agent choices. Rather than dictating the behavioral rules through expert opinion, these behavioral rules are revealed to researchers through empirical data. Second, choice models allow one to test the statistical significance of preferences for various attributes. This statistical estimation gives confidence to researchers about the importance that various attributes have at influencing choice decisions of agents. Third, choice models forecast probabilities of behaviors that are ideally suited for simulation modeling. Finally, choice models, which are consistent with random utility theory, allow one to estimate changes in economic value of behaviors from different management scenarios.

Despite its benefits, the use of choice models to guide agent behaviors is not without problems. It is difficult to accommodate complex processes such as learning and fatigue of agents within choice models. This difficulty may limit the ability of agents to make decisions from past experiences and efforts. The rigour needed to estimate choice models may induce costs that exceed the benefits of using this approach. Monitoring the tendencies of people or using expert opinion may be more appropriate to estimate choices for relatively trivial decisions such as deciding to stop on a hiking trail. Significant decisions such as visitation and timing of trips to recreational areas are likely well suited to choice models. Therefore, choice models provide a method to assist and not to replace the development of rules that govern agent behaviors.

CONCLUSION

Choice models allow one to embrace theories that describe decision-making processes of individuals and the abilities of researchers to model these processes. Choice models rely on two theoretical postulates. First, individuals are assumed to choose alternatives based on a utility maximization process. Second, researchers acknowledge that they cannot account for all the individual nuances that lead to utility from the alternatives. These theories along with measures of attributes of alternatives and choices of alternatives are sufficient to estimate choice models. Researchers and others can use these choice models to examine the potential implications of management scenarios on choice. For outdoor recreation behaviors, choice models can provide information on the amount, timing, location, and value of recreational activities.

Early choice models such as the conditional logit model were developed to reduce computational burden at the expense of making rigid assumptions about decision making. Continuing developments to choice models have re-oriented the focus of model development to reduce assumptions of decision-making. As Keane (1997) notes it is time that researchers fit statistical models to behavioral problems and not vice versa. Choice model developments have great potential to accommodate variations in individual tastes, substitution among alternatives, choice sets, participation and site choices, and dynamic aspects of decision-making. These developments provide ways for researchers to heed the call from Keane.

Choice models like every other model are a simplification of reality. The assumptions that one makes to simplify reality have the potential to introduce bias. As discussed earlier, choice models assume that individuals are maximizers while most researchers acknowledge that people are satisficers.

The vast developments to choice models seem to open limitless opportunities for researchers to estimate models. However, incorporating complexities into choice models is not a simple task. Often constraints on data, computing power, and understanding of behavioral processes limit the complexity that researchers may build into choice models. For example, the unobserved utility term receives a lot of attention for GEV, mixed logit, and even some choice set models. These models provide opportunities to account for preference heterogeneity, complex patterns of substitution, choice set development, models of participation and site choice and dynamic decision-making. As Train (2003, p.145) eloquently notes "There is a natural limit on how much one can learn about things that are not seen". This statement is a caution to researchers to understand and acknowledge that choice models cannot account for all aspects of reality. However, researchers can develop choice models to capture aspects of behavioral processes that are most important to a given choice context.

Despite these important limitations and

considerations for estimating choice models, individuals who use simulation approaches to study outdoor recreation behaviors can benefit from choice models. Hybrid choice and agent based models may benefit from the statistical rigour and theoretical foundation from choice modeling and the flexibility and ability to uncover emergent behaviors from agent based models. These hybrid models should continue to receive attention by both choice model and agent based model researchers.

REFERENCES

Adamowicz, W.L., Louviere, J.J., & Williams, M. (1994). Combining stated and revealed preference methods for valuing environmental amenities. *Journal of Environmental Economics and Management, 26*, 271-292.

Ajzen, I. (1985). From intentions to actions: A theory of planned behavior. J. Kuhi, & J. Beckman (Editors), *Action-control: From cognition to behavior* (pp. 11-39). Heidelberg: Springer.

Akabua, K.M., Adamowicz, W.L., Phillips, W. E., & Trelawny, P. (1999). Implications of realization uncertainty on random utility models: the case of lottery rationed hunting. *Canadian Journal of Agricultural Economics, 47*, 165-179.

Ben-Akiva, M.E., & Bierlaire, M. 1999. Discrete choice methods and their applications to short-term travel decisions. R.W. Hall (Editor), *Handbook of transportation science: International series in operations research and management science*. (Vol 23, pp. 5-34). Norwell, MA: Kluwer.

Ben-Akiva, M.E., Bolduc, D., & Walker, J. (2001). *Specification, identification, & estimation of the logit kernel (or continuous mixed logit) model*. Unpublished manuscript.

Ben-Akiva, M.E., & Lerman, S.R. (1985). *Discrete choice analysis: Theory and application to travel demand*. Cambridge, MA: MIT Press.

Ben-Akiva, M.E., & Morikawa, T. (1990). Estimation of switching models from revealed preferences and stated intentions. *Transportation Research, Part A, 24*(6), 485-495.

Bhat, C.R. (2002). Recent methodological advances relevant to activity and travel behavior analysis. H.S. Mahmassani (Editor), *Perpetual motion: Travel behavior research opportunities and application challenges* (pp. 381-414). New York: Pergamon.

Bhat, C.R. (2003). Random utility-based discrete choice models for travel demand analysis. H.K. Goulias (Editor), *Transportation system planning: Methods and application* (pp. 1-30). CRC Press.

Boxall, P.C., & Adamowicz, W.L. (2002). Understanding heterogeneous preferences in random utility models: a latent class approach. *Environmental and Resource Economics, 23*, 421-446.

Cascetta, E., & Papola, A. (2001). Random utility models with implicit availability/perception of choice alternatives for the simulation of travel demand. *Transportation Research, Part C, 9*, 249-263.

Daly, A.J., & Zachary, S. (1978). Improved choice models. D.A. Hensher, & Q.Dalvi (Editors), *Determinant of travel choice* (pp. 335-357). Sussex, UK: Saxon House.

Dillman, D.A. (2000). *Mail and internet surveys: The tailored design method*. Toronto, ON: John Wiley & Sons.

Dubin, R.A. (1995). Estimating logit models with spatial dependence. L. Anselin and R.J.G.M. Florax (Editors). *New directions in spatial econometrics* (pp. 229-242). New York: Springer.

Ferguson, M.R., & Kanaroglou, P.S. (1997). An empirical evaluation of the aggregated spatial choice model. *International Regional Science Review, 20*, 53-75.

Fishbein, M., & Ajzen, I. (1975). *Belief, attitude, intention and behavior*. Reading, MA: Addison-Wesley.

Gilbride, T.J., & Allenby, G.M. (2004). A choice model with conjunctive, disjunctive, and compensatory screening rules. *Marketing Science, 23*(3), 391-406.

Haab, T.C., & Hicks, R.L. (1997). Accounting for choice set endogeneity in random utility models of recreational demand. *Journal of Environmental Economics and Management, 34*, 127-147.

Hausman, J.A., & Wise, D.A. (1978). A conditional probit model for qualitative choice: Discrete decisions recognizing interdependence and heterogeneous preferences. *Econometrica, 46*(2), 403-426.

Heckman, J.J. (1981). Statistical models for discrete panel data. C.F. Manski, & D. McFadden (Editors), *Structural analysis of discrete data with econometric applications* (pp. 114-175). Cambridge, MA: MIT Press.

Hensher, D.A, Rose, J.M., & Greene, W.H. (2005). Applied choice analysis: A primer. New York: Cambridge.

Hicks, R.L., & Strand, I.E. (2000). The extent of information: Its relevance for random utility models. *Land Economics, 76*(3), 374-385.

Hunt, L.M., (2007). Examining state dependence and place attachment within a recreational fishing site choice model. *Journal of Leisure Research, 27,* 832-847.

Hunt, L.M., Boots, B.N. & Boxall, P.C. (forthcoming). Predicting fishing participation and site choice while accounting for spatial substitution, trip timing and trip context. *North American Journal of Fisheries Management.*

Hunt, L.M., Boots, B.N. & Kanaroglou, P.S. (2004). Spatial choice modeling: New opportunities to incorporate space into substitution patterns. *Progress in Human Geography, 28*(6): 746-766.

Hunt, L.M., Kushneriuk, R., & Lester, N.L. (2007). Linking agent-based and choice models to study outdoor recreation behavior. *Forest, Snow and Landscape Research.* 81 (1/2), 163-174.

Jakus, P.M., & Shaw, W.D. (1997). Congestion at recreation areas: empirical evidence on perceptions, mitigating behavior and management preferences. *Journal of Environmental Management, 50,* 389-401.

Keane, M.P. (1997). Current issues in discrete choice modeling. *Marketing Letters, 8*(3), 307-322.

Lancaster, K. (1966). New approach to consumer theory. *Journal of Political Economy, 74,* 132-157.

Louviere, J.J., Hensher, D.A., & Swait, J.F. (2000). *Stated choice methods: Analysis and applications.* New York: Cambridge University Press.

Louviere, J.J., & Woodworth, G. (1983). Design and analysis of simulated consumer choice or allocation experiments: An approach based on aggregated data. *Journal of Marketing Research, 20,* 350-367.

Manski, C.F. (1973). *The analysis of qualitative choice.* Unpublished doctoral dissertation, Department of Economics, Massachusetts Institute of Technology, Cambridge, MA.

Manski, C.F. (1977). The structure of random utility models. *Theory and Decision, 8,* 229-254.

McFadden, D. (1974). Conditional logit analysis of qualitative choice behavior. P. Zarembka (Editor), *Frontiers in econometrics* (pp. 105-142). New York: Academic Press.

McFadden, D. (1978). Modeling the choice of residential location. A. Karlqvist, L. Lundqvist, F. Snickars, & J.W. Weibull (Editors), *Spatial interaction theory and planning models* (Vol. 3pp. 75-96). Amsterdam: North Holland Publishing.

McFadden, D. (2000). *Disaggregate behavioral travel demand's RUM side: A 30-year perspective.* Unpublished manuscript.

Morey, E.R. (1999). Two RUMs uncloaked: Nested logit models of site choice and nested-logit models of participation and site choice. J.A. Herriges, & C.L. Kling (Editors), *Valuing recreation and the environment: revealed preference methods in theory and practice* (pp. 65-120). Northhampton, MA: Edward Elgar.

Morey, E.R., Rowe, R.D., & Watson, M. (1993). A repeated nested-logit model of Atlantic salmon fishing. *American Journal of Agricultural Economics, 75,* 578-592.

Parsons, G.R., & Hauber, A.B. (1998). Spatial boundaries and choice set definition in a random utility model of recreation demand. *Land Economics, 74*(1), 32-48.

Parsons, G.R., Jakus, P.M., & Tomasi, T. (1999). A comparison of welfare estimates from four models for linking seasonal recreational trips to multinomial logit models of site choice. *Journal of Environmental Economics and Management, 38,* 143-157.

Parsons, G.R., Massey, D.M., & Tomasi, T. (1999). Familiar and favorite sites in a random utility model of beach recreation. *Marine Resource Economics, 14,* 299-315.

Peters, T., Adamowicz, W.L., & Boxall, P.C. (1995). Influence of choice set considerations in modeling the benefits from improved water quality. *Water Resources Research, 31*(7), 1781-1787.

Phaneuf, D.J., Kling, C.L., & Herriges, J.A. (1998). Valuing water quality improvements using revealed preference methods when corner solutions are present. *American Journal of Agricultural Economics, 80,* 1025-1031.

Provencher, W., & Bishop, R.C. (1997). An estimable dynamic model of recreation behavior with an application to Great Lakes angling. *Journal of Environmental Economics and Management, 33,* 107-127.

Schuhmann, P.W., & Schwabe, K.A. (2004). An analysis of congestion measures and heterogeneous angler preferences in a random utility model of recreational fishing. *Environmental and Resource Economics, 27,* 429-450.

Simon, H.A. (1955). A behavioral model of rational choice. *Quarterly Journal of Economics, 69,* 99-118.

Skov-Petersen, H. (2005). *Feeding the agents – collecting parameters for agent-based models.* Proceeding for Computers in Urban Planning and Urban Management 2005. London. Worl wide web publication http://128.40.111.250/cupum/searchpapers/papers/paper60.pdf accessed April 12, 2007.

Swait, J.F. (1994). A structural equation model of latent segmentation and product choice for cross-sectional revealed preference choice data. *Journal of Retailing and Consumer Services, 1*(2), 77-89.

Swait, J.F., Adamowicz, W.L., & van Bueren, M. (2004). Choice and temporal welfare impacts: Incorporating history into discrete choice models. *Journal of Environmental Economics and Management, 47*(1), 94-116.

Thurstone, L. (1927). A law of comparative judgment. *Psychological Review, 34,* 273-286.

Train, K.E. (1999). Mixed logit models for recreation demand. J.A. Herriges and C.L. Kling (Editors), *Valuing recreation and the environment* (pp. 121-140). Northampton, MA: Edward Elgar.

Train, K.E. (2003). *Discrete choice methods with simulation.* New York: Cambridge University Press.

Walker, J. (2002). The mixed logit (or logit kernel) model: Dispelling misconceptions of identification. *Transportation Research Record, 1805,* 86-98.

Wen, C.H., & Koppelman, F.S. (2001). The generalized nested logit model. *Transportation Research, Part B, 35,* 627-641.

Williams, H.C.W.L. (1977). On the formation of travel demand models and economic evaluation measures of user benefit. *Environment and Planning A, 9,* 285-344.

CHAPTER 12

MODELING INDIGENOUS CULTURAL CHANGE IN PROTECTED AREAS

Laura Sinay
Carl Smith
R.W. (Bill) Carter

Abstract: Minimizing environmental impact, damage to physical resources and interpreting resources for visitors, has been the focus of management in protected areas at both natural and cultural sites. However, there are less tangible aspects of protected areas that need to be managed. Culture has physical or material elements as well as socio-structural and ideational elements. Protecting the culture of traditional peoples is not straightforward because many aspects of culture are intangible, subjective and difficult to measure. This chapter outlines both a framework and modelling approach for assessing cultural impact and predicting the efficacy of cultural change management strategies. The framework is based on one that has been used widely in environmental change management – the Pressure-State-Response (PSR) framework. The modelling approach reported is based on the application of Bayesian Belief Networks. Bayesian Belief Networks provide a range of benefits when modelling cultural change. A study was conducted using the Bayesian Belief Network modeling approach to cultural change management in the Juatinga Ecological Reserve, Brazil. This modelling approach was found to be extremely valuable in the establishment of cultural change indicators, and the assessment of the relative impact of pressures (tourism being one) on culture and the efficacy of management action.

Keywords: Bayesian Belief Networks, culture, protected area management, indicators, modelling

INTRODUCTION

This chapter presents a framework and a modeling approach for assessing cultural impact and predicting the efficacy of cultural change management strategies. A Bayesian Belief Network modeling approach has been applied to cultural change management on a visited protected area inhabited by traditional people. From empirical study, practical indicators and measurements to forecast cultural change were identified. These data, moderated by Delphic review (see Dalkey and Helmer, 1963; Richey et al., 1985), formed the basis for the model. Thus, a procedure is presented for collecting, organizing and analysing data that can be used to model the process of cultural change and define effective strategies to manage this change.

Much of the discourse on cultural

change has been descriptive and explanatory, with few attempts to be predictive. Where indicators of and buffers to change are identified (see Bleasdale and Tapsell, 1996), they tend to be only post-event assessable. The need for a tool with strong predictive power is fundamental to cultural impact assessment and the rationale behind this developmental work.

CULTURE AND CULTURAL CHANGE

In this chapter the word culture is used to describe the totality of mental and physical reactions and activities that characterize the behavior of individuals in relation to their natural environment, to other groups, to members of the group itself and to themselves (Boas, 1965; 1966; Campbell, 1967; Mitchell, 1979; Lett, 1987). This includes all knowledge, beliefs, morals, laws, art and customs, and other expressions and habits acquired by people as members of a society, as well as products of human activity and their role in the life of groups. It refers to the perceptions and standards by which people see cultural resources, traditions and expressions. It is a trait possessed by every person, learnt from living within a community; it is a framework for interpreting and responding to experiences of the world. It is a way of life (Linton, 1945); a blueprint for living (Rogers *et al.*, 1988). Culture is shared between members and is expressed by similar patterns of behavior. However, there is never complete sharing and individuals have a unique appreciation of the group culture and knowledge to contribute to it (Rogers *et al.*, 1988).

Cultures are distinguished by material and non-material elements. Non material elements are the thinking patterns that define standards for deciding what is right, how one feels about it and how to go about it (Goodenough, 1981). This affects how world events are perceived and interpreted as well as the subsequent response (Carter, 2000). Each thinking pattern is unique, thus societies perceive, interpret and respond in different ways to the same event (Matta, 1981). Cultural thinking patterns are based on:

- beliefs - symbolic statements about reality;
- values - symbolic statements about what is right and important; and
- norms - symbolic statements about what is acceptable behavior for community members (Fishbein and Ajzen, 1975).

Beliefs, values and norms contribute to build social structures, which promote stability and protect societies from anarchy (Campbell, 1967). In turn, these bring well-being to the community and the individual. These non-material or ideational elements of culture direct and restrict the possible range of responses a cultural group will have to a stimulus (Levi-Strauss, 1963-1976; Campbell, 1967; Matta, 1981). Within the environmental context (see Steward, 1979), ideational elements will influence and define a culture's material and observable elements or cultural expressions. These strengthen the identity of a society and buffer it from change (Carter, 2000). The material elements of culture, in contrast to non material elements, are easily observable (Carter, 2000). Thus, cultural differences and similarities can be perceived, especially through expressions, traditions and resources (Carter, 2000).

By inventions and discoveries, individuals are continually adding knowledge to culture. Hence, cultures are constantly changing (Levi-Strauss, 1963-1976; Boas, 1965; Steward, 1979; Goodenough, 1981; Matta, 1981; Laraia, 1989). Environmental conditions and social systems may influence or prevent inventions and discoveries as well as the adoption of innovation (Boas, 1966; Campbell, 1967; Steward,

1979; Cavalli-Sforza and Cavalli-Sforza, 1995). Therefore, environmental conditions and socio-cultural institutions are permissive or prohibitive of cultural change (Steward, 1979). However, they tend not to be deterministic of how change will occur, because between physical environments and human responses, there is always a cultural pattern to drive change (Steward, 1979). In addition, between the social system and human response there are always limitations imposed by the environment.

Cultural change is the process by which alterations occur in the structure and functioning of a culture (c.f. social change as defined by Rogers *et al.*, 1988). Of relevance here, is that cultural change will be reflected by altered cultural expressions, traditions and individual and collective perceptions of cultural resources. Change is inherent in culture. It occurs through a process generally described as socialisation (Rogers *et al.*, 1988; Haralambos *et al.*, 1996), where culture is transmitted to and learned by associates. The process of transmission and learning is always within the culture through experiences of individuals, but may be driven by experiences of other cultures.

Social systems establish the grounds on which changes can take place. Usually, cultural changes are acceptable if they bring more benefits than harm to the individual and community (theory of social exchange); and if they do not confront the social representations of the group; the concepts, statements, explanations, personal constructs and mental models common to a group (Farr and Moscovici, 1984).

The process of cultural change can be divided into internal changes that relate to inventions and discoveries that originate and are transmitted among one cultural group, and external changes that result from contact with other cultural groups (Matta, 1981). In this case, one or both groups will adopt an invention or discovery identified by the other group (Levi-Strauss, 1963-1976; Goodenough, 1981).

Since some elements of a culture are more easily perceived than others, then perception of change in a cultural expression does not necessarily equate to cultural change. Carter (2000) first discussed cultural change using a multi-dimensional model, in which the core of the model represents a community's non-observable or ideational elements and linked observable elements or expressions external to this core (see also Carter and Beeton 2004 and Figure 1). In this representation, numbers indicate different cultural expressions, lines between numbers represent links bound by tradition and lines between the core and the numbers represent the time allocated by the community to performing the tradition and the importance attributed to the expression for community and personal well-being. These links to the core are also bound by, and reinforced by tradition. As the time, or the importance, allocated by a community to performing a tradition, declines or increases, so does the link between the core (the ideational dimensions) and the expression via a change in tradition, changing the shape of the model and the overall expression of the culture. Inevitably, this alters the link to other expressions. These concepts are presented in Figure 1 for a hypothetical culture from three perspectives in two moments: a pre-change state at time t^0 and a post-change state at t^{0+1}, where time spent on tradition (5), and its importance, increases.

From perspectives A and B, the change in expression 5 can be identified, although it is more evident from perspective A. From Perspective C, the change is hidden and not observable. It is masked by other expressions. In all cases however, the increase in time spent on tradition 5 or the importance attributed to it, would change the length of all its links (but not un-linked

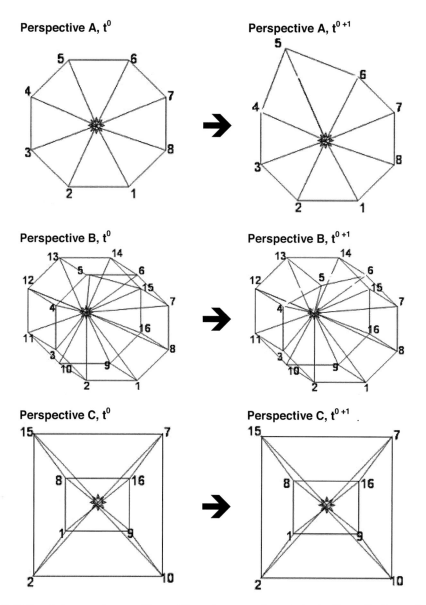

Figure 1. Cultural change observed from different perspectives (adapted from Carter and Beeton, 2004)

expressions), changing the shape of the overall model of the culture. Since the interval and available time are constant, then, despite some expressions being buffered by their importance, ultimately the time spent on one or more expressions must decrease, with an ultimate decrease in the importance attributed to them. Hence, a change in one expression ultimately has a 'ripple effect' throughout the model demanding individual and community adjustment to maintain the dynamic equilibrium that is the culture.

The model indicates that:

- the view gained from where an observer stands limits their ability to perceive cultural expressions, links, changes and ultimately cultural change, and

- all expressions that define a culture are vulnerable to change given a change to one, and the ability to absorb change will depend on the number of links and community capacity to absorb or limit the ripple effect.

Hence, all expressions have a role in maintaining cultural stability and integrity.

TRADITIONAL COMMUNITIES

Today, there are about 300 million traditional people occupying 20% of the world surface, many living in protected areas (Kemf 1993b; The Office of the High Commissioner for Human Rights 2006). They inhabit all continents (except Antarctica) and most terrestrial ecosystems (Martin 1993). Each has their traditional religion, cuisine, language, laws, and way of explaining life (Martin 1993) and are specifically identified as, for example, Aborigine, Maori, Inca, Pataxó, Yanomami, Guarani, Quilombola or Caiçara. They often live in small communities, based on some form of kinships linkage (Jones et al. 1992; Diegues 1994; Cavalli-Sforza and Cavalli-Sforza 1995). Many normally make decisions by consensus and do not have any centralised state organisation (Jones et al. 1992; Diegues 1994; Cavalli-Sforza and Cavalli-Sforza 1995). Some live wholly or partly on wild produce, gathering plants, hunting or fishing (Jones et al. 1992; Diegues 1994). They tend to have low levels of production and low inputs (Jones et al. 1992; Diegues 1994). A major point of difference from other cultural groups is their largely self-sufficient, closed economy (Jones et al. 1992; Diegues 1994). Most traditional communities remain separate from urban industrial civilization, maintaining their own systems of thought and values (Jones et al. 1992; Nogara and Diegues 1994). This includes a profound involvement with the land in which they live, which is often held to be sacred (Roberts 1878; Kemf 1993b; 1993a; Ramos 1993; Diegues 1994; Cavalli-Sforza and Cavalli-Sforza 1995; Ribeiro 1997). Worldwide pressures influencing cultural change in traditional people include loss of land due to western economic activities, encroachment (Jones et al. 1992), protected areas (Lado 1992; Diegues 1994), introduction of diseases (Jones et al. 1992), stereotyping (Evans-Pritchard 1989), racism (Roberts 1978), religion (Wilson et al. 1996; Ribeiro 1997), education (Archibold and Davey 1993), laws (Kemf 1993b) and tourism (Xie 2003).

While cultural change is an inherent characteristic of culture and inevitable (Carter 2000), globalization forces are affecting multiple expressions concurrently. The web of cultural expression is under constant stress with no time to stabilize and communities are not in control of the change process. Cultures collapse and are absorbed into the dominant culture and individuals are dislocated spatially, socially and culturally. For traditional peoples under globalization pressures, few individuals have prospered, and the majority suffer land alienation, high rates of unemployment, extreme poverty, oppression and violence (Jones et al. 1992). From egalitarian and social justice perspectives, understanding and managing cultural change is important to avoid these social problems and inequities. It is equally important for humanistic objectives of maintaining cultural diversity and avoiding homogeneity. But for tourism, it is an economic and sustainability issue, because indigenous culture is a major product for tourism world-wide and tourism itself may be a major cause of indigenous cultural change.

A BRIEF DESCRIPTION OF THE CASE STUDY AREA

Cultural change was examined in the Juatinga Ecological Reserve, Brazil, which is located in the extreme south of Rio de

Janeiro State. The reserve was established on 30 October 1992 by the State Act 17,981 to protect 8000 hectares of Atlantic Forest (SEMADS, 2001) and the 1500 Caiçaras living within its boundaries (SOS Mata Atlantica 2002; Silveira and Brandão 2005).

The Caiçara culture was formed by the union of indigenous people and fugitive slaves (Africans), pirates and colonizers. The main difference between the Caiçara people and the rest of Brasilian society is that their isolation has led to the retention of traces of the indigenous and African culture. With this, the environment in which they have lived for centuries is conserved.

Ciranda and Folia dos Reis are traditional celebrations of the Caiçaras. During these celebrations, Caiçaras dance and sing. In 2005, these celebrations were rare. Instead, Caiçaras were singing evangelic songs and no longer dancing. Caiçaras have also adopted evangelic dresses. Men wear long pants and long sleeved shirts (even on hot days of 44°C), and women wear long skirts and do not cut their hair. There are, however, some Caiçaras who are Catholic. These men tend to wear board shorts and the women cut their hair.

Because of their isolation, Caiçaras had limited access to technology (including energy and telecommunications) or markets for their produce or purchases. Consequently, they were self-sufficient, living a subsistence existence through fishing, gardening, hunting and gathering. During the 1970s, the construction of a national road (BR 101) facilitated access to the reserve by tourists from São Paulo and Rio de Janeiro via Paraty; about two hours away (by boat) from the Reserve. Since then, tourism started to be part of Caiçara life. In 2005, the Reserve received around 9000 tourists, almost 30 times more than in 1985. Some villages have received steady slow growth; others have experienced exponential growth and collapse as visitors become dissatisfied with the decline in the attractiveness of specific areas (Figure 2).

METHODS

Figure 3 summarises the overall process used to develop a cultural change model. The main steps in the process involved (a) identifying the indicators of cultural change, the reasons for change (pressures) and the possible actions required to manage change, (responses) (b) structuring these factors

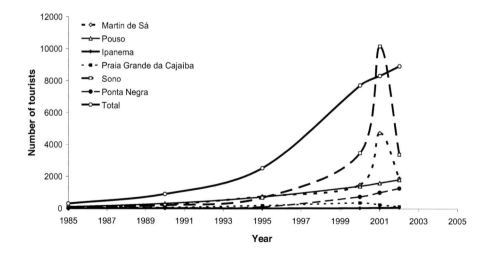

Figure 2. Tourism growth at indicative villages in Juatinga Ecological Reserve (1985 to 2004)

into an influence diagram, (c) developing a predictive model from the influence diagram, and (d) using the model for analysis. The detail of each of these steps is outlined in the following sections.

FIELD SURVEY

A survey was conducted within Juatinga Ecological Reserve to identify how people perceive and describe the Caiçaras cultural change process, the factors influencing change and the elements of Caiçaras culture that are resistant to change. Caiçaras, tourists, members of Rio de Janeiro State Forest Institute (IEF) and a local NGO (Verde Cidadania) were interviewed. The data collected were organized into:

- indicators: how people perceive and describe cultural change in the Caiçaras communities;
- pressures: factors influencing change; and
- responses: management interventions that may affect the extent of change.

BUILDING AN INFLUENCE DIAGRAM

The State-Pressure-Response (PSR) framework (Figure 4) was used to organise the survey outcomes into a model structure. This framework links pressures and responses to the current state of the environment, which is measured using indicators. The PSR framework has been used for environmental change management by most OECD countries and the World Bank (OECD, 1993). In the context of cultural change modeling, state of the environment indicators were substituted with state of culture indicators.

The resulting cultural change influence diagram is shown in Figure 5. Two points in time were used in the influence diagram as references for assessing the relative change in culture over the short term – 2008 to 2010. Pressures listed down the left-hand side of the diagram influence one or more indicators at 2008. Indicators at 2010 are replicas of the indicators at 2008 and are influenced by the corresponding 2008 indicator plus one or more responses listed down the right-hand side of the diagram. In other words, the model assesses the state of culture at 2008 given existing pressures, and the state of culture at 2010 given the indicator levels at 2008 and the

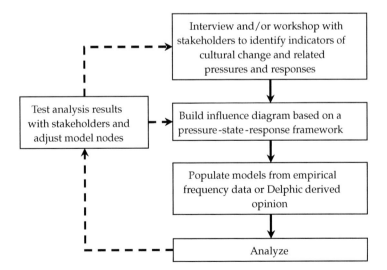

Figure 3. Process for building a model for assessing cultural impact and predicting the efficacy of management strategies

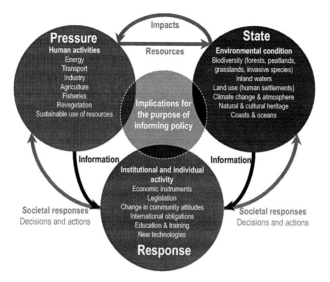

Figure 4. The pressure-state-response framework (Smith, 1999)

responses implemented during the period 2008 to 2010.

To accommodate multiple opinion about the influence of pressures and responses on cultural change indicators at 2008 and 2010, a *Different Perceptions* component was added to the influence diagram (Figure 5). This allows the different opinions to be selected in the model, thereby modifying model behavior according to the perceptions or beliefs of different groups of people. Weights were used to accommodate differences in the importance placed on cultural change indicators by groups of people (such as Caiçaras, tourists, members of Rio de Janeiro State Forest Institute (IEF) and a local NGO). The weight for an indicator was calculated as the percentage of people within a group observing change in that indicator, normalised by the percentage of people within that group observing change in all other indicators. For example, the weight given to Indicator A for Caiçaras would be: Percentage of Caiçaras observing a change in Indicator A, divided by, the sum of the percentages of Caiçaras observing change in all other indicators.

BUILDING A PREDICTIVE CULTURAL CHANGE MODEL

A Bayesian Belief Network (BBN) (Jensen, 2001) was used to convert the cultural change influence diagram into a predictive cultural change model. BBNs use conditional probabilities to quantify the cause and effect relationships within an influence diagram (Cain, 2001; Nadkarni and Shenoy, 2004). Insert Table 1 is an example conditional probability table (CPT) that relates *Number of tourists* to *Number of Caiçaras surfing*. The first row in this table stores the probabilities that *Number of Caiçaras surfing* will *Decline more than 20pc, Decline less than 20pc, etc* , given that the *Number of tourists Declines more than 50pc*. Figure 6 is the BBN showing this relationship, which consists of two *nodes* representing variables, each with possible *states*, and an arrow or *link* representing cause and effect. In this example, the arrow is from *Number of tourists* (known as the parent node) to *Number of Caiçaras surfing* (known as the child node) because the belief is that *Number of tourists* influence *Number of Caiçaras surfing*.

The probabilities required to populate

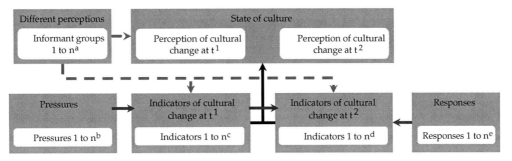

Figure 5. Structure of an influence diagram for cultural change

CPTs within a BBN can come from a number of sources, including frequency data obtained from survey or monitoring, probabilities obtained from other models or probabilities elicited from people. In the case of the cultural change model developed for Juatinga Ecological Reserve, probabilities were elicited from people because data from which probabilities could be obtained were not available. Four respondents were selected to provide probability estimates for indicator levels at 2008 and 2010 under different pressure and response scenarios. These respondents were selected based on their experience with Caiçaras and the Juatinga Ecological Reserve.

A five point percentage scale of improvement or decline was used as the basis for probability estimates (Table 2). Except for one indicator (*Number of Caiçaras surfing*), a 20% improvement target was set for all cultural change indicators as the maximum expected level of improvement. Numerical values were also assigned to each of the states included within the indicator scale. A numerical value of zero (0) was given to the best state in the scale (the Target state) because achieving this state would mean the least degree of cultural change. Within the cultural change model, the numerical values given to the states of an indicator are multiplied by the weight assigned the that indicator. The weighted sum of all cultural change indicators then becomes the *Perception of cultural change* (Figure 5). Hence, the weighted sum of indicators at 2008 becomes the *Perception of cultural change at 2008* and the weighted sum of indicators at 2010 becomes the *Perception of cultural change at 2010*.

Probabilities for the indicator levels were elicited from each of the four respondents, giving four different opinions about the influence of pressures and responses on

Table 1. Example of a CPT

Number of tourists	Number of Caiçaras surfing				
	Declines more than 20pc	Declines less than 20pc	Remain stable	Increase less than 50pc	Increases more than 50pc
Declines more than 50pc	0.49	0.23	0.18	0.09	0.03
Declines less than 50pc	0.37	0.19	0.18	0.15	0.12
Remains stable	0.23	0.15	0.21	0.21	0.21
Increases less than 50pc	0.13	0.10	0.21	0.26	0.30
Increases more than 50pc	0.04	0.05	0.21	0.31	0.39

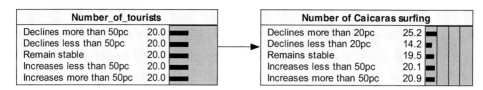

Figure 6. A simple BBN showing the relationship between number of tourists and number of Caiçaras surfing

Table 2. Five point scale used for cultural change indicators

State	Description	Numerical Value
Target	The indicator improves by 20% or more	0
Less than 20% improvement	The indicator improves by less than 20%;	0.001 to 0.25
Remains stable	The indicator does not change	0.251 to 0.5
Less than 50% decline	The indicator worsens by up to 50%	0.501 to 0.75
Greater than 50% decline	The indicator worsens by more than 50%	0.751 to 1

the cultural change indicators. During the probability elicitation process, a CPT calculator developed by Cain (2001) was used to (a) maintain logical consistency in the estimated probabilities, and (b) reduce the number of probabilities that had to be elicited to populate the BBNs. The CPT calculator works by selecting particular scenarios from a CPT. These are (a) the best-case scenario where all of the parent nodes are in the best state, (b) the worst-case scenario where all of the parent nodes are in the worst state, and (c) scenarios where only one parent node is not in the best state (see Table 3). Probabilities for these scenarios were elicited and the CPT calculator used to interpolate probabilities for all remaining scenarios in the CPTs.

As a final check, and to validate the probability estimates supplied by the respondents, a sensitivity analysis was performed using the cultural change model

Table 3. Example CPT where shaded rows are scenarios where probabilities are elicited first and used by the CPT calculator (in this example "Low" is the best state and "High" is the worst state for all pressures).

Pressures		Indicator 1 at 2008				
Pressure 1	Pressure 2	Target (20% Improvment)	Less than 20% Improvement	Stable	Less than 50% Decline	Greater than 50% Decline
Low	Low					
Low	Moderate					
Low	High					
Moderate	Low					
Moderate	Moderate					
Moderate	High					
High	Low					
High	Moderate					
High	High					

Table 4. Informants per group (%) observing cultural change indicators

Group	Number of Caiçaras wearing traditional clothes	Number of Caiçaras eating non traditional dishes	Number of Caiçaras emigrating	Number of Caiçaras surfing	Time spent by Caiçaras on traditional subsistence activities	Number of Caiçaras maintaining traditional beliefs and values	Time spent by Caiçaras performing traditional arts
Caiçaras (n=63)	4.8%	16%	0%	0%	68%	22%	35%
NGO & IEF (n=12)	0%	0	0%	8%	67%	25%	8%
Tourists (n=57)	1.8%	7%	3.5%	5%	39%	40%	11%

to determine the relative influence of pressures and responses on each of the cultural change indicators. Sensitivity analysis was performed by systematically changing the state within the *Pressure* and *Response* nodes within the model and observing the effects on each indicator. The results of this analysis were organized into tables and returned to the respondents for their review. Where the results did not match expectations, respondents were asked to adjust the CPTs accordingly.

RESULTS

Perceptions of cultural change

Data collected during the field survey revealed that there was a significant difference in the perception of cultural change between informant groups ($p^2 < 0.05$). For example, 35% of Caiçaras observed change in traditional arts, while only 11% of tourists observed this change. Hence, the change perceived depends on the observer (in this case the informant group). The percentage of informants within each group observing change for each indicator is summarised in Table 4. The normalised importance weights calculated for each indicator are summarised in Table 5.

CULTURAL CHANGE INFLUENCE DIAGRAM

The influence diagram of cultural change developed for the Juatinga Ecological Reserve is shown in Figure 7. Table 6 lists the pressures and responses that are linked to each of the cultural change indicators in the diagram.

Table 5. Normalized weights for cultural indicators

Group	Number of Caiçaras wearing traditional clothes	Number of Caiçaras eating non traditional dishes	Number of Caiçaras emigrating	Number of Caiçaras surfing	Time spent by Caiçaras on traditional subsistence activities	Number of Caiçaras maintaining traditional beliefs and values	Time spent by Caiçaras performing traditional arts	Weighted sum
Caiçaras (n=63)	0.03	0.11	0.00	0.00	0.46	0.16	0.24	1.0
NGO & IEF (n=12)	0.00	0.00	0.00	0.07	0.62	0.23	0.07	0.99
Tourists (n=57)	0.02	0.07	0.03	0.05	0.36	0.37	0.10	1.0

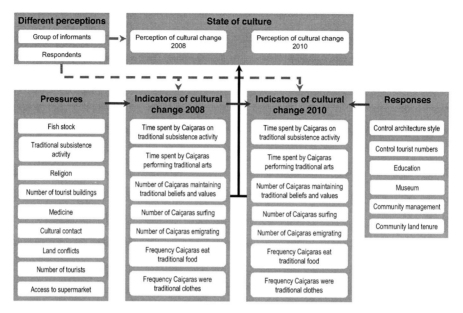

Figure 7. Structure of the influence diagram for cultural change at Juatinga Ecological Reserve

Table 6. Indicators of cultural change, pressures and responses identified by informants

Indicator	Pressures	Responses
Time spent by Caiçaras on traditional subsistence activities	Number of tourists Fish stock Restrictions to traditional subsistence activities	Control tourists numbers Community engagement on management
Time spent by Caiçaras performing traditional arts	Number of tourists buildings Religion	Architectural control
Number of Caiçaras maintaining traditional beliefs and values	Number of tourists Religion Cultural contact Time spent on traditional subsistence activities Medicine Number of Caiçaras emigrating	Education Community management Control tourists numbers Museum
Number of Caiçaras surfing	Number of tourists	None
Number of Caiçaras emigrating	Number of land conflicts and intensity of land speculation Religion Restrictions to traditional subsistence activities Number of tourists	Community control over land tenure
Number of Caiçaras eating non traditional dishes	Access to supermarket Number of tourists	Education
Number of Caiçaras wearing traditional clothes	Religion Number of Caiçaras watching TV Number of tourists	None

Table 7. Pressures and their states

Pressures	States
Access to supermarket	Difficult
	Stable
	Facilitated
Cultural contact	Declines
	Remains stable
Fish stock	Increases more than 50%
	Increases less than 50%
	Remains stable
	Declines less than 50%
	Declines more than 50%
Medicine	Mostly traditional
	Mostly western
Number of Caiçaras emigrating	Target (declines by more than 20%)
	Declines less than 20%
	Remains stable
	Increases less than 50%
	Increases more than 50%
Number of Caiçaras watching TV	Declines
	Remains stable
	Increases
Number of land conflicts and intensity of land speculation	Declines more than 50%
	Declines less than 50%
	Remains stable
	Increases less than 50%
	Increases more than 50%
Number of tourists	Declines more than 50%
	Declines less than 50%
	Remains stable
	Increases less than 50%
	Increases more than 50%
Number of tourists buildings	Declines more than 50%
	Declines less than 50%
	Remains stable
	Increases less than 50%
	Increases more than 50%
Religion	Traditional
	Other
Time spent on traditional subsistence activities	Target (declines by more than 20%)
	Declines less than 20%
	Remains stable
	Increases less than 50%
	Increases more than 50%
Traditional subsistence activities	Allowed
	Prohibited

The states assigned to each of the pressures in the influence diagram are listed in Table 7. Table 8 lists the states assigned to all other nodes in the influence diagram and the methods used to populate their probability tables.

CULTURAL CHANGE MODEL

Figure 8 is the complete cultural change model for Juatinga Ecological Reserve. Note that the worst state of all pressures and the best state of all responses have

Table 8. Nodes, their states and methods used to populate their probability tables

Nodes		States	Method used to populate probability tables
Pressures		See Table 7	Equal prior probability assigned to all states
Responses		True False	Equal prior probability assigned to all states
Different perceptions	Group of informants	Caiçaras Tourists NGO and IEF members	Populated with the percentage of informants interviewed per group, 43%, 47% and 10% respectively.
	Respondent	Respondent 1 Respondent 2 Respondent 3 Respondent 4	Equal prior probability assigned to all states
Perception of cultural change		Low (weighted sum of indicators is <0.33) Moderate (weighted sum of indicators is between 0.33 and 0.66) High (weighted sum of indicators is >0.66)	The type and importance of change depends on the observer (in this case the informant group). This difference in perception was incorporated into the BBN model using indicator weights. The CPTs for Perception of cultural change nodes were populated using the weighted sum of indicator levels. Below are the equations used to populate the Perception of cultural change nodes at 2008 and 2010. These equations incorporate the indicator weights for Caiçaras, Tourists and the NGO respectively: Caiçaras = 0.003* Number of Caiçaras wearing traditional clothes +0.11* Number of Caiçaras eating non traditional dish+0* Number of Caiçaras emigrating +0* Number of Caiçaras surfing +0.46* Time spent by Caiçaras on traditional subsistence activities +0.16* Number of Caiçaras maintaining traditional beliefs and values +0.24 *Time spent by Caiçaras performing traditional arts Tourists = 0.02 * Number of Caiçaras wearing traditional clothes + 0.07* Number of Caiçaras eating non traditional dish + 0.03* Number of Caiçaras emigrating + 0.05*Number of Caiçaras surfing+ 0.36 * Time spent by Caiçaras on traditional subsistence activities + 0.37* Number of Caiçaras maintaining traditional beliefs and values + 0.10 *Time spent by Caiçaras performing traditional arts NGO = 0 * Number of Caiçaras wearing traditional clothes + 0 * Number of Caiçaras eating non traditional dish+ 0 * Number of Caiçaras emigrating + 0.07 * Number of Caiçaras surfing + 0.62 * Time spent by Caiçaras on traditional subsistence activities + 0.23 * Number of Caiçaras maintaining traditional beliefs and values + 0.07* Time spent by Caiçaras performing traditional arts

Indicators of cultural change	Target	Probabilities elicited from experts (respondents)
	Improved up to 20% Remain stable Declined up to 50% Declined more than 50%	

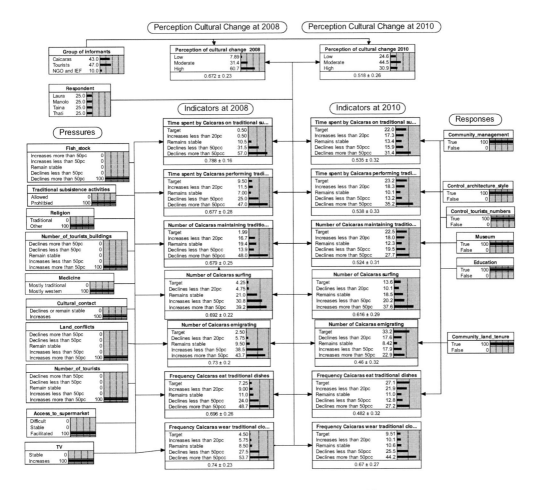

Figure 8. Cultural change model for Juatinga Ecological Reserve with no pressure or response scenario selected

been selected. Uniform probability distributions have been given to the opinion of Respondents (see Respondents node) who were the experts that provided probabilities for the cultural change indicators at 2008 and 2010. The probability distribution given to the *Group of informants* node represents the percentage of people interviewed in the field survey falling within each group.

Table 9. Pressures and responses ranked according to their influence on each cultural change indicator indicator

Indicator	Pressures and responses ranked according to their influence on each indicator	Change in the probability of the best state of each indicator due to a change in the state of the pressure or response from its best to worst state
Time spent by Caiçaras on traditional subsistence activities	1. Control tourists numbers	16.4%
	2. Number of tourists	11.4%
	3. Community engagement on management	9.3%
	4. Fish stock	4.4%
	5. Restrictions to traditional subsistence activities	3.3%
Number of Caiçaras performing traditional arts	1. Religion	18.6%
	2. Number of tourists buildings	13.2%
	3. Architectural control	7.4%
Number of Caiçaras eating non traditional dishes	1. Education	18.3%
	2. Number of tourists	10.7%
	3. Access to supermarket	2.7%
Number of Caiçaras dressing on a new style	1. Number of tourists	16.3%
	2. Religion	1.8%
	3. TV	1.5%
Number of Caiçaras emigrating	1. Community control over land tenure	22.8%
	2. Number of land conflicts and intensity of land speculation	14.6%
	3. Number of tourists	10.1%
	4. Restrictions to traditional subsistence activities	3.0%
	5. Religion	1.0%
Number of Caiçaras maintaining traditional beliefs and values	1. Number of Caiçaras emigrating	9.6%
	2. Education	7.6%
	3. Religion	6.5%
	4. Community engagement on management	4.4%
	5. Time spent by Caiçaras on traditional subsistence activities	3.3%
	6. Control tourists numbers	2.5%
	7. Cultural contact	2.2%
	8. Medicine	1.9%
	9. Number of tourists	2.5%
	10. Museum	1.2%

ANALYSIS

Assessing the Influence of Pressures and Responses on Indicators of Cultural Change

Table 9 summarizes the results of sensitivity analysis conducted on the indicators of cultural change. It ranks pressures and responses with respect to their influence on the best state (Target) of each of the cultural change indicators. These results suggest that tourism, including its related impacts and responses, has a relatively large influence on all indicators, except on the number of Caiçaras maintaining traditional beliefs and values.

Assessing the Resistance and Resilience of Culture

Table 10 summarizes the probability of Targets being achieved when all pressures are in their worst state and all responses are in place. That is, it represents the resistance (the ability of culture to absorb pressures) and resilience (the ability of culture to return to its original condition once

Table 10. Probability that the Target state of cultural change indicators will be met in 2008 and 2010

Indicator of cultural change	Probability of Target state being met in 2008	Probability of Target state being met in 2010	Change
Time spent by Caiçaras on traditional subsistence activities	0	22%	22%
Number of Caiçaras maintaining traditional beliefs and values	1.99%	22.5%	20.51%
Number of Caiçaras emigrating	2.5%	33.2%	30.7%.
Number of Caiçaras dressing on a new style	4.50%	9.5%	5.0%
Number of Caiçaras eating non traditional dishes	7.25%	27.1%	19.85%
Number of Caiçaras performing traditional arts	9.5%	28.2%	18.7%

Table 11. Probability of perception of cultural change at 2008 and 2010 under best and worst case scenarios

Scenario	Perception of cultural change 2008			Perception of cultural change 2010		
	Low	Moderate	High	Low	Moderate	High
Best case (all pressures and responses in the best state)	68.2%	23.81%	7.99%	61.20%	29.96%	8.84%
Worst case (all pressures and responses in the best state)	7.89%	31.41%	60.70%	10.70%	37.10%	52.20%

Table 12. Influence of pressures on the perception of cultural change at 2008

Pressure	Change caused by pressure in the probability of Perception of cultural change being Low at 2008
Tourism	22.90%
Policy	19.90%
Religion	10.60%
Fish stock	5.60%
Land conflicts	1.60%
Cultural contact	1.40%
Medicine	1.20%
Access to supermarket	0.90%
TV	0.00%

Table 13. Influence of responses on the perception of cultural change at 2010

Response	Change caused by response in the probability of Perception of cultural change being Low at 2010
Control tourists numbers	8.6%
Community management	4.6%
Education	3.9%
Control architecture	1.7%
Museum	0.3%
Community land tenure	0.2%

responses are implemented) of each expression of culture included in the model (see Carter 2000). These results suggest that time spent by Caiçaras on traditional subsistence activities is the only expression of culture that may not reach its target state by 2008 under maximum pressure. Therefore, it is the least resistant. However, if responses are implemented, it may recover by 2010. The least resilient aspect of culture is the *Number of Caiçaras dressing in a new style*.

Assessing the Influence of Pressures and Responses on the Perception of Cultural Change

Table 11 summaries the results of sensitivity analysis on the *Perception of cultural change*. The analysis was performed by selecting the best state followed by the worst state for all pressures and responses and recording the outcomes for *Perception of cultural change* under the best and worst case scenarios. The results indicate that, under the worst case scenario (all pressures and responses in the worst state), the *Perception of cultural change* is most likely to be high, however there is still a small probability that it will be low. In other words, a small probability that cultural change will go unnoticed.

Table 12 summarises the influence of individual pressures on the probability of the *Perception of cultural change* being Low at 2008. Table 13 summarises the influence of individual responses on the probability of the *Perception of cultural change* being Low at 2010. The results indicate that at 2008, *Tourism* and *Policy* are the most influential pressures (Table 12), while *TV*, *access to supermarket* and *medicine* have almost no effect on the *Perception of cultural change*. *Controlling tourist numbers* is the most influential response (Table 13), followed by *Community engagement on management* and *Education*. *Community land tenure* is the least effective response.

Assessing the Influence of Cultural Change Indicator Weights on the Perception of Cultural Change

As described in the methods, weights were used in the cultural change model to

Table 14. Probability of perception of cultural change at 2008 and 2010 under best and worst case scenarios using equally weighted cultural change indicators

Scenario	Perception of cultural change 2008			Perception of cultural change 2010		
	Low	Moderate	High	Low	Moderate	High
Best case (all pressures and responses in the best state)	55%	37.23%	7.77%	54.3%	37.89%	7.81%
Worst case (all pressures and responses in the best state)	7.69%	20.51%	71.8%	8.84%	37.86%	53.3%

Table 15. Influence of pressures on the perception of cultural change at 2008 using equally weighted cultural change indicators

Pressure	Change caused by pressure in the probability of Perception of cultural change being Low at 2008
Tourism	13.70%
Policy	12.60%
Religion	5.30%
Land conflicts	3.40%
Fish stock	1.20%
Access to supermarket	0.90%
Cultural contact	0.70%
Medicine	0.50%
TV	0.40%

accommodate the relative importance of each cultural change indicator to different groups of people (such as Caiçaras, tourists, members of Rio de Janeiro State Forest Institute (IEF) and a local NGO). However, this introduces the bias of these different groups into the model. To assess the influence of removing the weighting bias in the model, all indicators of cultural change were given equal weight.

Table 14 summarizes the probabilities of *Perception of cultural change* at 2008 and 2010 under best and worst case scenarios with equally weighted indicators. The results are similar to those obtained previously using weighted indicators (Table 11) in that, under the worst case scenario, the perception of cultural change is most likely to be high. However, the probability of the perception of cultural change being high is slightly increased. Under the best case scenario, the *Perception of cultural change* tends to be low, however, the probability of the perception of cultural change being low is slightly reduced.

Table 15 and 16 summarise the influence of individual pressures and responses on the probability of the *Perception of cultural change* being *Low* at 2008 and 2010 respectively using equally weight indicators. For pressures, the results are similar to those obtained using weighted indicators (Table 12) in that *Tourism*, *Policy* and *Religion* remain the top three most influential pressures. However, *Land conflicts* and *Access to supermarkets* become more influential when equal indicator weights are applied. For responses, *Controlling tourist numbers* remains the most influential response under equally weighted indicators (Table 16), yet *Community land tenure* jumps from the last (Table 13) to the second most influential response (Table 16).

DISCUSSION

Policy

Policy is a formal strategy established to address potential threats, such as tourism and land conflicts. It is normally seen to be a response, targeting maintenance of cultural integrity. Yet, at Juatinga respondents identified existing government policy as a pressure for change. This is because existing policy creates confusion about issues such as land tenure, tourism management and use of natural resources.

If policy was adequate then tourism

Table 16. Influence of responses on the perception of cultural change at 2010 using equally weighted cultural change indicators

Response	Change caused by response in the probability of Perception of cultural change being Low at 2010
Control tourists numbers	4.0%
Community land tenure	3.0%
Community management	2.0%
Control architecture	1.8%
Education	1.4%
Museum	0.2%

would be controlled so that its negative impacts are avoided, community would hold land tenure so that land conflicts would cease, restrictions to the use of natural resources could be constrained to limit change to traditional economy, and community would participate in the protected area management. Yet as policy is inadequate at Juatinga, tourism is the main cause of cultural change, the frequency and intensity of land conflicts is increasing, the use of natural resources is prohibited and local community does not participate in management.

For Juatinga Ecological Reserve current policy is perceived to be one of the strongest pressures on cultural change. Yet it could be the most effective response. Therefore, to avoid radical cultural change, an effective policy framework is essential and it must be beyond 'paper policy' to include proactive action.

Pressure assessment

Tourism and *Religion* were found to be the two most influential pressures on the *Perception of cultural change*. *TV* was found to be the least influential pressure. The order of the remaining pressures depended on the weights of importance associated to the indicators of cultural change. This suggests that the weights of importance only affect the ranking of pressures with moderate pressures, pressures with very low or very high influence are either so weak or so strong that their importance weights do not affect their rank influence on indicators.

Intervention assessment

Of the responses assessed, controlling tourist numbers was found to be the most influential response in managing cultural change. Yet, the tourism literature suggests that the impacts of tourism depends not only on the number of tourists, but also on the type of tourists and other tourism and community variables (see Carter 2000). In addition, tourist numbers are affected by, and affect, the type of tourists. Hence, controlling the type of tourists may be as effective as controlling numbers. This suggests that the informants approached during the field survey component of this study had limited knowledge of potential management options, and therefore it is necessary to include a broader range of experts to diversify the responses available for managing pressures such as tourism.

Indicators of cultural change

Most often, cultural change has been studied using anthropological approaches, through which changes are described, rather than measured. Carter (2000) and Carter and Beeton (2004) introduced the idea of measuring cultural change through the time spent by a community on cultural expressions and the number of community members performing it. This research builds on Carter's findings as it identifies, empirically, broad indicators of cultural expressions that can be measured through time (time spent on traditional subsistence activities and performing traditional arts), and through the number of community members performing the expression (eating traditional cuisine, dressing traditional clothes, emigrating and maintaining traditional beliefs and values).

This research uses the perception of cultural change as an instrument to assess cultural change. Yet, the perception of cultural change depends on the perspective of the observer, which may be from inside the culture (Caiçaras informants), from the edge (respondents with familial associations with the Caiçaras) or from outside (other informants and respondents). In addition, not all expressions of a culture can be observed from every perspective (see Figure 1). Therefore, it is possible that a relevant change may be occurring and few, if any, informants observe it. This was the case with emigration, which reduced in less than 10 years the Praia Grande da Cajaiba community from 24 to three families. If this tendency continues, there will soon be no community to give continuity to culture, imposing great risk for the survival of that culture. Only 1% of field informants identified emigration as an issue in this study. Thus, the weight of this indicator on the total perception of cultural change was under estimated.

This means that the percentage of informants identifying a change may not be a valid measurement of the importance weight of an indicator. Instead, it is necessary to interview a broader group of informants (including people with different experiences, such as experts on tourism and culture management) to guage the importance of indicators. The results of this study suggest that within the cultural change model developed, indictor importance weights do not affect the relative influence of strong and weak pressures and responses on the overall perception of cultural change. In other words, pressures and responses that are highly influential remain so independent of whether indictors are given deferential or equal importance weights. However, indicator importance weights may affect the influence of moderate pressures and responses with the model on the overall perception of cultural change. For example, when indictors were given differential weights in our model as compared to equal weights, controlling tourist numbers remained the most influential response while community land tenure went from the least to the second most influential response (see Tables 13 and 16). Further research is needed to identify the most appropriate method for setting indicator importance weights within the model.

Application to other areas

Cultural indicators used to describe cultural change, pressures influencing change and responses that can be put in place, tend to vary between cultures and the socio-environmental context. However, the participatory model building approach used in this study provides a generic method for understanding, modeling and managing cultural change. This chapter demonstrates how a model can be constructed with stakeholders to identify:

- which aspects of culture might change under specific pressures;
- which pressures are likely to cause the most significant changes; and
- which responses are most likely to mitigate change.

In situations where comprehensive data or monitoring records are available, these can be used to populate the probability tables within a change model to improve rigour and remove the vagaries of perception. It is unlikely, however, that any location will have a complete set of these data, so Delphic approaches to populating cultural change models will continue to be a major part of any cultural modeling process.

Creating a co-learning environment

About 130 people participated in the development of the Juatinga Ecological Reserve cultural change model. Yet, no person identified all pressures, indicators, responses and links. Each informant described the process from their own perspective, adding valuable knowledge to the development of the model. In this sense, the participatory process, necessary for the construction of the model, provided a platform for combining knowledge and experiences and an opportunity for co-learning and knowledge sharing, which is essential for the effective management of a protected area. The process of interviewing tourists and Caiçaras was an educational process. It focused attention on changes occurring in the community and may have provided impetus for the Caiçara community to consider what they can do to manage the change process themselves.

The need for adaptive management

The outputs of any modeling exercise are only estimates based on available knowledge or opinion, and uncertainty means the outcomes of management decisions or responses are difficult to predict. Therefore, cultural change management and modeling should not be a once-off activity. It should be a continuous process, learning from the successes and failures of previous management responses. This process is known as adaptive management (Holling, 1978; Bosch et al., 2003). The process is cyclic (Figure 9), with plans for achieving management objectives developed using current knowledge (such as that embedded in a model), and monitoring undertaken to track the success of implemented management actions (responses). Reviewing management outcomes contributes new knowledge, which can then be used to refine management plans and models for future implementation.

The benefits of adaptive management are twofold. It allows for decisions and action to be based on experience, and it engenders a culture of continuous improvement by consciously reflecting on previous

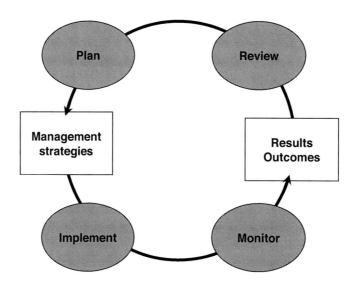

Figure 9. Generalized adaptive management cycle (Bosch et al., 2003).

management outcomes. The cultural change model outlined in this chapter is a tool to assist in the planning of cultural change management responses; however, its use should be embedded within an adaptive management process, so that continuous improvement of both the model and management decisions can occur.

One advantage of using BBNs as a modeling tool is that their underlying conditional probabilities can be updated over time using more informed opinions, frequency data collected via survey or monitoring. This allows for periodic update of a model based on prior outcomes from one or many localities, making them suitable for adaptive management. This is a particular advantage for cultural change modeling because many model assumptions are based on the opinion of people, which need to be adjusted or fortified using real-world observations.

Limitations and difficulties

The use of BBNs as a tool in cultural change modeling has the particular advantages of accommodating uncertainty and facilitating the integration of qualitative and quantitative variables. However, the need to use probabilities requires basic mathematical knowledge, which limits the variety of people that can be involved in the model building process. In the case of the Juatinga Ecological Reserve, it excluded the Caiçaras. Therefore, it is still necessary to develop techniques for capturing probabilistic data from illiterate people. Participatory rural appraisal techniques (Chambers, 1981; 1983; 1994; Mukherjee, 1993) may be one solution.

The other important limitation of BBN models is that they are acyclic. In other words, cycles or feedback loops cannot be accommodated. This presented some problems when developing the cultural change model for the Juatinga Ecological Reserve because field informants did describe cyclical process that affected culture. For example, according to informants, time spent on traditional subsistence activities affected traditional beliefs, and traditional beliefs affected the time spent on traditional subsistence activities.

Even though the Juatinga Ecological Reserve was created in 1992, prior to this study there were no established indicators with which to monitor cultural change, nor cultural management targets or management plans to protect Caiçara culture. Moreover, there were no empirical data relating to cultural change for protected areas elsewhere. This means that experiential knowledge and local opinion were the sole source of information available for developing the model. The model is, therefore, biased towards this opinion. However, the modeling framework was sufficiently flexible to accommodate variety of opinion, allowing multiple perspectives to be examined.

CONCLUSION

This chapter described a cultural change modeling approach that can be used to assess the impact of pressures on culture and predict the efficacy of cultural change management responses. The model building process involved stakeholder consultation and the integration of different perceptions of cultural change. A Bayesian Belief Network (BBN) was used as the modeling tool. Using a BBN had the particular advantages of (a) accommodating uncertainty, (b) accommodating multiple opinion about the relationship between variables, (c) facilitating the integration of qualitative and quantitative variables, and (d) facilitating sensitivity and scenario analysis.

This model demonstrates that tourism, policy and religion are the strongest pressures affecting the cultural integrity of the Caiçaras in the Juatinga Reserve. While tourism and religion bring benefits to the community that they desire, the model suggests that effective policy framework that seeks to maintain the cultural identity of the Caiçaras is essential to avoid radical cultural change. Whilst local communities should be involved in the development of such a policy, it will be necessary to include additional expertise (especially in areas such as tourism and cultural change) to diversify the range of management options considered. Policy will not be able, nor should it seek to avoid cultural change influenced by pressures, such as religion, cultural contact, use of western medicine, access to supermarket and TV. For social equity reasons, regulation of these pressures is likely to be untenable. Instead, policy should focus on controlling manageable impacts, such as tourism and land conflicts. Since policy will not control the effects of some pressures, then even under the best efforts, cultural change will occur, but more likely over a timeframe and in a manner that the Caiçaras might be able to manage through their own social system.

However, the greatest contribution of this research is its innovative approach to understanding, managing, forecasting and studying cultural change. The continuation of this research may lead to the identification of general theories relating pressures, responses and indicators of cultural change.

ACKNOWLEDGMENT

The research for this article was partially supported by the Brazilian agency Coordenação de Aperfeiçoamento de Pessoal de Nível Superior – CAPES, through the fellowship granted by the process 1672029, and the NGO Verde Cidadania.

LITERATURE CITED

Archibold, G. and Davey, S. 1993. Kuna Yala: protecting the San Blas of Panama. In Kemf, E. (ed) *The Law of the mother: protecting indigenous peoples in protected areas*. Sierra Club Books, San Francisco, pp. 52-60.

Bleasdale, S. and Tapsell, S. 1996. Saharan Tourism: Arabian Nights or Tourist 'Daze'? The Social-Cultural and Environmental Impacts of Tourism in Southern Tunisia. In Robinson, M., Evans, N. and Callaghan, P. (Eds), *Tourism and Culture: Towards the 21st Century, Conference Proceedings*, Vol. Tourism and Cultural Change, The Centre for Travel & Tourism in association with Business Education Publishers Ltd, Sunderland, pp. 25-48.

Boas, F. 1965. *The mind of primitive man*. Rev. edn., Free Press, New York.

Boas, F. 1966. *Kwakiutl ethnography*. University of Chicago Press, Chicago.

Bosch, O.J.H., Ross, A.H. and Beeton, R.J.S. 2003. Integrating Science and Management Through Collaborative Learning and Better Information Management. Systems Research and Behavioral Science. Vol. 20:107-118.

Cain, J. 2001. *Planning improvements in natural resources management: guidelines for using Bayesian networks to support the planning and management of development programmes in the water sector and beyond*. Centre for Ecology and Hydrology, Wallingford, Oxon.

Campbell, B.G. 1967. *Human evolution: an introduction to man's adaptations*. Heinemann, London.

Carter, R.W. 2000. *Cultural change and tourism: towards a prognostic model*. PhD Thesis, The University of Queensland, St. Lucia, Qld.

Carter, R.W. and Beeton, R.J.S 2004. A Model of Cultural Change and Tourism. *Asia Pacific Journal of Tourism Research*, Vol 9, No 4: 423-442.

Cavalli-Sforza, L.L. and Cavalli-Sforza, F. 1995. *The great human diasporas: a history of diversity and evolution*. Addison-Wesley, Reading, Mass.

Chambers R. 1981. Rapid rural appraisal: rationale and repertoire. *Public Admin Develop* 1981; Vol 1: 95–106.

Chambers, R. 1983. *Rural Development: Putting the Last First*. Longman, New York.

Chambers, R. 1994. Participatory Rural Appraisal (PRA): challenges, potential and paradigm. *World Development*, Vol 22, No 10: 1437-1454.

Dalkey, N. and Helmer, O. 1963. An experimental application of the Delphi Method of the use of experts. *Management Science,* Vol 93: 458 - 467.

Diegues, A.C. 1994. *O Mito Moderno da Natureza Intocada* (The Modern Myth of the Untouched Nature). USP, São Paulo.

Farr, R.M. and Moscovici, S. 1984. *Social representations.* Cambridge University Press; Paris: Editions de la Maison des Sciences de l'Homme, Cambridge, New York.

Fishbein, M. and Ajzen, I. 1975. *Belief, attitude, intention, and behavior: an introduction to theory and research.* Addison-Wesley, Reading, Mass, London; Sydney.

Goodenough, W.H. 1981. *Culture, language, and society.* 2nd edn., Benjamin/Cummings Pub, Menlo Park, California.

Jones, S., Martin, R.D. and Pilbeam, D.R. 1992. *The Cambridge encyclopedia of human evolution.* Cambridge University Press, Cambridge, 506 pp.

Kemf, E. 1993a. Aluna: the place where the mother was born. In Kemf, E. (Ed) *The Law of the mother: protecting indigenous peoples in protected areas.* Sierra Club Books, San Francisco, pp. 131-140.

Kemf, E. 1993b. *The law of the mother: protecting indigenous peoples in protected areas.* Sierra Club Books, San Francisco, 296 pp.

Lado, C. 1992. Problems of wildlife management and land use in Kenya. *Land Use Policy.* Vol. 9, No. 3: 169-184.

Laraia, R. 1989. *Cultura um conceito antropologico* (Culture an anthropologic concept). Jorge Zahar, Rio de Janeiro.

Lett, J. 1987. *The human enterprise: a critical introduction to anthropological theory.* Westview Press, Boulder; London..

Levi-Strauss, C. 1963-1976. *Structural Anthropology.* Translated from French by C. Jacobson and B.G. Schoepf. Basic Books, New York.

Martin, C. 1993. *The Law of the mother: protecting indigenous peoples in protected areas.* Sierra Club Books, San Francisco, p. xv.

Matta, R.D. 1981. *Relativizando: uma introdução à antropologia social* (Relativizing: an introduction to social anthropology). Vozes, Petropolis.

Mitchell, G.D. 1979. *A new dictionary of the social sciences.* Aldine, New York.

Mukherjee, N., 1993. Participatory Rural Appraisal: Methodology and Applications. Concept Publishing Company, New Delhi, India.

Nadkarni, S. and Shenoy, P.P. 2004. A causal mapping approach to constructing Bayesian networks. *Decision Support Systems,* Vol. 38, No. 2: 259-281.

Nogara, P.J. and Diegues, A.C. 1994. *O nosso lugar virou parque: estudo sócio-ambiental do Saco de Mamanguá-Parati-Rio de Janeiro* (Our place became a park: social and environmental study of Saco de Mamanguá-Parati-Rio de Janeiro), USP/Nupaub, São Paulo.

Ramos, A.R. 1993. Paradise gained or lost? Yanomami of Brazil. In Kemf, E. (Ed) *The Law of the mother: protecting indigenous peoples in protected areas.* Sierra Club Books, San Francisco, pp. 89-94.

Ribeiro, D. 1997. *Confissões* (Confessions). Companhia das Letras (Ed), São Paulo, Brazil.

Richey, J., Mar, B. and Horner, R. 1985. The Delphi Technique in Environment Assessment I: Implementation and Effectiveness, *Journal of Environmental Management,* Vol 21, 135 - 146.

Roberts, J.P. 1978. *From massacres to mining: the colonization of Aboriginal Australia.* CIMRA, London.

SEMADS, Secretaria de Estado de Meio Ambiente e Desenvolvimento Sustentável (State Secretary of Environment and Sustainable Development) 2001. *Atlas das unidades de conservação da natureza do Estado do Rio de Janeiro* (Atlas of protected areas in the Rio de Janeiro State), São Paulo, Metalivros.

Silveira, G.N.D. and Brandão, H.B. 2005. *Aspectos da gestao da Reserva Ecologica da Juatinga sob a luz da legislacao ambiental considerando a ocupacao por comunidades Caiçaras* (Aspects of the management of the Juatinga Ecological Reserve focusing on environmentals law and considering Caiçaras communities). Proceedings of 8th Brazilian Congress of Public Advocacy (Congresso Brasileiro de Advocacia Publica). Available at: http://www.ibap.org/.

Smith, C.S. 1999. *Assessing agricultural land management sustainability development and application of a method to a sugar producing area in North Queensland.* PhD Thesis, The University of Queensland, Brisbane.

Steward, J.H. 1979. *Theory of Culture Change: the methodology of multilinear evolution*. University of Illinois Press, Urbana.

The Office of the High Commisioner for Human Rights 2006. *The Rights of Indigenous Peoples*. United Nations. Accessed 3 November 2006. Available at: http://www.unhchr.ch/html/menu6/2/fs9.htm#intro,.

Wilson, J., McMahon, A. and Thomson, J. 1996. *The Australian welfare state: key documents and themes*. Macmillan Education Australia, South Melbourne.

Xie, P.F. 2003. The bamboo-beating dance in Hainan, China: authenticity and comodification, *Journal of Sustainable Tourism*, Vol. 11, No. 1, pp. 5-16.

CHAPTER 13

AGENT-BASED MODELING – VIEWS FROM THE MANAGEMENT PERSPECTIVE

Ulrike Pröbstl
Peter Visschedijk
Hans Skov-Petersen

Abstract: In many studies agent-based modeling (ABM) is presented as a new and highly valued tool Agent-based Modeling (ABM) by management agencies and identifies crucial factors that effect its adoption. Two projects with similar preconditions for the possible use of ABM provide the background for this evaluation. The evaluation is based on observation during a participatory project management process, combined with a review of the written records. Results of this analysis illustrate that the managers' attitudes, and pre-disposition to use ABM is influenced by many factors. When the affected species or the diversity of recreational use in the study area makes the management task very complex, the willingness to use ABM increases. The willingness to adopt ABM is also influenced by the availability of various databases, the expected costs and the amount of previous experience of managers with geographic information systems (GIS) and other technical management tools like GPS or modeling. If there is little experience with technical planning tools, then the willingness to use ABM is low. In the case studies all of which represent high use recreation areas the acceptance of management decisions by the public and various user groups, the amount of anticipated changes and the need for public relations were crucial factors in the implementation of ABM. Finally the increasing need to communicate the management goals in a participatory planning process enhances the success of using ABM. In summary, this new technology offers promising new directions, but further research will be needed to evaluate its positive effects on the planning process in the field of nature conservation.

Key Words: Management, simulation, GIS, Natura 2000, planning processes, participation, biodiversity

INTRODUCTION

The wealth of recent publications on methods and applications of Agent-based Modeling (ABM) generally conveys the message that this rather new method has quickly gained a high level of acceptance from resource and recreation managers or management agencies (Lawson et al. 2004, Van Wagtendonk 2004, Cole 2005, Jochem et al. 2006). This statement conforms to the thinking of those researchers involved in the development of these tools. However, seldom is that effectiveness of these methods assessed and discussed from a practitioner's perspective. According to (Gimblett 2002; Lawson et al. 2004b; Murdock 2004; Cole 2005; Henkens et al. 2006 and Manning 2007), ABM can be very helpful in three fields of application:

- Visitor information, guidance and information strategy;
- Nature conservation, especially wildlife disturbance; and
- Locating facilities

While ABMs have potential, it is important to evaluate their application to enhance nature conservation in protected areas that are intensively used by outdoor enthusiasts. This chapter explores the question of acceptance of the method by management agencies. More specifically, crucial factors for its adoption are identified, and subsequently used as a framework for a discussion and evaluation of the factors that actually influence the successful adoption of the method. This exploration is based on our collective experience during two EU funded projects under the interregional research and development fund (Interreg III B) dealing with Natura 2000 sites and their use by tourists and visitors (PROGRESS und AlpNatour, Pröbstl et al. 2007 and Visschedijk and Henkens 2002).

BACKGROUND AND FRAMEWORK

Two case studies were used to test the use of ABM as a tool in the management planning process in Natura 2000 sites. In order to understand the results of this study, the following discussion of the purpose of Natura 2000 and the role of management plans must first be explained.

The European Commission is establishing a network of biotopes and habitats – referred to as Natura 2000 – by decree of two directives, which all member states of the European Union must comply with.

Once established, adequate management shall ensure a favorable conservation status of the protected species, habitats and biotopes of European interest. In most cases once the sites are established and defined, the management plan is the required instrument to protect the Natura 2000 sites by integrating ecological, social and economic interests. Progress on these issues requires a solid database about ecological, local, social, and visitor information, as well as analytical tools that integrate these data effectively with the planning process. Furthermore the management plan should be developed in a collaborative planning process and involve all kind of local stakeholder groups (Pröbstl 2002; Kovac 2007).

Two research programmes funded by EU-grants, one in the Netherlands, Great Britain and France (PROGRESS) and one in Austria, Slovenia, Germany and Italy (AlpNaTour) are designing and assessing management plans in intensively used recreation and tourism areas. Both projects attempt to identify efficient tools for data collection, analysis, stakeholder involvement and management in different ecological environments (Pröbstl et al. 2005; Elmauer et al. 2005). In PROGRESS, the research team used the agent-based modeling to optimize the balance between recreation and nature and to prevent decline in quality of the protected sites. The applicability of GPS, aerial photos, and agent-based modeling-program MASOOR have been developed further and tested (Visschedijk and Henkens 2002). In a related initiative, agent-based modeling was also adapted to the alpine environment within a test site of AlpNaTour by the same team.

The research process used in both Interreg projects was composed of regular meetings of European steering committees, and the two first authors of this chapter. The following discussion is based on participatory observation during the project management process, in which the authors collectively assessed the planning process in retrospective, predominantly based on their personal recollection of the various stages of the process, combined with a review of the written records. It is the great advantage of this form of action research that it takes place in a real-world situation (Winter 1989; O'Brien 1998). The researcher, unlike in other disciplines, makes no attempt to remain objective, but openly acknowledges their bias to the other participants. The principle of reflective critique ensures people reflect on issues and processes and make explicit the interpretations, biases, assumptions and concerns upon which judgments are made. In this way, practical accounts can give rise to theoretical considerations. Furthermore it is of great value to understand decision-making processes when they are conceptualized in dialogues. It especially makes assessment of the contradictions both between many viewpoints (inter-discourse) and within a single viewpoint (intra-discourse) possible.

Since the action research is based on two projects only, and the number of managers who were consulted was rather limited and confined to participants from Germany, Austria, France, and the UK. Accordingly the discussion can only have qualitative character. Overall about 35 managers, members of the steering committee and local stakeholders were involved over the two-year period of the projects involved.

RESULTS

Site management in recreation areas is predominantly governed by cost effectiveness, efficient implementation of management and monitoring, and increasingly by concern about public acceptance and satisfaction (Farrell & Marion 2002; Queensland Parks and Wildlife Service 2000). The predominance of these three management concerns was apparent in both projects reported on here. In particular, as it turned out, the acceptance of management decisions by visitors was a major concern (see Figure. 1).

The search for an evaluative framework of ABM has lead to the identification of several factors that potentially influence the attitude to and acceptance of the method. These factors can conveniently be categorized according to the three themes presented in Figure 1. The first column in Table 1 identifies the factors that were deemed used to develop the application of ABM. Also seen in Table 1 is the evaluative continuum, describing conditions of high readiness of adopting ABM on one side, and conditions of low readiness on the other scale. This continuum is relevant for each of the factors.

COMPLEXITY OF MANAGEMENT TASKS

Over the past 10 years, ABM has been applied in recreation research for a variety of different purposes. The objective of the first applications in North America was to

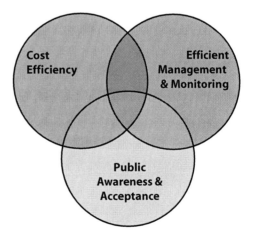

Figure 1. The most important themes influencing the choice of ABM from a management perspective, and decision-makers' attitudes towards it.

Table 1. Factors influencing Attitudes towards ABM

Factors influencing decision-making and attitudes / opinions	High readiness for applying AbM; (positive attitude towards)	Low readiness for applying AbM; (more negative attitude)
Complexity of management tasks	Highly complex management tasks, e.g. Integration of nature protection and recreation	Low level of management complexity, e.g. no comparison and trade-offs between competing land uses required
Size and type of area	Large sites and/or divers environment	Small and homogeneous environments
Diversity of factors determining visitor behaviour	Large number of factors influencing behaviour	Low number of factors influencing behaviour
Number of anticipated / planned changes	Large number of changes planned	Few changes anticipated
Management expererience with modelling	A lot of experience by managers in the field of GIS	Minor experience by managers with GIS
Cost	Low extra costs are to expect	Additional high costs must be expected
Data availablity	A lot of actual data are available	A weak and/or old database
Acceptance of management decisions	Problems with the acceptance of the decisions must be expected	The decisions will most likely be accepted by the public or different user groups
Planning process complexity	Complex participatory (communicative), button-up planning.	Rational, top down planning
Stakeholder diversity	Diverse sectors and professional leveles represented	Few sectors represented by similar professional backgrounds
Documentation of process and	Necessity of documentation and	Little or no requirements for

contribute to recreation management in general and to the management of the recreation experience in particular by modeling visitor flows and identifying potential conflicts (e.g. the capacity limits of campsites, more equal distribution of visitors across the existing infrastructure etc.) (Van Wagtendonk & Cole 2005; Daniel & Gimblett 2000). The problems were not very complex, and easy to be simplified as a model. It would have been quite possible to arrive at a management solution without ABM. Here, the relevance for practitioners was not the challenge for the management itself, but more the visualisation of problems and their possible solution to increase the acceptance.

Recent European applications on the other hand were less concerned with the optimization of the recreation quality, but the coordination and multi-objective optimization of recreation goals and conservation goals in the same land base (Henkens et al. 2003; Visschedijk et al. 2006).

These differences in the complexity and themes of application and management objectives are reflected in the comments and evaluation of the European managers who were interviewed for this project. ABM finds a high level of acceptance if it can serve as a decision support tool when the management tasks and challenges at hand are considered particularly complex. Complexity increases when both conservation and recreation concerns ought to be considered in the planning process. The various European applications, in the Alpine region as well as in France and England reveal that ABM is a highly effective and desirable tool for documentation of the recreational effects on protected species, both avifauna as well mammals. If several user groups and several protected species need to be considered concurrently, and the planning tasks are strict and clearly defined, then managers have a high level of acceptance of ABM tools. This is the case around planning processes of the new European protected areas strategy NATURA 2000 which is aiming at elimina-

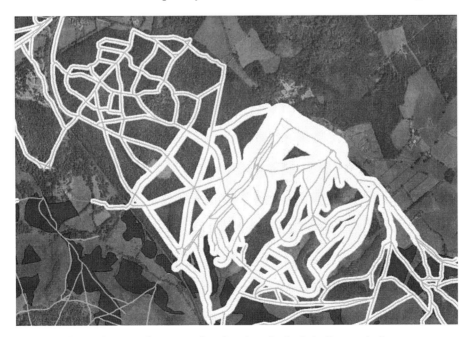

Figure 2. Use of ABM for complex recreational and ecological challenges in the same area.

tion of any further deterioration of the ecological conditions. On the other hand, if the conflict potential is low, for example if protected areas are already clearly delineated, or the recreational use to the area is fairly homogenous, then most managers, especially older ones with ample experience, felt that they can achieve acceptable or even superior results without ABM. Temporal complexity, as well as semantic and spatial complexity increases significantly the acceptance and willingness to cooperate during an ABM based planning process. Since the purpose of computer models is to simplify complex problems, some managers still feel that the long-term experience of managers who are very familiar with 'their' area might actually be undervalued vis-à-vis the increasing dependence on computer models especially against complex problems.

SIZE OF AREA AND HETEROGENEITY

The content and volume of management tasks also depends on the size of the area and its heterogeneity. In small areas with homogeneous structures the acceptance by management is lower than in heterogeneous structures, which might differ in terms of access or suitability for recreation. The test of MASOOR in an alpine environment showed that a topographically complex mountainous area constitutes challenges to its application, as the methods would need to be re-calibrated for complexity of terrain and other variables.

Following Visschedijk et al. (2006), the application of MASOOR in the alpine area requires it to be adapted to different ecological conditions and sites. An adaptation of the database to the model was necessary concerning the:

- Attributes of the paths (for ex. steep paths, only suitable for well equipped hikers),
- Different user groups,
- Access to the area (for ex. by cable car and by hiking),
- Influence on the protected habitat types,
- Different types of possible deterioration (different habitat types like alpine meadows or mountainous shrubs).

DIVERSITY OF FACTORS INFLUENCING VISITOR BEHAVIOR

One fundamental influence on the acceptance and willingness to apply an ABM depends on the additional factors that need to be considered in a decision making process. This concern is equal to the debate among modelers about the complexity of the model such as which decision processes should be considered and what type of hierarchical structures of spatial choices should be assumed such as the decision to go to the site, decisions where to enter the site, decisions follow a trail system or not). Typical questions need to address the structure of the model such as should accessibility and distance covered be the only or primary concern, or should these concerns be joined with the number and quality of encounters, and should a certain degree of importance also be assigned to the various points of interests. When preferences between various user groups differ, the judgments and opinions of the managers vary widely. They range from high acceptance ('the computer should be able to include and calculate these differences') to very low acceptance of its usability ('the necessity of a simplification of the model cannot simulate the reality in an adequate, sufficient way'). Overall, most managers indicated that if only a few behaviorally related variables influence the decision, then conflicts or management concerns can also be addressed without ABM (Skov-Petersen 2005).

NUMBER OF ANTICIPATED OR PLANNED CHANGES (STRATEGIC DIRECTIONS OF MANAGEMENT)

During the management planning processes and the discussion with local managers it turned out that the demand for new strategic directions in management and the number of possible changes are influencing the readiness to use ABM. This will be illustrated by the following two situations.

The first situation is about an attractive recreation area, in which a number of changes and improvements are planned to increase satisfaction of visitors and recreation suitability in general. One main purpose of the ABM should be to simulate the changing visitor composition of bicyclists, joggers and dog walkers so that management can work through several scenarios, and a monitoring system can be developed. Overall, the situation encompasses a long-term management question without any current conflicts or any fundamentally new recreation uses.

In the second situation it is assumed that an intensively used recreation area (such as Fontainebleau, France, New Forest, UK, or many alpine destinations already have significant challenges of visitor management and improvements of recreation experiences. But additional management concerns for protected areas have recently been increased as many of these traditional recreation areas have now been reported as NATURA 2000 areas. Since these areas are governed by a strict 'No net loss' policy, mandated by the Habitat and the Bird Directives of the European Commission, and must be reported to the EU, site management has changed significantly. While Situation 1 is mostly concerned with maintaining or slightly improving status quo, Situation 2 represents a complex situation that requires intensive planning and might be associated with significant changes for visitors.

Consequently, the readiness of ABM in Situation 1 is rather low, especially if new data would need to be collected for an ABM, which would incur additional cost and require more time. However, situation 2, which needs to consider the national duty to ensure or to develop a favorable conservation status for special species and habitats of European significance, cannot afford any experiments with negative results. These protected sites require careful planning, which takes the natural sensitivity as well as the various needs and the behavior of different user groups into account. Otherwise negative reactions from the visitors could endanger the actions that favor nature protection. Without ABM or another form of careful participatory planning, negative effects on the conservation goals, are more likely to occur since the acceptance and the efficiency of the measures will be lower. Often the process cannot be changed or repeated. A possible loss of habitats by intensive disturbance can – in most cases – not be compensated or redeveloped. This type of scenario increases the willingness of a manager to use ABM to aid decision-making.

DATA AVAILABILITY, COSTS, AND SYSTEM USER-FRIENDLINESS

Inevitably, an ABM constitutes a very focused inventory of the frequency of use of the study area, together with a range of additional variables describing and determining behavior. The two European funded projects, AlpNatour and PROGRESS, did not only require spatial data, but also data specific to the various user groups, and special biological data of the protected species and habitats.

The initial data collection for such a model is considerable, and frequently leads to the rejection of this modeling approach, simply due to the lack of resources. Even if most of the data are available, additional costs are inevitable because ABM requires data in a compatible format, which makes its application even more problematic. Finally, a specific ABM program is required, which is perceived as a particularly negative aspect by management. So far, ABM has not developed user (manager) friendly and cost efficient protocols, as each case study requires a separate and unique modeling approach. MASOOR has several applications in different landscapes with different user concerns, and a further modular development of these components might lead to such cost-efficient modules in the near future. When this occurs, the cost for data collection will be reduced, sufficient data will have been collected for typical conflict areas (such as urban areas) and more standardized tools that can be applied indifferent landscape types will be developed. Until now, most ABM has been primarily driven as research initiatives, financed by research projects, model initiatives and dissertations. Consequently little is known about the costs of applying this technology in practice. Both, proponents of ABM and land managers would be well served by making the 'true' costs much more transparent and realistic. At the same time, with more systematic ABM applications, economies of scale can eventually be achieved in the areas of data collection, and the calibration of models for various types of applications (high density or low density, various user mixes, types of terrain). In general it can be stated that the acceptance is higher when the database is sufficient and when the expenses for programming and adaptation are low or research funded.

EXPERIENCE AND PRIOR KNOWLEDGE

Experience and previous knowledge are important considerations when applying ABM. The extent of experience of managers especially in the field of geographic information systems (GIS), other technical management tools like GPS or modeling is a concern. If managers have little experience with technical planning tools, then the willingness to use ABM is low. Specific knowledge in the field of recreation and/or nature conservation is not as much a determining factor as was expected. The disposition is also influenced by the experience of the management in planning processes and new challenges in participatory planning (see below). The more experienced managers are, the more they see ABM as an attractive tool for these new demands. The collaborative approach that was taken between management and modeler was invaluable. On the other hand, the less experience managers have, the more challenging it is to initiate ABM. For a complete and continuous application of ABM, one would require a high level of diverse information (e.g. on the behavior of different visitor groups, spatial information and ecological data), and of particular importance would be some competency in the use of the final model (i.e. for the creation and interpretation of scenarios) among the end users (i.e. NGOs or management agencies).

ACCEPTANCE OF MANAGEMENT DECISIONS BY USERS AND THE PUBLIC

Decisions relating to visitor management can be made in such a way that is rarely perceived as being negative by the users or the population at large. Their acceptance depends on the type of management intervention, as well as the way it is communicated (Ammer and Pröbstl 1991). ABM will have a more important role to play when it is deemed necessary to introduce measures that might be considered controversial by at least some of the users. In such a situation ABM will provide the opportunity to simulate the pending restrictions.

The same is true for possible adaptive and preventive strategies, for which ABM can provide simulations and can be combined with a public relations campaign. Examples for such a strategy come from France and England, where the ABM was used to communicate use restrictions for the benefit of certain species in an attractive and easy to comprehend manner. The more controversial the management issue and the more likely it is to lead to conflict the more likely managers are to resort to ABM for the purpose of presenting the solutions as being derived from a scientific process. Computer simulation of recreational behavior has been reported to be a well functioning teaching and training aid (Manning and Potter 1984).

In the two case studies, the use of ABM enhanced the interest of the target groups and stakeholders. As described by several authors in the field of GIS, the presented spatial data and analytic results are perceived as 'truth' and stakeholders are less critical than the managers. The acceptance of the ABM by target groups and stakeholders was therefore an excellent tool to communicate the necessity of management, its aims and proposed measures. ABM increases the understanding of nature protection and the required behavior. The willingness to support the management by stakeholders or sport associations (e.g. horse back riders, hiking club) is influenced in a positive manner. Against this background and similar experiences in the two case studies we propose to use ABM especially in a participatory context and collaborative planning processes.

PLANNING PROCESS COMPLEXITY

Physical planning and the processes involved cannot be treated as a presumed 'gray box'. In spite of this, GIS models and ABM's often are intended for application in planning without further consideration of for instance the organizations or stakeholders, information flow, discourse and norms involved (Skov-Petersen 2002). Planning processes ranges largely from closed, apparently objective, rational planning, over incremental processes to planning where communication or participation is in focus (Hudson 1979). In rational planning the aim is more or less automated to identify a single optimal solution to a planning and management problem. If proper data and models are available, this goal can be achieved without further external involvement. Iterative planning is based on mutual trade off between involved stakeholders. The goal is to find a balanced solution that meets the needs and expectations of the stakeholders. Participatory planning is based on the idea of mutual learning through the planning process. In the context of planning outdoor recreation which is a highly diverse, multi scale, multi sector planning theme, it appears that application of ABM's has its highest potential in relation to participatory setting like those discussed in the present chapter.

STAKEHOLDER DIVERSITY

The transparency of ABM's is an obvious offset for communication between stakeholders. The more complex the stakeholder situation is the more likely application of simulation is to succeed. Participants in planning can be categorized along two axis'; whether A) the matters dealt with are concrete (as they often are in relation to management) or they are more abstract (as is more likely in relation planning) or whether B) participants are professional academics or have lay persons perspectives. This reveals four prototypic stake holders: A manager who's professional/concrete, a lay person who's non-professional/concrete, a politician who's non-professional/abstract and finally a planner being professional/abstract (Skov-Petersen 2002). The more likely a variety of these types are to participate in a planning exercise, which is typically seen in relation to recreational planning, the more need there will be for transparent communicative means, including ABM.

DOCUMENTATION OF THE PROCESS AND PARTICIPATORY PLANNING

ABM amounts to the presentation of large amounts of information in a succinct manner, enabling the visualization of spatial and temporal processes very effectively. For example, short clips illustrate the visitor's behavior, their distribution over time across the area. These video clips showing for example moving persons, different user groups and a divers spatial distribution, facilitate the understanding of complex problems and a huge amount of data. Therefore the method is well suited for the public presentation of visitor flows and visitor behavior. Visualization and animation are useful for all forms of public information campaigns, and are particularly relevant to convey management objectives to publish via (mass-) medias. All these arguments support the increasing application of ABMs. Figure 3 shows several images of visitor flows in a high alpine area: Arriving by foot or by cable car agents are moving around the high alpine management site.

The experiences and reaction of different user groups, the public and local stakeholders illustrates that ABM can also be used as a learning tool. A learning tool that can show that actual behavior and it's negative consequences on the environment and positive consequences on selected species, can occur if visitors will change their behavior in order to protect the environment. As a result, understanding and the efficiency of management can be enhanced. One way to accomplish this is to use ABM in interpretation centers, especially with the younger generation that is familiar with the concept of modeling through technology and computer games.

SUMMARY

Agent-based modeling has been used in Europe and North America in several recreation and visitor management applica-

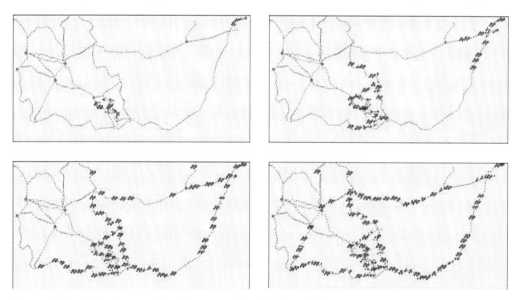

Figure 3. Illustrates the spatial and temporal distribution of hikers on the Schneeberg in the Austrian alpine area. In the south visitors are coming by cable car, in north east visitors hiking up

tions and its usefulness has been highlighted in several publications including chapters in this volume. While applications in the field of visitor information guidance and location of facilities that are often described as rather successful and well accepted by management, there are fewer experiences with more complex applications such as in the field of nature conservation. These latter types of applications are predominantly found in Europe where frequently recreation management and nature conservation concerns coincide in the same areas. The two European case studies document from a manager's perspective the advantages as well as disadvantages associated with this tool.

Even in EU-funded projects the increasing of costs due to data collection and modeling is strongly influencing the discussion. Furthermore, the likelihood of applying ABM depends on the availability of existing databases, and the experience of managers with geographic information systems (GIS), as well as other technical tools like GPS or modeling. If they have little experience with technical tools, then the willingness to use ABM is low. Therefore, in the future the cost efficiency of ABM must be illustrated right at the beginning considering reduced expenses for monitoring and management in the long term. It is time to discuss the 'true' costs much more transparent and realistically in the future.

Besides cost efficiency, effective management and monitoring are other key issues mentioned. If in a specific area of application, the affected species or the diverse recreational use makes the management task very complex, then the willingness to use ABM is increasing. In the two cases of highly used sites in the Alpine environment, as well as in large French and British forests, the importance of the anticipated changes influenced the manager's attitude towards ABM. However, due to the confidence of many managers into traditional planning methods and the personal skills of experienced, well-trained managers often disregard computer models. Future applications must prove to practitioners that they can contribute to the solution of complex situations and make management and monitoring more efficient.

Finally, the increasing need to communicate the management goals in a participatory planning process enhances the use of ABM. The visualization effects associated with ABM are important tools for the discussion with stakeholder groups. In the field of nature conservation and the protection of endangered species, the simulation of the disturbance over time can be extremely helpful. It contributes to the awareness of the local population and different stakeholders, especially concerning wildlife. If the model simulates the behavior of recreationists, the participants' willingness to cooperate is increasing. Overall, the contribution of ABM's to public awareness, a transparent planning processes and acceptance the strongest arguments for a further use and continued applications of this type of planning and management aids.

In summary, these new technologies offer promising directions, but significant amounts of further research will be needed to evaluate its positive effects on the planning and management processes in the field of nature conservation all over Europe.

LITERATURE CITED

Ammer U. and U. Pröbstl. 1991. Freizeit und Natur. Probleme und Lösungsmöglichkeiten einer ökologisch verträglichen Freizeitnutzung. Pareys Studientexte Nr. 72. Hamburg, Berlin.

Cole, D. 2005. Why model recreation use? In

Cole, D. (compiler). 2005. Computer simulation modeling of recreation use: Current status, case studies, and future directions. USDA, General technical report RMRS-GTR-143.

Daniel, T. C. & H. R. Gimblett. 2000. Autonomous Agent Model to Support River Trip Management Decisions in Grand Canyon National Park. International Journal of Wilderness -Special Issue on Wild Rivers. December 2000. Volume 6, Number 3. Pgs 39-42.

Elmauer, T., T. Knoll, U. Pröbstl U & W. Suske. 2005, Management planungen für Natura 2000 in Österreich, in: Naturschutz und Biologische Vielfalt, Heft 26, 269 -286.

Farrell T. A, J. L. Marion. 2002 The Protected Area Visitor Impact Management (PAVIM) Framework: A Simplified Process for Making Management Decisions. Journal of Sustainable Tourism. Vol.10. No.1, p31-51.

Gimblett, H. R. 2002. Integrating Geographic Information Systems and Agent-based Modeling Techniques for Simulation Social and Ecological processes. Oxford University Press. 2002. 327p.

Henkens, R. J., R. Jochem, D.A. Jonkers, J.G. de Molenaar, R. Pouwels, M.J. S.M. Reijnen, P. A.M. Visschedijk & S. de Vries. 2003. Analysis on the impact of recreation on breeding birds. Literature study and coupling of the models FORVISITS and LARCH (in Dutch). The Netherlands Environmental Assessment Office (MNP). Document 2003/29.

Henkens, R. J., R. Jochem, R. Pouwels, P.A.M. Visschedijk. 2006. Development of a zoning instrument for visitor management in protected areas. (Editors) D. Siegrist, C. Clivaz, M. Hunziker, S. Iten Proceedings of the third conference on Monitoring and Management of Visitor flows in Recreational and Protected Areas, Rapperswil, 2006. 513p.

Hudson, B. M. 1979. Comparison of current planning theories: Counterparts and contradictions. With comments by Thomas D. Galloway and Jerome L. Kaufman. Journal of American Planning Association. Vol. 45, number 4. 1979.

Jochem, R., R. Pouwels, P.A.M. Visschedijk. 2006. MASOOR: The power to Know - A story about the development of an intelligent and flexible monitoring instrument. (Editors). D. Siegrist, C. Clivaz, M. Hunziker, S. Iten Proceedings of the third conference on Monitoring and Management of Visitor flows in Recreational and Protected Areas, Rapperswil, 2006. 513p.

Kovac, M. 2007. Participation - Democratic Dealing With Problems In Natura 2000 In The Countries Of The Alpine Space. In: Pröbstl, U. Kovac, M., Knoll, T., Ruffini, F., Schneider, W., Martin, K. (Eds.) 2007. Tourismus in Natura 2000 Gebieten - Empfehlungen für die Managementplanung im Alpenraum. Tourism in Natura 2000 sites - Guidelines and Recommendations for the management planning in the alpine space, Haupt Verlag, Bern (in press).

Lawson, S., R.M. Itami, H.R. Gimblett & R. E. Manning. 2004. Monitoring and Managing Recreational Use in Backcountry Landscapes Using Computer-Based Simulation Modeling. In: (Editors) Sievänen, T., Erkkonen, J., Jokimäki, J., Saarinen, J., Tuulentie, S., Virtanen, E. (The Second International Conference on Monitoring and Management of Visitor Flows in Recreational and Protected Areas. June 16–20, 2004. Arktikum, Rovaniemi, Finland. p.107-114.

Lawson. S. A. M. Kiely, R. E. Manning. 2004b. Computer simulation as a tool for Developing Alternatives for Managing Crowding at Wilderness Campsites on Isle Royale. In: (Editors) Sievänen, T., Erkkonen, J., Jokimäki, J., Saarinen, J., Tuulentie, S., Virtanen, E. The Second International Conference on Monitoring and Management of Visitor Flows in Recreational and Protected Areas. June 16–20, 2004. Arktikum, Rovaniemi, Finland. p. 115-120.

Manning, R.E. and F. I. Potter. 1984. Computer simulation as a tool in teaching park and wilderness management. Journal of Environmental Education. 15(3): 3-9.

Manning R. E., 2007. Parks and Carrying Capacity: Commons Without Tragedy. Island Press Washington, Covelo. London.

Murdock, E., 2004. Understanding Recreation Flow to protect Wilderness Resources at Joshua Tree National Park, Carlifornia. In: (Editors) Sievänen, T., Erkkonen, J., Jokimäki, J., Saarinen, J., Tuulentie, S., Virtanen, E. The Second International Conference on Monitoring and Management of Visitor Flows in Recreational and Protected Areas. June 16–20, 2004. Arktikum, Rovaniemi, Finland. p.121-128

O'Brien, R. 1998. An Overview of the Methodological Approach of Action Research. Faculty of Information Studies, University of Toronto, Toronto

Pröbstl, U., P. Sterl, V. Wirth, V. (ed.), 2005. Tourismus und Schutzgebiete - Hemmschuh oder Partner? Tagungsband, Universität für Bodenkultur, Wien, 28 S.

Pröbstl, U., 2002. NATURA 2000 - The acceptance of the European directives in communally and privately owned forests, in Barros, S. (ed) Collaboration and Partnership in Forestry, proceedings of the IUFRO Division 6 Conference in Chile, Valdivia

Pröbstl, U., M. Kovac, T. Knoll, F. Ruffini, W. Schneider & K. Martin. (Eds.), 2007. Tourismus in Natura 2000 Gebieten - Empfehlungen für die Managementplanung im Alpenraum. Tourism in Natura 2000 sites - Guidelines and Recommendations for the management planning in the alpine space, Haupt Verlag, Bern (in press).

Queensland Parks and Wildlife Service. 2000. Benchmarking And Best Practice Program User-Pays Revenue, Report for Australian and New Zealand Environment and Conservation Council (ANZECC) http://www.environment.gov.au/parks/publications/best-practice/user-pays/pubs/user-pays.doc, access 30.7.2007

Skov-Petersen, H. 2002. The role of Geographic information technology in physical planning. In Skov-Petersen, H. GIS, accessibility and physical planning – exemplified by models of recreational behaviour. Ph.D. thesis. Institute of Geography. University of Copenhagen.

Skov-Petersen, H. 2005. Feeding the agents – Collection parameters for agent based models. 9'th Conference on Computers in urban planning and urban management (CUPUM) 29/6-1/7 2005. London

Van Wagtendonk, J.W. and Cole, D.N. 2005. Historical development of simulation models of recreation use. In Cole, D. (compiler), 2005. Computer simulation modeling of recreation use: Current status, case studies, and future directions. USDA. Forest service. Rockey mountain research station. General technical report RMRS-GTR-143. 75 pp.

Van Wagtendonk J.W., 2004, Simulation modeling of visitor flows: where have we been and where are we going? In: Sievänen, T., Erkkonen, J., Jokimäki, J., Saarinen, J., Tuulentie, S., Virtanen, E. (eds) 2004 Policies, Methods and Tools for Visitor Management, Finish Forest research Instituts, Saarijärvi, p. 129-136

Visschedijk, P.A.M. & R.J.H.G. Henkens. 2002. Recreation monitoring at the Dutch Forest Service. (Editors). A. Arnberger, C. Brandenburg, A. Muhar. Monitoring and Management of Visitor flows in Recreational and Protected Areas. Institute for Landscape Architecture and Landscape management Bodenkultur University Vienna. 485p.

Visschedijk, P.A.M., U. Pröbstl, R. Henkens. MASOOR in the Alpine areas: Agent based modeling as a tool for the management planning in Natura 2000 sites. (Editors). D. Siegrist, C. Clivaz, M. Hunziker, S. Iten Proceedings of the third conference on Monitoring and Management of Visitor flows in Recreational and Protected Areas, Rapperswil, 2006. 513p.

Winter, R. 1989. Learning from Experience- Principle and Practice in Action-Research. The Falner Press, Philadelphia

CHAPTER 14

LINKING ECOLOGICAL AND RECREATION MODELS FOR MANAGEMENT AND PLANNING

Rogier Pouwels
Rene Jochem
Jana Verboom

Abstract: Most nature areas (e.g. national parks, regional parks, wildlife sanctuaries, etc.) are open for recreation. In these areas, recreation can be an extra stress factor for animal populations. Increased recreation pressure decreases the probability of long-term population persistence. Recreation goals and biodiversity goals need to be well balanced: what is the recreation pressure an area can support, and how should the recreation be distributed spatially and temporally in order to achieve both recreation and biodiversity goals. In this chapter we demonstrate how a linkage between an ecological and a recreation model can help managers in finding this balance, using a case study.

Keywords: Biodiversity, Framework, Birds, Disturbance, Nature reserves

INTRODUCTION

Many nature areas in Northwestern Europe are open for recreational use. Visitors enjoy restorative health benefits of contact with nature and they experience many other valued aspects of visiting the countryside such as tranquility, open space, fresh air, unpolluted waters and scenery (Natural England 2006, Natuurmonumenten 2006). Health programs are set up to stimulate more people to visit nature areas (e.g., Natural England 2006) and managers must accommodate an increasing number of visitors. The policy of opening nature areas for recreation can conflict with the policy of protecting species in these areas (Drewitt 2007). In England, the Countryside and Rights of Way Act 2000 (CRoW) integrates freedom of rambling with protecting biodiversity. This integration should be evidence-based instead of believe-based (Bathe 2007). While the last decade has witnessed significant research on the impact of recreational activities on biodiversity, there is a need for more research to balance or integrate recreation and biodiversity (Sutherland *et al.* 2006, Sutherland 2007, Haider 2006).

In industrialized countries like the Netherlands, the persistence of many populations in the landscape depends on nature areas. Next to fragmentation, eutrophication, desiccation and pollution, recreation can be an extra stress factor for these populations and can threaten their persistence. Many studies have stressed the negative effects of recreational disturbance on bird behavior, distribution and breeding success (Blanc et al. 2006, Gill 2007, Mallord et al. 2007). However from a conservation viewpoint, the impact at the population level is of paramount importance (Sutherland 2007). Modeling the consequences of alternative recreational access scenarios will help policymakers choose appropriate mitigation measures (Taylor et al. 2007, O'Connell et al. 2007, Mallord et al. 2007). These models should include a recreational as well as a conservation viewpoint (Sutherland 2007). The main questions that should be answered are: does the area fulfill the expectation of the visitors? What are the impacts of recreation on species persistence? And is the viability of a population affected by the impact?

Models should be seen as part of a conceptual planning/managing framework that includes both scientific and managerial perspectives (Haider 2006). Scientists need the managers to give their research more focus as much as managers need the empirical data of scientists to help them develop standards (Cole 2004). The framework integrates three dimensions for managing multifunctional land use problems, including goals, monitoring and design (Figure 1). These dimensions allow managers (or decision makers) to be fully aware of: (1) the desired future they wish to achieve; (2) the alternative routes to the future; and (3) consequences of those alternatives (Haider 2006). The framework implies that these dimensions can be independently considered. First, stakeholders negotiate goals with respect to biodiversity and recreation and come to operational management area objectives (goal setting dimension). Second, goal realization needs to be monitored as changes occur in the level and spatial/temporal distribution of recreational access that is provided for visitors (monitoring dimension). Finally, the development of future management plans should simultaneously consider dimensions of both nature and recreation (design and evaluation dimension). Models are a useful tool for this evaluation (Opdam et al. 2002).

The arrows within the triangle indicate information flows between the three dimensions. Establishment of a monitoring program should chart success in achieving management goals (e.g. Manning 2004). Analyses of monitoring data can also provide rules of thumb for goal setting (e.g. Fernandez-Juricic et al. 2005, Moran-Lopez et al. 2006). Monitoring data

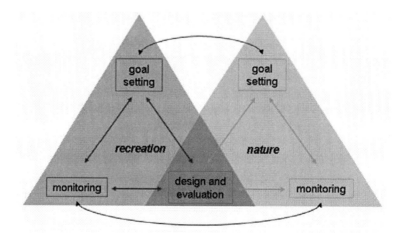

Figure 1. Planning framework for multiple land use in protected nature areas.

provides a basis for goal reassessment as well as defining design and evaluation strategies, which, in turn must be based on stated goals (e.g. Verboom *et al.* 2001, Opdam *et al.* 2002). Goals define results that should be produced and monitored in modeling the implementation of design strategies. Models often provide insight into what monitoring data are missing (Jochem *et al.* 2007).

The design dimension evaluates the effect of management plans toward realization of both recreation and nature goals. Plans for both nature enhancement and recreational access should be subjected to this evaluation. Few researchers have used models and monitoring data to integrate recreation and nature goals in evaluating future designs of an area, including Poe *et al.* (2006), Henkens *et al.* (2006) and Liley *et al.* (2006). In these models functional relationships are established between attributes managers can control and desired management goals.

Integration is possible only when the impacts of recreation and wildlife on each other are well understood. This is indicated by the arrow between the monitoring dimensions. However, monitoring is rarely integrated. Social scientists tend to monitor motivation and experience of visitors and ecologists tend to focus on recreation impact on animal behavior. Within the proposed framework, monitoring research should result in a description of the functional relationships between attributes that managers can control and the outcomes that managers seek (Cole 2004). Goal setting is rarely integrated. The inability to resolve the competing values of a diverse public (Cole 2004) make it difficult to establish agreed upon standards (Seidl and Tisdell 1999). Most so-called integrated goals are in fact nature goals that restrict recreation behavior within a certain distance or period of time from nests or colonies of sensitive birds (e.g. Moran-Lopez *et al.* 2006).

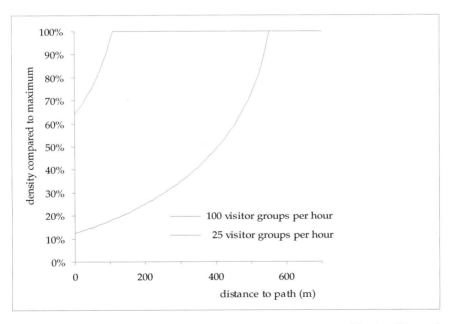

Figure 2. Schematic example of recreation impact on densities of nests of Curlew (Numenius arquata) (after Vos and Peltzer 1987).

RECREATION IMPACT

Research examining the impact of recreation on animals is diverse (Hill et al. 1997, Blanc et al. 2006). This research can be categorized into four types (Gill 2007): change in distribution; change in behavior; change in demography; and change in population size and persistence. First, densities of birds are lower near paths (Vos and Peltzer 1987, Van der Zande and Vos 1984, Yalden and Yalden 1990, Riffel et al. 1996, Miller et al. 1998, Langston et al. 2007, Mallord et al. 2007 and O'Connell et al. 2007). Vos and Peltzer (1987) also illustrate that the intensity of use increases the distance over which this reduction occurs (Figure 2).

Second, recreation has an impact on the escape behavior of animals like birds and mammals (Blanc et al. 2006, Gill 2007). The closer a visitor approaches, the higher the probability an animal will flee. Larger animals tend to flee at longer distances from an encounter with humans (Blumstein et al. 2005). Also the type of visitor and the way a visitor moves have effect on escape behavior. Ecotourists, hunters, browsers and visitors with dogs have a higher impact than cyclists or walkers. Unpredicted movement patterns of visitors and changes in speed or direction have a higher impact (Blanc et al. 2006). In periods when the escape behavior is an extra load on the scarce energy budgets of animals, recreation can have an impact on the survival of the animal or even the viability of the population. These periods are the winter, when food is scarce and energy costs are high (Gross-Custard et al. 2006, Stillman et al. 2007), and the spring, when energy budgets are directly linked to the number of offspring (Yalden and Yalden 1990, Murison et al. 2007, Langston et al. 2007). However, there is no guarantee that the behavioral response to disturbance is related to the population consequence, measured in terms of decreased reproduction or increased mortality (Gill et al. 2001).

Birds exhibiting an escape behavioral response might actually be moving to alternative breeding or feeding sites (Stillmann et al. 2007). Stillmann & Goss-Custard (2002) document the seasonality of escape behavior for Oystercatchers. In late winter, when energy demands are higher and food quality is lower, Oystercatchers respond less frequently to disturbance. Individual based models, consisting of fitness-maximizing individuals, are one means of linking disturbance induced behavioral responses to population consequences (West et al. 2002, Stillmann et al. 2007).

Thirdly, research shows that reproduction of birds is lower when recreation pressures are high (Van der Zande & Verstraal 1984, Bijlsma et al. 1985, Vos & Peltzer 1987 Gaddy & Kohlsaat 1987, Miller et al. 1998, González et al. 2006, Murison et al. 2007). Adult survival also decreases with higher recreational pressure (Gross-Custard et al. 2006). Ecotourism can also be a cause of reduction in survival (Müllner et al. 2004). There might be a feedback between impact on density and impact on reproduction. Only Mallord et al. (2007) found no impact on reproduction for the Woodlark (*Lullula arborea*) in plots with and without recreation.

Fourth, lower densities, lower reproduction success and higher mortality rates might lead to lower survival rates of populations. Depending on the spatial and temporal characteristics of the impact on patch size, habitat quality and other stress factors the populations might go extinct (Blanc et al. 2006). There is little research on the impact of recreation on changes in population size or persistence. Mallord et al. (2007) modeled the consequences of several access scenarios for Woodlark populations in southern UK and found that compared to the current situation the same number of people distributed evenly across all sites leads to a major negative impact

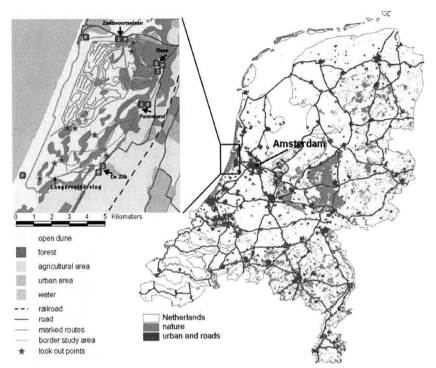

Figure 3. 'Amsterdamse Waterleiding' dune area near Amsterdam. Black arrows (left figure) are entrance points.

on the population. In the case study presented in this chapter we also use a model to translate the impact of recreation on the population size and persistence.

CASE STUDY

The study area is a dune area located near Amsterdam (the Netherlands) that is heavily used for recreation (Figure 3). It is called the 'Amsterdamse Waterleidingduinen' and with an area of 3500 ha it is one of the largest dune areas in the Netherlands. In 1998, a total of 723,000 visitors used the area (Jaarsma & Webster 1999). Besides its functions for protecting biodiversity and ensuring recreational opportunities, the area is used by City of Amsterdam for producing drinking water. Yet its main function is protecting the lowlands against the sea.

In this case study, we illustrate the integrated use of an agent based model to simulate recreational behavior and an individual based model to simulate response of a population of avian species to alternative management scenarios (Figure 4). The recreational simulation model is MASOOR (Jochem et al. 2007) and the population model is METAPHOR (Vos et al. 2001, Verboom et al. 2001).

We chose an individual based population model because it allows us to translate the disturbance impact of recreational use on individual members of an indicator species to a population level. The combination of both models allowed us to evaluate alternative access scenarios from a conservation viewpoint and a recreational viewpoint.

In the case study, the Skylark (*Alauda arvensis*) was chosen as an indicator species. The Skylark is a species that showed a decline of 60% in the Netherlands during the last decade. In dune areas, this decline is even larger (Van 't Hoff 2002). The population decline in the dune areas is due to increased recreation pressure and habitat quality decline. One of the reasons is a change in vegetation structure brought about by a decline of rabbits. The numbers of Skylark and Northern Wheatear (*Oenanthe oenanthe*) show a correlation to the number of rabbits (Koning & Baeyens 1990).

The case study models three scenarios for management of the dunes, including: no recreation; the current pattern of recreational use; and a zoning scenario. In the zoning scenario, the central part of the nature area was closed for recreation. For each scenario both nature indicators (percentage of occupied patches and total numbers of Skylark) and recreation indicators (overall recreation density and total length of paths with low recreation densities) were assessed. The nature indicators were identified by the managers of the area. The recreation indicators were determined by the researchers after the study was conducted. We chose one indicator that relates to crowding and one indicator that relates to opportunities for visitors that seek tranquility. The exploratory study illustrates the integrated use of the two models to manage a nature area for both ecological values and high quality recreational experiences.

SIMULATION MODELS

MASOOR

MASOOR (Multi Agent Simulation of Outdoor Recreation) (see Jochem *et al.* 2007 for a complete description of the model) is a multi-agent recreational behavior simulation model. The model captures the transactional experience of different visitor types in natural areas containing a high density trail network.

The visitors are modeled as individual agents. Each visitor type is defined by specific behavior and goals. The visitors' behavior is contained in a hierarchical control system that provides a framework for identifying visit goals and constraints, defining path networks that enable goal attainment and specifying specific behavioral rules for navigating through the network in pursuit of goals. The benefit of a hierarchical control system in MASOOR is that it defines the symbolic landscape used by recreational visitors at multiple scales. Spatial characterization of these inputs is retained in a GIS framework, where, for example, the spatial configuration and actual attributes of the defined trails are stored as a fixed path network. The spatial

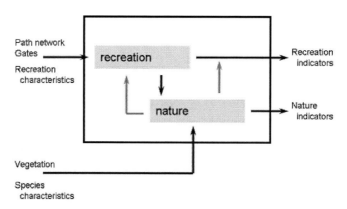

Figure 4. Instrument for evaluation of recreation and nature indicators. The inputs are GIS maps and process parameters. The outputs are tables and GIS maps for recreation indicators and nature indicators. The arrow between 'recreation' and 'nature' represents recreation impact. Grey arrows are not implemented in this case study.

and temporal outputs generated by MASOOR can be processed in GIS to provide the required output for the recreation impact.

PARAMETERS OF MASOOR CASE STUDY

The area has five main entrance points (Figure 3). In the case study, the entrance points are treated as exit points as well. This means that all visitors will make a circular trip. The distribution of visitors over the entrance points is based on counting from 1998 (Table 1) (Jaarsma and Webster 1999). Two types of visitors were defined and the mean duration of a trip is 2.5 hours (Bakker and Lengkeek 1999). The first type of visitors follows a marked route with an average length of 5 km and the second type of visitors follows randomly selected paths having an average length of 7.5 km. The standard deviation in trail length for both types of visitors is 1 km. For each scenario one model run with 50000 agents was done. There was no interaction between the agents.

Table 1. Distribution of visitors over the different entrance points (Figure 2) (From: Jaarsma and Webster 1999).

Entrance point	% of total visitors
Zandvoortselaan	15.5
Oase	41.6
Panneland	21.5
Zilk	20.6
Langevelderslag	0.7

METAPHOR

The main processes that determine the fate of a population are birth, death and exchange (dispersal). METAPHOR simulates these processes. METAPHOR is a spatially explicit, individual based model that simulates the dynamics of a metapopulation. A simulation starts with a given number of individuals of different age classes and sex categories in a specified number of patches. In the case study the simulations started with each patch filled to carrying capacity. Carrying capacity is a linear function of patch area, truncated to discrete numbers. METAPHOR follows the life history of each individual. The first simulated event is formation of breeding pairs and reproduction, followed by a mortality event. Individuals age and finally disperse through the landscape. Reproduction and mortality are density dependent and stochastic processes. The dispersal algorithm is spatially explicit (Verboom et al. 2001, Vos et al. 2001).

Each year, METAPHOR assesses the state of sub-populations, thus, the state of the metapopulation. The results are, for example, the persistence probability of the metapopulation and the mean densities in the sub-populations. METAPHOR can be used to simulate the effects of changes in landscape pattern as well as processes within the metapopulation. Because each individual is assigned to a specific location the model is able to translate local impacts (recreation disturbance) into results on the landscape level (persistence). METAPHOR is a flexible tool that can easily be adjusted to meet the requirements of new applications.

Comparable models are ALEX (Possingham & Davies 1995), ALMASS (Topping et al. 2003) and RAMAS (Akçakaya 2000), which all have in common that they can be used for population viability analysis (PVA, see Brook et al. 2000) and differ in the exact formulation of density dependence, population structure, and dispersal algorithms. METAPHOR has detailed and realistic algorithms for these features, but lacks the interaction between species that characterizes ALMASS.

Parameters METAPHOR case study

In the case study, we used the model for evaluating the persistence of the Skylark

(*Alauda arvensis*). The Skylark breeds in open vegetation without shrubs or trees (Topping *et al.* 2005, Beintema *et al.* 1995). Species density is directly correlated with the presence of more open landscapes (Van 't Hoff 2001). These conditions were used to select suitable vegetation types from a local land cover map (Van Til and Mourik 1999). We categorized three types of suitability: optimal, sub-optimal and marginal habitat. In optimal habitats, densities can reach 60 breeding pairs per square kilometer (Cramp *et al.* 1998, Teixeira 1979). In sub-optimal habitats, the densities where set at 30 breeding pairs per square kilometer while in marginal habitats the densities where set at 15 breeding pairs per square kilometer. Because Skylarks forage within 300-350 meters of their nest (Cramp *et al.* 1988), all patches within 300 meters of one another belong to the same local population. The dispersal capacity of the Skylark is estimated at 10 kilometers and 90% of all dispersal events are assumed to remain within this distance (Pouwels *et al.* 2002). All patches in the study area lie within this threshold distance and form one metapopulation. The main parameters for reproduction, mortality and dispersal are based on Cramp *et al.* (1988) and Beintema (1995) (Table 2). Mortality rates and reproduction used by Topping *et al.* (2005) for Skylark are within the ranges set for high and low densities in METAPHOR.

For each scenario, 100 replica runs were generated. Results were collected between years 150 and 250. It is expected that after 150 years the metapopulation in METAPHOR achieves a balance (Vergeer 1997).

RECREATION IMPACT: LINKING MASOOR AND METAPHOR

MASOOR and METAPHOR are linked by the impact of visitors on two model parameters: density and reproduction. In their study, Vos and Peltzer (1987) related the impact of recreation on birds to the number of people on a path at the tenth busiest day in the year (mostly a sunny weekend day in spring). The result of MASOOR is translated into this output. The number of visitors on each path was scaled to 4500 visitors for the total area (the estimated number on the tenth busiest day). This resulted in visitor densities per path segment per hour. The recreation impact is calculated by using buffer zones. The width of the disturbance zones depends on the visitor densities (Vos and Peltzer 1987, see Figure 2).

For the Skylark, Vos and Peltzer (1987) found a reduction of more than 50% in a zone of 40 meter when visitor densities were 5 groups per hour and a reduction of 100% in a zone of 40 meter when visitor densities were 20 groups per hour. These parameters were extrapolated to disturbance zones (Table 3). These parameters were used to define a buffer zone surrounding each path segment. Within the buffer zones, the density was reduced by 50%.

Little is known about the reduction in reproduction success for Skylark. For Stonechat (*Saxicola torquata*), the percentage of nests that were abandoned in disturbed zones is 75% and for Curlew (*Numenius arquata*) the percentage is 64% (Vos and Peltzer 1987). Based on expert

Table 2. Main yearly parameters for the Skylark in METAPHOR.

Parameter	Probability
mortality low densities	0.2
mortality high densities	0.4
standard deviation mortality	0.05
reproduction low density	0.75
reproduction high density	0.35
standard deviation reproduction	0.1
fraction of juveniles that disperse	0.7
fraction of adults that disperse	0.1

Table 3. Size of disturbance zone produced by varying sized groups of visitors used in the models.

Number of groups	Disturbance zone (m)
0-1	30
2-5	60
6-15	100
16-30	200
31-60	300
61-100	400
>100	600

judgement, we chose a low reduction in reproduction of 25% in disturbed zones. Because the impact is expressed as visitor impact for one day and the viability of the Skylark is simulated over 100 years the time scale of MASOOR differs greatly from the time scale of METAPHOR. MASOOR uses discrete event simulations (see Jochem et al. 2007) and METAPHOR uses discrete time steps (Vos et al. 2001).

RESULTS

In the area, 11 populations can be distinguished. Some of the populations are completely within disturbance zones of recreation, while others are more or less disturbance free (Figures 5a and 5b). In the zoning scenario, all paths near population 5, 6, 7 and 8 and two paths through population 4 have been closed (Figure 5b).

In the current situation, 35% of the habitat is disturbed while in the zoning scenario 28% is disturbed (Table 4). The simulation results show that the impact of zoning is most pronounced in protecting. In the current situation the disturbed area of population 4 is twice as large as in the 'zoning scenario'. This results in a less stable population with an average of 14 breeding pairs (Table 4). Because population 4 is the largest population in the metapopulation, a reduction in this population is expected to have the largest effect

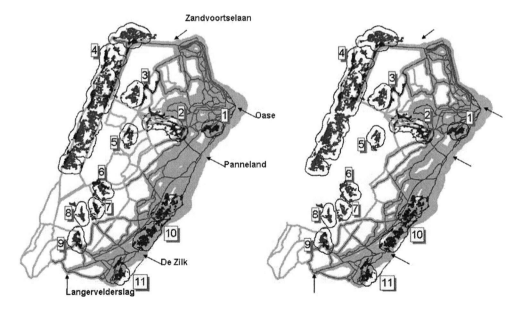

Figure 5a-b. Predicted zones disturbed by visitors in the scenario 'Present situation' (a) and 'Recreation zoning plan' (b). The disturbance zones along the foot paths are shown (light red). Entrances are indicated by black arrows. The populations are divided in a disturbed part (dark red) and an undisturbed part (dark green).

Table 4. Results of the simulations for three scenarios.

	no recreation	current situation	zoning scenario
% disturbed habitat	0%	35%	28%
viable metapopulation	yes	yes	yes
population 4 (bp)	41	14	44
average total population (bp)	83	33	82
standars error total population	1.14	1.14	1.10

Note: Number of runs was 100. Population 4 corresponds with population 4 in figures 6a-6b. 'bp' is number of breeding pairs.

on the metapopulation (Vos et al. 2001, Verboom et al. 2001, Opdam et al. 2003).

In the zoning scenario, the number of breeding pairs were not significantly different from the scenario without recreation. Also the nature indicators (percentage of occupied patches and total numbers of Skylark) of the zoning scenario are the same as in the scenario without recreation (Table 4). The current situation has lower values of number of breeding pairs in the total population. Extinction frequency did not differ significantly between the three scenarios. All scenarios result in a viable metapopulation.

For the scenario without recreation, no recreation indicators can be determined by MASOOR. Compared to the current situation, the number of visitor groups per path length increased in the zoning scenario by almost 20 percent (Table 5). In comparison with the current situation, the length of paths in the zoning scenario containing less than 1 visitor per hour (i.e. tranquil paths) decreased by almost 40 percent.

DISCUSSION

The results show that the zoning scenario achieves almost maximum achievable nature values in the case study area. As a next step, a scenario with zoning in population 4 only might be considered. Within a metapopulation the largest population is very important for the persistence and occupation of all the subpopulations (Verboom et al. 2001, Opdam et al. 2003). At the same time, the effects of the zoning scenario seem to produce a noticeable decline in the quality of recreational experience as compared with the current situation.

Choosing between the current situation and the zoning scenario is a political choice. If specific objectives or thresholds for nature indicators and recreation indicators that are agreeable to concerned stakeholders, scenario results can be explicitly compared. The existence of agreed upon objectives and thresholds may help managers resolve what might otherwise be considered an intractable situation.

This chapter illustrates how models can be part of the framework presented in Figure 1. Local knowledge and monitoring data increase the quality and credibility of the results from the models. Stakeholder involvement is important not only at the

Table 5. Recreation indicators derived from MASOOR for the three scenarios.

	no recreation	current situation	zoning scenario
visitor groups/km	-	3.7	4.4
tranquil paths (km)	-	62.6	38.4

start (scoping / goal setting) and at the end (evaluation) of the planning process (Bentrup 2001), but also during the spatial analysis of scenarios (Johnson & Campbell 1999). Since citizens participating in planning processes will not support what they do not understand (Theobald et al. 2000), decision support tools for choosing common and measurable goals need to be easily understood. The effectiveness of conservation management is thought to be closely linked to adaptive management processes that empower stakeholders, rather than by "ever-more precise techniques for prioritizing elements of nature" (Knight et al. 2006).

The results of the simulation models are useful in communication with stakeholders. Managers can show clearly what the effects of different scenarios are for attainment of both habitat goals and recreation goals. In an iterative and reflective process, indicators for nature and recreational quality can be compared across scenarios. Communicative action among participating stakeholder groups allows eventual construction of a consensus policy on recreational access. Managers of three large dune areas in and around the case study area put effort in the further development of MASOOR (Jochem et al. 2007). This was mainly because the model showed its usefulness in communication with stakeholders. See Jochem et al. (2007) for a discussion on this issue.

Euler stated: "Give me five parameters and I will draw you an elephant; six, and I will have him wave his trunk". This quotation (in Mollison 1986) illustrates the pitfalls of model parameterization and calibration and is often used as a criticism of using models. However spatial models may be the only objective tools for scenario studies. Translating scenario studies into model parameters can simulate effects of, for example, changes in land-use. While the exact quantitative model outcomes sometimes have high levels of uncertainty, when used for comparing scenarios the results are more robust (Verboom and Wamelink 2005). For example, in applying the NTM model, Schouwenberg et al. (2000) illustrate that the model output had a large uncertainty for a single prediction, but when scenarios were compared the uncertainty was much smaller. The best alternative predicted by the model is likely to be the best one in real life (Verboom and Wamelink 2005). It is in the comparative evaluation of scenarios that integrated use of models such as MASOOR and METAPHOR may have its greatest utility. Managers faced with the task of accurately estimating outcomes of specific scenarios may find use of the models more problematic.

Models are inherently sensitive to modification of input parameters. The predicted numbers of Skylark in the scenario of the current situation are 3 times higher then the actual numbers over the last years (personal communication with Antje Ehrenburg). Although the simulation defines the metapopulation as viable, a small change in parameter settings can cause much lower estimates of breeding pairs. The lower numbers in the current situation could also be attributable to other stress factors not considered in the simulations. A land cover map from 1990 was used, possibly overestimating open areas and carrying capacity. Successional changes in vegetation of some open dune areas may have lead to the production of smaller patches and lower estimates of breeding pairs. In the case study, the Skylark populations in the Amsterdamse Waterleidingduinen are considered as a metapopulation on its own. In the agricultural areas to the East there is a large population of Skylarks. Because this population is declining also, the number of dispersal events to the dune area is difficult to estimate, but the large population will have at least a small positive effect on the actual population in the dune area.

Although both recreation and ecological models are being developed further, more effort should be put in the combination of these models. The integrated models should organize ecological, managerial and recreational information in a related manner, and subcomponents of the respective models should influence each other (Haider 2006). The output of recreation models should allow managers to place value on different aspects of the recreation experience as well as impact of recreation on nature. Also the impact of nature (development) on recreation should be implemented (grey arrows in figure 3). Do people appreciate the typical song of the Skylarks on sunny days? Visitors have strong preference for the preservation of species richness and scenic beauty. However, when there is a trade-off between these benefits visitors chose their favourite scenery at their favourite recreation site and prefer management options, which preserve biodiversity at other sites in an area (Horne et al. 2005). Visitors don't like to see changes in their own backyard.

In the case study no clear recreation goals were stated by the managers. In northwestern Europe, recreation goals are rarely specified. However, clear goals and thresholds are needed when future planning scenarios are to be comparatively evaluated and when recreation goals and nature goals need to be optimized. Integration of recreation and nature functions can not be accomplished in the absence of explicitly stated goals. We think that the lack of clear goals is one of the reasons why modelling frameworks are not more widely used in Europe. Another main reason is the complex private and public ownership of nature areas in Europe (Haider 2006).

Future research should focus on the integrated monitoring and modelling of recreation and nature (Haider 2006, Sutherland 2007). Simultaneous consideration of both recreation and nature goals might lead to construction of scenarios not immediately apparent when only one set of values is considered (Önal and Yanprechaset 2007). General rules of thumb for predicting recreation impact should be developed (Sutherland 2007). Also there are still knowledge gaps of recreation impact on the population level, measures that can reduce human impact and the combination of recreation and nature at the scale of landscape planning (Blanc et al. 2006, Sutherland 2007).

ACKNOWLEDGEMENTS

We thank dr. M. Hootmans and drs. A. Ehrenburg (Gemeentewaterleidingen Amsterdam) for comments and discussion. Prof P. Opdam, dr. C. Vos, dr. H. Baveco, dr. A. Beintema, drs. H. Henkens and drs. H. Kuipers (Alterra) contributed to the model development and the simulations. David Pitt and Ramona van Marwijk contributed to the outline of the framework. David Pitt also made editorial improvements.

LITERATURE CITED

Akçakaya, H.R. 2000. Population viability analyses with demographically and spatially structured models. Ecological Bulletins 48:23-38.

Bakker, J.G. & J. Lengkeek. 1999. [In Dutch] Monitoringsonderzoek recreatie Amsterdamse Waterleidingduinen – Deel II: Onderzoek naar beleving, recreatiegedrag en routepatronen van de bezoekers in 1998 en 1999. Wageningen Universiteit en Research Centrum (Nota Vakgroep Ruimtelijke Planvorming nummer 78). Wageningen.

Bathe, G. 2007. Political and social drivers for access to the countryside: the need for research on birds and recreational disturbance. Ibis 149 (s1): 3-8

Beintema, A., O. Moedt & D. Ellinger. 1995. [In

Dutch] Ecologische atlas van de Nederlandse weidevogels. Schuyt. Haarlem.

Bentrup, G. 2001. Evaluation of a collaborative model: a case study analysis of Watershed planning in the Intermountain West. Environmental Management 27: 739-748

Bijlsma, R.G., R. Lensink & F. Post 1985. [In Dutch] De boomleeuwerik (Lullula arborea) als broedvogel in Nederland in 1970-1984. Limosa 58: 89-96.

Blanc, R., M. Guillemain, J-P. Mouronval, D. Desmonts and H. Fritz. 2006. Effects of non-consumptive leisure disturbance to wildlife. Rev. Écol. (Terre Vie) 61: 117-133.

Blumstein, D.T., E. Fernández-Juricic, P.A. Zolnner and S.C. Garity. 2005. Inter-specific variation in avion responses to human disturbance. Journal of Applied Ecology 42: 943-953.

Brook, B.W., J.J. O'Grady, A.P. Chapman, M.A. Burgman, H.R. Akçakaya and R. Frankham. 2000. Predictive accuracy of population viability analysis in conservation biology. Nature 404: 385-387

Cole, D.N. 2004. Monitoring and management of recreation in protected areas: the contributions and limitations of science. In: Sievänen, T., J. Erkkonen, J. Jokimäki, J. Saarinen, S. Tuulentie & E. Virtanen (eds.). Policies, methods and tools for visitor management – proceedings of the second International Conference on Monitoring and Management of Visitor Flows in Recreational and Protected Areas, June 16–20, 2004, Rovaniemi, Finland. pg: 10-17.

Cramp, S., D.J. Brooks, E. Dunn, R. Gillmor, J. Hall-Craggs, P.A.D. Hollom, E.M. Nicholson, M.A. Ogilvie, C.S. Rosselaar, P.J. Sellar, K.E.L. Simmons, K.H. Voous, D.I.M. Wallace & M.G. Wilson. 1988 Handbook of the birds of Europe, the Middle East and North Africa : the birds of the Western Palearctic. Vol. 5: Tyrant flycatchers to thrushes. Oxford, New York. Oxford University Press.

Drewitt, A.L. 2007. Birds and Recreational Disturbance. Ibis 149 (s1): 1-2

Farrell, T.A. and J.L. Marion. 2002. The Protected Area Visitor Impact Management (PAVIM) framework: a simplified process for making management decisions. Journal of Sustainable Tourism 10(1): 31-51.

Fernández-Juricic, E., M.P. Venier, D. Renison and D.T. Blumstein. 2005. Sensitivity of wildlife to spatial patterns of recreational behavior: a critical assessment of minimum approaching distances and buffer areas for grassland birds. Biological Conservation 125: 225-235.

Gaddy, L.L.& T.L. Kohlsaat 1987. Recreational impact on the natural vegetation and avifauna and herpetofauna of south Carolina Barrier Islands. Journal of Natural Areas 7: 55-64.

Gill, J. A. 2007. Approaches to measuring the effects of human disturbance on birds. Ibis 149(s1): 9-14.

Gill, J. A., K. Norris and W.J. Sutherland. 2001. The effects of disturbance on habitat use by black-tailed godwits Limosa limosa. Journal of Applied Ecology 38(4): 846-856.

González, L.M., B.E. Arroyo, A. margalida, R. Sánchez and J. Oria. 2006. Effect of human activities on the behaviour of breeding Spanish imperial eagle (*Aquila adalberti*): management implications for the conservation of a threatened species. Animal Conservation 9: 85-93.

Gross-Custard, J.J., P. Triplet, F. Sueur and A.D. West. 2006. Critical thresholds of disturbance by people and raptors foraging wading birds. Biological Conservation 127: 88-97.

Haider, W. 2006. North American Idols: Personal Observations on Visitor Management Frameworks and Recreation Research. In: Clivaz, S.D., C. Clivaz, M. Hunziker & S. Iten (eds.) Exploring the Nature of Management. Proceedings of the Third International Conference on Monitoring and Management of Visitor Flows in Recreational and Protected Areas. University of Applied Sciences Rapperswil, Switzerland, 13-17 September 2006. Rapperswil. pg: 16-22

Henkens, R.J.H.G. 1998. [In Dutch] Ecologische capaciteit natuurdoeltypen I: Methode voor de bepaling van effect recreatie op broedvogels. IBN-report 363. Alterra, Wageningen.

Henkens, R.J.H.G., R. Jochem, R. Pouwels and P.A.M. Visschedijk. 2006. Development of a Zoning Instrument for Visitor Management in Protected Areas. In: Clivaz, S.D., C. Clivaz, M. Hunziker & S. Iten (eds.) Exploring the Nature of Management. Proceedings of the Third International Conference on Monitoring and Management of Visitor Flows in Recreational and Protected Areas. University of Applied Sciences Rapperswil, Switzerland, 13-17 September 2006. Rapperswil. pg: 238-240

Hill, D., D. Hockin, D. Price, G. Tucker, R. Morris and J. Treweek. 1997. Bird disturbance: improving quality and utility of disturbance research. Journal of Applied Ecology 34: 275-288.

Horne, P., P.C. Boxall and W.L. Adamowicz. 2005. Multiple-use management of forest recreation sites: a spatially explicit choice experiment. Forest Ecology and Management 207: 189-199.

Jaarsma. C.F. en M.J. Webster. 1999. [In Dutch] Monitoringsonderzoek recreatie Amsterdamse Waterleidingduinen – Deel III: Analyse van het recreatiebezoek en het recreatieverkeer in het jaar 1998. Wageningen Universiteit en Research Centrum (Nota Vakgroep Ruimtelijke Planvorming nummer 79). Wageningen.

Johnson, B.R., and R. Campbell. 1999. Ecology and participation in landscape-based planning within the Pacific Northwest. Policy Studies Journal 27: 502-529.

Knight, A.T., R.M. Cowling, and B.M. Campbell. 2006. An operational model for implementing conservation action. Conservation Biology 20:408-419.

Koning F.J. and Baeyens G. 1990. [In Dutch:] Uilen in de duinen. Stichting Uitgeverij KNNV / Gemeentewaterleidingen Amsterdam, Utrecht / Amsterdam.

Langston, R. H. W., D. Liley, G. Murison, E. Woodfield and R.T. Clarke. 2007. What effects do walkers and dogs have on the distribution and productivity of breeding European Nightjar Caprimulgus europaeus? Ibis 149(s1): 27-36.

Liley, D., R.T. Clarke, J.W. Mallord and J.M. Bullock. 2006. The effect of urban development and recreational access on the distribution and abundance of nightjars on the Thames Basin and Dorset Heaths. Unpublished report [Draft 6th December 2006]. Footprint Ecology / Natural England.Mallord, J. W., P. M. Dolman, et al. 2007. How perception and density-dependence affect breeding Woodlarks Lullula arborea. Ibis 149(s1): 15-15.

Mallord, J.W., P.M. Dolman, A.F. Borwn and W.J. Sutherland. 2007. Linking recreational disturbance to population size in a ground-nesting species. Journal of Applied Ecology 44: 185-195.

Manning, R. 2004. Carrying Capacity of parks and wilderness. In: Harmon, D., B.M. Kilgore and G.E. Vietzke (eds.) Protecting Our Diverse Heritage: The Role of Parks, Protected Areas, and Cultural Sites. Proceedings of the 2003 George Wright Society / National Park Service Joint Conference. Hancock, Michigan: The George Wright Society.

Miller, S.G., R.L. Knight and C.K. Miller. 1998. Influence of recreeational trails on breeding bird communities. Ecological Applications 8(1) 162-169

Mollison, D. 1986. Modelling biological invasions: change, explanation, prediction. Phylosophical Transactions of the Royal Society of London B 314: 675-693

Murison, G., J.M. Bullock, J. Underhill-Day, R.H.W. Langston, A.F. Brown, and W.J. Sutherland. 2007. Habitat type determines the effects of disturbance on the breeding productivity of the Dartford Warbler Sylvia undata. Ibis 149(s1): 16-26

Natural England. 2006. Strategic Direction 2006-2009). website: http://www.naturalengland.org.uk/pdf/about/Natural_England_Strategic_Direction.pdf

Natuurmonumenten. 2006. website [Dutch]: http://www.natuurmonumenten.nl/natmm-internet/vereniging/natuurmonumenten.htm.

O'Connell, M. J., R. M. Ward, C. Onoufriou, I.J Winfield, G. Harris, R. Jones, M.L. Yallop and A.F. Brown. 2007. Integrating multi-scale data to model the relationship between food resources, waterbird distribution and human activities in freshwater systems: preliminary findings and potential uses Ibis 149(s1): 65-72.

Önal, H. and P. Yanprechaset. 2007. Site accessibility and priorization of nature reserves. Ecological Economics 60: 763-773.

Opdam P., R. Foppen, C.C. Vos. 2002. Bridging the gap between empirical knowledge and spatial planning in landscape ecology. Landscape Ecology 16: 767-779.

Opdam, P.F.M., J. Verboom and R. Pouwels. 2003. Landscape cohesion: an index for the conservation potential of landscapes for biodiversity. Landscape Ecology 18, 113-126.

Poe, A., R.H. Gimblett, M.I. Golstein and P. Guertin. 2006. Evaluating Spatiotemporal Interactions between Winter Recreation and Wildlife Using Agent-Based Simulation Modeling on the Kenai Peninsula, Alaska. In: Clivaz, S.D., C. Clivaz, M. Hunziker & S. Iten (eds.) Exploring the Nature of Management. Proceedings of the Third International Conference on Monitoring and Management of Visitor Flows in Recreational and Protected Areas. University of Applied Sciences Rapperswil, Switzerland, 13-17 September 2006. Rapperswil. pg: 306-307

Pouwels, R., M.J.S.M. Reijnen, J.T.R. Kalkhoven and J. Dirksen. 2002 [In Dutch] Ecoprofielen voor soortanalyses van ruimtelijke samenhang met LARCH. Alterra-rapport 493. Alterra, , Wageningen.

Possingham, H.P. and I. Davies. 1995. ALEX: A Model For The Viability Analysis Of Spatially Structured Populations. Biological Conservation 73 (2): 143-150.

Riffell, S. K., K. J. Gutzwiller and S.H. Anderson. 1996. Does repeated human intrusion cause cumulative declines in avian richness and abundance? Ecological Applications 6(2): 492-505.

Seidl, I., and C.A. Tisdell. 1999. Carrying capacity reconsidered: from Malthus' population theory to cultural carrying capacity. Ecological Economics 31: 395.408.

Schouwenberg, E.P.A.G., H. Houweling, M.J.W. Jansen, J. Kros and J.P. Mol-Dijkstra. 2000. Uncertainty propagation in model chains: a case study in nature conservancy. Alterra report 001. Alterra, Wageningen.

Stillman, R.A. and J.D. Goss-Custard. 2002. Seasonal changes in the response of Oystercatchers *Haematopus ostralegus* to human disturbance. J. Avian Biol. 33: 358–365.

Stillman, R. A., A. D. West, R.W.G. Caldow and S.E.A. le V. dit Durell. 2007. Predicting the effect of disturbance on coastal birds. Ibis 149(s1): 73-81.

Sutherland, W.J. 2007. Future directions in disturbance research. Ibis 149 (s1): 120-124

Sutherland, W.J., Armstrong-Brown, S., Armsworth, P.R., Brereton, T., Brickland, J., Campbell, C.D., Chamberlain, D.E., Cooke, A.I., Dulvy, N.K., Dusic, N.R., Fitton, M., Freckleton, R.P., Godfray, H.C., Grout, N., Harvey, H.J., Hedley, C., Hopkins, J.J., Kift, N.B., Kirby, J., Kunin, W.E., MacDonald, D.W., Markee, B., Naura, M., Neale, A.R., Oliver, T., Osborn, D., Pullin, A.S., Shardlow, M.E.A., Showler, D.A., Smith, P.L., Smithers, R.J., Solandt, J.-L., Spencer, J., Spray, C.J., Thomas, C.D., Thompson, J., Webb, S.E., Yalden, D.W. & Watkinson, A.R. 2006. The identification of one hundred ecological questions of high policy relevance in the UK. J. Appl. Ecol. 43: 617–627.

Taylor, E. C., R. E. Green, and J. Perrins. 2007. Stone-curlews Burhinus oedicnemus and recreational disturbance: developing a management tool for access. Ibis 149(s1): 37-44.

Teixeira, R.M. 1979. [In Dutch] Atlas van de Nederlandse broedvogels. Deventer. De Lange van Leer b.v.

Theobald, D.M., N.T.Hobbs, T. Bearly, J.A.Zack, T. Shenk, and W.E. Riebsame, 2000. Incorporating biological information in local land-use decision making: designing a system for conservation planning. Landscape Ecology 15: 35-45.

Topping, C.J., R.M. Sibly, H.R. Akçakaya, G.C. Smith and D.R. Crocker. 2005. Risk assessment of UK Skylark populations using life-history and individual-based landscape models. Ecotoxicology 14: 925-936.

Topping, C.J., T.S. Hansen, T.S. Jensen, J.U. Jepsen, F. Nikolajsen and P. Odderskær. 2003. ALMaSS, an agent-based model for animals in temperate European landscapes. Ecological Modelling 167: 65–82

Van der Zande, A. N. and P. Vos. 1984. Impact of a semi-experimental increase in recreation intensity on the densities of birds in groves and hedges on a lake shore in the Netherlands. Biological Conservation 30(3): 237-259.

Van der Zande, A.N. and T. Verstrael 1984. Impacts of outdoor recreation upon nest-site choice and breeding success of the kestrel (Falco tinnunculus) in 1975-1980 in the Netherlands. In: A.N. van der Zande (ed.), Outdoor recreation and birds: conflict or symbiosis; Impacts of outdoor recreation upon density and breeding success of birds in dune and forest areas in the Netherlands. Dissertatie, Universiteit van Leiden, Leiden, Pp 130-150.

Van Til, M. and J. Mourik. 1999. [In Dutch] Vegetatie en landschap van de Amsterdamse Waterleidingduinen. Architectura & Natura. Amsterdam.

Van 't Hoff, J. 2002. [In Dutch] Veldleeuwerik *Aulada arvensis*. In: SOVON Vogelonderzoek Nederland. 2002. Atlas van de Nederlandse broedvogels 1998-2000. Nederlandse Fauna 5. Nationaal Natuurhistorisch Museum naturalis, KNNV Uitgeverij & European Invertebrate Survey-Nederland, Leiden. pg: 312-313

Verboom, J., R. Foppen , J.P. Chardon, P.F.M. Opdam en P.C. Luttikhuizen. 2001. Introducing the key patch approach for habitat networks with persistent populations: an example for marshland birds. Biological Conservation. Vol 100 (1). pp. 89-100.

Verboom, J. and W. Wamelink. 2005. Spatial modeling in landscape ecology. In: Wiens, J.A. and M.R. Moss (eds.). Issues and perspectives in landscape ecology. Cambridge University Press, Cambridge. 79-89

Vergeer, P. 1997. Changes in landscape: how do species react?. Student report. IBN-DLO, Wageningen.

Vos, C.C., J. Verboom, P.F.M. Opdam and C.J.F. Ter Braak. 2001 Towards ecologically scaled landscape indices. American Naturalist 157: 24-51.

Vos, P. & R.H.M. Peltzer 1987. [In Dutch] Recreatie en broedvogels in heidegebieden: Strabrechtse en Groote Heide. Bos en recreatie 15, Afdeling Sociologisch Onderzoek t.b.v. Natuur en Landshcap, SBB Utrecht.

West, A.D., J.D. Gros-Custard, R.A. Stillman, R.W.G.Caldow, A.E.A. le V. dit Durell and S. McGrorty. 2002. Predicting the impacts of disturbance on shorebird mortality using a behaviour-based model. Biological Conservation 106: 319-328

Yalden, P.E. and D.W. Yalden 1990. Recreational disturbance of breeding golden plovers (Pluvialis apricarius). Biological Conservation 51: 243: 262.

CHAPTER 15

MASOOR: MODELING THE TRANSACTION OF PEOPLE AND ENVIRONMENT ON DENSE TRAIL NETWORKS IN NATURAL RESOURCE SETTINGS

Rene Jochem
Ramona van Marwijk
Rogier Pouwels
David G. Pitt

Abstract: MASOOR (Multi Agent Simulation of Outdoor Recreation) is a multi-agent recreational behavior simulation model designed for front country situations containing high density trail networks. It was developed in collaboration with researchers at Wageningen University and Research Center and natural resource managers in the Netherlands, France and Great Britain. The model uses object oriented programming language to capture the transactional experience of natural areas by different types of visitors. A hierarchical control system provides a framework for identifying visit goals and constraints, defining path networks that enable goal attainment and specifying specific behavioral rules for navigating through the network in pursuit of goals. Application of the model to a Dutch national park with 1750 agents in a path network containing 2150 path segments finds a respectable correlation (r=0.68) with patterns of actual behavior as recorded from the geographic position system (GPS) tracks of 309 visitors. Future directions for conducting research into the transactional experience of recreational environments, refining MASOOR and similar models and managing recreational resources are discussed.

Key words: recreation simulation, MASOOR, transactional experience of environment

INTRODUCTION

The development of models for managing the transaction of people and environment in a manner that accommodates increasing human use while also maintaining ecological integrity is a growing area of recreation resource management research (see also Gimblett 2005). For example, the establishment of Natura 2000, a network of natural habitats distributed among the countries of the European Union, created a political need to systematically assess the effectiveness of landscape management plans for maintaining biological diversity in portions of this network that are also subject to intensive recreational and tourism activity (Visschedijk et al. 2005). A major concern in modeling the transaction of recreational visitors with natural settings is optimizing the production of high quality outdoor recreational experiences and ecosystem values and services (e.g. biological diversity and water quality) in the landscape (Lawson 2006)

Early development of these models focused on maintaining the ecological integrity of landscapes in the face of increasing pressures for recreational use with foci on measuring the effects of recreational behavior on critical habitats, water quality and soil resources (Goodwin 2000; McCool & Lime 2001). Consideration of experiential quality focused on collection of visitor data through interviews and visitor counts. As described elsewhere in this volume (Pouwels et al. 2008), the 21st Century witnessed the evolution of spatially explicit models capable of relating management actions concerning the geographic distribution and density of recreational use to metapopulation dynamics and persistence of sensitive plant and animal species. This required development of a visitor use model that was in spatial scale and resolution comparable to the ecological models (Pouwels et al. 2008). Early visitor use simulation models replicated the spatial and temporal distribution of visitor use patterns as they relate to an area's natural and developed environments. They allowed managers to visualize the spatial and temporal implications of different strategies to direct visitor use patterns toward or away from key environmental resources (Cole 2005).

More recently, computer models began focusing on the effects of visitor use on experiential quality. Models such as RBSim 2 (Itami & Gimblett 2001; Lawson et al. 2002), Wilderness area Simulation Model (Wagtendonk 2004), Extend (Lawson et al. 2003), and GCRTSim (Roberts et al. 2002) have been developed to assist managers establish the optimum level of use of a wilderness area with regard to crowding. These models have been applied to back country locations in Australia and USA that have an extensive or moderate path density. As a result, they often use 'typical travel itineraries' (Arrowsmith & Chhetri 2003) that involve a fairly limited range of itinerary options.

In front country recreational settings that are characterized by more intensive land use patterns and higher population densities (e.g. the Netherlands), the range of travel pattern itineraries for accomplishing a specific goal is large. Consequently, path density is higher, and recreational use simulation models must be able to accommodate multiple path opportunities for accomplishing trip goals. Even in relatively remote recreational settings in the Netherlands, for example, the higher density of paths that characterizes Dutch nature areas offers the visitor several possibilities to reach a 'goal' (which can be a specific facility or just avoiding crowds).

This paper describes the development of MASOOR (Multi Agent Simulation of Outdoor Recreation), a visitor simulation model designed for front country situations containing high density trail networks. After describing the process used to create MASOOR, the paper explores the conceptual frameworks used to characterize human-environment transactions as well as to model visitor use patterns in a high-density trail network. The structure and dynamics of the MASOOR model are discussed. Finally, an application of the model to understanding visitor use patterns of hikers in Dwingelderveld National Park in the Netherlands is presented. The paper concludes by suggesting future directions for development of simulation models such as MASOOR.

CREATION OF MASOOR

The development of recreation resource management decision support systems requires a dynamic interaction between the person building the model and the eventual end user of the model. The model becomes far more useful and more frequently used if it is socially constructed within the community where it will be used. Technical information needed for the construction of the model gains utility in eventual model application if it is related to organizational routines and practices of the end user's application of the model. In other words, iterative and reflective communication between the builder and the user of the model during the process of construction and testing is more likely to produce a model that is not only better understood by the user but also already incorporated into user behavior related to resource management (Innes 1996).

Recognizing the value of a communicative action approach to model construction, MASOOR was constructed through a series of workshops between researchers at the Wageningen University and Research Center and personnel from several agencies responsible for managing recreational use in Dutch natural areas. Early outcomes from these workshops included recognition that both model builders and end users were interested in constructing a model that could characterize spatial and temporal patterns of visitor use. This characterization needed to enable examination of both the ecological footprint of recreational use as well as the implications of use on the quality of visitor experiences. Both parties agreed that the model needed to have a sufficient spatial and temporal scale and resolution to permit the characterization of ecological and experiential consequences of management actions implemented in discrete geographic settings and time periods.

Special needs were identified by both model builders and model users. As is discussed in subsequent sections of this paper, model builders were interested in two sets of scientific criteria. The first of these relates to advancing the state-of-the-art of multi-agent modeling of visitor behavior in the dense path networks that characterize natural area recreation in the intensively used landscapes of Western Europe. The second scientific criterion relates to constructing a simulation model that is based on a transactional framework in which environment provides a cognitive framework for action toward accomplishment of recreational objectives and satisfaction of human needs. In this sense, scientific interest resides in developing criteria to define with conceptual, spatial and temporal accuracy and precision how different types of visitors interpret, relate to and behave in various environmental settings.

Managers were especially interested in the model's ability to consider landscape elements over which they could exert influence. The geographic scale of management criteria ranges from local to regional. At the local scale, managers are interested in the effects on recreational behavior of trail design alternatives (e.g. path surface and width) and geographic positioning of elements such as benches, picnic tables, trails, parking spaces and various types of recreational attractions and interpretative information. At the regional scale, managers may be concerned with the visitor behavioral implications of actions taken to alter hydrologic regime or dispersion of plant and animal communities in attempts to accomplish ecological restoration objectives.

MASOOR evolved through an iterative process in which the intuition of model builders was informed by the professional insights of natural area managers and experts in the field of recreation resource management as well as consultation of scientific literature inside and outside the field of recreation. While the model is under continual development and refinement, it has been applied in several European settings to understanding the implications of various scenarios of visitor use. It has proved useful in estimating the implications of spatially and temporally specific visitor use management actions. Most importantly, managers find themselves able to integrate model output into the process of constructing policy choice among stakeholders and policymakers (see also Cole 2005).

CONCEPTUAL FRAMEWORKS USED IN CONSTRUCTING MASOOR

In this section of the paper we will describe the conceptual frameworks that informed construction of the MASOOR model. A discussion of the transactional approach to examining human-environment interaction, which was used as a framework for characterizing recreational behavior in natural settings, is succeeded by an elaboration of spatial modeling. An integration of the two frameworks provides a basis for subsequent description of model structure and operation.

Transactional Understanding of Recreation Behavior in Natural Settings

Nature of the Transaction. Recreational behavior can be viewed as a transaction of visitors with attributes of environment (Pitt 1989; Zube et al. 1982). From a transactional perspective, recreational visitors are surrounded and engulfed by environment and they constantly receive multi-sensory cues about environmental content and structure (Ittelson 1978). Visitors use both the content of this information as well as its organization in the environment to interpret meaning in the environment (Kaplan & Kaplan 1982; Kaplan & Kaplan 1989). This information evokes autonomic affective response (Bourassa 1990; Ulrich 1986; Zajonc 1980), which demonstrably affects human understanding of and performance in natural environments (Hartig 1993). In this sense, visitors are active participants within rather than passive observers of recreational environment (Ittelson 1978). Visitors process environmental information relative to its ability to satisfy a wide range of personal needs and to enable accomplishment of goals and plans (Cohen 1979; Pearce 1982; Yiannakis & Gibson 1992). Goal-seeking individuals and groups evaluate environment based on its salience to desired outcomes and in terms of its congruence or ability to afford behavioral opportunities for realizing desired goals (Gibson 1986; Stokols 1978).

Dimensions of the Transaction. Operationalizing a transactional framework requires specification of three sets of parameters (Pitt 1989). As noted above, multi-sensory **attributes of environment** must be assessed from the standpoints of both their content and their spatial organization (Ittelson 1978; Kaplan & Kaplan 1982; Kaplan & Kaplan 1989). Variability in **participant attributes** (e.g. socio-economic status, life cycle/style, mobility impairments, cultural/ethnic characteristics, childhood experiences, personality traits and value/belief systems) affects the production of transactional outcomes. A person returning to the same environment in a different **context of participation** will experience varying outcomes. Differences may relate to *psychological context* (e.g. emotional disposition, motivation, skill, knowledge), *social context* (e.g. group composition, referent social groups, encounters), *physical context* (e.g. mode, speed, direction, position), and temporal context (e.g. integration of specific events in linear time).

The dynamic interaction of environmental attributes, visitor characteristics and contextual dimensions in the production of transactional outcomes is illustrated in Figure 1. None of the three sets of inputs to this production process is more important than the others. The outcome of the

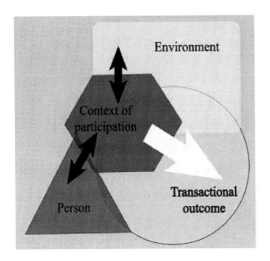

Figure 1. The production of transactional outcomes

transaction is a function of what lies outside the visitor, what lies inside the visitor and how the visitor engages environment (Meinig 1979).

Spatial Modeling Framework

Probabilistic Simulation Models. One way to approach spatial modeling of recreational behavior is to use probabilistic simulation models, which are based on a representative sample of visitor trip itineraries. Visitors' trips are then modeled based on the probability of a visitor selecting a single trip itinerary out of the entire sample (Manning et al. 2005). Advancements in the development of geographic information system (GIS) and geographic positioning system (GPS) technology make creating trip itineraries of visitors easier and more reliable. However, probabilistic simulation may not be an appropriate way to model behavior in new recreational settings or in existing settings where management policies may introduce new travel networks, delete or alter existing travel networks, or where behavior may change due to changes in recreation mode or mix of recreation types (Manning et al. 2005).

Individual Agent-Based Models. To create Spatial Decision Support Systems that maintain validity outside the current and fixed environmental domain of an existing trail network, Individual-Based Models (IBM) have been used. IBM's have been successfully applied to spatially-explicit modeling of ecological phenomena and are capable of modeling variation among individuals and interaction between individuals (Slothouwer et al. 1996). IBM's have been used in ecological studies to model population dynamics (Verboom 1996) and animal movement or dispersion (Vos et al. 2002). A form of individual-based modeling is "Agent-Based modeling" in which the basic unit of activity is the "agent". Usually agents explicitly represent individual actors in the situation being modeled (Schelhorn et al. 1999). In socio-economic applications, there is a considerable interest in the use of agent-based techniques for spatial analysis, geocomputation and the development of Spatial Decision Support Systems (Haklay et al. 2001). An example with considerable links to recreation behavior simulation is STREETS (Schelhorn et al. 1999), which was built to understand the movement of people in town centers. Itami and Gimblett (2001) have shown that agent based systems are also suitable for simulating recreation behavior by building RBSIM.

Multiple Agent-Based Models. The challenge of simulation modeling is to capture essential behavior of the system being modeled. In outdoor recreation, this means capturing and representing the characteristics of the physical environment and modeling the behavior of multiple visitors as they interact with the environment and with each other (Cole 2005; Manning et al. 2005). For the domain of MASOOR, we selected a Multi-Agent System (MAS) linked to a Geographical Information System (GIS). The agent's artificial intelligence is situated in a movement module that is multi-phasic and dynamic. Like the STREETS Model, the movement model of MASOOR is based on a Hierarchical Control System. MASOOR can interpret the 'world' at different scales in order to make agents autonomously navigate a given recreational track network.

Multi-Agent Systems evolved from fields like robotics, psychology, ethology, biology and artificial life to characterize the processes by which two or more agents interact among themselves and with their environment while accomplishing a complex task. Given the fact that recreational way-finding involves the transaction of individual recreationists with biophysical and cultural dimensions of the environment as well as with other visitors, an MAS-based architecture provides an ideal structural framework for simulating recreational behavior. The MAS framework allows for interaction among different visitor groups as well transaction with a complex environment.

Cognitive Agents. The agents in MASOOR are mobile, cognitive and spatially aware, since one of their primary tasks is navigation in a representation of an actual spatial environment. An important distinction between agent-based models is the extent to which agent behavior is determined by limited local information, or by overall knowledge of global outcomes. This distinction is closely related to that between reactive and cognitive agents (Ferber 1994). Reactive agents usually respond in a stimulus-response mode to individual events in their environment, with little capacity to consider a specific event in its larger spatial context. Cognitive agents, in contrast, incorporate some model of their world into their decision-making framework (Schelhorn et al. 1999). The agents of MASOOR respond to specific events but also have knowledge of the total environment in which they operate.

Integration of Transactional and Spatial Modeling Frameworks

Overall Model Structure. As illustrated in Figure 2, MASOOR operates in three distinct stages. During the *first stage*, management and scientific criteria must be specified as inputs to the model. As noted earlier, management criteria may be local in geographic scale (e.g. characterization of path design) or regional (e.g. characterization of hydrologic regime or vegetative community structure through which the trail passes). Feedback from visitor monitoring studies and manager expertise is also used to define the types and numbers of visitors that access the natural area from specific entry points. Scientific criteria must be specified to describe relationships between visitor types, preferences and expected behavior in the natural area. In the design criteria specification stage, the model user is defining parameters of the environment within which visitor behavior will occur and writing rules of allowable behavior for specified types of visitors.

As will be further discussed in the next section of this paper, the *second stage* involves the actual modeling of visitor behavior within a specified natural area. Management criteria inputs specify dimensions of the environment in which visitor behavior occurs. Spatial characterization of these inputs is retained in an ArcGIS™ framework, where, for example, the spatial configuration and actual attributes of the defined trails might be stored as a fixed path network shape file. Similarly, management criteria inputs specify the magnitude of the visitor population entering a natural area. In conjunction with scientific criteria, management inputs also provide a basis for defining typologies among the population of visitors entering an area.

The *third stage* of MASOOR operation involves generation and interpretation of

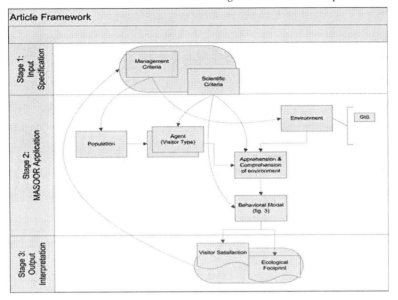

Figure 2. Overall structure of MASOOR model.

output, which includes a temporally dynamic representation of the movement of different visitor types within a specified path network (see Figure 4). To the extent that the distribution and persistence of plant and animal species is sensitive to the spatial and temporal distribution of different visitor types (Pouwels & Vos 2001), MASOOR output can be used to examine ecological footprint of recreational use. Using 'dose-effect' research strategies (Galloway 2002) in which an increasing dosage (i.e. increasing temporal/spatial densities of recreational use) are examined in terms of their effect (i.e. impact on biodiversity), managers can examine the implications of alternative strategies to manipulate the temporal and spatial dispersion of visitors.

From scientific criteria and understandings, rules are developed to define how various visitor types apprehend and understand the biophysical reality of a natural area. The actual construction of visitor behavior by MASOOR is illustrated in Figure 3 and will be discussed in more detail in a subsequent section of the paper.

Visitors as Agents in a Multi-Agent System. The development of the MASOOR model incorporated many dimensions of the transactional paradigm of human-environment relationships. The population of agents visiting a specific natural area is segmented into visitor groups based on both managerial insight and scientific investigation of visitor behavior. Numerous studies link typologies of behavioral preferences and other attitudes with socio-demographic characteristics (Johns & Gyimothy 2002). Recently, researchers have related visitor types to trail typologies (Farías et al. 2005). The implications of spatially explicit behavior (e.g. velocity of tourist travel, human way-finding logic and crowd avoidance) have yet to be considered in typological investigations of recreational behavior in natural areas (O'Connor et al. 2005). In the context of MASOOR, visitors become individual agents in a multi-agent system. Agent types are assigned role characteristics and behavioral characteristics. Role characteristics relate to types of behavior pursued by an agent type; for example, agents may be assigned roles within a family group or as dog owners, joggers, cyclists, etc. Agent types can be given the capacity to affect the behavior of one another. For example, the asymmetrical technological antipathy associated with encounters between motorized and non-motorized visitors can be simulated by assigning asymmetrical relationships to the condition existing when agents from non-motorized roles encounter motorized agents (Adelmann et al. 1982; Blahna et al. 1995; Catton 1983; Graefe & Thapa. 2004; Jackson & Wong 1982; Watson et al. 1994). In addition, specific ecological footprints can be assigned to varying agent roles.

Behavioral characteristics define the detailed behavior of agents, and they define factors such as intended trip speed and duration. Within an agent type, speed is kept constant and is, for each agent type, randomly drawn from a statistical distribution. The intended duration of the trip is also held constant within visitor type but assigned randomly to an agent type based on a statistical distribution. The intended duration assigned to an agent type can vary from the actual duration. The actual duration of an agent's trip is dependent on the configuration of the network and the cumulative path network choices an agent has made. Some agent types can be programmed to consider specific attractions in planning their activity schedule while other types might visit the area with no intention to visit a particular attraction. In both cases time is the main component of the activity schedule.

Representation of Environment. Agents encounter an environment that is value free and assigned no specific meaning. Agents construct meaning in the context of their spatial/temporal encounters with environment as well as with other agents experiencing the same environment. In MASOOR, the environment is modeled in an extendable and rich geographic information systems (GIS) framework. A number of data representations are used.

Path networks are stored as *vector data* that are attributed in terms of surface type. Path surface preferences are assigned to agents based upon manager experience and scientific understandings of visitor preferences. The path network is used in the generation of agent routes. Network segments are also attributed for their accessibility to the public based upon management decision-making. Management scenarios may be simulated by altering the accessibility of individual path segments to some or all types of agents for specified periods of time.

Attractions, such as refreshment concessions, viewpoints, visitor centers, historical places, etc., are represented as *point data*. Attraction points are assigned varying levels of attractiveness to the schedule of agent types using the path network. Attractions compete based on their relative ability to fulfill agent goals and closeness to agent location at a specific point in time. Areas that are attractive to various agent types are represented as *raster data*. Like point attractions, attractive areas within a raster framework have varying abilities to draw agents based upon the schedule of attractiveness that a particular type of area holds for a specific agent type.

Agent Engagement of Environment. Agent engagement of environment involves an environmental scan that positions the individual geographically and temporally in the natural area and apprises the agent of available opportunities for realization of defined goals. The engagement process also informs agents of social context, allowing them to react to the presence of other nearby visitors based on rules for interaction among the various role and behavioral characteristics assigned to agent types. Agents are also able to process an encounter at a specific point in space and time relative to their previous encounters during the trip. This iterative process leads to apprehension and comprehension of the spatial structural of environment in terms of opportunities and constraints for goal attainment that are afforded by the biophysical and cultural environment as well as the social and behavioral context of engagement (Kaplan & Kaplan 1989).

Spatial Behavior in Natural Areas. The interplay of mind and environment is essential to understanding human behavior in natural areas (Kaplan & Kaplan 1983). Understandings of both object and space as well as the position of objects in space are essential to perceive and give meaning to the environment. The idea of a mental model or cognitive map that is profoundly influenced by properties of the environment (ibid.) seems intuitively promising. A cognitive map is a personal representation of the environment that we experience. The term 'cognitive map' was introduced by Tolman (1948) who, based on observations of rats searching for food in a maze, used it to describe the representation of the environment that a rat, person or any other agent consults in order to guide intelligent behavior towards a goal. In formulating a cognitive map, processes such as homing, piloting, and chunking take place.

'Homing' or path integration is a person-based procedure in which the traveler constantly updates position with respect to a home base by semantically integrating time and motion as a journey proceeds. Under these circumstances the traveler, at any time in the route, should be able to turn and point in the direction of home base and to estimate the line distance that must be traveled to get there (Golledge 2002). Homing is a prerequisite for gaining spatial orientation with respect to an environment and subsequently finding one's way toward accomplishment of established goals.

The strategy known as 'piloting' makes use of landmarks that act as environmental clues that may be received in sequence. In a familiar environment, travel over specific routes is traced into a pattern of neural synapses in the brain that constitute formation the cognitive map. Retracing the route from multiple directions facilitates the accomplishment of route learning (Golledge 2002).

'Chunking' (Allen 1982) involves subdividing or chunking down planned, or experienced routes to aid memorization. When a significant directional change occurs, or a particular landmark is reached, a new pattern or trend is established that integrates combinations of distinct landmarks and direction (Golledge 2002). Repeated encounters with this trend lead to the creation of 'cognitive schemata' that assist in understanding the network's spatial configuration and in the recognition of generic layout. In the natural environment, these schemata can be based on slopes, gradients, or watercourses to build a configurational or layout picture of the environment (Golledge 2002).

In MASOOR, the cognitive map is constructed and stored on the so called 'Blackboard', where an agent stores information about an environment that is gathered by multi-sensory processes. In essence, the cognitive rules of homing, piloting and chunking provide a compass that allows characterization and remembrance of the spatial positioning in the environment.

Recreational behavior in natural areas can be segmented into distinct temporal phases or contexts, including: entry, immersion and exit (Borrie & Birzell 2001; Borrie & Roggenbuck 2001). Expectations, motivations, and the meaning of specific encounters with both biophysical and social/behavioral dimensions of environment will vary along the temporal continuum that is inherent in entering, immersing and leaving a natural area. MASOOR accommodates the multi-phasic nature of visitor behavior in natural areas by allowing an agent's behavior pattern along a path network to be segmented into entry, immersion and exit phases based on spatial and temporal positioning relative to points of network access and egress. During the first phase (entry), agent behavior is modeled toward moving away from the car park and finding the location of attractive areas that will lead to accomplishment of defined goals. Towards the core of the recreational area, agents move into a browsing behavior where solitude, attractiveness of the track and the environment might play a more important role in navigation. After a while, when trip goals have been attained or the time allotted for the journey is reached, the agent/visitor is 'homing' in on its exit point.

Segmenting the agent's behavioral track has practical advantages for the design and operation of MASOOR. Path segmentation enhances the ability to replicate agent behavior within the network. In the "object-oriented" environment within which MASOOR exists, path segmentation also facilitates alteration of behavioral rule definition as the internal state of an agent changes (Gamma et al. 1994).

DESIGN AND OPERATION OF THE MASOOR BEHAVIORAL MODEL

Within MASOOR, the path network is divided physically into a series of segments that begin and end with a node. Path choice events occur on the concluding node of each path segment and are based on the behavioral model described below. Nodes represent opportunities to change not only physical characteristics of the trip (e.g. direction and speed of travel) but also phase of the trip (i.e. entry, immersion, exit) as well as role and behavioral characteristics of the agent. Rules of interaction among agents can also be altered at each network node. The element in Figure 2 that is labeled as 'Behavioral Model' is the heart of MASOOR's path choice decision-making as it is this component that actually operates at each path node and assigns agents to specific path segments at discrete temporal intervals. Figure 3 further elaborates upon the behavior model component of MASOOR, using a Discrete Event Simulator (DES) framework to make temporally discrete path-choice decisions that are based on local, relational, and global structural characteristics of the paths.

Model users specify criteria for each agent relating to role and behavioral characteristics as well as rules of interaction among agents. These inputs define goals and preferences of agents, specifying the nature of their desired behavioral outcomes. Input criteria also specify the nature and location of points and areas in the path network environment that have varying abilities to assist agents in realizing desired goals and satisfy needs. Environmental information may also be used to specify constraints on attainment of agent goals and satisfaction of agent preferences. Based upon these input specifications and information stored on the Blackboard of each agent, a Hierarchical Control System, involving a Planner, Navigator, Pilot and Chooser assigns agents to specific path segments at discrete temporal intervals. Each of these elements is further described below. The dynamics of object oriented design of the MASOOR behavioral model are illustrated in Figure 3.

Following the transition during the 1980's from procedural programming to object oriented programming (OOP), modelers envision a set of well-defined objects that possess methods and properties. For example, a 'recreational visitor' can be treated as an 'object' who is 'selecting' (method) a new 'path'. The 'path' (object) can be 'occupied' (property) by other 'visitors' (objects). It is the most common way models are built in engineer-

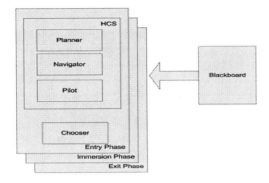

Figure 3. Behavioural model component to MASOOR.

ing and science, including ecology and the social sciences. Following the parameters of the OOP programming paradigm, MASOOR code is easily maintained, flexible and can be extended to changes in the research or management environment.

Hierarchical Control System

The behavioral brain of the system by which agents navigate through the path network was adapted from artificial intelligence robotics, in the form of a nested Hierarchical Control System (HCS) (Kronreif & Furst 2001). This system is divided into three hierarchically nested sub-modules: Planner, Navigator and Pilot. Temporal resolution and spatial scale among these three modules increases in specificity as among the modules from Planner to Pilot. The transition from Planner to Navigator and Pilot also reflects a refinement of specificity through which plans and preferences are evaluated relative to a series of alternatives from which a concrete course of action is selected. Final preferences are combined using Multi Criteria Analysis to assess preferences in a precise or fuzzy way.

Hierarchical relationships among both the structure and function of these three modules allow construction of a highly modular framework that simulates behavior at multiple temporal and geographical scales. The different levels interact through the agent's Blackboard where the Planner defines the goals, the Navigator sets the direction and the Pilot steers. As illustrated in Figure 3, the multi-phasic and dynamic nature of MASOOR allows construction of a HCS for each phase of a recreational visit. Phase change can be designated by specifying temporal/spatial criteria regarding location of trail segment nodes relative to defined points of network access and egress. As agents cross phase-defining trail segment nodes, their preferences, roles and behavioral characteristics can be automatically reassigned from one trip phase to another. Movement from one phase to another, as defined either by spatial or temporal criteria, allows the multiple HCS contained in MASOOR to interact in a manner similar to the temporal variability of actual visitor behavior.

Planner. The Planner's role is to take care of overall time constraints and in some cases to determine which attractions to visit. The main task of the Planner is to keep track of time. Each phase of the trip has a duration which is a given percentage of the total duration. The Planner initiates phase changes in order to get to a succeeding phase when the time for the current phase has passed.

Navigator. The Navigator sets the agent's heading in events of homing so that it does not deviate too far from the target destination when an attraction or exit gate is required. It also can create a 'map' towards a destination by use a shortest distance/time algorithm.

Pilot. At the lowest level of navigational reasoning, the Pilot module mediates between all feasible choices of path segment combinations within the trajectories defined by Planner and Navigator. Humans are capable of a variety of methods of wayfinding, depending on the modes of information available to them and considerations of efficiency and aesthetics (Cornell & Heth 2000; Cornell et al. 2003; Golledge 1999). The Pilot assumes that independent behaviors along discrete path segments can be sequenced to create complex behaviors resulting in attainment of agent goals. The Pilot decomposes the complex actions needed to accomplish agent goals into independent behaviors that are evaluated according to both motivations and constraints. Motivating and constraining behaviors occur concurrently and independently, and constraints have the capacity to overrule fulfillment of motivations.

Motivation behaviors result in the specification for an agent of a preference value for a specific path segment alternative. Preference values are calculated for each path segment alternative occurring at a specific path node. As described in Table 1, the set of possible Pilot motivation behaviors that can currently be used in defining path segment preference value includes:

- Global heading: Visitors arrive at a nature area with specific goals in mind. The location of desired destinations that correspond to goal attainment is maintained.
- Path type: Individual visitor types are assigned preferences for different types of path design and surfaces.
- Homing direction: This criterion relates to direction of the exit gate. Paths that point toward the gate receive higher preference values than those leading away from the exit, especially as the visitor enters the exit phase of the experience.
- 'Chunking direction': Also known as local heading, this criterion gives preference to path segments that continue an established direction of travel. It is based on the fact that visitors often prefer to divide their route into longer and more identifiable stretches.
- Path segment history: This algorithm gives higher preference to new previously unvisited path segments. It plays an important part in creating 'round trip' itineraries.
- Shortest distance: This is a basic algorithm in wayfinding. Presented with path segments whose completion requires variable time allotments, the agent prefers the segment requiring the least amount of time. This algorithm can be used to navigate to an attraction point or back to the exit gate.
- Crowding: Agent types can be specified as having variable preferences for the presence of other visitors within a defined viewshed. The crowding algorithm scans alternative path segments for the presence of other agents within a defined distance. Paths containing more agents receive a lower preference value.

Table 1. Weighting of Various Motivation Behaviors in MASOOR

Motivation behavior	Entry	Immersion	Exit	Direct exit
Global heading	10	0	0	0
Path type	3	3	1	0
Homing direction	0	0	7	10
'Chunking' direction	3	3	0	0
Path segment history	1	3	1	0
Shortest distance	0	0	5	1
Crowding	2	1	0	0

Note: 0 = Motivation does not influence spatial behavior of visitor.
    10 = Motivation highly influences spatial behavior of visitor.

The framework could easily be extended to include preference values that consider agent preferences for marked versus unmarked trails as well as terrain features (e.g. length and steepness of slope). Values shown in Table 1 indicate weighting coefficients for each of the possible Pilot motivation behaviors. A value of '0' indicates that MASOOR excludes that behavior from its decision-making calculus.

Limiting behaviors exclude certain alternatives because they conflict with the goals set by the Planner. For example, the consequence of selecting a long path segment alternative may lead the agent too far from its goal and result in a trip that exceeds time constraints established by the Planner, which is not allowed in MASOOR's logic. Such an alternative would consequently not be pursued, regardless of its preference value.

Comparison of motivation behaviors relative to limiting behaviors occurs for each agent as it evaluates alternatives at every path segment node. An algorithm assesses the preference values and constraints associated with the path segment alternatives available at the node corresponding to the agent's path network location. As more criteria are added to the definition of preference value, the path choice model has potential for more accurately approximating the richness of decision-making that occurs in visitor behavior along a path segment.

To calculate the combined preference for each alternative path segment, the Pilot uses multiple criteria analysis. The criteria included in the algorithm's reasoning can be weighted in importance to create a composite preference value that reflects an agent's predisposition toward consideration of the varying preference criteria. Preference criteria weighting schema can be specified independently for various agent types and for different phases of the visitor experience. Current applications of MASOOR do not make full use of the potential for structuring motivation behaviors in a hierarchical fashion.

Chooser

The Pilot's reasoning should result in an action, so the Chooser identifies the next target in the route of the agent. By definition, the target must be one of the alternative track segments present at the node corresponding with an agent's current location.

MASOOR uses two types of chooser behavior. An 'optimal' chooser logic ranks path segment alternatives according to their composite preference values and selects the alternative that *maximizes* preference values. In instances where multiple alternatives all present equal preference value, choice is resolved by randomly selecting one of the alternatives.

Humans often behave in ways that do not maximize benefits. For example, visitors are often inclined to take routes that *satisfice* (Simon 1978). They are 'good enough' , 'seems simple', or 'get me there any way'(Golledge 2002). Furthermore, most cognitive maps are incomplete or often distorted because of incomplete information (Evans & Pezdek 1980). Cognitive maps may become further distorted because they often require mental rotation, alignment, and matching as well as scale transformations when used in actual wayfinding situations (Golledge 2002; Kitchin 1994; Portugali 1996). To implement the principle that our environmental knowledge is gained 'incidentally' and represents 'naive' or 'common sense' environmental knowing and is usually, fuzzy, inaccurate, and error prone (Golledge 2002), MASOOR incorporates a 'fuzzy' chooser for path alternative choice decision-making in the entry and immersion phases of the visit. The 'fuzzy' chooser uses the preferences of the alternatives as a distribution to select randomly a preferred track segment. Tracks with a higher preference have a higher chance of being selected. The fuzzy chooser also compensates for variability in behavior among members of the same visitor type.

During the exit phase of the trip, the MASOOR Chooser applies both fuzzy and optimal logic in trip selection. During the early stages of the exit phase, trip selection is based on application of fuzzy logic. However when the agent exceeds the time allocated to the trip by the Planner, the Chooser uses an optimal logic that finds the most direct shortcut to the exit point.

Blackboard

In human terms, the Blackboard constitutes the sensory organs (ears, eyes), 'backpack' (compass, maps, watch, etc.), memory and physical state of the visitor. The Blackboard records alternatives, preferences and attractions experienced by the agent as it navigates through a path network. The Blackboard observes directions and distances and is able to keep track of time during the agent's transactions with the path network. The Blackboard's 'backpack' can reflect a real backpack with physical maps, compasses and watches. But it also symbolizes the sensory/intuitive counterparts of the backpack, maintaining sensitivity to direction, cognitive maps and time. Both the sensory and the 'backpack' representations of the Blackboard can be seen as a façade to the environment and time. The 'memory' of the agent is like an internal notebook, keeping track of the exit time established by the Planner, its current location, where it has been, what the location of its current goal is, etc. The physical state of the agent is based on speed, group size and the type of the agent, (e.g. horse rider, walker, cyclist, dog owner, etc.).

In Object-Oriented Programming terms, the Blackboard is implemented as a Façade, a unified interface to a set of subsystems. It defines a higher-level interface that make all the subsystems easier to use (Gamma et al. 1994). The use of façades is not only good Object-Oriented design practice, but it also plays an important role in communication with non-modeling behavioral scientist.

Within MASOOR the object model of the network is based on a forward star network with associated attributes and methods for calculating travel time and distances across the network. Research has demonstrated that the Forward and Reverse Star Representation is the most efficient among all existing network data structures for representing a network (Ahuja et al. 1993; Zhan 1997).

APPLICATION

MASOOR has been applied to simulate hiker behaviour in Dwingelderveld National Park (NPD), the Netherlands (see also Elands & Marwijk 2005). This Dutch nature area of 3,700 ha was chosen because of its recreational attractiveness and ecological quality. The area contains the largest wet heath land area in Northwest Europe (1550 ha). The heath land area is bordered by forest (2000 ha). The NPD receives at least 1.6 million visitors yearly. It is a typical Dutch nature recreation area with an extensive recreational network for both short strolls (60 km marked trails <7 km) and long walks, for cycling ('normal', racing, ATB) and horse riding. Research on visitor countings has shown that on an ordinary Sunday at least 7,000 people visit the National Park (Visschedijk, 1990). Half of those are walking, and the average group size is two. This means that MASOOR have to deal with 1,750 agents, or visitor groups within the simulation. The agents started from one of the nine parking places in the area and were free to move about the 2150 path segments in the network.

The outcomes of the simulation study

have been compared to actual visitor behaviour. In May and August 2006 hikers within NPD were asked to carry a GPS during their walk. A total of 309 valid tracks were collected at five (of the in total nine) parking places (see also Marwijk et al. 2007).

Figure 4a shows the outcome of the simulation by MASOOR, while Figure 4b shows the actual distribution of visitors measured by GPS. To make a valid comparison, Figure 4a shows only the outcome of agents starting from the parking places where the GPS study was carried out. We have deliberately chosen not to rerun the simulation for only those 5 parking places, because that would leave out the crowding effects of visitors from the other four parking places.

Comparing Figures 4a and 4b, we can clearly see popular routes that are marked trails. However, while in MASOOR marked trails in general are appealing to walkers, in reality walkers have preferences for specific marked trails. This can be seen clearly at the letters A and C in both figures. At both places a marked trail is situated and MASOOR suggests that visitors frequently use these trails (see Figure 4a). In reality these two trails are not often walked (See Figure 4b). An explanation might be that trail A is passing only through forest, while other trails reach the heath land providing a more appealing vista of the open heath land. A reason why marked trail C is not often chosen, is that walkers have to cross a road before they start walking this trail. However, since walkers do cross the road at other places in the network to start marked trails, a more valid reason might be the location of a unique attraction in the marked trail north of route C, starting from the same parking place. Walkers more often chose to walk this route to a popular destination known as 'Lake in the Clouds'.

A second striking difference is that agents in MASOOR use almost all of the paths in the path network, while actual walkers tend to walk on a limited number of paths (see B in both figures). This means that some paths are more appealing to walkers than others. An environmental analysis of actual behaviour is needed to give insight into this matter.

This comparison shows that MASOOR is not yet able to capture the richness of the environment. Not all unpaved small paths are the same. For some reason certain paths are more appealing. Not all marked trails of the same length are similarly attractive: trails that lead to unique places or pass different landscape types are preferred over other trails. However,

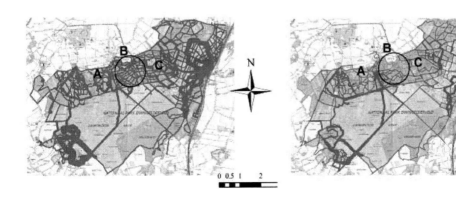

Figure 4. Comparison of MASOOR output (left) and GPS tracks (right).

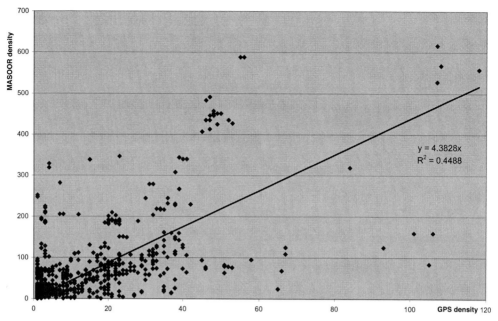

Figure 5. Correlation between MASOOR output and GPS tacks.

the correlation between the modelled and the actual visitor behaviour is r=0.68 (see Figure 5).

FUTURE DIRECTIONS

Several implications for future directions in modeling human-environment transactions of recreational visitors with natural resource settings emanate from this discussion. These implications relate to future directions for research into human-environment transactions in natural settings, the development of models to simulate this behavior as well as the application of this work to natural resource management.

Implications for Transactional Research

Displacement. In the past three decades, a significant amount of recreational research has focused on the so-called displacement phenomenon. The displacement hypothesis suggests that recreational visitors not finding an appropriate type or quality of transactional outcome in a particular setting go elsewhere. They are replaced by visitors who are satisfied with outcomes offered at the setting. Over time, this pattern leads to a general decrease in the quality of the recreational experience as visitors having higher evaluative standards are displaced by individuals with lower standards (for literature review see: Graefe et al. 1984; Manning 1986; Shelby & Heberlein 1986; Stankey & McCool 1984; Vaske et al. 1980).

Translating empirical relationships between dimensions of environment or participatory context and experiential quality into inferences about displacement requires time-series rather than cross-sectional studies (Kuentzel & Heberlein 1992; Shindler & Shelby 1992), which are difficult, costly and time-consuming (Robertson & Requla 1994). They also presume an ability to locate the displaced visitor (Nielsen & Endo 1977). Lacking this ability, modelers have to make assumptions about threshold conditions that are likely to produce displacement. Accurate incorporation of displacement concepts into a simulation modeling framework becomes problematic.

In a sense, the problems of encapsulating the perspectives of displaced as well as actual visitors means that modeling efforts lack the ability to represent the potential range of experiential quality associated with a recreational resource. If the perspectives of displaced visitors can not be located or are too costly to incorporate into modeling strategies, management actions emanating from modeled patterns of actual behavior will never be able to capture the perspective of 'paradise lost' that may be embodied in the experiences of the displaced visitor. In high density settings, such as the Netherlands, there may be many recreational visitors whose perspectives will never be incorporated into models that reflect the behavior of only current participants. Finding means of capturing and considering the experiences of displaced visitors is a significant issue in any form of recreation research that is based entirely on cross-sectional data.

Incorporating the Context of Participation in Modeling. The dimension of context of participation is fundamental to developing a transactional understanding of recreational experiences in natural areas. Creating profiles of visitor types and rules of visitor behavior, which are necessary inputs in the modeling process, need to accommodate the implications of contextual differences in understanding the production of human-environment transactional outcomes.

Two implications of this requirement affect the conduct of person-environment research related to recreational engagement with natural settings. First, more basic research is needed to better understand exactly how psychological, social, physical, and temporal dimensions of context influence the production of experiential quality. Second, applied research is needed to better understand the implications of context on the production of transactional outputs in a specific geographic and temporal setting. Standardized 'rapid assessment' systems are needed that will efficiently provide information to accurately characterize the nature of transactional outcomes in a specific natural area (e.g. national, state or regional park). A protocol also needs to be developed to monitor changes in the production of transactional outcomes as changes occur in the nature of visitors entering a recreational area, as the context in which these visitors engage the environment changes over time and as the environment is affected by resultant use levels.

Adaptation of Space Syntax Technology as an Approach to Modeling Implications of Environment on Recreational Behavior.

From a transactional perspective, most of the studies used in generating models such as MASOOR focus on predicting recreational behavior as functions of visitor characteristics and various dimensions of context of participation. When considered, environment is conceptualized as being impacted by behavior. Research examining the role of environmental affordances (Gibson 1986) in facilitating recreational behavior is less prevalent.

Advances in space syntax research (Bafna 2003; Hillier 1996; Penn 2003) offer a framework for examining environmental attributes as affordances for visitor behavior in a dense recreational trail network. Space syntax research examines relationships between the human inhabitation of environment and the spatial configuration of the environment being inhabited (Bafna 2003). The unit of analysis in space syntax research can be defined as the path segment, a section of the path network between two decision points (e.g. selecting a path at a junction in the path network). Configurational relationships among path segments within a network can be examined on a local level by examining the number of other segments to which a specific segment is directly connected. The global integration of a segment within a network is defined by the number of directional changes required to access a particular segment. Intensity of segment use is directly proportional to increasing connectivity among segments and inversely proportional to the number of directional changes required to access a segment (Bafna 2003; Chang 2002; Joseph & Zimering 2007). In studies of walking in elderly communities, space syntax methods have been used to relate different patterns of walking by various types of residents to the spatial structure of the path network within the community and to the physical attributes of path segments as well as the environment traversed by the network (Joseph & Zimering 2007). Space syntax has also proven useful in modeling behavior in various types of urban environments (Chang 2002; Kim & Penn 2004).

The application of space syntax methods to recreational research in dense path networks allows the explanation of path selection behavior to be partitioned among parameters related to visitor characteristics, social and psychological characteristics of the context of path use and physical and spatial characteristics of the path network and surrounding environment. Coupled with the application of geographic positioning system technology to record actual patterns of path behavior, application of space syntax methods provides an opportunity to develop calibrating heuristics that model visitor use of a path network on the basis of empirical studies conducted in the context of a specific environment.

Implications for Development of Simulation Models

Integration with Ecological Footprint Modeling. The spatial and temporal outputs generated by MASOOR are well suited to defining ecological footprints associated with different patterns of recreational use (see Pouwels 2007 within this volume for an application of MASOOR to the modeling of ecological footprint). Several studies document sensitivity of plants and animals to disturbance regimes generated by recreational use (eg. Vos & Peltzer 1987; Yalden & Yalden 1990). However, research relating spatial and temporal patterns of recreational use to biological diversity is complex (Duim & Caalders 2002). Not all types and intensities of recreational use have the same effects on habitat disturbance, and not all species respond in the same manner. Fluctuations in biological diversity often follow longer time cycles that may have little to do with patterns of recreational use. The effective use of models such as MASOOR in evaluating the ecological footprint of recreational use requires additional research that calibrates the existence of specific use patterns with the persistence of specific guilds of species.

Displacement. A correlate of the displacement hypothesis discussed above suggests that even within the context of a single visit, displacement of intention and motivation may occur. Differences between expected and encountered conditions, whether attributable to physical, social or psychological dimensions, may cause a shift in visitor intention and motivation (Shelby & Heberlein 1986). For example, a person intending to follow a marked trail from a visitor center may find conditions too wet or too crowded to continue and make a bee-line back to the trip origin. Based on surveys of initial intention, such a visitor may be incorrectly identified and modeled as marked trail user. In fact, the visitor belongs to both the marked trail and the social trail visitor groups, but it is difficult to model exactly when and where the shift in recreational intention occurs.

Furthermore, it is impossible to ascertain the threshold conditions in the environment and the context of participation that produced the shift in visitor intention and, thus, the need for reclassification.

Need for Manager-oriented Model. In its current form, use of MASOOR requires a reasonably high level of computer literacy. For example, specification of input parameters requires the ability to write computer programming script. Yet field managers react favorably when presented with model output, and they often see immediate applications for MASOOR to help them resolve complex management issues.

In a sense, the state of the art of recreation simulation modeling can be compared to early efforts to integrate geographic information systems (GIS) technology into natural resource management. Early GIS technology required a high level of sophistication to input, analyze and represent geographic data and the hardware requirements were often cost prohibitive for use in a field management context. Advances in computer technology, the growth of satellite technology permitting real-time monitoring of earth system processes and development of sophisticated geographic positioning technology, the advent of internet technology and the development of internet-based GIS systems such as MapServer™, Google Earth™, ArcGIS Server™ and ArcWeb Services™ now make it possible for field managers to access geographic information using hand-held Personal Digital Assistants (PDA) and inexpensive laptop computers having internet access. The development of decentralized but linked, user-friendly, menu driven systems to access web-based GIS data and technology makes it possible for field managers to integrate GIS into routine natural resource management decision-making. There is a similar need to design less-technically oriented and more user-friendly versions of MASOOR and other simulation models that can be operated with information technology expertise and systems commonly available in field offices of management agencies.

Applicability Across Spatial Scales. MASOOR currently operates most efficiently at the landscape scale (i.e. at scales ranging between 1:10,000 and 1:50,000). Its multi-agent and scalar capabilities allow it to keep track of up to 5000 agents operating in dense trail networks composed of less than 5000 path segments. As the spatial extent of the environment being modeled increases, the number of user groups, participatory contexts, path segments and trail types increases, making model operation more complex and cumbersome. Information needed to specify input parameters may not be available at scales beyond the landscape. While travel to and from recreational settings has been modeled at the regional scale (Skov-Petersen 2002; Vries et al. 2004), it is especially difficult to conceptualize regionally scaled recreational behavior. Furthermore the scale and resolution of regional databases and models is too coarse to provide detailed understandings of human-environment transactions in specific landscapes. Thus, there are limitations on the geographic scale at which simulation models like MASOOR can be applied. It is more likely that MASOOR will use output from regional origin-destination models to generate input parameters relating to visitor numbers at specific settings.

Implications for Natural Resource Management

Modeling Visitor Behavior Versus Visitor Experience. MASOOR has proven quite useful in modeling actual behavior of visitors in dense path networks. However, in its current format, MASOOR does not directly address the issue of experiential quality generated by the transaction of recreational visitors with a natural area. Indirectly, the implications of various dimensions of environment and context of participation as well as visitor characteristics are factored into MASOOR through the behavioral rules that have been written for different visitor types. For example, the visitor sensitivity to crowding may be incorporated by defining a behavioral rule that is based on encounters with other visitors within a specified viewshed of agent types. However, the model produces geographic patterns of visitor movement based on the validity that underlies the specification of behavioral rules for agent behavior. Translation of model output into implications for experiential quality requires careful interpretation by model users.

Stimulating Stakeholder Conversation. One of the values of simulation models such as MASOOR is that users can actually see visitors moving across a dense recreational trail network. These models provide real-time simulations of the consequences for visitor density of changes in the composition and distribution of the recreational resource. Recreational managers are eager to play out various scenarios in which they simulate long-contemplated but unrealized changes in path network layout (e.g. alteration of start points, limitations on trail access for certain types of visitors, removal of path segments from a trail network). They see these models as an opportunity to ask the question of 'what if I pursued this management action?' (Klosterman 2001) without having to incur the potential costs of failure.

The ability to generate and evaluate management scenarios offers the prospect of constructing management policies that will prove to be more sustainable. Visitor simulation models provide feedback on geographic patterns of visitor density resulting from management action. This feedback can be used as input into models examining the ecological footprint recreational use. In addition, they provide opportunity to broaden the participatory basis of management decision-making. Assuming development of a version of MASOOR that can be operated in field offices by local area managers, various stakeholder groups could participate in an iterative, reflective and communicative action approach to the design and evaluation of alternative management scenarios. Such conversation is more likely to lead to the generation of locally-generated policy that more closely reflects the perspectives of concerned stakeholders. In short, it is more likely to result in policy that can be agreed upon, adhered to and enforced by all concerned stakeholders (Innes 1996), providing a basis for resolving one of the dilemmas in managing a common property resource such as a publicly-held natural resource recreation area (Hardin 1968).

Additional Support for Model Refinement and Adoption. Technological innovation can be classified into several categories, including: a) that which has been developed and adopted and is currently in use within mainstream society; b) that whose development is complete and is ready for adoption and use in mainstream society; c) that being developed and tested in terms of its societal applications; d) that which has been conceptualized but requires refinement and further development before it becomes useful to society; e) that being conceptualized; and f) that about which nobody has even dreamed (Solow 1974). MASOOR falls into the third class of technological innovation. Some of the basic concepts embedded in the development of MASOOR have been outlined in this chapter. A prototype of the model is operational and its application to recreation resource management is being tested in the Netherlands, France, the United Kingdom and Austria. The model needs refinements related to refining specific behavioral rules of the model for such environmental dimensions as slope, path segment length, view characteristics and the nature and location of specific visitor attractions as well as calibrating behavioral rules during different phases of the visitor experience. In addition, agencies adopting simulation modeling into their culture will need to provide technical support to model users. As noted earlier, there is also a need to develop a less technical, user-friendly version of MASOOR that can be integrated into the management culture of a natural resource agency's local field office.

Institutional Development That Encourages Model Use. Beyond additional financial support, adoption and use of models such as MASOOR in recreational resource management will also be enhanced creation of an institutional culture that fosters use of the models. Currently, the use of simulation models is, for most managers, an idea that is outside the realm of their day-to-day routines. In large part, they lack the technical sophistication to operate the model. Often the agencies for which they work lack the capacity to provide information technology (IT) support for innovative managers seeking to experiment with model application. A parallel can again be drawn between the adoption of GIS technology in natural resource management and the use of simulation modeling technology. As natural resource agencies created IT divisions within their central offices, technical expertise was applied to the final refinement of GIS technology relative to an agency's mission. IT expertise became more readily available to support and encourage the application of GIS technology to routine issues encountered by line management staff in the field. Adoption of models such as MASOOR by field staff will be facilitated by the development of a similar intuitional framework and support system related specifically to simulation modeling.

Cooptation by Managers. A final consideration in the application of visitor simulation models to manage recreation resources is its possible cooptation by managers. The impact of technological advancements on society has more to do with how they are applied rather than with dimensions of their specific structure or function. It is conceivable that managers might use the 'gee whiz' dimension of watching multiple agents move about a dense path network in multiple stakeholder discussions to advocate for scenarios that best serve their own agenda rather than agency mission or stakeholder interests.

LITERATURE CITED

Adelmann, B. J., Heberlein, T. A. & Bonnickson, T. M. 1982. Social psychological explanations for the persistence of a conflict between paddling canoeists and motorcraft users in the Boundary Water Canoe Area. *Leisure Sciences* 5, 45-61.

Ahuja, R. K., Magnanti, T. L. & Orlin, J. B. 1993. *Network Flows: Theory, Algorithms and Applications.* . Englewood Cliffs, NJ: Prentice Hall.

Allen, G. L. 1982 The organization of route knowledge. *New Directions for Child Development* 15, 31-39.

Arrowsmith, C. & Chhetri, P. 2003. Port Campbell National Park: Patterns of Use. A report for the development of a visitor typology as input to a generic model of visitor movements and patterns of use: Prepared for Parks Victoria Visitor Research. Melbourne: RMIT University.

Bafna, S. 2003. Space Syntax: A Brief Introduction to Its Logic and Analytical Techniques. *Environment and behavior* 35, 17-29.

Blahna, J. D., Smith, S. K. & Anderson, A. J. 1995. Backcountry llama packing: Visitor perception of acceptability of conflict. *Leisure Sciences* 17, 185-204.

Borrie, W. T. & Birzell, R. M. 2001. Approaches

to measuring quality of the wilderness experience. In *Visitor use density and wilderness experience: proceedings; 2000 June 1–3; Missoula, MT. Proc. RMRS-P-20* (ed. W. A. Freimund & D. N. Cole), pp. 29-38. Ogden: Department of Agriculture, Forest Service, Rocky Mountain Research Station.

Borrie, W. T. & Roggenbuck, J. W. 2001. The Dynamic, Emergent, and Multi-Phasic Nature of On-Site Wilderness Experiences. *Journal of Leisure Research, Vol. 33 No. 2*

Bourassa, S. C. 1990. A paradigm for landscape aesthetics. *Environment and Behavior* 22, 787-812.

Catton, W. R. J. 1983. Social and behavioral aspects of carrying capacity in natural environments. . In *Human Behavior and Environment: Advances in Theory and Research. Vol. 6, Behavior and the Natural Environment* (ed. I. Altman & J. F. Wohlwill), pp. 269-306. New York: Plenum Press.

Chang, D. 2002. Spatial Choice and Preference in Multilevel Movement Networks. *Environment and behavior* 34, 582-615.

Cohen, E. 1979. Rethinking the sociology of tourism. *Annals of Tourism Research* Jan/Mar 1979, 18-35.

Cole, D. N. 2005. *Computer Simulation Modeling of Recreation Use: Current Status, Case Studies, and Future Directions*. Fort Collins, CO: U.S. Department of Agriculture, Forest Service, Rocky Mountain Research Station.

Cornell, E. H. & Heth, C. D. 2000. Route learning and wayfinding. In *Cognitive maps: Past, present, and future* (ed. R. Kitchin & S. Freundschuh), pp. 66-83. London: Routledge.

Cornell, E. H., Sorenson, A. & Mio, T. 2003. Human Sense of Direction and Wayfinding. *Department of Psychology, University of Alberta*.

Duim, R. v. d. & Caalders, J. 2002. Biodiversity and tourism: impacts and interventions. *Annals of Tourism Research* 29, 743-761.

Evans, G. W. & Pezdek, K. 1980. Cognitive mapping: Knowledge of real-world distance and location information. *Journal of Environmental Psychology* 6, 13-24.

Farías, E. I., Grau, H. R. & Camps, A. 2005. Trail Preferences and Visitor Characteristics in Aigüestortes i Estany de Sant Maurici National Park, Spain. *Mountain Research and Development* 25, 51–59.

Ferber, J. 1994. *Simulating with reactive agents*. Many Agent Simulation and Artificial Life: IOS Press.

Galloway, G. 2002. Psychographic segmentation of park visitor markets: evidence for the utility of sensation seeking. *Tourism Management* 23, 581-596.

Gamma, E., Helm, R., Johnson, R. & Vlissides, J. 1994. *Design Patterns: Elements of Reusable Object-Oriented Software*. Professional Computing Series: Addison-Wesley.

Gibson, J. J. 1986. *The Ecological Approach to Visual Perception*. Hillsdale, NJ: Lawrence Earlbaum Associates.

Gimblett, R. 2005. Modelling Human-Landscape Interactions in Spatial Complex Settings: Where are we and where are we going?

Golledge, R. G. 1999. Wayfinding Behavior: Cognitive Mapping and Other Spatial Processes. *Psycoloquy* 10.

Golledge, R. G. 2002. Human Wayfinding and Cognitive Maps. In *Colonisation of Unfamiliar Landscapes: The Archaeology of Adaptation*. (ed. M. Rockman, Steele, J.). Baltimore: The Johns Hopkins University Press.

Goodwin, H. 2000. Tourism, national parks and partnerships. In *Tourism and National Parks: Issues and Applications* (ed. R. W. Butler & S. W. Boyds). New York: Wiley.

Graefe, A. & Thapa., B. 2004. Conflict in natural resource recreation. In *Society and Natural Resources: A Summary of Knowledge* (ed. M. Manfredo, J. J. Vaske, B. Bruyere, D. Field & P. Brown), pp. 209-224. Jefferson, MO: Modern Litho.

Graefe, A. R., Vaske, J. J. & Kuss, F. R. 1984. Social carrying capacity: An integration and synthesis of twenty years of research. *Leisure Sciences* 6, 395-431.

Haklay, M., Schelhorn, T., O'Sullivan, D. & Thurstain-Goodwin, M. 2001. "So Go Down Town": Simulating Pedestrian Movement in Town Centres. *Environment and Planning B: Planning and Design* **28**, 343-359

Hardin, G. 1968. The tragedy of the commons. *Science* 162, 1243-1248.

Hartig, T. 1993. Nature experience in transactional perspective. *Landscape and Urban Planning* 25, 17-36.

Hillier, B. 1996. *Space is the Machine*. Cambridge, UK: Cambruidge University Press.

Innes, J. E. 1996. Planning Through Consensus Building: A New View of the Comprehensive Planning Ideal. *Journal of the American Planning Association* 62, 460-472.

Itami, R. M. & Gimblett, H. R. 2001. Intelligent recreation agents in a virtual GIS world. *Complexity International* 8.

Ittelson, W. H. 1978. Environmental perception and urban experience. *Environment and Behavior* 10, 193-213.

Jackson, E. L. & Wong, R. A. G. 1982. Perrceived conflict between urban cross-country skiers and snowmobilers in Alberta. *Journal of Leisure Research* 14, 47-62.

Johns, N. & Gyimothy, S. 2002. Market Segmentation and the Prediction of Tourist Behavior: The Case of Bornholm, Denmark. *Journal of Travel Research* 40, 316-327.

Joseph, A. & Zimering, C. 2007. Where active older adults walk. *Environment and Behavior* 39, 75-105.

Kaplan, R. & Kaplan, S. 1982. *Cognition and environment: Functioning in an uncertain world*. New York: Praeger.

Kaplan, R. & Kaplan, S. 1989. *The experience of nature: A psychological perspective*. Cambridge: Cambridge University Press.

Kaplan, S. & Kaplan, R. 1983. *Cognition and environment: functioning in an uncertain world*. Michigan: Ulrich's Bookstore.

Kim, Y. O. & Penn, A. 2004. Linking the Spatial Syntax of Cognitive Maps to the Spatial Syntax of the Environment. *Environment and behavior* 36, 483-504.

Kitchin, R. M. 1994. Cognitive maps: what are they and why studied them? *Journal of environmental psychology* 14, 1-19.

Klosterman, R. E. 2001. The What If? Planning Support System In *Planning Support Systems: Integrating Geographic Information Systems, Models and Visualization Tools* (ed. R. K. Brail & R. E. Klosterman), pp. 263-284. Redlands, CA: ESRI Press.

Kronreif, G. & Furst, M. 2001. TCP-Based communications for task oriented programming and controle of hetrogenous multi-robot- systems. In *International Symposium on Robotics*. Seoul, Korea.

Kuentzel, W. E. & Heberlein, T. A. 1992. Cognitive and behavioral adaptations to perceived crowding: A panel study of coping and displacement *Journal of Leisure Research* 4, 377-393.

Lawson, S. R. 2006. Computer Simulation as a Tool for Planning and Management of Visitor Use in Protected Natural Areas. *Journal of Sustainable Tourism* 14, 200-217.

Lawson, S. R., Itami, B., Gimblett, R. & Manning, R. E. 2002. Monitoring and Managing Recreational Use in Backcountry Landscapes Using Computer-Based Simulation Modeling. In *Workshop: Travel Simulation Modeling for Recreation Planning*, pp. 106-113. Rovaniemi, Finland: Finnish Forest Research Institute.

Lawson, S. R., Manning, R. E., Valliere, W. A. & Wang, B. 2003. Proactive monitoring and adaptive management of social carrying capacity in Arches National Park: an application of computer simulation modeling. *Journal of Environmental Management* 68, 305-313.

Manning, R. E. 1986. *Studies in outdoor recreation: a review and synthesis of the social science literature in outdoor recreation*. Corvallis, Oregon: Oregon State University Press.

Manning, R. E., Itami, R. M., Cole, D. N. & Gimblett, R. 2005. *Overview of Computer Simulation Modeling Approaches and Methods* Computer Simulation of Recreation Use: current status, case studies and future directions.

Marwijk, R. v., Elands, B. & Lengkeek, J. 2007. Experiencing nature: The recognition of the symbolic environment within research and management of visitor flows. *Special Issue of Forest Snow and Landscape Research* in review.

McCool, S. F. & Lime, D. W. 2001. Tourism carrying capacity: Tempting fantasy or useful reality? *Journal of Sustainable Tourism* 9, 372-388.

Meinig, D. W. 1979. *The Interpretation of Ordinary Landscapes: Geographical Essays*. New York: Oxford University Press.

Nielsen, J. M. & Endo, R. 1977. Where have all the purists gone? An empirical investigation of the displacement process hypothesis in wilderness recreation. *Western Sociological Review* 8, 61-75.

O'Connor, A., Zerger, A. & Itami, B. 2005. Geo-temporal tracking and analysis of tourist movement. *Mathematics and Computers in Simulation* 69, 135-150.

Pearce, P. L. 1982. *The social psychology of tourist behavior*. New York: Pergamon Press.

Penn, A. 2003. Space Syntax And Spatial Cognition: Or Why the Axial Line? *Environment and Behavior* 35, 30-65.

Pitt, D. G. 1989. The attractiveness and use of aquatic environments as outdoor recreation places. In *Public Places and Spaces* (ed. I. Altman & E. H. Zube), pp. 217-254. New York Plenum Press.

Portugali, J. 1996. *The construction of cognitive maps*. Dordrecht, The Netherlands: Kluwer publishers.

Pouwels et al. 2008. *Using Agent based models for optimizing recreation and biodiversity*. This Book.

Pouwels, R. & Vos, C. C. 2001. Recreatie en biodiversiteit in balans: een ruimtelijke benadering van functiecombinaties. Wageningen: Alterra.

Roberts, C. A., Stallman, D. & Bieri, J. A. 2002. Modeling complex human–environment interactions: the Grand Canyon river trip simulator. *Ecological Modelling* 153, 181–196.

Robertson, R. A. & Requla, J. A. 1994. Recreational displacement and overall satisfaction: A study of central Iowa's licensed boaters. In *Journal of Leisure Research 26(2), Second Quarter*, vol. 2007.

Schelhorn, T., O'Sullivan, D., Haklay, M. & Thursaint-Goodwin, M. 1999. Agent-Based Pedestrain Model, Center for advanced spatial analysis working paper series. *Paper 9, UCL*.

Shelby, B. & Heberlein, T. A. 1986. *Carrying Capacity in Recreation Settings*. Corvallis: Oregon State University Press.

Shindler & Shelby, B. 1992. Management implication of displacement and product shift: A panel study of Rogue River floaters. In *Abstracts of the 4th North American Symposium on Society and Natural Resources held in Madison, Wisconsin, May 17-20, 1992*.

Simon, H. A. 1978. Satisficing and the one right way. In *Humanscape: Environment for People* (ed. S. Kaplan & R. Kaplan), pp. 127-131. North Scituate, MA: Duxbury Press.

Skov-Petersen, H. 2002. GIS-based modeling of car-borne visits to Danish forests. In *Monitoring and Management of Visitor Flows in Recreational and Protected Areas Conference Proceedings* (ed. A. Arnberger, C. Brandenburg and A. Muhar), pp. 233-239. Vienna January 30-February 2, 2002: Institute for Landscape Architecture and Landscape Management, Bodenkultur University.

Slothouwer, R. L., Schwartz, P. A. & Johnson, K. M. 1996. Some guidelines for implementing spatial explicit, individual-based ecological models within location-based raster GIS. In *Third international conference integrating GIS and environmental modeling*. January 21-25, Santa FE.

Solow, R. M. 1974. The economics of resources or the resources of economics. *American Economic Review* 64, 1-14.

Stankey, G. H. & McCool, S. F. 1984. Carrying capacity in recreational settings: Evolution, appraisal and application. *Leisure Sciences* 6, 453-473.

Stokols, D. 1978. Environmental Psychology. *Annual Review of Psychology* 29, 253-295.

Tolman, E. C. 1948. Cognitive maps in rats and man. *Psychological Review* 55, 189-208.

Ulrich, R. S. 1986. Human responses to vegetation and landscapes. *Landscape and Urban Planning* 13, 29-44.

Vaske, J. J., Donnelly, M. P. & Heberlein, T. A. 1980. Perceptions of crowding and resource quality by early and more recent visitors. *Leisure Sciences* 3, 367-381.

Verboom, J. 1996. Modelling fragmented populations: between theory and application in landscape planning, vol. PhD: Wageningen, The Netherlands: Forestry and Nature Research.

Visschedijk, P. A. M., Probstl, U. & Henkens, R. 2005. MASOOR in the Alpine Areas: Agent Based Modelling as a tool for management planning in Natura 2000 sites. In *MMV*, pp. 416-417. Rapperswil, Switzerland.

Vos, C. C., Baveco, H., Chardon, P. & Goedhart,

P. 2002. The role of habitat heterogeneity on dispersal in agricultural landscapes. *Landscape Ecology (submitted)*.

Vos, P. & Peltzer, R. H. M. 1987. Recreatie en broedvogels in heidegebieden: Strabrechtse en Groote Heide. In *Bos en recreatie 15*. Utrecht: Afdeling Sociologisch Onderzoek t.b.v. Natuur en Landschap, SBB

Vries, S. d., Jellema, A. & Goossen, M. 2004. FORVISITS: Modeling visitor flows at the regional level. In *Modeling visitor flows at the regional level. Policies, Methods and Tools for Visitor Management. Proceedings of the Second International Conference on Monitoring and Management of Visitor Flows in Recreational and Protected Areas* (ed. T. Sievanen, J. Jokimaki, J. Saarinen, S. Tiilentie & E. Virtanen), pp. 77-83. Rovaniemi, Finland, June 16-20, 2004.

Wagtendonk, J. W. v. 2004. Simulation modeling of visitor flows: where have we been and where are we going? In *Policies, Methods and Tools for Visitor Management: Second International Conference on Monitoring and management of Visitor Flows in Recreatonal and protected Areas, June 16-20, 2004* (ed. T. Sievänen, J. Erkkonen, J. Jokimäki, J. Saarinen, S. Tuulentie & E. Virtanen), pp. 129-136. Rovaniemi, finland: Finnish forest Research Institute.

Watson, A. E., Niccolucci, M. J. & Williams, D. R. 1994. The nature of conflict between hikers and recreational stock users in the John Muir Wilderness. *Journal of Leisure Research* 26, 372-385.

Yalden, P. E. & Yalden, D. W. 1990. Recreational disturbance of breeding golden plovers (Pluvialis apricarius). *Biological Conservation* 51, 243-262.

Yiannakis, A. & Gibson, H. 1992. Roles tourists play. *Annals of Tourism Research* 19, 287-303.

Zajonc, R. B. 1980. A one-factor mind and emotion. *American Psychologist* 35, 151-175.

Zhan, F. B. 1997. Three fastest shortest path algorithms on real road networks: data structures and procedures. . *Journal of Geographic Information and Decision Analysis* 1.

Zube, E. H., Sell, J. L. & Taylor, J. G. 1982. Landscape perception: Research, application and theory. *Landscape Planning* 9, 1-33.

CHAPTER 16

APPLYING AGENT-BASED MODELING FOR SIMULATING SPRING BLACK BEAR HUNTING ACTIVITIES IN PRINCE WILLIAM SOUND, ALASKA

Spencer Lace
Randy Gimblett
Aaron Poe
Dave Crowley

Abstract: Black bear harvest levels have increased rapidly during the past 10 years in Prince William Sound (PWS), Alaska. Alaska Department of Fish and Game (ADF&G) has found a 100% increase in reported bear harvest between 1995 and 2001. In regulatory year 2001/02, this area reached a record of 436 bears taken which was approximately 25% more than any other Game Management Unit in Alaska. The Chugach National Forest which manages the vast majority of the land surrounding PWS was in need of assessing the spatiotemporal distribution of the spring black bear harvest with hopes of assessing its overlap with other recreational groups. This study combines Geographic Information Systems (GIS) with existing standardized harvest datasets and an agent based modeling approach to analyze complex, spatially dynamic patterns of black bear hunting in PWS. This study illustrates that human use simulation modeling, driven by a harvest record dataset, can inform decision making leading to proactive management of human-landscape interactions and enhance long-term management of harvested wildlife populations.

Keywords: Agent-Based Modeling, Recreation, Wilderness, Long-Term Monitoring, Simulation, Wildlife Management, Human-Landscape Interactions

INTRODUCTION

Black bear harvest levels have increased rapidly during the past 10 years in Prince William Sound (PWS). Harvest in this area is managed by the Alaska Department of Fish and Game (ADF&G) as a single hunt unit. Unit 6d includes approximately four million acres of coastal, temperate rainforest. An ADF&G status report for 6d found a 100% increase in bear harvest between 1995 and 2001. In 2002, this area reached a record of 435 bears taken which was approximately 25% more than any other Game Management Unit (GMU) in Alaska. This harvest level represents an unknown number of individual hunting parties using the shoreline of PWS. A hunt success of ~ 50% was documented by unguided hunters out of the port of Valdez during the late 60s (McIlroy

1970). It is not certain if this success rate is applicable to current harvest practices and locations throughout PWS but its reasonable to assume that the number of reported kills could represent as much as 50% of the actual harvest effort in PWS (McIlroy 1970). With the opening of the Anton Anderson Memorial Whittier Tunnel in June of 2000 it is thought that the majority of current bear hunting use enters the Sound through Whittier and likely occurs generally in the western and northern parts of the Sound (pers. com. Crowley, ADF&G).

At the time of writing black bear harvest season is open from September 1 – June 30 (subject to change in 2006 from September 1 – June 10), although the majority of harvest (~80%) occurs in during May and June (ADF&G harvest data, 1989-2002). In this six-week period, bear hunters likely become the most prevalent recreation use group on the shoreline of PWS. The Chugach National Forest (CNF) hosts a vast majority of these hunting parties and yet knows almost nothing specific related to their use patterns. This timing overlaps with an increasingly popular early kayaking and pleasure boating season. During the spring hunting harvest season black bears concentrate their activity along shorelines (ADF&G 1982; McIlroy 1970), often with beach or estuary characteristics that are also attractive to kayak campers and other shoreline recreation uses. The USFS has received several reports of user conflicts in the western Sound between bear hunting groups and other non-harvesting use of the shoreline during late May and early June. Conflicts are exacerbated by the practice of bear baiting. Using this technique hunters establish a bait station by placing bear attractants, to attract bears to a specific location. Many bait stations are established on beaches that may also be used by non-harvest users. Though actual reports of conflict have been limited the potential for severity of each individual conflict between these user groups is a concern to both Forest and ADF&G managers. To reduce potential conflict, ADF&G has closed bear baiting in two popular recreation areas in Western PWS, Blackstone Bay and Harriman Fiord. In addition, Alaska hunting regulations prohibit baiting within one mile of any dwelling structure or developed campsite.

During 2004 CNF permitted nine hunting guide services that were authorized for total of 790 user days in western PWS alone. In addition an unknown number of water taxi services specialize in supporting bear hunting groups throughout PWS but these numbers are not tracked by USFS. The CNF also permits another 24 non-harvest oriented commercial operators for shoreline activities in Western PWS. With more harvest and non-harvest commercial operations being permitted, increasing private recreation, and new CNF efforts to manage shoreline camping areas, it is critical to have a solid understanding of both the spatial and temporal patterns of activities related to the spring black bear harvest. The application of relevant technologies to the management issues of black bear harvest and other human uses in Prince William Sound (PWS) represents a new opportunity. This paper explores the use of **RBSim**2 (Itami et al. 2003), a spatial agent-based simulation model in conjunction with a harvest database containing location information for bear kill sites in the area to analyze complex, spatially dynamic patterns of black bear hunting in PWS. This study evaluates through simulation the existing patterns of hunter use over the season, identifies peak visitation periods in the study area, evaluates the durations and destinations of hunter choice, assesses commercial and non-commercial hunting activities and provides some guidance in describing the density of harvest per management of capacity areas managed by the CNF in PWS.

STUDY AREA

The region of interest is the western part of PWS that is primarily accessible from Whittier, Alaska. The study area is shown in Figure 1. This area includes 42 Uniform Code Units (UCU) areal units of the Game Management Unit (GMU) 6D listed. It also includes sixteen Capacity Analysis areas (CAs), which encompass one or more UCUs. Campsite and attraction facilities fall within what the CNF refers to

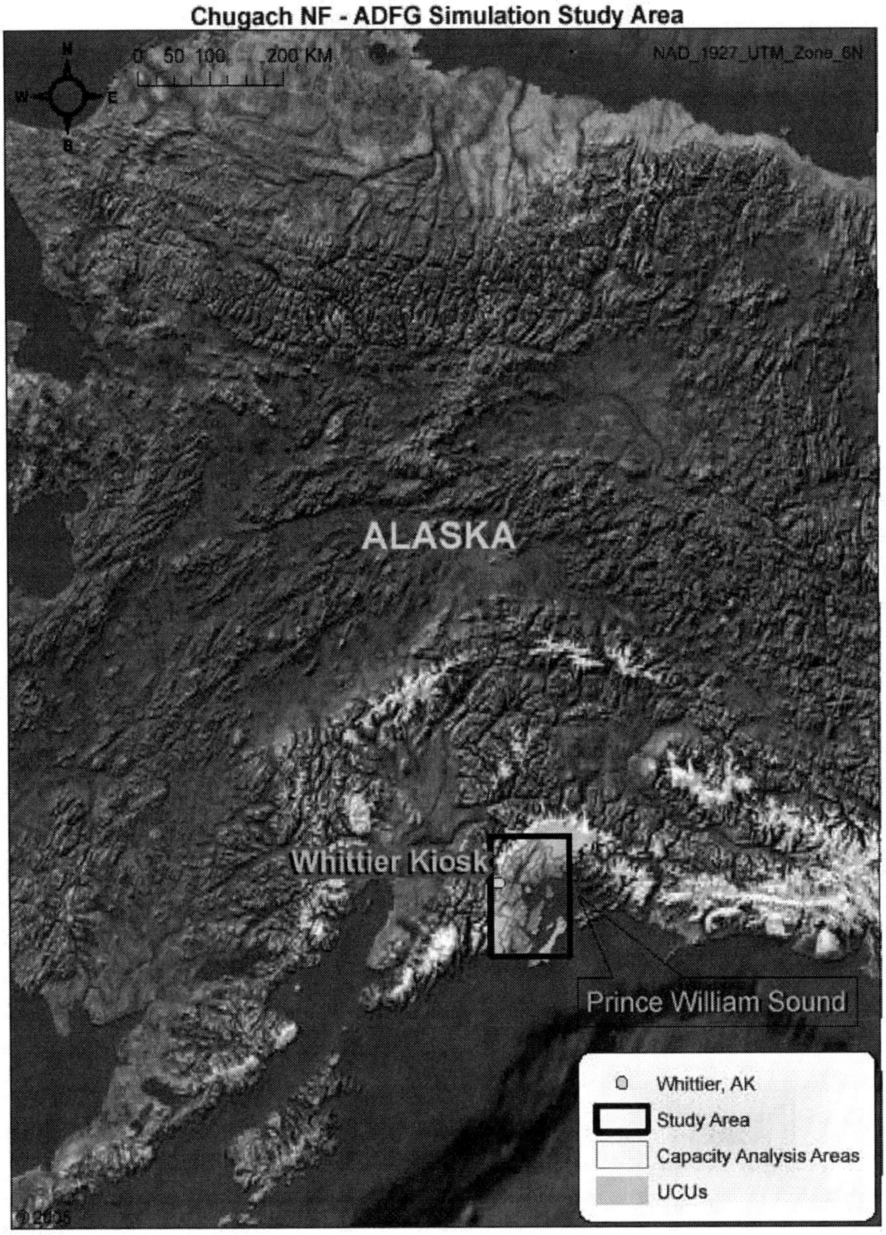

Figure 1. Context Map for the Study Area.

as capacity areas. CAs derived using watershed associations are used to summarize recreation activity on CNF lands in western PWS. The Capacity Analysis areas are much larger and cover a water body while the ADF&G units are related more specifically to the land. ADF&G Game Management units are used to summarize harvest records The area defined by this study boundary in Figure 1 is approximately 17529 km2 (or 6768 mi2), while the land area contained in this boundary is 5513 km2 (2129 mi2). The shoreline in the study region is roughly 2574 km (or 1599 miles). The coastal slopes may rise abruptly from sea level to mountain peaks at 2300 – 5600 feet (700 – 1700 m) above sea level (McIlroy 1970). Where freshwater glacial streams discharge into the water body, alluvium forms tidal marshes up to several acres in size (McIlroy 1970). These and other shoreline areas plus the water routes connecting them comprise the actual study area, which is described more specifically in the habitat capability model section.

SIMULATING HUMAN BEHAVIOR IN RECREATION ENVIRONMENTS USING RBSIM

The simulation framework referred to as the Recreation Behavior Simulator (RBSim) (Itami et al. 2004). RBSim is a computer simulation tool, integrated with a Geographic Information System (GIS) that is designed as a general management evaluation tool in any human/landscape interaction setting where humans travel on a linear network. This software uses geographic data, imported into RBSim where the simulation analyst then sets other variables in the management model. RBSim uses the concept of 'Agents'. An agent is embedded with behavior characteristics, contains a mode of travel and a set of rules how it interacts with other agents and the landscape. Unique to this study is the use of a rule-based agent approach. Rule-based simulation approaches provide agents with their wayfinding logic and intelligence through a set of user definable rules. Analytical Hierarchy Process (AHP) is used in the way-finding logic to weight the importance of various attraction factors for each agent type that are associated with features attributed in the network. These attraction factors in conjunction with behavioral rules provide the agent with the capability to make travel and destination or attraction decisions during the course of the simulation (See Itami et al. 2004).

METHODS

This project addresses the need to improve the understanding of the complex patterns of hunter use in Prince William Sound (PWS) with simulation modeling. ADF&G and CNF provided a harvest database containing location information for bear kill sites by hunters in the area. This data includes the home city of each hunter, the duration of each hunt, mode of travel, kill date, and the UCU in which each bear was taken. Trip itineraries were constructed from this harvest data for 1995-2004 only those hunts assumed to orginate from Whittier, based on kill location and the home city of each hunter were included. In addition, GIS data was used to develop a habitat model that explicitly identifies and ranks value of spring habitat for black bears and accessibility to hunters in PWS. Criteria defined by Suring et al. (1988) were used identify and rank the value of habitat used by black bears during spring in PWS. A habitat capability model developed by Poe (2005) identifies the following biophysical features: (1) estuaries; (2) south, east, and west facing avalanche slopes; (3) muskeg forest; (4) beach fringe; (5) grasslands; and (6) old growth forest. The resulting coverage of habitat suitability for bears is summarized for each ADF&G management Uniform Code Unit

(UCU) polygon. Bear habitat is then evaluated and ranked for accessibility to hunters as a function of terrain slope and shoreline type. The resulting data layer represents areas attractive to hunting parties on a scale of one to ten. Ten is consider the best quality site. Figure 2 illustrates the overall analysis of shorelines and Figure 3 illustrates specific sites or referred to as facilities included on the simulation network.). For more detailed information on the habitat model see Gimblett & Lace (2005).

Using n=1199, or 75 percent valid bear harvest records for the time period to construct trip itineraries, the bear harvest capability model and likely travel route patterns of hunters linked to campsite and

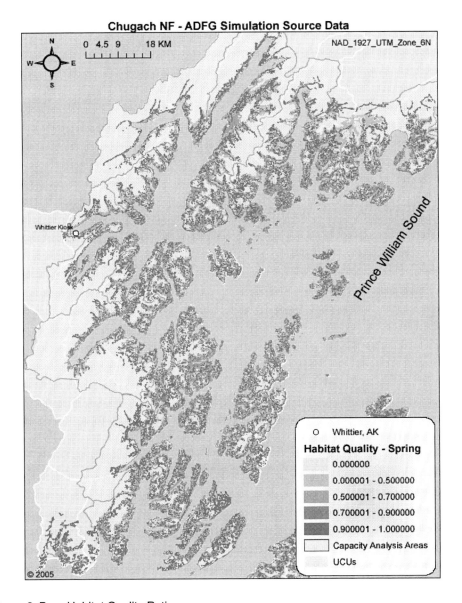

Figure 2. Bear Habitat Quality Ratings.

bear quality ranking nodes, a simulation model was built to analyze available harvest data from ADF&G Unit 6D from 1995-2004 to predict and quantify areas favored by black bear hunters and define the seasonal patterns of harvest effort in PWS. Imperative to this model was the development of rules that define hunter behavior in the simulation. If the hunter simply left Whittier, went to a capacity area, killed bear and returned on the same day that would be simple to model and no need for way-finding and intelligent rules. However, in the case of bear hunting in PWS, if a bear was reported being killed on the last day of a four day trip, little is known about the behavior of hunters during the first three days. In order to simulate hunter behavior the agents need the ability to randomly roam up and down the estuaries starting on day one, selecting day use and overnight (campsites) based on attraction values until day four when the kill is simulated the boat leaves the capacity area and returns to the origin of the trip. What is of interest to CNF is how the hunter is using PWS prior to the kill date. Where are they likely to camp? Are they having encounters with others? What might be the prime sites that both bears and hunters are attracted to? In order to construct the rules, consultation with biologists and other federal agency personnel who have knowledge of hunter behavior was sought. From these experts several assumptions were made to construct the rules. They are as follows:

- Majority of harvesters in western PWS enter through and return to the port of Whittier.
- Guided operations in western PWS are primarily live-aboard boats, displacement and semi-planing hull boat, supported by short range skiffs.
- Access through Whittier for private harvesters is primarily small boat, likely planning hull.
- After killing a bear, harvesting users terminate their trip and return to Whittier.
- Trip starts on kill date minus the reported length of trip and trip ends on kill date plus one.
- Harvesters in western PWS are primarily attracted to quality spring bear habitat (defined by PWS spring bear habitat model) for harvesting activity for search efforts and focus their efforts on accessible beach types.
- Small boat based and taxi drop-off harvesters will be attracted to campsites in close proximity to quality spring bear habitat and live-aboard harvesters will be attracted to anchorages in quality habitat.
- Harvest/search activity occurs along a shallow depth contours within sight of beach, crossing open water only when needed, trace shoreline for opportunities.
- Hunting parties focus their search efforts during morning and evening hours. The simulation imposes this with facility durations.

A linear network provides connectivity between the campsite and attraction facilities, known as nodes. The network is comprised of links with correct topology for agent travel between nodes. In addition, campsite and attraction facilities fall within what the CNF refers to as capacity areas (See Figure 4). Capacity Areas (CAs) were derived using watershed associations and are used to summarize recreation activity on CNF lands in western PWS. ADF&G Game Management units are used to summarize harvest records. The Capacity Analysis areas are much larger and cover a water body while the ADF&G units are related more specifically to the land. For the simulation, ADF&G units

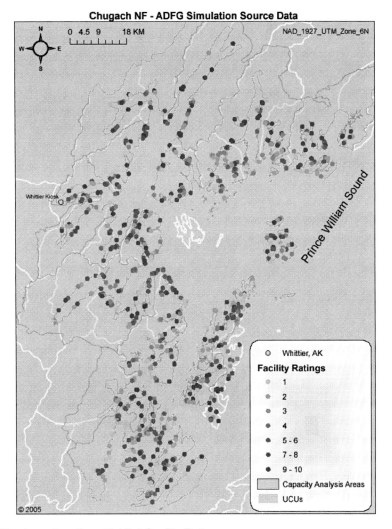

Figure 3. Sites based on Bear Habitat Quality Ratings.

and CAs were used as a basis for analysis. When a hunter agent leaves Whittier, arrives and then enters into a capacity area that they are permitted to hunt in, RBSim employs rules then to represent the behavior of the agent for the remainder of the trip. They randomly select campsites, roam up and down the estuary scouting for bears, camp at night and repeat this behavior until the kill date. Random stops, durations or time spent while stopped, travel speeds etc. all can be traced while the simulation is running. RBSim uses the concept of a locale or a group of nodes on a network to represent a basic unit of measure in this case a capacity area. Once in the locale the agents rules turn on they determine their own behavior.

GENERATING CONFIDENCE INTERVALS FOR SIMULATION AND MODEL VERIFICATION

Two important facets of any simulation modeling effort is to first generate confidence intervals around the simulation models and undertaking some form of model verification or validation. Any model requires verification and validation according to Law & Kelton (2000). The procedure of actual results verification was undertaken in this study. It compares the simulation model results with values from the real world system (Law & Kelton, 2000). In the case of the bear harvest model the simulation is repeatedly using a different set of random numbers for each replication to determine the number of replications of the model that need to be run to acquire a 95% confidence level of reliability. The simulation was run for the entire period for which data was available from 1995-2004. During this period the 1199 trips are simulated using the trip itineraries generated in the previous section. Standard practice is to run the simulation for many replications and calculate the confidence intervals for the key output variables (in this case, daily node and link

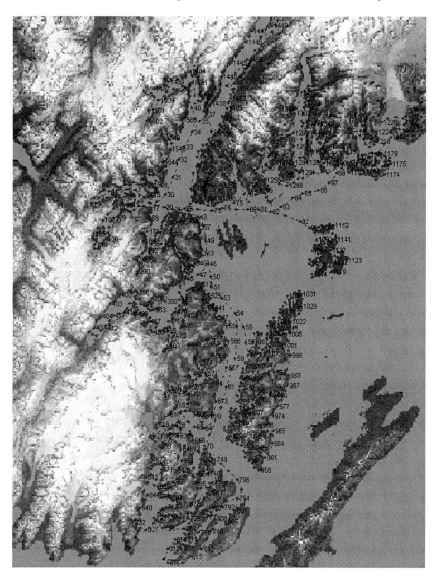

Figure 4. Travel routes, campsites, bear habitat sites used in Simulation.

use and node and daily link encounters). This analysis was performed at a 95% level of reliability and with 40 replications and a user defined confidence half interval of one hunting party. The results of this analysis showed that for node use, link use, and overnight node encounters the required 95% confidence intervals could be achieved within the 20 replications. For link encounters however, because of the higher impact of small variations on encounters, 95.7% of the links produced 95% confidence intervals within 18 replications, another 2.82% requiring an estimated 19 to 30 replications and the remaining 1.5% requiring an estimated 31 to 109 replications to achieve 95% confidence intervals. So the results were averaged and then analyzed based on a 20 replication simulation.

There were two forms of verification run on the model to ensure both the simulated trips were arriving within specified time periods to the locales or capacity areas in other words how accurate were the trips in getting to their destinations and secondly, once in the capacity areas, were the rules firing correctly and is reasonable behavior being obtained from these agents. Looking at the differences in these actual trips versus simulated trips gives "face validity" to the model since the assumptions embodied in the rules are being followed and pattern of behavior is near expectations. Face validity is a term used by Law & Kelton (2000) to describe reasonableness: when results are consistent with perception of typical recreation behavior. The degree of variation also is within reason for the experiment as this depends on many other factors, namely the travel network and encounters between agents both of which impact the outcome of the simulation run.

Figure 5 provides an evaluation of the comparison of the arrival patterns of the simulated trips to the actual trips obtain from ADF&G and converted into trip itineraries. 84% of the simulated trips arrived at the CAs within specified time periods. While random starts and stops were observed along the way, a high proportion

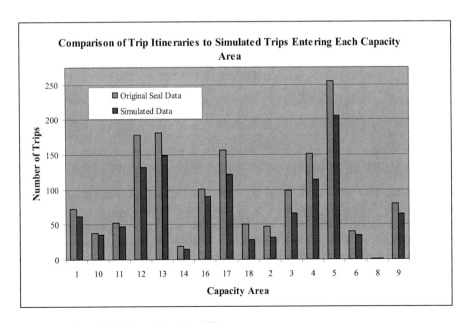

Figure 5. Comparing Original to Simulated Trips.

of the trips did what they were programmed to do. Those trips that did not perform adequately were identified and adjusted to improve the performance of the simulation.

The second form of verification was performed to evaluate if the agent rules were performing accurately and providing the intended behavior. This was done using a series of prototype agents, or sample trips, where one trip was run by itself and inspected. This verification then consists of tests on the outputs to ensure the rules are firing in sequence and the relative outcomes of the agents are as desired. The rules used in this project are a simple set that keep the agents active for the duration of the trip. The trip length is known and assumed to start the reported number of days minus one day before the kill date, and the trip ends one day after the kill. Hence the number of facilities visited by the agent is not dictated by the rules, rather is a function of the length of the trip. The facility durations are not fixed either; the time spent at each destination is determined by the software as part of a distribution. In short, the trips do not result in the same facilities being visited for the same duration. As expected the agents tend to disperse and utilize different facilities for variable amounts of time. Yet there is a consistent pattern established by the rules and range of durations allowed at each destination. For example, in a multi-day trip, more than three sites will be used even though there are only three rules. The maximizing behavior of the agent makes it accomplish as many rules as possible, while it cannot exceed the length of stay at each site. The facilities and the order they are used is not predetermined as these are chosen during the run according to the agent's rule set. For these reasons the agent trips result in unique itineraries with a consistent pattern or use. It can be concluded from this analysis that the trips, while having a high amount of variation in site selection do perform the behaviors as intended. No clandestine or unintended behaviors were observed.

RESULTS AND DISCUSSION

While there are many results that can be presented from the simulation, this paper will focus on those that answer the following questions:

- What was the average use per day at locations within PWS?
- What was the distribution of hunters in the CAs?
- What is the bear harvest density per CA?
- How are hunter use days distributed by site facilities in PWS?
- What are the most popular facilities frequented by hunters for each of the CAs?
- Did hunters respond to ranked quality bear habitat sites?
- What are the differences between commercial and non-commercial trips entering each CA?
- What was the distribution of hunters by travel mode entering each Locale or CA?
- What was the total number of hunter days for each UCU?

WHAT WAS THE AVERAGE HUNTER USE PER DAY AT LOCATIONS WITHIN PWS?

A primary measure is node use, and "nodes" include the destination facility points. Node use (in days) is best regarded as events, defined as any day that at least one agent visited the node. For example, in Node 1 is the Whittier Harbor, and it was used over 1800 times in the nine year period by black bear harvesters (nine years has

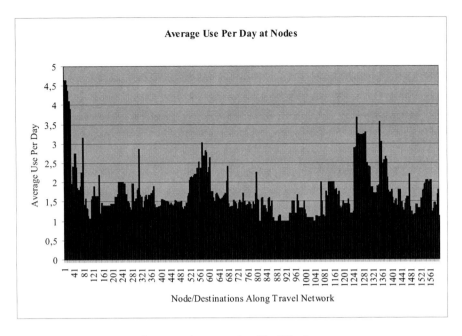

Figure 6. Average Hunter use days per day at nodes (Facilities).

3285 days/540 days 'in-season' = May through June or 60 days of spring hunting season per year). Also other main destination nodes were used almost as heavily. Further out on the network, destination nodes were used by half as much, with high use around Node 1241 at 200 days in the nine-year period. Other less used nodes were well below 50 days in ten years. One Hunter each day would give such quantities, but actually more than one hunter may have used a node. Figure 6 shows the average use per day for each node. Again at Node 1241 a high average number of use days are around 3.5. These averages are generally below 2 per day for the nine years. Both of these figures provide a view of the distribution of hunter use.

Figures 7 shows the average daily use spatially and the high points in the graphs can be found with large diameter dots. The underlying capacity analysis area and UCU is identified later. From this figure it becomes very apparent which of the facilities are most heavily used through the period for which the data represents but reveals very little about which facilities those areas represent.

A 'visitor use day' is a measure of the sum of the number of visits that occur each day at each site. Visitor use days are generally much higher than the number of distinct days the site was visited. A maximum user days escalates to 747 for the nine-year period. It is not permissible to compare this to the actual number of discrete days in the time period as above, since many of these user days occur on the same calendar day. This total could be compared to other popular recreation areas total user days. Yet, each user day does not account for the entire number of hours on a day. While some do as in the case of overnight, 24-hour camping, the rest are less than a full day's use. This changes the quantity of the actual use a site receives. For instance, a site that was used 2 days (either one Hunter for 2 days or two hunters on the same day) was really used for less than 48 hours. A "use" is cre-

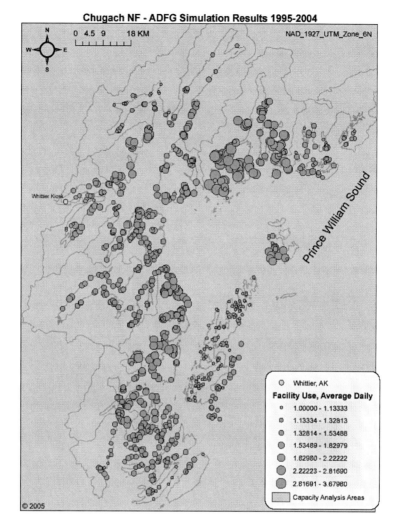

Figures 7. Simulation results of Hunter Average Daily use per day at Facilities.

ated by the event of a hunter reaching a site, not by how long the activity takes place. So the duration spent at a node is more revealing of the actual level of hunter use. It is found by subtracting the minimum time stamp from the maximum time stamp of the agent's itinerary. Figure 8 has the duration at sites in hours for the study period. This map has much of the same pattern as daily use maps but with more distinction between nodes and regions. Using both maps helps to assess the pattern of use: did a facility tend to be used very little on many different days, or all day for fewer days? For example, Knight Island receives more intensity of use possibly due to greater distance from the origin of Whittier, but fewer uses than closer CAs.

To develop a strategy for sampling trip itineraries for each day of a simulation, it would be necessary to define of "pool" of trips with common patterns of destinations over the course of a year. If itineraries are changing through the year the pool of trips must change to reflect changing patterns of use. The idea of a sampling pool of trips is to capture the variability of trips that may occur on any given day of the year. If the pool is limited to only the trip itineraries sampled for the corresponding day, the effect is to "overfit" the simulation model to the sample population. This may increase the statistical validity of the resulting simulation, but lowers its utility from a management perspective since the model is not representative of the true variability of trip itineraries selected by hunters on any given day.

Given that this model was built with a limited dataset, future considerations should be given to sampling to improve the reliability of the model and ultimately the predictability and usefulness as a management tool.

WHAT WAS THE DISTRIBUTION OF HUNTERS IN THE CAS?

Knowing the total number of hunter and visitor use days in PWS provides critical information useful to CNF. But more

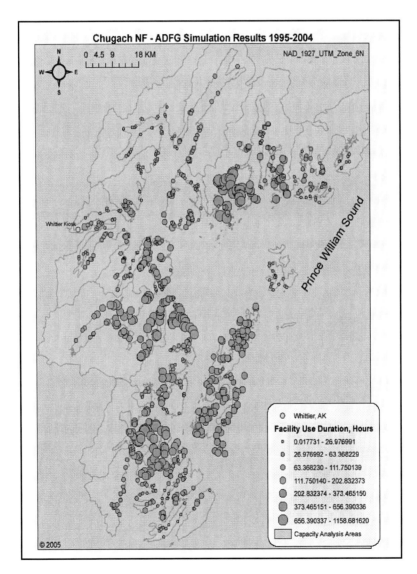

Figures 8. Simulation results of Duration per day at nodes.

importantly, the number of hunters using each of the CAs provides managers with more information to undertake more site specific and targeted management of these areas. An assessment of each node within each CA was undertaken, and is subsequently summed to provide an overall assessment of use. This analysis can be done year by year but for this study the analysis is done as a total for the nine-year period. The amount of use ranges widely from over 10,000 hunter use days to less than 2,000 (See Figure 9). CAs 5 (6,000 – 10,000 hunter days), 12, 13, 16 and 17 (2,000- 6,000 hunter days) are the most heavily used in the study area. CAs 5 (325-456 days), 12, 13, 16 and 17 (215-325 days) reveal the longest durations spend hunting in the study area (See Figure 10).

WHAT IS THE BEAR HARVEST DENSITY PER CA?

Of interest to wildlife biologists and managers is the number of bears from these CAs that are taken over the period for which data was available. Using maps of CA use duration and number of parties to CA (number of kills), assessment of the overall time spent harvesting could be made. Figure 11 provides an assessment of the number of black bears harvested from each capacity unit. As expected, there should be consistency with the analysis above that revealed that CAs 5, 13, 12, 17, 4 and 16 had the highest amount of time spent and hunter use days with total number of bears harvested. We see similar patterns in Figure 11.

HOW ARE HUNTER USE DAYS DISTRIBUTED BY SITE FACILITIES IN PWS?

Interesting to this study was the examination of how hunter use days were distributed among facilities. This analysis comes directly out of the simulation since very little is directly known about hunter behavior once they arrive into a CA. As

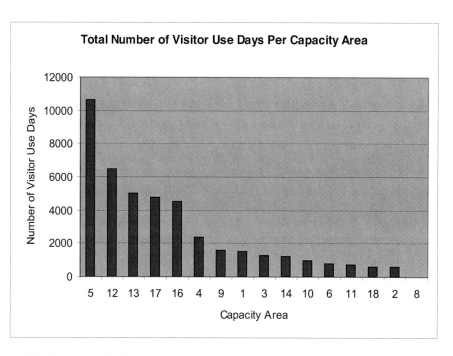

Figure 9. Simulation results for the total number of Visitor Use days per CA.

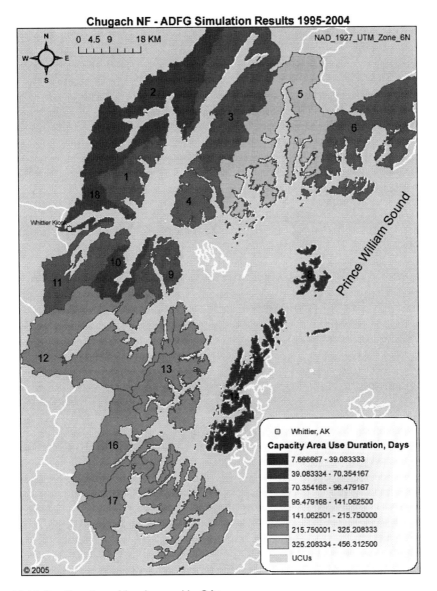

Figure 10. Visitor Durations (days) spend in CAs.

outlined earlier, each of the CAs have a number of facilities identified in addition to the bear habitat quality rating. But what was not known before this study was how often these facilities were visited and which ones were visited the most (See Figure 12).

Depending on the length of trip, many of the trips are required to stay in an overnight facility, either at anchorages on the water or campsites on the shore. All other types are less preferred. Since most trips were boat-based, the anchorage overnight facility resulted in higher uses. While not being a precedent setting finding, it does suggest for managers the need for more focused management on the shores to accommodate the annual hunting activities. It also provides information that could be used in conjunction with other

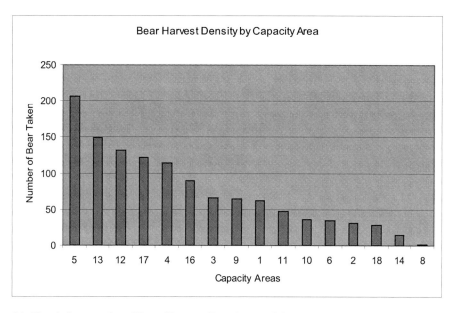

Figure 11. Simulation results of Bear Harvest Density per CA.

recreation use information to identify potential conflict area. While it is important to identify the range of use levels at site attractions, the more important questions revolved around where hunter use levels tend to occur and is there a correlation with the most highly ranked bear habitat sites.

WHAT ARE THE MOST POPULAR FACILITIES FREQUENTED BY HUNTERS FOR EACH OF THE CAS?

To provide some insight into how the most popular facilities are distributed across the study area an analysis was undertaken summarizing facility use by CA. As seen in Figure 13, CAs 5, 12, 13, 16, 17 & 4 consistently have the highest number of total visits and trends in bear habitat and overnight facility. Again CA 5 stands out.

WHAT ARE THE DIFFERENCES BETWEEN COMMERCIAL AND NON-COMMERCIAL TRIPS ENTERING EACH CA?

A question that is prevalent in many outdoor recreation studies is the evaluation of commercial and non-commercial use. What is the percentage of each? Do they conflict with each other? And most importantly are they appropriate for the area?

Figure 14 provides an assessment of Commercial and Non-commercial for each locale entry. In other words, each trip that entered into a locale or CA was kept track of, summarized and displayed in Figure 14. Locale entries 1 (CA 18), 13 (CA 52) reveal some commercial use but dominated mostly by non-commercial activity. Locale entries 12 (CA 49), 68 (CA 16), 71 (CA 17) & 78 (CA5), reveal equal if not dominant commercial versus non-commercial use. In other words in these four areas there is a significant amount of reported commercial activity. The later three, 5, 16 & 17 in earlier analysis are not only the sites most frequently visited by commercial activity, but also are the most frequented overall, account a high percentage of the areas where the most bears are harvested and where the duration of stay and the most overnight activity occurs.

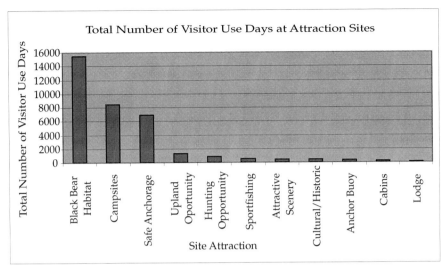

Figure 12. Simulation results of Hunter Use days and Site facilities.

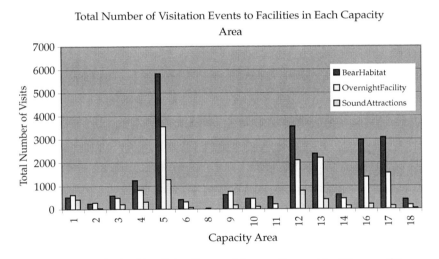

Figure 13. Simulation results for identifying the most frequently used facilities per CA.

SUMMARY OF FINDINGS

There are several interesting findings from the simulation outputs based on the bear harvest record dataset, logical assumptions and rules derived from experts. They are as follows:

- Peak hunter use days range between 315 and 503 on the main travel routes to less than 26 days in remote areas.
- Cumulative use of routes over the nine year period ranges from 1015 to 2338 visitor or hunter use days.
- Average hunter use per day for each node was approximately 3.5. The averages generally fall below 2 per day for the nine years.
- The amount of hunter use days ranges widely from over 10,000 to less than 2,000, depending on the capacity area. Capacity areas 5 (6,000 – 10,000 hunter

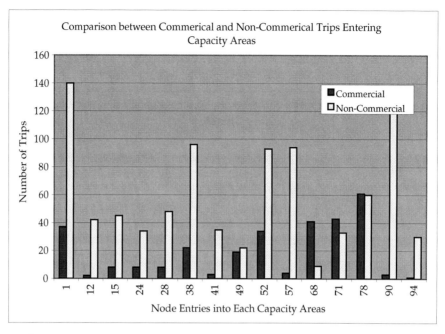

Figure 14. Simulation results comparing Commercial versus Non-Commercial trips entering each CA.

days), 12, 13, 16 and 17 (2,000- 6,000 hunter days) are the most heavily used in the study area.

- Capacity areas 5, 12, 13, 16 & 17 consistently have the highest number of total visits and trends in bear habitat and overnight facilities.
- Capacity area 5 receives the most harvest use days and number of bear taken.
- Private boats account for approximately 75% of the travel modes entering into the capacity areas and Water Taxis an additional 15% with the remaining group being hunters hosted by a registered guide. While water taxi services specialize in supporting bear hunting groups throughout PWS, they are certainly no where near as high as private boats but do account for a significant number of trips entering PWS and in contrast to registered guides are not tracked by the CNF.
- While there is a significant amount of commercial use in PWS, capacity areas 5, 16 & 17 are not only the sites most frequently visited by commercial activity, but also are the most frequented overall, accounting for areas where the most bears are harvested, the longest duration of stay and where the most overnight activity occurs.
- While there is no empirical evidence at the onset of this study as to other recreation groups using the area, anecdotal information does point to the potential for conflicts with bear hunters.

CONCLUSIONS

There is a growing body of research focused within the context of human-environment interactions. This work examines the need to develop a comprehensive and empirically based framework for linking the social, biophysical and geographic disciplines across space and time. This paper has outlined a prototype bear harvesting model using a spatial agent based modeling (ABM) approach. While the ABM approach has been successfully used in many situations, each one has its unique character. Unique to this study has been the use of a rule-based approach, using a minimal amount of empirical data on hunter movement patterns and behavior. Bear harvest data served as a beginning point to develop this model, but to have a high degree of reliability and sophistication in the model and associated outputs, a representative valid sample of data should be captured. This is not just about acquiring more data but more focused quality data that represents the spatial and temporal variability of bear hunting trips into PWS.

ACKNOWLEDGEMENTS

The authors would also like to thank those folks on the Chugach National Forest who have contributed background information, knowledge, data and funding for this prototype study.

LITERATURE CITED

Alaska Department of Fish and Game. 1982. Black bear movements and home range study. ADF&G Federal Aid in Wildlife Restoration Report. Juneau, Alaska 73 pp.

Gimblett, H.R. & S. Lace. 2005. Spring Black Bear Harvest Simulation Modeling in Prince William Sound, Alaska. Final Report Prepared for the Glacier Ranger District. Chugach National Forest. Girdwood, Alaska. 65pgs. http://www.srnr.arizona.edu/~gimblett/PWS_Report_2005_Final.pdf

Itami, R, R. Raulings, G. MacLaren, K. Hirst, R. Gimblett, D. Zanon, P. Chladek. 2003. RBSim 2: Simulating the complex interactions between human movement and the outdoor recreation environment. Journal for Nature Conservation. 11, pgs. 278-286.

Law, A. M. and W. D. Kelton. 2000. Simulation Modeling and Analysis, Third Edition, McGraw Hill, Boston.

McIlroy, C.W. 1970. Aspects of the ecology and hunter harvest of black bear in Prince William Sound. M.S. Thesis. University of Alaska, Fairbanks. 69 pp.

Poe, A. 2005. GIS model for Evaluating Bear Habitat Quality. Unpublished Document. Chugagh National Forest. Girdwood, AK 99587

Suring, L. H., E. J. Degayner, R. W. Flynn & T. M. McCarthy. 1988. Habitat capability model for black bear in southeast Alaska. Unpublished report, USDA Forest Service, Alaska Region, Juneau, Alaska. 34 pp.

Suring, L. H., K. A Murphy & Ali Iliff. 2001. Human Use and Wildlife Disturbance establishing the baseline for management in western Prince William Sound, Alaska. Unpublished Report, USDA Forest Service, Chugach National Forest, Anchorage, Alaska. 14pp.

CHAPTER 17

BALANCING INCREASED VISITOR USE AND WILDERNESS CHARACTER USING AGENT-BASED SIMULATION MODELING IN MISTY FJORDS NATIONAL MONUMENT

Rachel Nelson
Randy Gimblett
Chris Prew
Chris Sharp

Abstract: This chapter describes the development and application of a computer-based simulation model of recreational use in Misty Fjords National Monument (MFNM) Tongass National Forest, Alaska, USA. This study used monitoring data collected by observation during summer 2004 to build an agent based model and uses average daily encounters to estimate the impact of increasing visitor use levels on wilderness management quality objectives established for these backcountry sites. The results of the study demonstrate conceptually how simulation modeling can be used as a tool for understanding existing visitor use patterns within the MFNM Area to compare those estimates to established wilderness quality standards. This study also identifies and discusses potential challenges of applying computer simulation to backcountry recreation management and provides recommendations for further research to address these issues.

Keywords: Recreation Planning, Computer Simulation, Visitor Monitoring, Wilderness Management

INTRODUCTION

Federal Wilderness legislation emphasizes that Wilderness areas should be managed to provide recreational visitors with opportunities for solitude or a primitive and unconfined type of recreation. With the renewed popularity of outdoor recreation in backcountry settings, wilderness managers are being challenged to provide unconfined recreation opportunities while protecting wilderness character: the bio-physical, experiential, and symbolic ideals unique to Wilderness lands.

Wilderness recreation management decisions are complex and need to be carefully monitored to ensure the decisions are realistic, practical, and effective. Misty Fjords National Monument (MFNM) in the Tongass National Forest is not exempt from these types of management strategies. MFNM is an extraordinary landscape of extremely high scenic and recreational value, contained within the National Monument Wilderness Land Use Designation (LUD) The intent of this LUD is to maintain an enduring wilderness resource while providing for public access and uses consistent with the Wilderness Act of 1964, the Alaska National Interest Lands Conservation Act of 1980 (ANILCA), and the Misty Fjords National Monument proclamation of 1978. Specific goals for Misty Fjords existence are: *to serve the scientific purposes of preserving a unique ecosystem and the remarkable geologic and biological objects and features it contain… and to provide a high degree of remoteness from the sights and sounds of humans, and opportunities for solitude and primitive recreation activities consistent with wilderness preser*vation (USDA 1997).

Under the LUD are carefully constructed objectives to manage recreation activities to meet the appropriate levels of social encounters, on-site developments, methods of access, and visitor impacts indicated for the Recreation Opportunity Spectrum (ROS) class. The ROS standards being applied to MFNM are designated as Primitive, Semi-Primitive Non-Motorized, and Semi-Primitive Motorized while recognizing exceptions due to ANILCA (See Table 1). Management leeway is given in these guidelines, as the actual authorized number can be larger or smaller based on a site-specific analysis.

Table 1. Maximum Recreation and Tourism Development Generally Allowed by LUD for National Monument Wilderness *

Recreational Opportunity Setting Classification	Flight-Based Sightseeing (number of landings per site per day)	Social Encounters (number of parties encountered per day)
Primitive	3	3
Semi-Primitive Non-Motorized	6	6
Semi-Primitive Motorized	6	6

*Actual number authorized could be larger or smaller depending on site-specific analysis

Common to the Alaskan landscape and way of life is the floatplane. Floatplane access to MFNM is the traditional and predominant form of travel. Commercial tours provide direct access to MFNM and vary depending on availability of time and recreation objective. Some visitors want to experience the MFNM by air on short tours while others prefer longer stays to fish on freshwater lakes or rendezvous with a charter fishing boat for a day on the water. The commercial trips that land do so on either salt or freshwater. Since the Tongass National Forest does not have any jurisdiction to manage salt water, they are primarily concerned with freshwater landings and their effect on wilderness quality. While minimal regulation occurs on salt water, all freshwater landings on lakes fall within the jurisdiction of the Tongass National forest and require permits. While it has never been quantitatively measured, it is thought that there may be adverse impacts of floatplane landings on the wilderness experience of others camped within site and sound of freshwater lakes. An evaluation of the spatial and temporal patterns of freshwater landings is imperative to assessing whether wilderness quality objectives established by the Tongass National Forest are realistic. Even though the management of saltwater landings does not fall within the jurisdiction on the Tongass, the frequency of landings and overflights may also be having adverse impacts on the recreation experience of visitors in kayaks or other modes of travel in these areas.

The current standards established for flight-based sightseeing landings allowed per day are three for the primitive ROS standards and six for the other ROS classes. When a site is defined as a fixed point such as a lake or cove, then the number of lake and saltwater landings in the Monument often exceed the maximum allowable standard. The ROS standards for social encounters are defined as less than three parities per day for Primitive areas and less than six for the other ROS classes.

Recent research suggests that computer-based simulation modeling is an effective tool for helping to address the challenges associated with managing visitor use in backcountry and wilderness settings (See Cole 2005). Simulation modeling can also be used to monitor the condition of "hard to measure" indicator variables (Lawson et al. 2003). For example, how many encounters do backpacking visitors have with other groups per day while hiking? How many nights do visitors camp within sight of other groups? In addition, simulation modeling can be used to test the potential effectiveness of alternative management practices in a manner that is more comprehensive, less costly, and less politically risky than on-the-ground trial and error (Lawson & Manning 2003a)

This paper describes the development and application of a computer-based simulation model of recreational use in Misty Fjords National Monument (MFNM) Tongass National Forest, Alaska, USA.

The objectives of this paper are to:

- Describe the development of a methodology for monitoring visitor use levels for all pertinent salt water, freshwater and shoreline modes of recreation use (i.e. Floatplanes, Sea kayaks, cruise ships, charter fishing etc.) during the peak season in 2004;
- Develop a simulation to evaluate visitor use levels on both salt and freshwater sites;
- Evaluate existing management policies and standards established for Misty Fjords National Monument to determine if they are being exceeded and/or appropriate for maintaining a high quality wilderness experience.

Figure 1. View into Nooya Lake

STUDY AREA

Located in southeast Alaska, Misty Fjords National Monument is 22 miles east of Ketchikan and about 680 air miles from Seattle. Created on December I, 1978 by presidential proclamation, the Monument encompasses 2,294,343 acres within the Tongass National Forest. In 1980, Congress passed the Alaska National Interest Lands Conservation Act (ANILCA) and designated all but 151,832 acres as Wilderness. Monument status protects ecological, cultural, geological, historical, prehistorical, scientific, and wilderness values. Remote and wild, the monument is a mountainous, nearly untouched coastline, characterized by deep saltwater Fjords and is home to mountain goats, brown and black bears, and a host of fish and marine mammals (See Figure 1).

The Behm Canal, a deep, long waterway of the northeastern Pacific Ocean, leads to the heart of the Monument. Places such as Walker Cove and Rudyerd Bay, characterized by rock walls jutting 3,000 feet above the ocean, provide flightseers, cruise ship passengers, boaters, and hikers with photographic opportunities. Commercial services are allowed, under permit, to provide basic public services in accordance with Monument designations. The Forest Service and the State of Alaska manage fish and wildlife habitat cooperatively. Management concerns are apparent as very little recreation use actually occurs on the land due to the steep, inaccessible terrain. Since there is limited regulation of air or water access, encroachment of tourists from anchored ships and commercial flights on wildlife is of serious concern.

MONITORING PATTERNS OF VISITOR USE IN MISTY FJORDS NATIONAL MONUMENT

In the summer 2004 a combination of field observations, visitor use surveys, interviews, and simulation was used to understand recreation use in MFNM. Results from the 2002 study aided in identifying critical, high use areas in MFNM that would subsequently be targeted for detailed monitoring in 2004. The Misty Core Area (MCA) is centered in the Rudyerd/Walker Cove area. Since a component of this study was to address the appropriateness of the wilderness quality objectives, fixed-point monitoring sites were carefully selected to observe all pertinent salt water, freshwater and shoreline modes of recreation use (i.e. floatplanes, sea kayaks, cruise ships, charter fishing etc.). Since permitted use by the Tongass National Forest (FS) only exists on those lands that fall into the national forest jurisdiction, those sites that were freshwater lakes were targeted. In addition, since the original simulation study identified critical areas that received above average amount of visits on salt water, it was important to capture them in the sample so they could be used to validate the original simulation results. In addition, the 2002 study (Gimblett 2002), identified the heaviest or peak visitations periods that coincided with the months of June, July and August. So the monitoring study in 2004 was established to capture visitor use patterns in the Misty core area during these peak visitation periods.

All monitoring work was done by sea kayak using both students from the University of Arizona as well as sea kayak rangers employed through the FS. Since kayaks were being used to monitor these remote areas, the monitoring locations were selected according to ease of access by the observer, clarity of viewshed for determining vessel routes, and significance based on outlined critical use areas from the 2002 study. Each monitoring location, fixed point, origin, and destination point was given a unique acronym. Fixed points were used to record the origin and destination time and location of observed travel routes. Fixed points that were monitored or observed from monitoring locations are identified in Table 1. The monitoring locations were allocated accordingly to observe fresh and saltwater landings and over flights. Table 2 provides a breakdown of those monitoring allocations. Data sheets were developed to record all spatial and temporal aspects of the observed vessels within the Misty core region. These included observation time, fixed point at which the vessel's time was recorded, a travel origin location, a travel destination location, and a vessel type. If known, the type of use: commercial; administrative; or private, and the company name were recorded. Vessels that were observed more than once during the observation period were tracked with a unique identifier to better describe their route.

Using teams consisting of both Forest Service Employees and University of Arizona graduate students, field observations were taken from June 19th 2004 to August 16th 2004 (59 days). To ensure statistically valid and representative sample, weekdays and weekends were consistently sampled throughout the study. In addition, two teams operated simultaneously, equally dividing up their observation periods between both saltwater and freshwater location.

A total of n=3649 trips were documented by the field crew over the sampling period. A aggregation of trips per month ranging from June (n=1051) to August with (n=979) and the peak month of July with (n=1819). A total of 36 observation days, or 66% of those days were sampled through these months. A total of 619 hours of actual observation time occurred throughout those 80 days dispersed over 20 sights. The average number of hours of observation at these locations at any point in time was between 10 to 14 hours. It is apparent that July is the peak month in MCA. An evaluation of these figures suggests that the sampling plan adequately captured peak periods in the MCA.

In addition to the observation data collected, administrative trips were obtained from the FS through radio communication logs. All observation and trip data was entered into a Microsoft Access database. Travel origins, destinations, and monitoring location codes corresponding with each trip were recorded and entered both into a database and on a travel network. The travel networks were either trip routes or flight paths that were captured through in-depth interviews with pilots and FS personnel, from those utilized in the 2002 study, and by using GPS tracking on random kayaks and boats. The information was either entered into Arc Map or converted and geo-referenced to the existing data. Both travel routes and destinations were used to construct the travel network used in the baseline simulation.

Although the total number of trips is important to obtain a representative sample of visitor use occurring in the MCA, equally important is how those trips are spatially and temporally distributed across the fixed points. For the purposes of travel simulation the sampling design must incorporate

Table 2. Fresh and Salt Water Monitoring Location

Freshwater Sites
CHE L- Upper Cheecats Lake, ELL L- Ella Lake, GOA L- Big Goat Lake, MAN L- Manzanita Lake, MZI L- Manzoni Lake, NOO L- Nooya Lake, PUN L- Punchbowl Lake, WIL L- Wilson Lake.
Saltwater Sites
CHE C- Cheecats Cove, EDD R- New Eddystone Rock, ELL B- Ella Bay, MAN B- Manzanita Bay, NOO C- Nooya Cove, PUN C- Punchbowl Cove, RUD M- Mouth of Rudyerd, RUD N- Rudyerd Narrows, WAL H-Head of Walker, WAL M- Mouth of Walker, WIN I- Winstanley Island

both spatial and temporal variations in the population (See Chapter 21 for complete description of sampling techniques).

SIMULATION MODELING OF VISITOR USE PATTERNS IN MFNM

The Tongass National Forest has the responsibility to improve the visitor experience and protect the ecological integrity of MFNM. This project addresses the need to improve the understanding of the complex patterns of visitor use with simulation modeling. The Tongass National Forest has already invested in data collection of visitor use. The use of simulation is a logical step in the overall effort to develop a comprehensive understanding of the relationship between the spatial-temporal patterns of visitor use and goals for wilderness management.

With this in mind a simulation was constructed to examine baseline or "existing conditions" of travel patterns in MFNM directly from the observation data collected in the summer 2004. This simulation forms the basis of a series of subsequent simulations that can be used to explore a wide range of management issues. The following management questions direct this study:

- What is the existing pattern of visitor use?
- How are visits distributed on Salt versus freshwater destinations?
- Are the wilderness management objectives appropriate for MFNM?

This simulation project uses the Recreation Behavior Simulator (RBSim), a computer simulation tool designed for evaluating linear recreation systems and developing management alternatives (See Itami et al. 2004 for details on RBSim). The MFNM model is a trace simulation of existing conditions. It is comprised of the following information has been used to model visitor use:

- Observation data collected from fixed point monitoring in the MCA - Summer 2004
- Trip Itineraries were created from these observations as input into the simulation
- The landscape model consisted of GIS flight and travel routes for the study area and known destinations extract from the fixed point observation data

The travel network was comprised of 47 Nodes and 124 links. Maximum travel speeds (speed limits) were assigned to

Table 3. Arrival distributions for encounters for the Sampling Period

Week	Monday	Tuesday	Wednesday	Thursday	Friday	Saturday	Sunday	Sum
23	0	3	6	6	5	0	2	22
24	6	1	6	7	1	0	0	21
25	1	7	5	7	2	12	9	43
26	31	126	190	205	3	5	4	564
27	11	88	298	304	31	0	0	732
28	0	43	51	125	48	5	4	276
29	4	4	11	166	118	152	111	566
30	169	46	7	9	5	2	0	238
31	3	8	21	241	101	9	0	383
32	7	8	107	164	57	1	1	345
33	2	2	114	175	104	35	59	491
34	108	4	2	1	1	0	1	117
35	0	2	0	1	1	2	0	6
36	2	1	0	0	0	0	0	3

each link of the network. All trip destinations were loaded into ArcView GIS and then loaded as a point file in the RBSim travel network. This process assigns a unique ID to each destination that is then used to cross-reference destinations each trip was documented passing and traveling towards. It is assumed that visitors are taking the shortest travel time path between destinations. This assumption allows RBSim to automatically generate the travel path from one destination to the next.

Itineraries were created from the 3849 observed trips itineraries described above. Trip itineraries were formatted by GeoDimensions Pty Ltd so they could be read into RBSim. RBSim simulates existing or baseline conditions by modeling and simulating trip itinerary data directly. This type of simulation is referred to as a trace simulation. The trace simulation minimizes random variability of probabilistic simulations and represents a "true" realization of the field-collected data in the form of dynamic simulation. The results of the trace simulation examine interactions overtime.

VERIFICATION AND VALIDATION OF MODEL OUTCOMES

Building valid spatial dynamic models has posed many challenges to researchers. Spatial simulation models use random numbers to generate input variables such as arrival times, durations at destinations, and trip selection. Itami et al. (2005) have examined this problem and suggest that:

> "It is not recommended to draw any conclusions from the output of a single replication of a simulation model since it represents only one realization of a stochastic process. This is a common characteristic of all simulation models however spatial simulations have added complexity because output measures such as number of visits or average duration at nodes are measured across many different links and nodes."

This study implements two methods for verification of simulation outputs. The first examines the problem of generating confidence intervals and verification of simulation model and outputs. We use a method used by Itami et al (2005) where "one observation per replication is generated (usually the mean value of the performance indicator)." The simulation is replicated a number of times to generate confidence intervals that meet a certain reliability (usually 0.90 or 0.95) and a given accuracy (measured in the same units as the performance indicator). A method is implemented in this simulation using the performance indicator use days for links and nodes. Link and node use days are the number of parties visiting each link or node for each day of the simulation.

While this step indicates how many replications to develop a reliable and accurate simulation, there is still a question as to whether the outputs from the simulation are representative or what has been or could be observed in the world. A second method is implemented to determine how representative the simulation is of the observed data in this study, an analysis is paired t-test was performed including a Pearson correlation to compare the differences in the use days to seventeen destinations where both observations were obtained and simulated trips visited. In addition, a sensitivity test is performed on three of the seventeen nodes to evaluate how much improvement can be obtained in the overall simulation results.

RESULTS OF EVALUATING CONFIDENCE INTERVALS AROUND SIMULATION RUNS

The simulation is run for the entire season of data collection from June 1, 2004 until August 31, 2004 (59 days). During this period 3849 trips are simulated using the trip itineraries generated from the observation data. Generally, the baseline trace simulation requires fewer replications than a probabilistic simulation in order to generate stable results because random variation is minimized. Standard practice is to run the simulation for 5 replications for baseline trace simulations and calculate the confidence intervals for the key output variables (in this case, node and link use). This analysis was performed at a 95% level of reliability and with 5 replications and a user defined confidence half interval of + or − 1 visitor. The results of this analysis showed that for node use and link use, the required 95% confidence intervals could be achieved within the 5 replications. For link use however, because of the higher impact of small variations on use levels, 95.7% of the links produced 95% confidence intervals within 5 replications, another 2.82% requiring an estimated 6 to 10 replications and the remaining 1.5% requiring an estimated 11 to 104 replications to achieve 95% confidence intervals. As this analysis shows, it is possible to achieve completely reliable simulation results based on an unrepresentative sample of the population. As a result of this analysis the simulation was run and results evaluated based on 10 simulation runs.

SIMULATION VERIFICATION A COMPARISON OF MODEL RESULTS TO OBSERVATION DATA

Before any analysis can be performed and conclusions drawn from the outputs of the simulation, it is imperative to ensure that the simulation results are representative of what has been or could be observed in the real world The observed data collected during the Summer 2004 monitoring period was used. A paired t-test was performed including a Pearson correlation to compare the differences in the use days to seventeen destinations where both observations were obtained and simulated trips visited. A correlation coefficient of .592767, (P = -0.07, T stat = -3.4E-16) reveals that there is approximately 60% correlation between the simulated visits and observed visits in the field. Given that spatial simulation models use random numbers to generate input variables (arrival times, durations at destinations, and trip selection etc.) this could be considered a reasonable correlation. However after further examination of the observation data, it is apparent that the data is collected from fixed points, providing detailed information at those specific points during specific times. This causes a certain amount of speculation as to where trips originate and the location of their final destination. Simulated trips on the other hand, leave a specific origin, but freely traverse the network on route to the desired destination. While each destination keeps track of a visit once a trip arrives, any node that happens to fall on the network that the trip passes through, on route to the destination will also record a visit. This in itself is a significant benefit of simulation because it provides information on destinations in the landscape where monitoring data on use levels did not occur. Use numbers are hard to valid if no data has been collected, but even more problematic is attempting to compare observation to simulated outputs as a measure of model verification. This tends to skew the comparative results as much higher visits are recorded at some nodes on the network then were actually observed in the field.

To test this idea, a sensitivity test was performed to evaluate how much higher a correlation could be obtained between simulation outputs in comparison to the observed data (See Figure 2). Three of the known destinations that fall on the network that are not major destinations but happened to exhibit exceptionally high user levels as compared to observed days, identified in Figure 3 (Cheecats Cove, Upper Cheecats Lake and Winstanley Island) were removed and the paired t-test was rerun. The difference in the mean number of visits significantly increases the

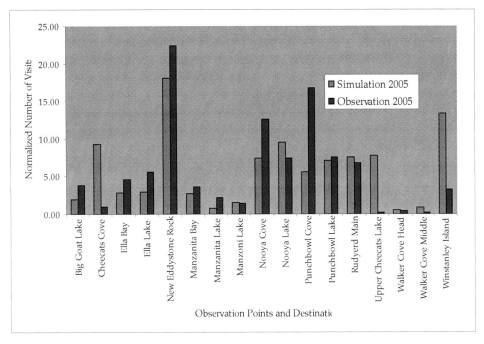

Figure 2. Comparison of the 2004 Simulation Results and Observation Data

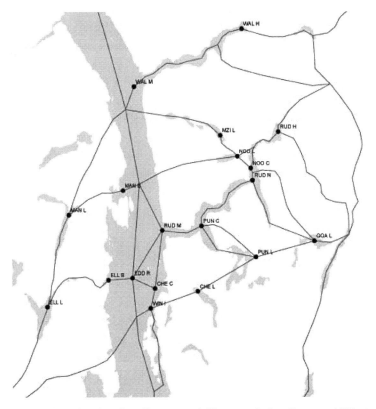

Figure 3. Map showing the alignment of Cheecats Lake, Cove and Winstanley Island

correlation coefficient to 0.86435, (P = -0.05, T stat = -2.10497) revealing approximately 86% correlation between the simulated visits and observed visits in the field.

This analysis has demonstrated the importance of both performing a verification test on the simulated data to ensure that there is adequate correlation with what is observed in the field. While both 60% and 86% correlation suggest that the simulation is performing as expected, more refinement of the model parameters needs to be undertaken to better align the observed data with the simulated outcomes before a probabilistic model can be developed. However, this analysis does illustrate the true benefit of simulation for acquiring information about recreation use in areas where monitoring does not occur.

WHAT ARE THE EXISTING PATTERNS OF USE?

All analyses on simulated and observed data are based on visitor use days. A visitor use day is the number of days a site was visited by one party through the sampling and simulation period. Table 4 provides an assessment of simulation outputs for nodes and destinations. Three specific measures are undertaken to assess recreation use patterns. The number of use days or number of days that the destination is visited, total use or number of trips that actually visited the destination and the average use per day for each of the nodes.

Since all trips originate and return to Ketchikan, it is apparent that the total use would be the highest of any node or destination. Notable other sites are Nooya, Cheecat, Punchbowl and Ella Lakes with Rudyerd head, main and north receiving a moderate number of visits (See Figures 4 & 5). Figure 6 illustrates the number of visitor use days. Some of these nodes such as Entrance, Winstanley Island and Eddystone rock have an above average number of use days. These nodes are not necessarily destinations but happen to sit on

Figure 6 provide an evaluation of the link use days and average daily use. There are two major routes or corridors being used for access in to Misty. Routes 116, 100, 117 and 122 (flight routes) receive the highest number of use days and average daily

Table 4. Assessment of Simulation Outputs for Nodes/Destinations

NodeLabel	NodeID	Use Days	Total Use	AvgUsePerDay
KTN	1	76	7109	93.54
NOO L	3	46	2155	46.85
CHE L	13	45	1410	31.33
PUN L	14	41	1139	27.78
CHE C	39	38	3186	83.84
RUD H	11	37	435	11.76
PUN R	21	33	628	19.03
NOO C	24	32	1632	51.00
PUN C	19	32	376	11.75
RUD M	40	32	1873	58.53
RUD N	22	31	989	31.90
ELL L	56	24	1205	50.21
GOA L	9	20	268	13.40
WAL L	30	20	35	1.75

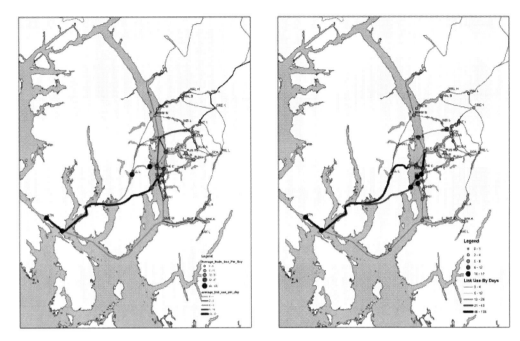

Figures 4 & 5. Average and Use Days at Nodes/Destinations

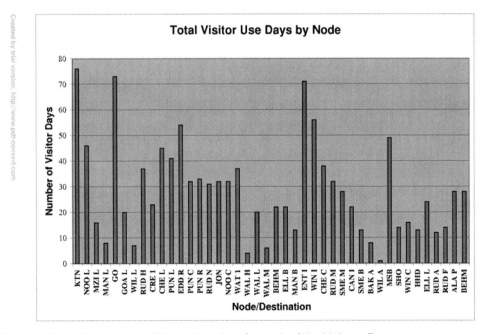

Figure 6. Simulation outputs of Visitor Use days for each of the Links or Routes

use of between 90 to 93 days. The second set of popular routes 48, 49, 50 and 52 are the water routes. All these routes can be seen in Figure 9. Route 100 is closely associated with the entrance or exit to Ketchikan. Route 117 (90.95) is a flight route leading to Ella Lake while route 100 is the water route associated with water travel. Link 50 for example with average use days at 64 is close to Entrance Island. These values, while not extremely high do reveal the frequency of use of these routes.

WHAT ARE THE NUMBER OF VISITOR USE DAYS SPENT AT DESTINATIONS ON SALT WATER?

Table 5 provides a summary of the number of visitor use days for destinations that fall in salt water. The sites that have the highest number of visitor use days are Ketchikan (all trips start and finish here), New Eddystone rock and Winstanley Island (because these points are along the route that a high percentage of the trips pass by and those are floatplanes in the air or kayaks on the water), Rudyerd Main and Nooya Cove again areas that are passed by along route to some specific destination like Rudyerd Float. Interestingly, Rudyerd float is not as frequently visited, as was the case in the 2002 simulation. This can be explained by the fact that in 2004, visits to the float were actually observed and documented. In the 2002 study, trip logs reported their destination was to the float, but many never stopped, just flew around Rudyerd on route to some other destination or were simply used as a convenient destination to report. When compared to the observed data the differences observed in Table 16 are mostly those sites that were not monitored as fixed points with observation. These are the points that were not observed during the summer 2004 but were points along the network the simulation documented trips passing by on route to the destinations. This is an important observation as it reveals the true predictive power of using the simulation to model recreation visits as sites where no data has been collected.

WHAT ARE THE NUMBER OF VISITOR USE DAYS SPENT AT DESTINATIONS ON FRESH WATER?

Table 6 provides a view of the total use days at freshwater sites. Nooya and Upper Cheecats Lakes followed closely by Punchbowl Lake received the most recreation use. Like the saltwater sites, floatplane landings on these lakes account for approximately 68% of the use. Passengers from kayaks, cruise ships, pleasure motor, and tour boats also access the lakes.

ARE THE WILDERNESS STANDARDS PROPOSED IN THE MANAGEMENT PLAN FOR THE MCA BEING EXCEEDED?

An analysis of average daily encounters from the simulation outputs can be seen in Table7. When the average daily encounters are combined with the ROS setting classes it is apparent that all four destinations (Nooya, Goat, Cheecats and Punchbowl Lake) all exceed the 3 parties encountered per day ROS social setting standards. Ella Lake is the only Non-Primitive or Semi-Primitive site that exceeds the social setting standard of less than six parties encountered per day. The results of this analysis suggest that either encounter levels are too high for the freshwater lakes or the ROS social setting standards need to be more reevaluated and ultimately aligned to more accurately reflect what is occurring. Further work needs to be undertaken to determine these standards.

CONCLUSIONS

Analysis of the temporal distribution of observations has shown that there is a need for survey designs that are specific to the needs of building and validating travel simulation models. A solid monitoring program was established to collected data at twenty fixed observation points. While a trace simulation has successfully been built to replicate both the trip itineraries extracted from travel logs (2002) and a site specific observation plan (2004) has been constructed to capture and simulate more detailed data on visitor use, it has been

Table 5. Salt Water Results from Simulation

Name	Use Days	AP	CH	CS	FS	HC	KY	PL/M	PL/S	TR
Bakewell Arm	8	6	0	0	0	0	1	1	0	0
Behm Canal	45	17	2	2	3	0	2	10	3	6
New Eddystone Rock	134	49	3	13	3	1	19	20	7	19
Nooya Cove	80	28	4	14	0	1	6	10	6	11
Punchbowl Cove	60	24	3	1	1	3	14	7	3	4
Rudyerd A	12	10	0	0	0	0	1	0	1	0
Rudyerd Head	51	35	0	0	0	0	0	6	1	9
Rudyerd Main	81	31	6	10	1	0	1	12	6	14
Rudyerd Float	77	27	4	14	0	1	5	10	6	10
Smeaton Bay	13	8	0	0	0	0	4	1	0	0
Walker Cove Head	6	3	0	0	0	0	0	3	0	0
Walker Cove Middle	9	5	0	0	0	0	1	3	0	0
Walker Cove	51	35	0	0	0	0	0	6	1	9
Wilson Arm	1	0	0	0	0	0	1	0	0	0
Winstanley Island	143	42	7	10	1	7	34	20	9	13

Table 6. Fresh Water Results from Simulation

Name	Use Days	AP	CH	CS	FS	HC	KY	PL/M	PL/S	TR
Big Goat Lake	21	20	0	0	0	1	0	0	0	0
Ella Lake	32	18	0	0	0	0	10	2	0	2
Manzanita Lake	8	8	0	0	0	0	0	0	0	0
Manzoni Lake	16	15	0	0	0	0	0	0	0	1
Nooya Lake	102	44	4	14	0	1	7	12	6	14
Punchbowl Lake	76	32	3	1	1	5	19	8	3	4
Upper Cheecats Lake	83	37	3	1	1	5	20	8	4	4
Wilson Lake	8	7	0	0	0	1	0	0	0	0

Table 7. Average Daily Encounters at Freshwater Lakes

Name	AvgDailyEncounters	Ros Class
NOO L	12.55	Non-Primitive
MZI L	0.58	Primitive
MAN L	0.77	Non-Primitive
GOA L	3.90	Non-Primitive
WIL L	0.03	Non-Primitive
CHE L	3.35	Non-Primitive
PUN L	13.32	Non-Primitive
ELL L	8.02	Non-Primitive

impossible to build a probabilistic simulation that can be used in a generalized way to generate management advice (other than encounter levels) and formulate alternative management scenarios. Continued data collect or monitoring is crucial that is well structured to capture the spatial variability of trips throughout the season.

The demonstration of methods for generating a probabilistic simulation from observation and travel logs shows, the potential of value adding simulation technology to the Tongass National Forest's current data store. This potential can be realized by developing protocols for collecting more detailed trip itineraries and a focused effort to integrate visitor monitoring and simulation into the planning and management process.

FUTURE IMPLICATIONS/RESEARCH

This paper only addresses the use levels within Forest Service jurisdictional areas, specifically lake landings. The simulation itself tracks use levels throughout the saltwater, freshwater, and land, and air regions. Clearly, restrictions on freshwater landing could increase use on saltwater locals if overall use was not decreasing. If the management aim is to limit the sights and sounds in the entire wilderness area, and the encircled landscape, then this study identifies and lays groundwork for collective management of the area.

In the short term the Tongass National Forest should:

- Continue to work towards a standardized set of data collection and monitoring procedures that are comprehensive and cost effective.

- Explore new technology for collecting data and analyzing data on visitor flows. Innovations such as radio frequency identification tags (RFID) provide inexpensive methods for collecting accurate data on large numbers of pedestrian and travel movements.

- Combining technologies such as pedestrian counters and traditional social survey techniques can provide the comprehensive information on visitor behavior.

- Integrate models of visitor flow, wildlife and environmental impact within the context of planning and management In the longer term:

- Continue to apply simulation in a diversity of environments to demonstrate the advantages of using visitor use monitoring and simulation in the planning and management process.

- Continue to coordinate at the national level to incorporate visitor monitoring and simulation in the planning and management process.

ACKNOWLEDGEMENTS

This project was developed in close cooperation with the Ketchikan-Misty Fjords Ranger District of the Tongass National Forest. Many thanks to Karen Brand of the Ketchikan Misty Fjords Ranger District for their support, cooperation and expertise in the developing and implementing the monitoring program in Misty Fjords Summer 2004, Leta Sharp as well as the kayak rangers working with the FS, who together formed the core of the monitoring team and who did outstanding job in capturing a representative sample of visitors using the MCA.

LITERATURE CITED

Daniel, T., & Gimblett, R. 2000. Autonomous agents in the park: An introduction to the Grand Canyon river trip simulation model. *International Journal of Wilderness, 6(3), 39-43.*

Gimblett, H.R. 2002. Modeling the Impacts of Recreation Visitation on Misty Fiords National Monument, Tongass National Forest. 2002. Final Report. U.S. Fish and Wildlife Service. http://www.srnr.arizona.edu/~gimblett/Misty_final_report1.pdf.. December 2002.

GeoDimensions Pty Ltd. 2005. Pattern of Human Use Simulation in Canada's Mountain Parks. Final Report Submitted to Parks Canada. March, 2005.

Gimblett, R., Richards, M., & Itami, R. 2000. RBSim: Geographic simulation of wilderness recreation behavior. *Journal of Forestry, 99(4), 36-42.*

Itami, R., Raulings, R., MacLaren, G., Hirst, K., Gimblett, R., Zanon, D., & Chladek, P. 2004. RBSim 2: Simulating the complex interactions between human movement and the outdoor recreation environment. *Journal of Nature Conservation, 11(4),* 278-286.

Itami, R.M., D. Zell, F. Grigel & H.R. Gimblett. 2005. Generating Confidence Intervals for Spatial Simulations: Determining the Number of Replications for Spatial Terminating Simulations. Paper presented at International Congress of Modeling and Simulation (MODSIM05). Melbourne, Australia, December 12-15, 2005.

Lawson, S. & Manning, R. 2003a. Research to inform management of wilderness camping at Isle Royale National Park: Part I – descriptive research. *Journal of Park and Recreation Administration, 21(3),* 22-42.

Lawson, S. & Manning, R. 2003b. Research to inform management of wilderness camping at Isle Royale National Park: Part II – prescriptive research. *Journal of Park and Recreation Administration, 21(3),* 43-56.

Lawson, S., Manning, R., Valliere, W., & Wang, B. 2003. Proactive monitoring and adaptive management of social carrying capacity in Arches National Park: An application of computer simulation modeling. *Journal of Environmental Management, 68,* 305-313.

Manning, R., R. M. Itami, D. Cole & H. R. Gimblett. 2005. Overview of Recreation Simulation Modeling Approaches and Methods. Pgs. 11 – 15. In: Cole, David N. (compiler). Computer Simulation Modeling of Recreation Use: Current Status, Case Studies, and Future Directions. Gen. Tech. Rep. RMRS-GTR-143. Ogden, UT: U.S. Department of Agriculture, Forest Service, Rocky Mountain Research Station. September 2005. Pgs 75.

USDA- United State Department of Agriculture. Forest Service Alaska Region. Tongass National Forest Land and Resource Management Plan. R10-MB-338-CD. May 1997.

Van Wagtendonk, J. 2003. The wilderness simulation model: A historical perspective. *International Journal of Wilderness, 9 (2),* 9-13.

CHAPTER 18

LEVEL OF SUSTAINABLE ACTIVITY: MOVING VISITOR SIMULATION FROM DESCRIPTION TO MANAGEMENT FOR AN URBAN WATERWAY IN AUSTRALIA

Robert M. Itami

Abstract: Visitor pattern of use simulations are an effective tool for describing and human movement in a variety of environments (Cole, 2005). Visitor simulations studies use paradigms such as discrete-event simulation, cellular automata and multi-agent simulation. However, regardless of the methodology used, the outputs are always the same – a quantitative description of movement patterns. Whereas the quantitative analysis of visitor flows is fundamental to a better understanding of the complex interactions of human use with the environment, the impacts on the quality of experience and behaviour can only be discovered through social science methods that elicit responses from users about their expectations, attitudes, preferences and behavioural responses to visitor densities, queuing times, flow rates, and the capacity of facilities. This chapter presents a decision-making framework called "Level of Sustainable Activity"(LSA) that links the outputs of agent based river traffic simulations to social and environmental performance objectives. The method is adapted from the US Federal Highway Administrations "Level of Service" for traffic capacity. However the LSA framework links user estimates of traffic density to quality of service objects and a risk management framework to identify social and environmental risk factors. The results of the method are then used to interpret simulations of existing and projected use for making management decisions.

Key Words: River Traffic Simulation, River Traffic Management, Level of Sustainable Activity, RBSim, Pattern of Use Simulation

INTRODUCTION:

Visitor pattern of use simulations are an effective tool for describing and quantifying the distribution, density, speed, and flow patterns of human movement in a variety of environments from wilderness back-country settings to highly urbanised high use settings (Cole 2005). The technical development of special purpose simulators for recreation environments continues using simulation paradigms including discrete-event simulation, cellular automata and multi-agent simulation. However, regardless of the methodology used to simulate visitor pattern of use, the outputs are always the same – a quantitative description of movement patterns. Whereas the quantitative analysis of visitor flows is fundamental to a better understanding of the complex interactions of human use, the impacts of these patterns on the quality of experience and behavior can only be discovered through social science methods that elicit responses from users about their expectations, experiences, attitudes, preferences and behavioral responses to visitor densities, queuing times, flow rates, the distribution of destinations, and the capacity of facilities. *Only by linking the social and environmental implications to the flow patterns generated by human pattern of use simulations can we begin to manage the quality of experience for visitors.*

This issue is at the heart of a complex management problem in Melbourne, Victoria Australia. The Melbourne Waterways Committee commissioned a study in 2005-2006 to determine the traffic capacity of the Maribyrnong and Yarra Rivers (see figure 1) to develop a traffic management plan on the basis of the current level of river traffic and the projected traffic for the next 5 and 10 year periods. The urgency for this study is prompted by existing conflicts between commercial passenger ferries and rowing and canoeing, and the increasing commercial and recreational traffic in the shipping zone. On top of this is the Melbourne Docklands development, which will create marinas for 700 to 1000 new private and public berths in the heart of the study area (see figure 1).

Earlier consultancies had established projected growth rates for commercial and recreational traffic, however a defensible method for determining river capacity for

Figure 1. The "2 Rivers" study area –Looking south, downstream to Yarra River to the left, Victoria Harbour and the Docklands development center right, and Bolte bridge and the Port of Melbourne in the background flowing into Port Philip Bay on the horizon

the various forms of traffic had not been determined. The underlying assumption had been that some single metric like "maximum number of vessels per hectare" could be established to determine the overall capacity of the river system. This definition, however does not recognize the very different physical operating characteristics of a rowing skull compared to a passenger ferry, or the quality of experience required for passive recreation versus competitive training. It is clear a robust defensible way of defining river capacity that takes into account river characteristics, competing users, vessel types, and physical infrastructure had to be developed.

STUDY METHOD

The Yarra and Maribyrnong Rivers and Port Phillip Bay in Victoria, Australia are complex environments. Located in an urban environment, their physical characteristics are diverse. Large shipping docks, commercial tourist berths, boat launching ramps and active recreation clubs support a diverse range of river traffic from cargo ships to commercial tourist operators, private motorized craft and rowers. These users share use of a restricted area of water and naturally have their own, sometimes-divergent views about how they and others they share the river with should be managed.

River traffic management must:

- Balance the competing demand of a diverse set of different types of users
- Maintain and enhance the significant commercial values of the water in terms of the operation of the Port of Melbourne and commercial tourist operators who operate throughout the study area
- Consider the safety and quality of experience of the many recreational users of the rivers and bay.
- Consult and incorporate the views of the many organizations involved in managing traffic on the two rivers and bays.

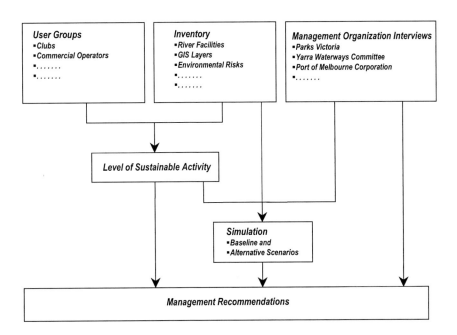

Figure 2. Overview of the Two Rivers Management Plan Methodology

Table 1. This table shows the relationship between Level of Sustainable Activity, Traffic Density and Quality of Service. As traffic densities increase, Quality of Service decreases.

Level of Sustainable Activity	Traffic Density	Quality of Service
A	Low Density	Highest
B	-	-
C	-	-
D	-	-
E	High Density	Lowest

A multi-faceted approach is required to meet these requirements. Figure 2 below describes the approach taken in the Two Rivers Traffic Management Plan.

The Two Rivers Traffic Management Plan consists of four main components:

1. Inventory. A detailed inventory of the physical characteristics of the waterways. Geographic data and locations of all major facilities was recorded and mapped in Geographic Information System (GIS). The inventory included a reconnaissance level assessment of erosion risk of river banks.

2. Management and User Group Interviews Various management organisations and User groups were interviewed in detail to obtain their views on river traffic management and the issues important to them. Feedback from interviews was used in the following ways:

   a. To define key traffic issues relating to quality of service, safety, and environmental impacts, conflicts between user groups and current traffic management issues.

   b. As an input to the simulation scenarios in the form of estimates of future traffic volumes and new future facilities;

   c. As an input to the simulation scenarios in the form of alternative management actions to be tested.

3. Level of Sustainable Activity Workshop and Interviews. The Level of Sustainable Activity (LSA) component uses the statistics and information collected in the inventory component to formulate a series of densities for motorized and non-motorized river traffic for each river management zone. Five LSA levels designated with the letters A to E define the relationships between traffic density and quality of service. LSA A is the lowest density and the highest quality of service and LSA E is the highest density and the lowest quality of service.

4. LSA focus group workshops then collected the preferences and views of the main user groups of the waterways about existing and future traffic levels and river facilities. The LSA workshops asked river users to rate existing peak and maximum tolerable traffic levels on the water. The LSA component results were used to

   a. Define existing LSA levels for each River Management Zone at peak periods of use

   b. Define maximum LSA levels while still maintaining safe navigation and quality of service objectives

   c. Existing and maximum LSA for competing users

   d. Define impacts and risks to quality of experience, safety, environment

   e. Suggest means of mitigating or managing these impacts and risks.

5. Simulation. The fourth component of the Two Rivers Traffic Management Plan involved building a series of simulations of existing and forecast future

traffic levels. The outputs of the simulations gave a spatial view of the changes in traffic densities and volumes from 2005 to 2010 and 2015. The Level of Sustainable Activity results were a crucial input to evaluating the simulations. Two alternative management options of river closures and speed limit changes were tested in the simulation process. The simulation was built using a special purpose recreation behavior simulator RBSim (Itami et al., 2004).

This chapter will focus on the Level of Sustainable Activity framework as applied to the analysis of river traffic management using the results of a single river management zone (Marina Transit Zone) as an example.

LEVEL OF SUSTAINABLE ACTIVITY FOR RIVER TRAFFIC

The Level of Sustainable Activity (LSA) concept is a generalization of the Level of Service concept developed by the Transportation Research Board (2000). River capacity is different for each user group and varies in relation to river geometry, vessel characteristics, the provision of facilities, and the interaction with other users. River traffic management must therefore be based on a comprehensive framework that integrates all the relevant factors in a format that is easy for users and decision makers to understand and that can be adapted to a wide range of environments and travel modes.

LSA can be thought of as a scale of end-user experience. Each river zone has a range of service levels defined for each vessel type ranging from very low levels of use, with minimal environmental and social impacts to high-density use with high levels of user interaction, higher levels of potential environmental and social impacts, and more intensive facility and management requirements.

The LSA concept integrates:
- Physical characteristics of the river, including navigable depth, width, and bank erosion potential. This information was available by taking measurements from aerial photographs, bathometric surveys of river depth and from on-site survey of bank conditions for each river management zone.
- Physical characteristics of different vessel types, their stopping distance and safe passing distance and speed (described in more detail below)
- Defining traffic densities for each LSA (A through E) for motorized and non-motorized vessels
- User preferences for levels of use for specific activities in specific river zones.
- User attitudes toward competing traffic Safety, Environmental and Social risk factors relating to increasing use densities.
- Suggestions from users on management options for dealing with the above risks.

Characterizing River Management Zones

The LSA framework requires that the environment that is under evaluation is homogenous in terms of physical configuration, mix of users and facilities. Previous studies of the 2 Rivers study area had formulated a sensible classification of the rivers and bay into 7 river management zones. These zones were homogenous for the characteristic required for the LSA framework. However, the Port Zone and the Bay Zones were not included in the LSA study due to lack of time and resources and the predominance of large commercial shipping in these zones. Each river zone is characterized according to total area in hectares, length and the average width of

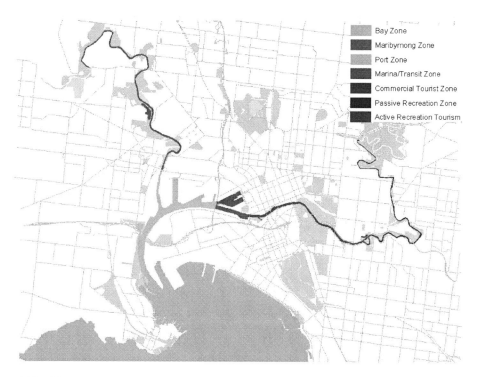

Figure 3. The "2 Rivers" study area – Port Phillip Bay to the south and the 7 river management zones.

the navigable area. This information is measured from aerial photography and bathometric surveys. These dimensions are then used to calculate traffic densities at different use levels. In addition existing facilities relating to river use were inventoried for each River Zone

An important factor in the environmental impact of the river is the bank erosion caused by wash from boats. The risk of erosion is related to bank conditions and the type and speed of vessels. Based on existing information on bank protection, evidence of bank slippage and expert opinion, the banks are rated on a relative scale from low to high for erosion risk. The risk factor is then compared to different levels of use by vessel type to determine the erosion hazard and mitigation options. Erosion risk is a highly complex and technical issue that varies significantly in the study area. Therefore, a qualified fluvial geomorphologist made an expert assessment of erosion risk.

Physical Characteristics of Vessels

Characteristics of the vessels shown in were collected from operators, manufacturer's specifications and from expert feedback.

These vessels were then grouped into two major subgroups to simplify the presentation of traffic to users. The two groups are:

- Motorized Vessels
- Non-Motorized vessels (rowing shells and canoes)

Commercial Shipping and Passenger Cruise ships were not included in the LSA since their movements are strictly regulated and they always have right of way because of their size, weight, limited maneuverability and the danger they pose

Table 2. Vessel classes in the Level of Sustainable Activity Study

Motorized Vessels • Ferry / Transport Boats • Cruising / Function Boats • Water Taxis • BBQ Boats • Small powered speed boats	• Large cruising motor boats • Victorian Water Police • Parks Victoria Boats • Customs Boats • Melbourne Port Boats • Fishing Boats
Non-Motorized Vessels • Rowing shells • Canoes • Sail boats Other non powered boats	
Commercial Shipping (not included in the LSA evaluations) • Bulk Carriers • Container ships • Passenger Cruise ships • Tug boats and Pilot boats	

for other vessels. Table 2 shows vessels included in the Motorized and Non-Motorized Vessel classes.

Establishing LSA Levels

In order to establish the traffic density at each of 5 LSA levels (A through E). The following factors have to be considered:

1. The type of vessel including its length, width (including oars for rowing), power and speed (translated into safe stopping distance with reverse thrust for powered boats)
2. The navigable width of the river
3. Navigation rules, in particular, staying to the right of on-coming traffic.

Operating characteristics of vessels is affected by speed, length, width, waterline depth and power. These factors affect the ability of the operator to avoid collisions with other vessels. In the LSA framework a set of equations were used to calculate stopping distance with reverse thrust (see Figure 4). These equations were run on a range of vessels with different weights, at the range of speeds allowed in each river zone. These calculations provide a safe distance in front of the vessel that allows the vessel to avoid collisions with statio-

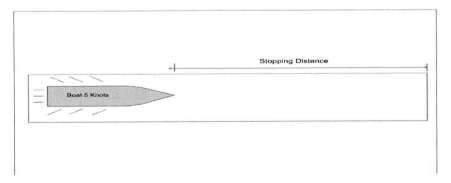

Figure 4. A safe operating envelope was calculated based upon the stopping distance of the vessel under maximum reverse thrust.

nery craft in the path of the boat traveling at a given speed.

The problem is complex because of the highly variable operating characteristics of the different vessels, however the problem, in the case of the 2 Rivers study could be simplified because the speed limit in each of the river zones studied was set at 5 knots/hour (9.26 km/hr). Speeds were determined by taking the minimum of the speed limit for the river as compared to the maximum speed of the vessel. A second simplification was to average the stopping distances of different vessels within the motorized and non-motorized vessel classes to create "typical" stopping distances for the evaluation.

A computer program was written to generate random patterns of boats separately for motorized traffic and rowing and paddling. The program incorporates the physics equations with spacing rules and navigation rules. The program works by randomly selecting locations and sizes for each vessel, if the new location meets the rules for navigation, safe operating envelope and spacing rules for horizontal and longitudinal proximities to other boats and the riverbank, then the boat is allocated to that position. If any of the rules fail, then the program generates another random location and tests for the same rules again. As the river segment fills up the rules are systematically relaxed until it is no longer possible to allocate any more boats to the river segment at this point the river is "jammed" and the program stops.

Random allocation produces similar densities at the lower levels for both motorized and non-motorized vessels. As densities increase however differences in stopping distances cause greater differences in the density levels of the two classes of vessels as shown in Table 3. Note that LSA levels A through C increase at a geometric rate (1, 2 and 4 vessels per hectare). LSA level D category however between Rowers/Canoeists and Motorized vessels. As Level D represents the "safe operating envelope" for a hypothetical "average" vessel in this zone the difference is attributed to differences in size, weight, speed and power. LSA level E represents the river zone with maximum number of vessels with all constraints relaxed. Level E therefore represents a crowded traffic level where vessels would need to slow down to navigate safely with operators having to be constantly vigilant to avoid collision with other vessels.

Focus groups to determine LSA level preferences

The preferences of different river users are key component of the management of river traffic and the definition of river capacity. Each river user-type (rowers, commercial tour operators, water taxis,

Level of Sustainable Activity	Rowers/Paddlers		Motorized Commercial and Recreational		Quality of Service
	Area/Boat	Boats/Ha	Area/Boat	Boats/Ha	
A	10,000 m²	1	10,000 m²	1	Highest
B	5,000 m²	2	5,000 m²	2	
C	2,500 m²	4	2,500 m²	4	
D	769 m²	13	1,250 m²	8	
E	625 m²	16	714 m²	14	Lowest

Table 3. Level of Sustainable Activity vessel density definitions for Rowers/Canoeists and Motorized vessels. LSA level A is the lowest density and the highest quality of service. LSA level E is the highest density and lowest quality of service. This table allows the manager or analyst to link outputs from the river traffic simulation to quality of experience or river users.

and ships) has different requirements in terms of safety, ability to perform their intended activity, and level of satisfaction based on the mix and density of vessels sharing the river zone. In the LSA framework two situations were addressed:

- Activity *within* a single user group (e.g. Rowers in relationship to rowers or commercial vessels in relationship to other commercial vessels) and
- Level Sustainable Activity *between* different users (e.g. rows in relationship to commercial vessels).

For each of these categories there is also difference in expectations during busy times of the day and season and during off-periods. A reliable way of making these judgments was needed.

First, users were divided into focus groups with no more than 12 participants in each focus group. The main categorizations were:

- Large Commercial Operators
- Small Commercial Operators
- Rowers and Paddlers

The large and small commercial operators were asked to evaluate LSA levels in three river zones they typically use:

- Marina Transit Zone
- Commercial Zone
- Active Recreation Zone

Rowers were similarly asked to evaluate LSA levels only in the river zone they used, there for the rowers were divided into three groups:

- Maribyrnong Rowers (Maribyrnong Zone)
- Rowing Clubs downstream of Herring Island (Active Recreation Zone)
- Rowing Clubs Upstream of Herring Island (Passive Recreation Zone)

The reason for separating out users in this fashion was to avoid conflict during workshops (it was known from previous questionnaires that there was considerable animosity between commercial operators and rowers), to ensure users were concentrating on their quality of service issues and when there was disagreement on LSA evaluations, consensus could be achieved through discussion between peers in a congenial environment. Government officials were excluded from observing or participating in the focus group meetings.

In order to elicit responses from the focus groups on LSA levels, plan view representations of each of the 5 LSA levels generated by the computer program described earlier, were overlaid on aerial photos to provide users with an accurately scaled image of each LSA level for motorized and non-motorized vessels. Examples of the LSA levels for the Marina Transit Zone are shown in Figure 7 through Figure 11.

The users were then asked to identify the peak period of use by season, day of week and hours in the day. Once the peak use period(s) was defined, the users were then asked to identify LSA levels for the following conditions:

- Current LSA during peak periods for OTHER users during YOUR peak period
- Maximum tolerable LSA during YOUR peak period
- Maximum tolerable LSA for OTHER users during YOUR peak period

The users were allowed to discuss differences in their evaluations and then through group discussion came up with a consensus for each of the 3 evaluations. This process defines the current peak period densities, the maximum LSA that users are willing to accept for their own group and for competing users.

After the above LSA levels were determined the focus groups were asked to identify risks to safety, quality of experience, and impacts on environment as traffic increased or exceeded the maximum

RESULTS AND DISCUSSION – COMMERCIAL ZONE

The LSA framework was succesful defining river capacity from a user's perspective as well as stimulating focussed discussion on the implications and responses to increasing traffic as traffic volumes exceed accepted levels of traffic. To demonstrate the findings from the LSA framework described above, the results from the Commercial Zone will be discussed. The Commercial Zone has a mix of recreational and Commercial traffic with many conflicts between the two user groups, it therefore is a good test of the LSA framework, to identify the issues, river capacity and user responses to increasing levels of use over time.

Figure. 6 shows the upstream half of the Commercial Zone, this area is a major hub for river tourist activity with commercial ferries, tour boats and restaurant boats originating in this zone. In addition, rowers from the adjacent Active Recreation Zone share the zone especially near Princes Bridge where rowers make a U-turn to return upstream on training runs. Elite rows also traverse through this zone enroute to the Marina transit zone downstream where there are long straight stretches of river favoured for race training by this group.

The Commercial Tourist Zone hosts a high level of traffic volumes mostly due to the high intensity tourist-oriented business along the south bank of the Yarra River and the high levels of pedestrian traffic generated by the zone's proximity to Melbourne's CBD. Many commercial passenger services originate or terminate in this zone. The zone also acts as a transit corridor for rowers and motorized recreation boats. Rowers in the Active Recreation Zone also turn downstream of Princes Bridge near the Southgate commercial berths causing conflict during the busy tourist season. Bank erosion is low risk in this zone because the banks are fully lined with concrete or stonewalls.

The Commercial Tourist Zone is currently used predominantly by commercial passenger services. With many visitor attractions including Southgate, Melbourne Aquarium, Crown Casino, the

Table 4. Commercial Zone Characteristics

Description	Value
Typical River Width in Zone	75-100 m
Zone Length	1.9 km
Area of Navigable Water	16 ha
Bank Erosion Risk Rating	Low
Major Features / Destinations	South Wharf Function Centre
	Polly Woodside / Melbourne Maritime Museum
	Melbourne Convention Centre
	Melbourne Exhibition Centre
	Crown Casino and Entertainment Complex
	Melbourne Aquarium
	Southgate and Commercial Tourist Berths
	Flinders Street Station
	Flinders Walk Landing
	Banana Alley Wharf
	Enterprise Wharves
	Princes Bridge

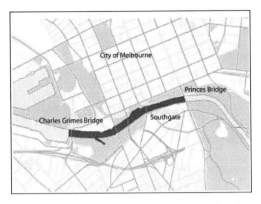

Figure 5. Location map for Commercial Tourist Zone, Melbourne CBD directly north.

Melbourne Exhibition Centre and Polly Woodside, this zone currently has a high level of pedestrian use and provides a strong market for commercial operations. The zone has heavy traffic not only because of the demand, but also because the zone is relatively small. Even if current traffic levels are within acceptable levels for users, the restricted navigable area means that there is little excess capacity.

Commercial Tourist Zone Level of Sustainable Activity Results

Table 4 shows the physical characteristics of the commercial zone along with the major tourist destinations accessible from the Yarra River.

Figure 7 through Figure 11 shows LSA levels A through E for motorized traffic in the Commercial Zone. A similar set of exhibits were generated showing the five rowing LSA classes. These were used in

Figure. 6. Upstream section of the Commercial Zone looking upstream on Yarra River. Princes Bridge near the top of the photo marks the boundary with the Active Recreation Zone. Southbank development to the right and Flinders Rail Station to the left and Sandridge rail bridge at the bottom of the photo. This area is a popular tourist zone with many restaurants and sidewalk cafes as well as close proximity to downtown Melbourne, the arts district and major sporting facilities.

Figure 7. Level A Level of Sustainable Activity Motorized Boats Commercial Zone Flinders Station rail yard to top of picture, Southbank at the bottom, Princes Bridge to the right and Queens bridge on the left (Sandridge Bridge is not shown in this image).

Figure 8. Level B Level of Sustainable Activity Motorized Boats Commercial Zone

Figure 9. Level C Level of Sustainable Activity Motorized Boats Commercial Zone

Figure 10. Level D Level of Sustainable Activity Motorized Boats Commercial Zone

Figure 11. Level E Level of Sustainable Activity Motorized Boats Commercial Zone

the LSA focus groups for Commercial Operators and for Rowers.

Table 5 shows the results of the LSA focus group with commercial operators. It shows this group has low tolerance for mixed traffic with rowers and that the zone is nearing maximum capacity.

Commercial operators also noted problems with poor night lighting at Spencer Street Bridge and on river bends. Also of concern was the training for new vessel captains. With the complexities of tidal effects, night conditions, river flows, flooding and competing river traffic it can take a year to "learn the ropes".

Conflicts with other river users include "unpredictable behavior" and speeding by motorized recreation vessels, unexpected turning by rowers, "verbal abuse" by rowers and coaches and rowing instructors often being unaware of safety issues with mixed traffic. However commercial operators generally agreed that all users have legitimate use of the river.

COMMERCIAL TOURIST ZONE SIMULATION RESULTS

Table 6 shows total hourly traffic for the peak use day simulated for 2005, 2010 and 2015 within the Commercial Tourist Zone. Traffic volumes more than double between 2005 and 2010 and then level off approach-

Table 5. Results of LSA focus group evaluations by Commercial Operators

Vessel Type	Current LSA at peak periods	Maximum tolerable LSA	Management Implications
Commercial	C+	C-D	Near Capacity
Rowers	A+	A+	Low tolerance to rowers

ing 2015 as marina capacities are reached. Densities of vessels increase rapidly in this zone because of its small area (16.27 Hectares). By 2010 densities typically reach LSA Level B at 11am and by 2015 LSA Level C is reached. In reality peak LSA levels are likely to be higher than those shown by the simulations. Areas with commercial berths and many bridges would be likely to be associated with localized traffic congestion making the LSA levels higher.

Figure 12 shows the pattern of use by travel mode. The early morning traffic is primarily elite rowers passing through the area to train between Charles Grimes Bridge and Bolte Bridge. Later in the morning the dominant traffic becomes commercial passenger and motorized recreation vessels. Commercial traffic declines sharply in the mid-afternoon, increases in the after-work hours and gradually declines into the evening.

Figure 13 shows that by 2010 motorized recreational traffic is the dominant use in the Commercial Tourist Zone due to the large number of Docklands berths now available.

Increased motorized traffic in 2010 will impact upon rowing.

Figure 14 shows the 2015 traffic pattern of use that was established in 2010 continuing. Again morning rowing extends later into the morning and motorized recreation and commercial passenger services show steady increases in volume.

Table 6. Hourly traffic volumes and densities for the Commercial Tourist Zone with projections for 2010 and 2015. Densities are vessels per hectare.

Hour	2005	2010	2015	2005 Density	2010 Density	2015 Density
6:00	2	3	5	0.12	0.19	0.28
7:00	6	10	14	0.37	0.65	0.84
8:00	8	20	26	0.52	1.22	1.58
9:00	11	29	38	0.70	1.79	2.35
10:00	16	34	48	1.01	2.07	2.98
11:00	18	46	52	1.10	2.81	3.18
12:00	15	38	43	0.93	2.30	2.64
13:00	19	39	43	1.20	2.39	2.64
14:00	19	40	43	1.16	2.47	2.65
15:00	11	26	32	0.69	1.61	1.97
16:00	13	29	35	0.82	1.78	2.15
17:00	20	33	34	1.20	2.01	2.09
18:00	17	31	32	1.06	1.88	1.94
19:00	12	24	28	0.75	1.45	1.70
20:00	11	22	26	0.65	1.33	1.62
21:00	5	18	22	0.33	1.09	1.32
22:00	8	19	19	0.51	1.17	1.17

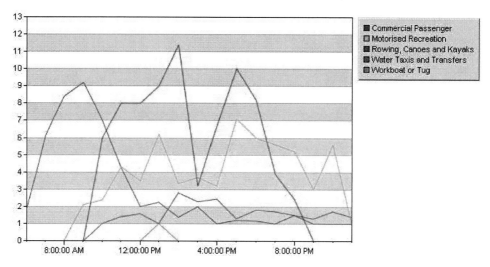

Figure 12. 2005 hourly traffic for Commercial Tourist Zone by Travel Mode

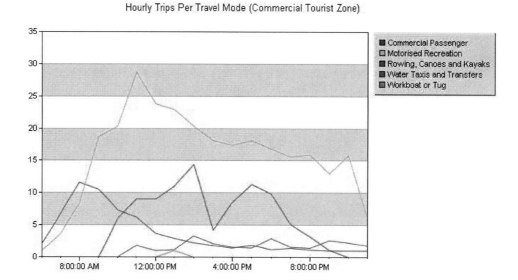

Figure 13. 2010 hourly traffic for Commercial Tourist Zone by Travel Mode

COMMERCIAL TOURIST ZONE TRAFFIC MANAGEMENT ISSUES

Motorized recreation is the biggest challenge for traffic management in the Commercial Tourist Zone. The high variability in the skills and experience of recreational boaters, the narrowness of the channel, and the complex movement patterns of commercial passenger vessels transiting to and from berths increase the risk of near misses and collisions. Commercial operators in this zone reported that there would be a lowering in quality of service for passengers in the form of delays and waiting times and higher stress on captains of vessels due to the often-unpredictable behavior of recreational boaters if current peak period LSA levels are exceeded.

Options for management include improving user education, compliance with navigation and speed rules, or limiting motorized recreation traffic by restricting access during peak hours of commercial use. Other mechanisms include instituting a "no overtaking" rule in this zone. This would have the affect of generating single file traffic, minimising the number of manoeuvres that would generate cross traffic and near misses.

Given that the current peak periods exceed the LSA, river traffic in this zone needs to be monitored carefully over the next few years including regular consultation with key user groups. The monitoring is to be targeted at identifying key high risk behaviors. In consultation with the commercial operators and other peak user groups develop vessel operating and zoning rules to reduce the likelihood and consequences of an incident occurring.

COMMENTS ON LSA FOCUS GROUPS

Since the methods used in this study are novel, it is worth making comments on the effectiveness of using user focus groups for evaluating LSA levels for river traffic. The major points are listed below:

LSA framework is easy for users to relate to.

Plan view-images of LSA levels overlaid on aerial photos is an effective method for eliciting responses from river users.

User's were able to make quick, spontaneous judgments of peak period LSA levels and were able to come to consensus quickly after brief discussion.

Users were similarly able to make judgments of LSA levels for other users

Users were able to make judgments relating to maximum tolerable LSA levels for their own use and the use levels for other users.

Discussion of LSA levels generated discussions relating to safety, quality of service and the nature of conflicts between different users. This information when linked to the LSA levels helps define the rationale for defining river capacity level for each group and between groups. This information is

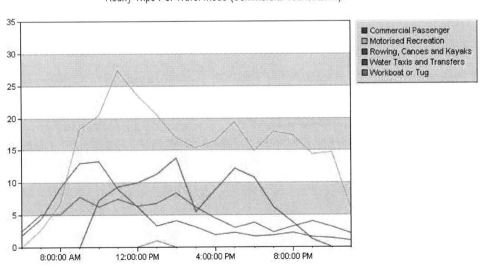

Figure 14. 2015 hourly traffic for Commercial Tourist Zone by Travel Mode

supported by information from face to face interviews.

The LSA framework is a valid method of determining River Traffic Capacity from a user's perspective.

USER BASED LSA RATINGS COMPARED TO SIMULATED LSA

Another aspect of the river management study methodology is the duo use of user-based LSA evaluations and computer simulated river traffic densities to inform management decisions. Ideally there should be good coincidence between these two evaluations, however there are differences, which are discussed here.

Generally there is good coincidence between user evaluations of river traffic densities at the lowest end of the scale (LSA A 0-1 boats per hectare), however as densities increase the user estimates of density are generally higher than the simulated densities. This discrepancy can be caused by the following factors:

- Simulated traffic volumes are under-estimated
- User estimates of traffic volumes are over-estimated
- User estimates are based on key locations and times within the zone when traffic is especially dense.

In reality the discrepancy is most likely due to a combination of the above factors. The simulation estimates of density are based on 4 sample counts per hour. These counts are then averaged and divided into the hectares for the river zone. This means traffic is averaged over time and space. The implications of this is that if the heaviest traffic occurs, say early in the hour, then this peak density is averaged with the remaining three counts for the hour which would lower the hourly density for the zone. Also if use is concentrated in a specific location in the zone, say at commercial berths or at boathouse launch sites, the high densities at these locations are averaged across the entire zone.

When users were asked to evaluate LSA levels they were instructed to give ratings during their "peak period of use". This would naturally prompt them to imagine the "worse case scenario" or the busiest times. For rowers on the Yarra in the Active Recreation Zone, for instance this would be during the early morning training periods when school rowers and elite rowers are all launching at the same time. For commercial operators in the Commercial zone, this would likely be at commercial berths. This would mean that users over estimated average densities for the zone.

The exact reasons for the discrepancy would have to be determined through follow up observations during busy periods. However the above explanation is probably close to the mark. The issue is what impact this discrepancy has on traffic management decisions. First it should be noted that the user LSA evaluations are seen as relative to the current situation. In other words, they are making judgments about traffic densities relative to the quality of service they desire. In this context the user LSA ratings can be used as a benchmark against current conditions. The likely impact of increases in traffic on users can then be estimated based on their assessment of current LSA levels and peak periods and the maximum tolerable LSA for the river zone. The simulation results may then be interpreted against these relative evaluations. The issues raised by users relating to boating safety and impacts on quality of service still provide valuable guidance to management decision-making, even if the exact evaluations of traffic density are in-accurate.

CONCLUSIONS

The LSA framework has proven to be a robust way of obtaining user-based definitions of river capacity, quality of service and the consequences of increasing traffic. The quantitative definition of LSA developed for river traffic was useful in interpreting the results of river traffic simulation. The use of focus groups to illicit current and maximum acceptable LSA levels for mixed traffic was efficient, easy to employ and stimulated positive and constructive feedback on the current conditions, desired conditions and the consequences of exceeding desired LSA levels.

The LSA framework will be developed further in other environmental contexts and other recreational settings and should be applicable across a broad range situations where there are competing demands for limited environmental resources.

Further research needs to be done to get better coincidence between LSA levels identified in workshops and traffic densities generated in simulations. However the flexibility of the framework for handling mixed and competing uses in a wide range of environments supports its further refinement and development.

ACKNOWLEDGEMENTS

I would like to thank Parks Victoria and especially David Ritman and Paul Dartnell who provided intelligent guidance and feedback throughout the project. Lindsay Smith and Glen Maclaren with Ennoscapes and Environmental Systems Solutions respectively for their assistance in data collection, analysis and visualization for the project. Colin Arrowsmith , Department of GeoSciences at Royal Melbourne Institute of Technology for data collection and analysis of river traffic monitoring data, Simon McGuiness of RM Consulting Group who acted as facilitator at the LSA focus group meetings.

LITERATURE CITED

Cole, David N., comp. 2005. *Computer simulation modeling of recreation use: current status, case studies, and future directions*. Gen. Tech. Rep. RMRS-GTR-143. Fort Collins, CO: U.S. Department of Agriculture, Forest Service, Rocky Mountain Research Station.

Itami, R., R. Raulings, G. MacLaren, K. Hirst, R. Gimblett, D. Zanon, P. Chladek 2003. *RBSim 2: Simulating the complex interactions between human movement and the outdoor recreation environment*. In Journal of Nature Conservation, Vol 11 pp278-286, Urban and Fischer.

Transportation Research Board 2000. *Highway Capacity Manual 2000*. National Research Council, National Academy of Sciences, Washington D.C.

CHAPTER 19

MONITORING AND SIMULATING RECREATION AND SUBSISTENCE USE IN PRINCE WILLIAM SOUND, ALASKA

Phillip Wolfe
Randy Gimblett
Laura Kennedy
Robert Itami
Brian Garber-Yonts

Abstract: This chapter outlines methods and results of a that study that employs survey and simulation data to reveal patterns in the spatial and temporal distribution of visitors across the Prince William Sound (PWS), Alaska. This study employs simulation to analyze the potential interactions between humans and wildlife directly relates to the mission of the recovery of the Sound from the Exxon Valdez Oil Spill. Five species were analyzed (Bald Eagles, Black Oyster Catchers, Harbor Seals, Cutthroat Trout & Pigeon Guillemot) to determine the interaction of recreational activities on known nesting sites of these species. To evaluate potential impacts, the number of visits and nesting sites per acre, duration of visit and the type of travel mode coinciding within these areas by season were combined to evaluate the potential threat or impact that is occurring in the Sound. This speculative analysis is provided in this paper.

Keywords: Computer Simulation, Recreation Behavior, Spatial and Temporal Patterns, Human/Wildlife Interactions

INTRODUCTION

PWS is a large marine environment characterized by its remoteness and pristine beauty that gives recreational users a sense of wilderness and the opportunity to participate in many unique activities. However, recreation in Alaska is expected to increase substantially over the next 15 years (Bowker 2001), and the effects this may have on the Sound are not yet known.

What is known however is that access to PWS changed considerably when the Whittier Tunnel opened to automobile traffic in the year 2000. This has allowed residents from Anchorage and its surrounding communities, to access the sound directly by car. Previously, visitors were required to board a train to reach Whittier. Traffic counts maintained by the Alaska Department of Transportation (ADOT) indicate a 48% increase in automobile traffic from 2001 to 2005. Furthermore, since ADOT tracks use of the Whittier tunnel based on various vehicle classes, we are able to estimate the number of vehicles that pass through the tunnel that are likely associated with recreation activities. Class B1 is defined as "Passenger vehicles pulling trailers, except as otherwise specified." During the five-month summer recreation season the number of vehicles in this class passing through the tunnel increased from 4,133[1*] to 8,483. This is an increase of over 200%. It is unclear if use of the Whittier tunnel will continue to increase, however this change in the access conditions to the Sound has had substantial impacts on the ability of recreation users to access the Sound and is one of the impetuses for the PWSHUS.

The second notable change occurred in 2006, one year after data for this study were collected, when the Alaska Marine Highway Department based one of its high speed ferries in Cordova. This marked the first time daily trips were scheduled to both Whittier and Valdez. Prior to this, ferry service to and from Cordova was limited to a few times a week on a conventional ferry. A high-speed ferry is capable of sailing the routes in half the time of a conventional ferry. The reduced travel times and daily service are expected increase the number of people visiting Cordova. Additionally outdoor recreation by Alaskan residents is expected to increase by 30% from 2000 to 2020 (Bowker, 2001).

When the *Exxon Valdez* (2004) oil tanker grounded on Bligh Reef on March 24th 1989 and proceeded to spill approximately 11 million gallons of Prudhoe Bay oil, the ecosystem of PWS was altered in numerous ways. Nearly two decades later the impacts of this ecological disaster are still evident and continue to be studied. As a result of a $900 million civil settlement with Exxon, the Exxon Valdez Oil Spill Trustee Council (EVOS) was established. Their goal is to oversee restoration of the ecosystem. They recognize a number of species that were negatively impacted by the spill. Ongoing monitoring of the species has occurred in order to determine recovery rates. EVOS' annual reports categorize affected species into four groups; Recovered, Recovering, Not Recovered and Recovery Unknown. They note that the recovery of certain species is important to both the commercial fishing industry and Alaskan residents who depend on these species for subsistence purposes (EVOS 2004). This study focuses on the proximity of human use to five species; bald eagle (Haliaeetus leucocephalus) nests sites, Black Oystercatcher (Haematopus bachmani) nest territories, harbor seal (Phoca vitulina) haul-out sites, Pigeon Guillemot (Cepphus columba) nests sites and cutthroat trout (Oncorhynchus carki) streams.

Critical to this study is the use of simulation to evaluate mode of travel, destination, duration and frequency of visit, which are useful in the study of human impacts on wildlife (Cole 2004) and the impacts that human activities have on wildlife and their associated habitats (Steidl & Powell 2006). Cole (2004), Blahna & McCool (2004) and Manning & Lime (2000) concur that recreation use is dispersed unevenly across the landscape. These types of dispersed use patterns make it difficult to uniformly manage for visitors. However, it does emphasize the need to more effectively collect spatial and temporal visitor data to empirically derive distributions across a landscape. While much is still not known about these patterns of use there is an extensive body of research on the biophysical impacts by visitors to sites. Manning and Lime (2000) suggest "that increasing wilderness use

---

[1*] Prior to June 2002 ADOT did not track use of the in the B1 category. . As a result we can only calculate vehicle numbers for June through September for 2002 and May through September for 2003 through 2005. This is not expected to have a substantial impact on the calculations and results presented in this section.

can and often does increase impacts in the form of damage to fragile soils and vegetation, and crowding and conflicting uses." However, the critical point at which a threshold is reached between visitor use levels and biophysical impact is not known. Cole (1993) and Blahna & McCool (2004) have both illustrated that the relationship is not linear. In fact, what they have shown is that biophysical impact occurs at a much higher rate as visitor use levels increase and at some point, use levels taper off showing very little increase in impact. Cole (2004) suggests that a "relatively infrequent and small amounts of use can cause substantial impacts." Hammitt and Cole (1998) suggest that at "low frequencies, small differences in use frequency can result in substantial differences in the amount of impact. At high use frequencies, even large differences in use frequency typically results in minor differences in impact."

This study is the first of many to determine appropriate and acceptable levels of visitor use in PWS, with the ultimate goal of maintaining wilderness quality and ecological protection throughout the Sound. The identification of essential factors to monitor is critical to provide necessary information for managing quality wilderness experiences. As a beginning point, it is important to consider the work by Cole (2004) who has concluded that to *"manage visitor use such that biophysical impacts are minimized, managers must attempt to minimize both the area of impact and the intensity of impact per unit area."* He goes on to say that the *"primary factors that influence intensity of impact are frequency of use, type and behavior of use, season of use and environmental conditions. The area of impact is primarily a result of the spatial distribution of use."* Cole (2004), Gregorie and Buhyoff (1999) suggest that any type of sampling should be designed to capture the amount, types and intensity of the human activity as well as how the activity varies spatially and temporally. Survey research and computer simulation can be used to derive these important parameters (Gimblett et al. 2007; Itami et al. 2003; Lace et al. 2007; McVetty et al. 2007).

OBJECTIVES

This study asks two broad questions: First, is it possible to determine the types, distribution and seasonal variation of human use in the Sound. This question provides a seasonal (Early, Core, Late) evaluation of where visitors are dispersing across the Sound. While the total number of trips has been shown to have some correlation to impact, this study evaluates the duration of trips, the total number of trips and the travel mode. Second the potential impacts of recreation users on identified wildlife sites be evaluated and quantified? The analysis of the potential interactions between humans and wildlife directly relates to the mission of the recovery of the Sound from the Exxon Valdez Oil Spill. Four species were analyzed (Bald Eagles, Black Oyster Catchers, Harbor Seals, Cutthroat Trout & Pigeon Guillemot) to determine the interaction of recreational activities on known nesting sites of these species. Of particular interest is not only the overlap between certain modes of travel and wildlife sites, but the duration and frequency of human interaction in these areas. To evaluate potential impacts, the number of visits and nesting sites per acre, duration of visit and the type of travel mode coinciding within these areas by season were combined to evaluate the potential threat or impact that is occurring in the Sound. This speculative analysis is provided in this report.

METHODS FOR SAMPLING VISITORS AND COLLECTING DATA ABOUT VISITOR PATTERNS

Development of the survey instrument

began in May of 2004, which was created to capture the behavioral characteristics and spatial/temporal patterns of visitor use in the Sound. By September of the same year initial testing of the survey instrument was done in Whittier, Valdez and Cordova. Extensive communication with resource managers of the Chugach National Forest identified specific issues they would like the study to focus on, including the dispersal of human use throughout the Sound, encounter levels of visitors, the identification of areas where human/wildlife interactions occur, and the ability to identify locations where visitors access land. In March of 2005 the revised survey instrument was reviewed by several focus groups in Anchorage and Cordova. Final revisions to the survey instrument were completed in time for sampling to commence on May 1st 2005.

Three distinct recreation user groups were also identified during the scoping/pilot study—paddlers, cruisers and fishers/hunters. The paddling group used non-motorized boats, primarily sea kayaks, on the Sound. Both the cruisers and fishers/hunters used motorized boats and were distinguished by the vessels' ability to provide overnight accommodations. The cruiser group used sailboats and motorized boats with overnight facilities on-board. This group generally spent several days in the Sound per trip and typically participated in fishing as a secondary activity. The final group, fishers/hunters, included users whose primary purpose was to participate in fishing and/or hunting, although fishing is the dominant activity of the two. Fishers/hunters' vessels generally did not include overnight accommodations. The trip length for this group varied dramatically. A majority of the trips were assumed to be short day trips, however some groups spend several days either fishing and/or hunting in the Sound.

The survey instrument consisted of two main portions: standard text based questions which addressed the behavior attributes of the users including the location, time of entry and exit, duration of visit, number in party and other contextual questions, as well as a diary/trip itinerary portion which used a map to focus on the specific trip taken by each user. Three survey instruments were used for each of the three user groups. The map based portion of the survey included a 17" by 22" map of Prince William Sound on which users were asked to document their trip by tracing the route of their travel in the Sound. They were also asked to number, record the date & time of arrival, the length of stay, and list the activities in which they participated at each place they stopped along their travel route. Additionally, they were asked to identify two of their sites as primary sites (P1 and P2) about which they were asked subsequent questions in the remaining text portion of the survey.

The sample design was randomized and stratified by port, day of week, and time of day. The three different sampling periods were morning (6:00-11:30), midday (11:30-17:00) and evening (17:00-22:30). These 5.5 hour periods were selected to coincide with the tunnel schedule for automotive access to Whittier. Further modifications of the sample design were necessary for the ports of Cordova and Valdez to coincide with the ferry schedule for transporting the survey crew covering these two ports. Sampling occurred between May 1st and September 30th of 2005. Data collection took place for 60 days in Whittier, 42 days in Valdez, and 48 days in Cordova for a total of 150 days of sampling. Surveys were distributed through intercepts of the next available and willing recreation user. Brief interviews were conducted in order to identify each potential respondent's user group, as well as allow the potential respondents to ask questions about the study. Supplemental information about the respondents was collected during the interview (detailed below). During the interview, survey packets were distributed to willing participants. Survey packets consisted of the survey, a prepaid and preaddressed envelope, and a postcard for entry into a prize drawing. A cover letter was also included in the survey packet that outlined information about the study as well as provided necessary information regarding participants' rights. A reminder postcard was sent two-weeks after distribution of the surveys.

Two two-person crews distributed the surveys. One crew was based at the Glacier Ranger District office in Girdwood, Alaska, and was responsible for collecting data on users accessing the Sound through Whittier. The second crew was based at the Cordova Ranger district in Cordova, Alaska. This crew was responsible for collecting data on users accessing the Sound through Cordova and Valdez. Due to staffing limitations the first 6 weeks and the final 4 weeks of the season only allowed for one member of each team to administer surveys.

Survey distribution and visitor interviews took place at each of the three harbors. Each member of the crew circulated independently throughout the harbor, along the floats, as well as around the boat ramp and staging areas. Brief interviews were conducted in order to identify each potential respondent's user group, as well as allow the potential respondents to ask questions about the study. A survey was distributed per vessel with instructions that the survey was to be completed by the member of the party with the earliest birthday in the year. The instrument consisted of the survey, return addressed stamped envelope as well as an enrollment postcard for a prize drawing used as a participation incentive. This entire survey instrument was placed in a ziplock bag for protection from water and other elements.

In addition to distributing the survey, the crews collected and recorded several pieces of information from the intercepted individuals, including name, address, primary purpose of trip (cruising, paddling, hunting/fishing), vessel type, anticipated duration of trip, and whether respondents agreed or refused to accept and complete a survey. As much information as could be observed was collected and recorded from those who refused a survey.

In October 2005 the data entry portion of the project commenced. Two separate databases were created based on the survey instrument. A web-based interface was created for the entry of the map data. This was done to build trip itineraries for the simulation. The spatial database was completed by the middle of December 2005. A second database for the text based portion of the survey was developed, using Microsoft Access. Data entry of the text questions was completed in January 2006.

This project uses the Recreation Behavior Simulator (RBSim), a computer simulation tool first developed in 1996 (Gimblett 1998, Itami & Gimblett 1997). It was designed as a general recreation management evaluation tool for linear recreation systems. RBSim uses concepts from recreation research and artificial intelligence (AI) and combines them in a GIS to produce an integrated system for exploring the complex interactions between humans (recreation groups) and the environment (geographic space). Survey data combined with simulation was used in this study to address the two broad questions: Is it possible to determine the types, distribution and seasonal variation of human use in the Sound and can simulation be successfully used to evaluate potential impacts of recreation users on identified wildlife sites? Simulation provides the unique ability to provide information on important indicator variables as outputs used to addresses both of these questions. They are a season spatial and temporal dispersal use patterns, mode of travel, duration, intensity and frequency of human interaction in these areas.

In order to construct a simulation, a travel network including trip destinations and trip itineraries needed to be constructed. Figure 1 illustrates all trip routes that were collected from the survey diaries.

Trip diaries from a total of n=703 trips were digitized and shown in Figures 2 & 3. It is clear that there are many overlapping travel routes that would be impossible to use as part of the simulation. As a result, travel routes were aggregated to reflect the major patterns of movement across the Sound. Figure 2 illustrates the final travel network derived from all n=703 travel diaries. This aggregate process resulted in n=104 links that represent the simulation network which was used to summarize link encounters and visitor use levels across the Sound.

The next step was to spatially represent all the possible stops and destinations that were recorded in the trip diaries. A total of n=4125 points were entered and mapped across the Sound. Figure xx provides an illustration of all trip destinations. It is clear from Figure xx that the distribution of trips across the Sound is not homogeneous. Visitors distribute themselves in a non-uniform way, suggesting that some locations or areas are more heavily used then others. Figure 4 shows the relationship of the n=4125 destinations to the generalized travel network. It is apparent that to insert all n=4125 points on the network would be a nightmare and result in a dramatic increase in processing in simulation runtime. So there needs to be a more efficient method to link the destinations to the network for the simulation.

Figure 1. Travel Routes from Survey Diaries

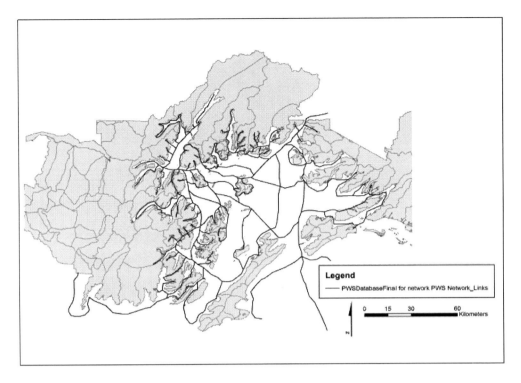

Figure 2. Final Aggregated Travel Routes from Survey Diaries

The Chugach National Forest (CNF) manages for a wide range of diverse, quality, recreational opportunities including the need to better disperse recreational use in response to increased user demands. In order to accomplish this management strategy, CNF employed the concept of a capacity analysis where areas were derived using watershed associations to summarize recreation activity on CNF lands. While this has been completed for the Western Sound, work is still underway to delineate capacity areas for the Eastern Sound. Since a significant portion of recreation use is water-based, with some interaction on the land, it is imperative that the capacity area analyses take both into account.

In addition, as recreation use becomes more dispersed across the Sound, the capacity analysis needs to more clearly reflect the diversity and intensity of visitor use. Given that the Eastern Sound had not yet completed their capacity analysis, an alternate strategy was required to spatially divide the Sound into more concentrated management zones that more accurately reflect the current and projected recreational use levels, intensities, durations of visit, travel modes and interactions with wildlife, such as suggested by Cole (2004) and Steidl and Powell (2006). Since this study was deployed to capture the spatial and temporal patterns of use, it was only logical to evaluate the data collected and try to determine appropriate capacity areas.

With this in mind a spatial clustering technique was applied to the data to aggregate these destinations into similar groups and physical features such as bays and estuaries that are commonly used throughout the Sound. Figure 5 shows that these destinations can be aggregated into areas that can ultimately serve as a beginning point for visitor management and open up discussions and future studies related to establishing realistic capacities. These areas

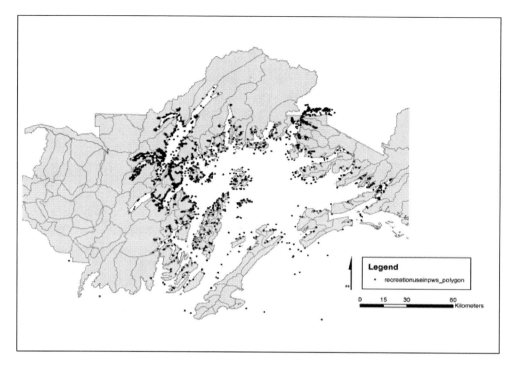

Figure 3. Reported Trip Destinations from Travel Diary Data

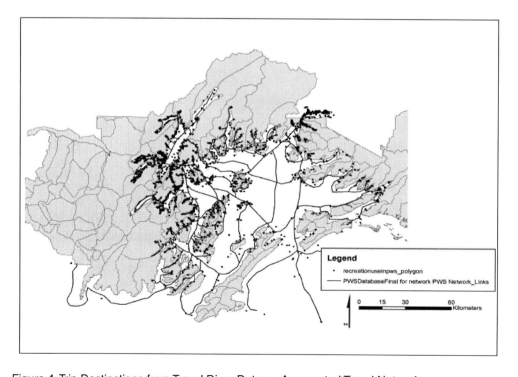

Figure 4. Trip Destinations from Travel Diary Data on Aggregated Travel Network

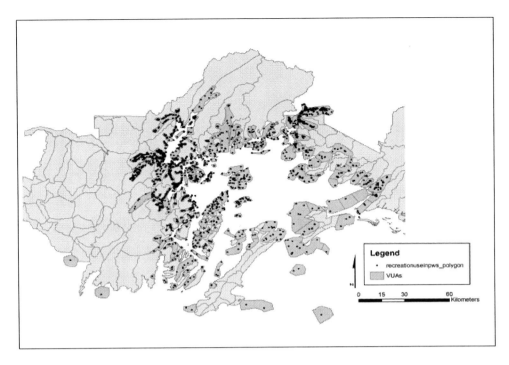

Figure 5. Clusters of Reported Trip Destinations Used to Derive Visitor Use Areas

can be disaggregated and then re-aggregated in the future into more realistic management zones. Figure 6 provide the mapped result of this analysis resulting in the creation of 152 Visitor Use Areas (VUAs). Much of the analysis in the second half of this report is done using these VUAs.

To simplify the number of destination points on the network, a single point was associated with each of the VUAs where the travel route dissected the VUA polygon. Figure 7 shows the relationship between the VUA points and the travel route. It is important not to lose the integrity of the original n=4125 destination survey points when using this aggregated network for the final analysis and results. Since this is the most up to date information about recreation visitation in the Sound, an identity function was used in ArcMap to relate the original destination points (those that fell within each of the VUA boundaries) to one of the n=152 points that are associated with each polygon and are on the topological network used in the simulation. Queries from the simulation database were done to summarize all n=4125 points to answer the variety of spatial and temporal research questions outlined earlier.

Trip itineraries were constructed from survey data once the travel network was condensed to represent all possible travel patterns and all original destinations from the travel diaries. Each trip itinerary has a travel mode, activity type, start and end data, travel destinations, trip durations, stay durations and number of visitors in the party. This information was subsequently used to construct each trip itinerary that was scheduled and run in the simulation. All spatial and temporal data were output from the simulation runs and queries were built to extract and analyze the data.

WHAT ARE THE SURVEY DISTRIBUTION AND RESPONSE RATES IN PWS?

Surveys were distributed from May 1

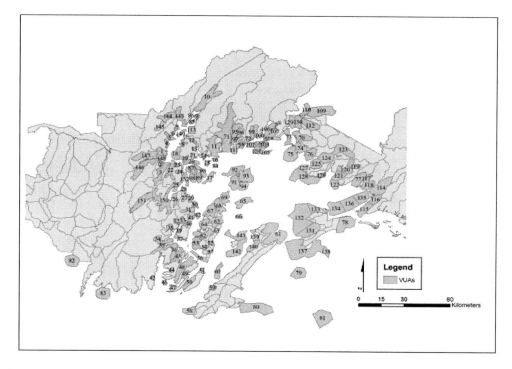

Figure 6. Aggregation of Reported Trip Destinations to Construct Visitor Use Areas

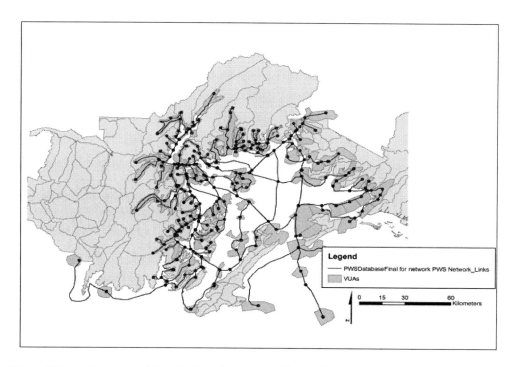

Figure 7. Travel Routes and Destinations Constructed for the Simulation

Table 1. Names of the 20 most "*prominent*" VUAs.

VUA	Name		
15	Dutch Group	116	Cordova
20	Fool Island	120	Parshas Bay
22	Culross Passage	126	Hells Hole
39	Dangerous Passage	134	Hawkins Island Cutoff
44	Hogg Bay	139	stockdale Harbor
50	Horseshoe Bay	143	Barry Arm
59	Hanning Bay	144	Harriman Fjord
66	Seal island	146	Blackstone Bay
78	Boswell Bay	147	Passage Canal
109	Port Valdez	150	Greystone Bay

through September 30 to coincide with the primary visitor season in Alaska. Sampling took place on a total of 150 days (60 in Whittier, 42 in Valdez and 48 days in Cordova). 2085 surveys were distributed during this time period. A total of 668 usable surveys were returned for an overall return rate of 32%. It should be noted that the traffic counts included vessels that were entering/exiting the harbor on short non-recreational trips (e.g. to check engines or to use the fuel dock).

Of particular note is the high response rate among paddlers relative to the other user groups (60% compared to 36-37%). Variation in response rate by location is also notable: paddlers responded at a much lower rate in Valdez (38%) than in Whittier (69%) and lower relative to hunter/fishers in Valdez (42%). This discrepancy is likely due to a larger fraction of paddling trips generated from Whittier by independent users rather than those using outfitter/guide services, and the relatively poor cooperation with the survey crew from the paddling outfitter services in Valdez.

Surveys returned from paddlers out of Cordova were not statistically representative of the population due to the poor return rate in this group (14% of 1 returned for every 7 distributed). This would be of greater concern if there was actually a significant level of paddling use generated from Cordova. Of greater concern is the low (24%) response rate for hunter/fisher surveys as this is currently the largest use category in Cordova. Further efforts to explain the low response rate are warranted.

SPATIAL AND TEMPORAL DISPERSAL OF VISITOR USE

The expectation that the level of use varies throughout the season was examined by breaking down the sampling period by month as well as into the early, core and late seasons. The following analysis in this report was undertaken for early, core and late seasons. However for purposes of this chapter all figures and tables will reflect on the core season. The section below reports the use, both in number of trips and number of stops, which occurred during the core season.

Table 2 provides an illustration of the dispersion of trips throughout the Sound by season. It is clear that a majority of the use (65%), measured in number of trips, did occur during the "Core Use Season" (June 15th through August 15th). Use in shoulder seasons dropped off considerably to 23% and below. When looking at use on a monthly basis, July appears to experience the most use with 31%. September saw the lowest use levels with only 5% of the trips.

WHAT IS THE TOTAL AMOUNT OF USE AT DESTINATIONS ACROSS THE SOUND?

Results of the simulation figure 9 provide a spatial view of the percentages of where the highest percentages of total trips actually went. VUA 147 (Passage Canal) had the

Table 2. Total trips, stops and VUAs visited throughout the sampling period

Season/Month	# of Trips	# of Stops	# of VUA's	Percent of trips
Entire Season	729	3989	149*	100%
May1- June 14	166	849	111	23%
June 15- Aug.15	475	2662	145	65%
Aug.16- Sep. 15	73	352	53	10%
May	109	564	84	15%
June	145	873	125	20%
July	227	1277	128	31%
August	193	983	97	26%
September	37	175	31	5%

* Three use areas were not visited- Please refer to section XX for a detailed description.

highest percentage of total trips with 29.3% followed by VUA 109 (Port Valdez) with 12.5% and VUA 22 (Culross Passage) with 5.3%. In addition, the question of where visitors stopped is important to understand as is the length of stay at each site. As the simulation data reveals, VUA 147 (Passage Canal) had 27.5% of total stops, VUA 109 (Port Valdez) with 12.9% and VUA 150 (Greystone Bay) with 3.8%.

WHAT IS THE SEASONAL (MONTHLY) DISTRIBUTION OF TRIPS ACROSS THE SOUND?

The percentage of total trips from the simulation output as expected is not evenly distributed. Over 66% of the use occurs during the Core season (from June 15 - August 15). An additional 24% can be found occurring in the Early season (from May 1 – June 14), followed by only 10% in the Late season (from August 16 – Sept 15). It is apparent from this analysis that during the early season the trips congregate closer to the major entry areas into the Sound (ie. Whittier, Valdez and Cordova). During the core season the use if more widely distributed across the sound. Passage Canal, Greystone, Culross passage and Blackstone notably with heavier use.

WHAT IS THE TOTAL NUMBER OF VISITS PER VISITOR USE AREA?

The first method of examining the dispersal of the recreational use within the Sound is by looking at the number and percentage of trips which occupied each Visitor Use Area. It should be noted that Visitor Use Areas147 and 109 correspond to Visitor Use Areas surrounding Whittier and Valdez respectively. These harbors were also the locations where most trips (466 from Whittier and 204 from Valdez) either originated or ended. (See Tables 3 & 4). This will be discussed in greater detail later on in the paper.

WHAT IS THE FREQUENCY OF USE AT THE VISITOR USE AREAS THROUGH THE SOUND?

As with the number of trips entering each Visitor Use Area, the numbers of stops are skewed due to the fact that each trip was given a stop for the beginning and end of trip. Thus we again find the number of stops in Visitor Use Areas 147 and 109 inflated due to the harbors being located within.

WHAT IS THE AVERAGE DURATION OF STAY PER VISITOR USE AREA?

A third way to look at the overall dispersal throughout the Sound would be to examine the amount of time spent in each Use Area (See Table 5).

Table 5 is similar to the previous two tables; however VUA 146 ranks higher with a higher cumulative number of hours

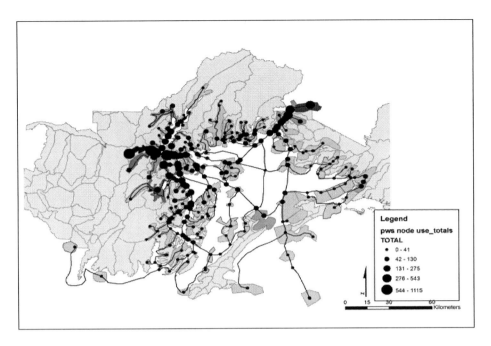

Figure 8. PWS Total Use at Destinations in the Sound

spent in it. This could be a result of several different factors, including a variety of different attraction points or campsites/mooring sites within this Visitor Use Area. (See Figures 9 & 10)

WHAT ARE THE INTENSITY LEVELS (DURATION + TOTAL USE) AT EACH VUA ACROSS THE SOUND?

Intensity is a measure derived to capturing both how long visitors are staying in particular locations but also the numbers of visits that are occurring. The hypothesis is that the greater number of visitors and the longer they stay the more potential for impact that will occur. So this measure simply indicates the intensity that humans focus in on certain visitor use areas in the Sound. Figure 11 provides a seasonal evaluation of levels of intensity in the Sound.

As one might expect, intensity levels change through the Sound by season. The western sound has greater levels of intensity during the early and core season which taper off in the late season. The eastern sound has a higher level of intensity during the late season and much less in the previous seasons. This can be explained by the heavy emphasis on fall hunting which brings a significant amount of use to the eastern sound area

WHERE IS THE GREATEST POTENTIAL FOR HUMAN IMPACT ON WILDLIFE TO OCCUR IN THE SOUND?

To evaluate the greatest potential for human impact on wildlife to occur in the Sound, an analysis was undertaken to examine the overlap between known locations of nesting sites and intensity of use of the sites from humans. This analysis at best is speculative as the inventory of nesting sites was undertaken in years prior to this study so the two data sets would be impossible to compare. However this analysis does provide a view of what interaction is possible and where further studies should focus to examine human/wildlife interactions.

For this analysis, data that was collected

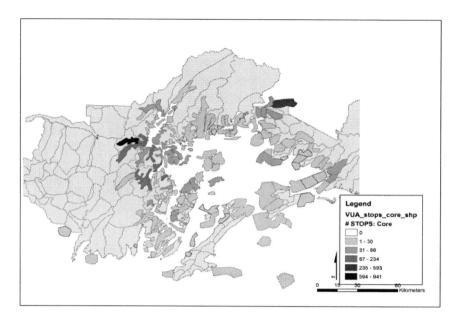

Figure 9. PWS Total Number of Stops during Core Season for Visitor Use Areas

prior to this study was used. This data was available for bald eagles, black oyster catchers, harbor seals, cutthroat trout and pigeon guillemot. In order to associate the impact which recreation use may have on wildlife sites a measure needed to be created. As a result the number of wildlife sites and the number of recreation based stops occurring in a Visitor Use Area was calculated on a per acre basis. The following figures and tables illustrate Visitor Use Areas containing the greatest number of wildlife sites on a per acre basis.

Bald eagle: The three areas with the highest density of Bald eagle nests also had the lowest density of recreational visitor stops per acre. VUA 120 (Parshas Bay) had the highest density with 3.9 bald eagle nests per acre, which corresponded to the lowest density, 0.5 stops per acre, by recreational visitors. This suggests there could be a negative relationship between eagle nesting sites and visitor use, but further studies need to be done to quantify this relationship (See Figure 12 & Table 6).

Table 3. Number and percentage of trips in each VUA Area

Rank	VUA	# of Trips	%
1	147	481	20.12%
2	109	212	8.87%
3	22	85	3.55%
4	25	64	2.68%
5	146	61	2.55%
6	148	60	2.51%
7	150	55	2.30%
8	1	52	2.17%
9	112	45	1.88%
10	116	45	1.88%

Table 4. Top Ten UA based on number of stops

Rank	VUA	# of Stops	%
1	147	1097	27.50%
2	109	516	12.94%
3	150	151	3.79%
4	22	130	3.26%
5	146	105	2.63%
6	25	94	2.36%
7	1	80	2.01%
8	116	79	1.98%
9	148	72	1.80%
10	112	61	1.53%

Table 5. Top Ten UA based on cumulative duration of stops

Rank	VUA	Hrs	%
1	147	6593	28.09%
2	146	3747	15.97%
3	109	2922	12.45%
4	150	1028	4.38%
5	49	639	2.72%
6	25	584	2.49%
7	22	562	2.39%
8	3	430	1.83%
9	143	413	1.76%
10	116	339	1.44%

Black oystercatcher: VUA 15 (Dutch Group) had the highest density of Black oystercatcher nests per acre (2.5 nests per acre) and had a recreational visitor density of 3.4 stops per acre, which was the second highest density of recreation use. VUAs 59 and 139 (Hanning Bay and Stockdale Harbor) both had 1.7 nests per acre and had very low densities of recreation use (0.6 and 0.5 stops per acre, respectively).

This overall trend seems surprising, since we would expect human presence to negatively impact the nesting behavior of these birds. However, this association could show that recreators are targeting areas known to have high densities of Black oystercatchers and that the impact of increased stops per acre is not yet high enough to have a negative impact. This is speculative, however, and further studies need (See Table 7).

Harbor seal: VUAs 39, 50, 44 (Dangerous Passage, Horseshoe Bay and Hogg Bay) had the highest densities of harbor seal haul out sites with 0.9, 0.6 and 0.5 sites per acre, respectively. There was no obvious relationship between haul out sites and density of recreational use (See Table 8).

Cut throat trout: VUAs 126, 134 and 78 (Hells Hole, Hawkins Island Cutoff and Boswell Bay) had the highest densities of cutthroat trout streams with 2.3, 1.9 and 1.0 streams per acre, respectively. There was no obvious relationship between

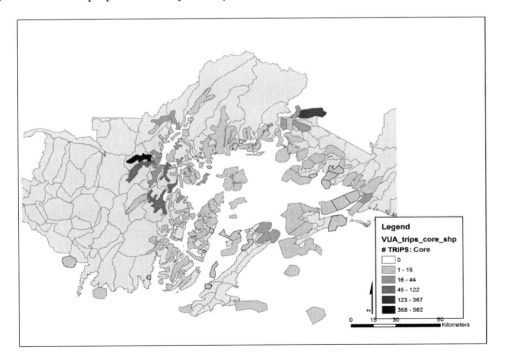

Figure 10. Total Number of hours (Duration) spent during Core Season in VUAs

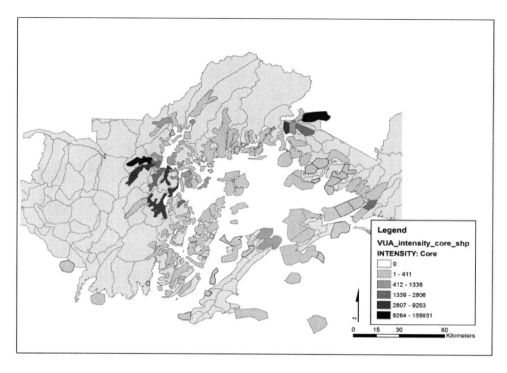

Figure 11. PWS Human Use Intensity at VUAs in the Core Season

stream density and density of recreational use (See Table 9).

Pigeon guillemot: VUAs 66, 20 and 15 (Seal Island, Fool Island and Dutch Group) had the highest densities of Pigeon guillemot nests per acre with 1.1, 1.1 and 0.8 nests per acre respectively. Recreation use was not extremely high or low in these three areas, which suggests that these data shows no relationship between nest sites and presence of recreators (See Table 10). Insert Table 10

CONCLUSIONS

The goal of this study was to use survey and simulation data to reveal patterns in the spatial and temporal distribution of visitors across the Prince William Sound (PWS). Critical to this study is the use of simulation to evaluate mode of travel, destination, duration and frequency of visit, which are useful in the study of human impacts on social and ecological systems. While these types of dispersed use patterns make it difficult to uniformly manage for visitors, it does emphasize the need to more effectively collect spatial and temporal visitor data to empirically derive distributions across a landscape.

This study is the first of many to determine appropriate and acceptable levels of visitor use in PWS, with the ultimate goal of maintaining wilderness quality and ecological protection throughout the Sound. This study was undertaken using two methodologies; survey and simulation. These two methods combined provide the Chugach National Forest with the best possible knowledge of the spatial and temporal distribution of recreational visitors in Prince William Sound.

Surveys were distributed in Whittier, Cordova and Valdez from May 1 – September 30, 2005, to coincide with the primary visitor season in Alaska. A total of 2,085 surveys were distributed during the

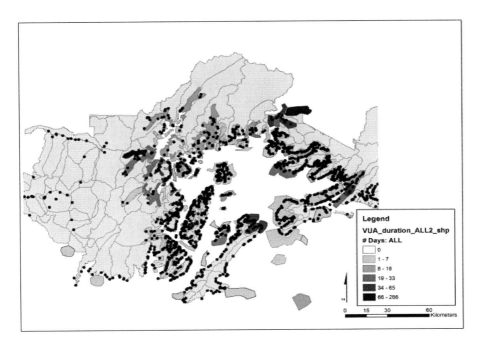

Figure 12. PWS Bald Eagle Nest Site and Visitor Durations at VUAs

Table 6 - An Evaluation of Bald Eagle Nest Sites in PWS

Use Area	Stop Per Acre	Bald Eagle Nest Sites Per Acre	May	June	July	August	September	Total Duration (hrs)
120	0.5	3.9	0	4	24	0	0	**28**
117	0.6	3.5	2	0	0	0	0	**2**
63	0.5	3.3	0	0	0	0	0	**0**
66	1.5	3.0	4	0	6	11	0	**21**
31	5.6	2.7	33	35	21	27	1	**117**
51	1.6	2.6	0	4	4	0	0	**8**
40	1.9	2.6	0	0	0	7	0	**7**
69	2.1	2.6	0	39	44	30	0	**113**
28	2.5	2.5	0	2	0	2	0	**4**
57	2.4	2.4	0	0	0	0	0	**0**

sampling period and 641 usable surveys were returned for an overall rate of return of 30.7%. Of the surveys returned over 67.7% came from Whittier; 25.4% from Valdez and 4.1% from Cordova. Of those, 67.6% consisted of hunters/fishers while 27.5% cruisers and 9.2% paddlers.

This study employs the concept of Visitor Use Areas (VUAs) which are more spatially explicitly derived management zones that more accurately reflect the current and projected recreational use levels, intensities, durations of visit, travel modes and interactions with wildlife. Since this study was deployed to capture the spatial and temporal patterns of use, it was only logical to evaluate the data collected and try to determine appropriate capacity areas. With this in mind a spatial clustering technique was applied to the data to aggregate these destinations into similar groups and physical features such as bays and estuaries that are commonly used throughout the Sound. Applying this pro-

Table 7 - An Evaluation of Black Oyster Catcher Nest Sites in PWS

Use Area	Stop Per Acre	BLOY Acre	May	June	July	August	September	Total
15	3.4	2.5	0	3	21	3	0	27
59	0.6	1.7	0	0	15	0	0	15
139	0.5	1.7	0	0	24	0	0	24
140	0.3	1.6	0	0	0	0	0	0
141	0.5	1.2	0	0	0	45	0	45
144	5.0	0.9	38	27	26	19	0	110
107	2.3	0.7	39	0	23	4	2	68
127	1.3	0.6	30	0	0	12	0	42
61	0.5	0.5	0	45	32	0	0	77
28	2.5	0.5	0	2	0	2	0	4

Table 8 - An Evaluation of Harbor Seals in PWS

Use Area	Stop Per Acre	Hseal Acres	May	June	July	August	September	Total Duration (hrs)
39	1.8	0.9	0	13	0	0	0	13
50	0.6	0.6	0	16	4	0	0	20
44	1.4	0.5	0	0	3	0	0	3
46	0.9	0.4	0	0	10	0	0	10
15	3.4	0.4	0	3	21	3	0	27
66	1.5	0.4	4	0	6	11	0	21
30	4.7	0.4	3	25	13	16	0	57
126	0.7	0.3	0	0	0	0	0	0
40	1.9	0.3	0	0	0	7	0	7
111	0.8	0.3	1	1	0	0	0	2

Table 9 - An Evaluation of Cutthroat Trout Streams in PWS

Use Area	Stop Per Acre	Cutthroat Trout Streams /Acre	May	June	July	August	September	Total Duration (hrs)
126	0.7	2.3	0	0	0	0	0	0
134	0.1	1.9	0	0	2	12	0	14
78	0.2	1.0	13	0	0	0	0	13
140	0.3	0.9	0	0	0	0	0	0
141	0.5	0.8	0	0	0	45	0	45
115	0.6	0.7	10	0	1	0	0	11
118	1.5	0.6	0	2	3	6	0	11
136	0.1	0.4	0	0	3	0	0	3
127	1.3	0.4	30	0	0	12	0	42
132	0.1	0.3	0	2	0	11	0	13

cess resulted in the discrimination of 152 Visitor Use Areas (VUAs).

Results of the survey conclude that of those Hunters/Fishers, that 26.88% - VUA 147, 14.49% - VAU 109 and another 3.5% at VUA 22. 11.04% of Cruisers visited VUA 147, while 10.37% - VUA 109 and 4.68% *at* VUA 116. As for Paddlers, 18.39% at VUA 147, 6.32% VUA 144, 5.75% - VUA 143 and 5.75 - VUA 146.

Over 92% identified the main purpose for use of the Sound as for recreation, 14% for subsistence use and 8% other. Of those that visited the Sound their major activity was Water based (74%) with 50% identifying fishing, 22% wildlife viewing, 2% pleasure boating. For those Land based (26%) activities, 14% day use, 10% camping/cabin use and 2% hunting

This study was the first of many to

Table 10 - An Evaluation of Pigeon Guillemot Nesting Sites in PWS

Use Area	Stop Per Acre	Pigeon Guillemot Sites Per Acre	May	June	July	August	September	Total
66	1.5	1.1	4	0	6	11	0	21
20	2.6	1.1	13	6	11	4	0	34
15	3.4	0.8	0	3	21	3	0	27
92	0.7	0.7	0	28	0	15	0	43
45	0.6	0.6	0	0	0	0	0	0
91	1.7	0.6	12	22	17	34	12	97
16	1.0	0.5	0	1	0	0	0	1
17	4.7	0.5	5	19	13	9	0	46
63	0.5	0.5	0	0	0	0	0	0
93	1.2	0.4	12	58	8	17	0	95

determine appropriate and acceptable levels of visitor use in PWS, with the ultimate goal of maintaining wilderness quality and ecological protection. The identification of essential factors to monitor is critical to provide necessary information for managing quality wilderness experiences. The simulation analysis in this study has been structured to provide answers to a number of questions. For example, what is the distribution of visitors currently using the Sound? This question provides a seasonal Early (from May 1 – June 14), Core (from June 15 – August 15) and Late (from August 16 – Sept 15) evaluation of where visitors are dispersing across the Sound.

To properly manage for visitor use in the Sound, it was imperative to understand the season of use or more generally when does the highest amount of visitor use occur. Over 66% of the use occurs during the Core season, 22% in the early season and only 8% in the late season. Additionally, how does the spatial distribution of use change across seasons? It is clear from the simulation data that the two areas that amassed the highest total stops were constant across all seasons (Passage Canal and Port Valdez). However, an examination of the distribution maps (pg. 47) shows recreation to have a far wider distribution in the core season than in the early and late seasons; the core season has more sites visited than both the early and late seasons.

Of interest to management was to identify where the highest frequency of visitor use occurs in the Sound. According to the survey data VUA 147 (Passage Canal) had the most trips and accounted for 20.12% of the total trips in the Sound followed by VUA 109 (Port Valdez) which accounted for 8.87% and VUA 22 (Culross Passage) which accounted for 3.55% of all trips. The simulation identified a similar trend: VUA 147 (Passage Canal) had the highest percentage of total trips with 29.3% followed by VUA 109 (Port Valdez) with 12.5% and VUA 22 (Culross Passage) with 5.3%. According to the survey data VUA 147 (Passage Canal) also amassed the highest percentage of stops in the Sound with 27.5% of total stops followed by VUA 109 (Port Valdez) with 12.9% and VUA 150 (Greystone Bay) with 3.8%. Simulation data was again very similar; VUA 147 (Passage Canal) had 27.5% of total stops, VUA 109 (Port Valdez) with 12.9% and VUA 150 (Greystone Bay) with 3.8%

Analysis of the potential interactions between humans and wildlife directly relates to the mission of the recovery of the Sound from the Exxon Valdez Oil Spill. Five species were analyzed (Bald Eagles, Black Oyster Catchers, Harbor Seals, Cutthroat Trout & Pigeon Guillemot) to determine the interaction of recreational activities on known nesting sites of these species. Of particular interest is not only the overlap between certain modes of travel and wildlife sites, but the duration and frequency of human interaction in these areas. To evaluate potential impacts, the number of visits and nesting sites per acre, duration of visit and the type of travel mode coinciding within these areas by season were combined to evaluate the potential threat or impact that is occurring in the Sound. This speculative analysis is provided in this report.

The three areas with the highest density of Bald eagle nests also had the lowest density of recreational visitor stops per acre. VUA 120 (Parshas Bay) had the highest density with 3.9 bald eagle nests per acre, which corresponded to the lowest density, 0.5 stops per acre, by recreational visitors. This suggests there could be a negative relationship between eagle nesting sites and visitor use, but further studies need to be done to quantify this relationship.

VUA 15 (Dutch Group) had the highest density of Black oystercatcher nests per acre (2.5 nests per acre) and had a recreational visitor density of 3.4 stops per acre, which was the second highest density of recreation use. VUAs 59 and 139 (Hanning Bay and Stockdale Harbor) both had 1.7 nests per acre and had very low densities of recreation use (0.6 and 0.5 stops per acre, respectively). This overall trend seems surprising, since we would expect human presence to negatively impact the nesting behavior of these birds. However, this association could show that visitors are targeting areas known to have high densities of Black oystercatchers and that the impact of increased stops per acre is not yet high enough to have a negative impact. There is little correlation between number of stops and the amount of time visitors spend at these sites. It appears in this initial analysis that there is little correlation between number of stops, duration of visit and nesting sites per acre. This is all very speculative, however, and further studies need to be undertaken to specially evaluate these trends. This is apparent in all the analysis undertaken to evaluate the impact of visitors on wildlife.

VUAs 66, 20 and 15 (Seal Island, Fool Island and Dutch Group) had the highest densities of Pigeon guillemot nests per acre with 1.1, 1.1 and 0.8 nests per acre respectively. Recreation use was not extremely high or low in these three areas, which suggests that these data shows no relationship between nest sites and presence of visitors.

VUAs 39, 50, 44 (Dangerous Passage, Horseshoe Bay and Hogg Bay) had the highest densities of harbor seal haul out sites with 0.9, 0.6 and 0.5 sites per acre, respectively. There was no obvious relationship between haul out sites and density of recreational use.

Finally, VUAs 126, 134 and 78 (Hells Hole, Hawkins Island Cutoff and Boswell Bay) had the highest densities of cutthroat trout streams with 2.3, 1.9 and 1.0 streams per acre, respectively. There was no obvious relationship between stream density and density of recreational use.

Finally, given this analysis of spatial and temporal patterns of visitor use, one might ask "*So what?*" How does this apply to the long-term recovery and apparent threats to indigenous wildlife and ultimate management of increased human use in the Sound? The answer to this question is more research. This study has provided critical, but baseline information on the current threats to the ecological integrity of the Sound from human intervention. The next step is to develop a framework, at an appropriate scale, to manage the future growth of recreation/visitor use in the Sound to provide for the ultimate protection of the resources and wildlife that live within its bounds. A more detailed study needs to be undertaken to evaluate a representative sample of these areas in conjunction with the constituents that frequent them to determine more realistic capacities, thresholds and displacement factors with consideration to resource crowding. It is only then, with this knowledge, that appropriate management strategies or prescriptions can be established and monitoring protocols initiated to balance increase use with the long-term protection and recovery of wildlife.

LITERATURE CITED

Blahna, D. & S. McCool. 2004. Managing Visitors in Wildland Settings. 2004. Paper presented at:Recreation Simulation Workshop. Anchorage, Alaska. May 17-18, 2004.

Bowker, J. M. 2001. Outdoor recreation by Alaskans: projections for 2000 through 2020. USDA Forest Service General Technical Report PNW-GTR-527.

Cole, David N. 1989. Wilderness Campsite Monitoring Methods: A Sourcebook. Res. Pap. INT-259. Ogden, UT: USDA Forest Service., Intermountain Research Station 57p

Cole, David N. 1993. Campsites in three western wildernesses: proliferation and changes in condition over 12 to 16 years. Res. Pap. INT-463. Ogden, UT: USDA Forest. Service., Intermountain Research Station 15p.

Cole, D. N. 2004. Monitoring and Management of Recreation in Protected Areas: the Contributions and Limitations of Science. In: Sievänen, Tuija, Erkkonen, Joel, Jokimäki, Jukka, Saarinen, Jarkko, Tuulentie, Seija & Virtanen, Eija (eds.). Policies, methods and tools for visitor management – proceedings of the second International Conference on Monitoring and Management of Visitor Flows in Recreational and Protected Areas, June 16–20, 2004, Rovaniemi, Finland.

Cole, D. (Compiler). 2005. Computer Simulation Modeling of Recreation Use: Current Status, Case Studies, and Future Directions. USDA Forest Service. Rocky Mountain Research Station. RMRS-GTR-143

Exxon Valdez Oil Spill Trustee Council. 2004. Then and Now- A Message of Hope: 15th Anniversary of the Exxon Valdez Oil Spill. *Exxon Valdez* Oil Spill Trustee Council, Anchorage, Alaska, USA.

Gimblett, H.R., R. Nelson, C. Prew & C. Sharp. 2007. Using Agent-Based Simulation Modeling in Misty Fjords National Monument: Balancing Increased Visitor Use and Wilderness Character. In Gimblett, H.R., H. Skov-Petersen & A. Muhar (eds) Monitoring, Simulation and Management of Visitor Landscapes. University of Arizona Press. (This Volume). 2008.

Gregoire, T.G., and G. J. Buhyoff. 1999. Sampling and Estimating Recreation Use. General Technical Report PNW-GTR-456. Portland, Ore.; U.S. Department of Agriculture – Forest Service, Pacific Northwest Research Station.

Hammitt, W.E., & Cole, D.N. 1998. *Wildland Recreation: Ecology and Management*. New York, NY: John Wiley.

Itami, R, R. Raulings, G. MacLaren, K. Hirst, R. Gimblett, D. Zanon, P. Chladek. 2003. RBSim 2: Simulating the complex interactions between human movement and the outdoor recreation environment. Journal for Nature Conservation. 11, pgs. 278-286.

Lime, D.W. et. al. 1978 An application of the simulator to a river setting. In: Shechter, M & Lucus, R.C. (1978) Simulation of recreational use for park andwilderness management. Baltimore, MD. John Hopkins University Press. 153-174

Manning, R.E. and D. W. Lime. Defining and Managing the Quality of Wilderness Recreation Experiences. USDA Forest Service Proceedings RMRS-P-15 Vol-4. 2000.

McVetty, D., F. Grigel, R.Itami & H.R..Gimblett. 2007. Simulating Patterns of Human Use in Canada's National Parks. In Gimblett, H.R., H. Skov-Petersen & A. Muhar (eds) Monitoring, Simulation and Management of Visitor Landscapes. University of Arizona Press. (Forthcoming). 2007.

Steidl, R. & B. Powell. 2006. Assessing the Effects of Human Activities on Wildllife. George Wright Society Forum. Volume 23, Number 2 (2006).

CHAPTER 20

ASSESSING THE RELIABILITY OF COMPUTER SIMULATION FOR MODELING LOW USE VISITOR LANDSCAPES

Brett C. Kiser
Steven R. Lawson
Robert M. Itami

Abstract: Previous applications of computer simulation modeling to outdoor recreation planning and management have generally done little to assess the reliability, or precision, of model estimates. This chapter examines the reliability of computer simulation estimates of wilderness solitude indicators that account for the timing and location of hiking and camping encounters in the backcountry of Great Smoky Mountains National Park. Thus, this chapter demonstrates procedures to assess the reliability of outputs from computer simulation models of visitor landscapes. Further, this study provides insights into the feasibility of generating visitor use-related model outputs for low use areas at managerially relevant levels of precision.

Keywords: Wilderness management, solitude, encounters, Great Smoky Mountains National Park, reliability, replications.

INTRODUCTION

The Wilderness Act of 1964 mandates that Congressionally designated wilderness areas in the United States should be managed to provide, among other qualities, "outstanding opportunities for solitude" to recreational visitors (Hendee & Dawson, 2002). To assist wilderness managers in meeting the mandates of the Wilderness Act of 1964 and related management objectives, several planning and management frameworks have been developed, including the Limits of Acceptable Change (LAC) (Stankey et al., 1985) and the Visitor Experience and Resource Protection Framework (VERP) (National Park Service, 1997). The process involved in these frameworks is similar and involves wilderness managers working with the public to define management objectives, indicators of quality, and standards of quality, and working with staff to develop an associated monitoring program. Indicators of quality are measurable, manageable variables that serve as proxies for broader management objectives (Manning, 2001). Standards of quality define minimum acceptable conditions of indicator variables, and must be quantifiable and measurable, time specific, and output oriented (Whittaker & Shelby, 1992). Perhaps the most commonly used indicator to operationalize the broader management objective of wilderness solitude has been the number of encounters visitors have with other groups (Dawson, 2004; Freimund, Peel, Bradybaugh, & Manning, 2003; Stewart & Cole, 2001). Recent studies have introduced indicators of wilderness solitude that account for the timing and location of encounters (Aplet, Thomson, & Wilbert, 2000; Hall, 2001; Saarinen, 1998).

Several studies in the field of outdoor recreation management and planning have used computer simulation modeling to demonstrate its utility as a tool to help managers monitor encounters and similar visitor use-related indicators of quality (Itami et al., 2003; Lawson, Itami, Gimblett, & Manning, 2006; Lawson & Manning, 2003). However, previous applications of computer simulation modeling to outdoor recreation planning and management have generally done little to assess the reliability, or precision, of model estimates. The reliability of computer simulation model estimates is a particularly important question because computer simulation modeling uses random numbers and/or empirical distributions to generate input variables (e.g., visitor arrival times, durations at destinations, travel routes, etc.) and therefore the estimates from a model vary across replications of the model. Consequently, conclusions should not be drawn from a single replication of a model (Law & Kelton, 2000). Thus, a significant issue within computer simulation modeling is identifying the number of simulation replications needed to estimate model outputs at a managerially relevant level of precision.

The authors are aware of only one study in which the reliability of visitor use-related estimates from a computer simulation model have been assessed, and this work was conducted in an area with very high levels of visitor use (Itami, Zell, Grigel, & Gimblett, 2005). In wilderness and related backcountry areas, the question of reliability is particularly pronounced because visitor use levels and inter-group encounters tend to be relatively low and even moderately imprecise estimates can lead to very different conclusions about the nature of visitor experiences. Furthermore, spatially complex simulations, such as those of recreational use of large, dispersed wilderness areas, contain added variability as it is often the goal to produce multiple outputs from the model simultaneously (e.g., use density and inter-group encounters on numerous trail segments and at several day use and/or overnight destinations). Thus, it is unclear whether computer simulation models can generate estimates of inter-group encounters and related outputs at a level of precision that is useful for management purposes in low use visitor landscapes, such as wilderness areas.

The purpose of the research presented in this chapter is to explore several questions concerning the reliability of computer simulation model estimates for monitoring wilderness solitude-related indicators of quality. In particular, can reliable estimates of solitude-related indicators be generated for low use recreation environments, such as backcountry and wilderness areas? Is there a spatial component to questions about the reliability of computer simulation estimates for low use visitor landscapes? That is, is it possible to generate estimates at a level of precision that is useful for management purposes for some, but not all locations within a low use recreation area (i.e., selected trails/trail segments and camping locations)? Similarly, can more precise estimates be generated for visitor-based outputs (e.g., average number of encounters per group per day) than for spatially-based outputs? The research presented in this chapter examines the reliability of computer simulation estimates of wilderness solitude indicators that account for the timing and location of hiking and camping encounters in the backcountry of Great Smoky Mountains National Park.

METHODS

Study Area

This study was designed to model visitor use and inter-group encounters in the Cosby and Big Creek areas of Great Smoky Mountains National Park. The study area is located in proposed wilderness in the northeast corner of the park and is used by day use hikers, day and overnight horseback riders, and backpackers, including Appalachian Trail thru-hikers (Figure 1). Over 85 miles of trails are located in the study area, including 16 miles of the Appalachian Trail (Figure 4.2). Four of the 6 campsites and all of the 4 shelters in this area of the park require visitors to obtain a reservation before visitors can camp overnight. Three of the shelters in the study area are located along the Appalachian Trail and these shelters receive most of the overnight use in the study area (National Park Service, 2005). There are multiple destination sites within the study area that are accessible within a relatively short day's hike, including several waterfalls that are within two miles of a parking lot and trailhead.

Data Collection

Two primary types of information about visitor use in the study area were collected to construct the computer simulation model in this study. First, information was collected about the amount of visitation to the study area. In particular, a combination of mechanical counters and direct observation was used to obtain counts of daily arrivals of day use visitors, by entry location into the study area, date, time of day, and type of visitor (i.e., hiker or

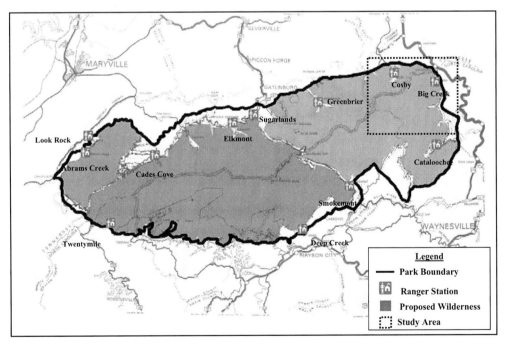

Figure 1. The proposed wilderness area of Great Smoky Mountains National Park, with study area marked (NPS, 1982).

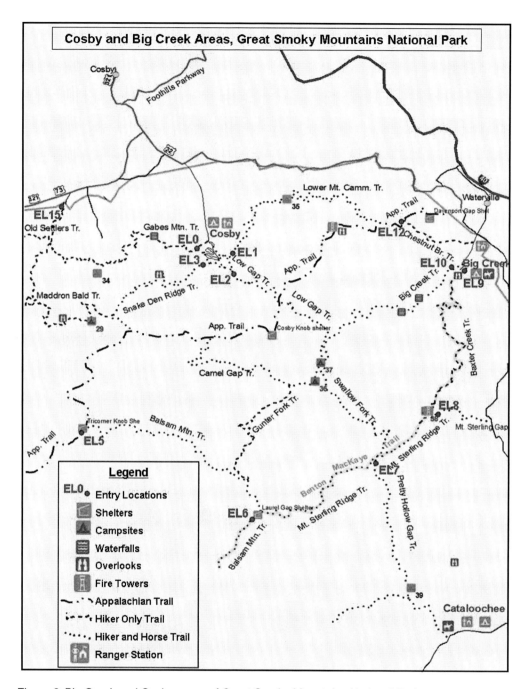

Figure 2. Big Creek and Cosby areas of Great Smoky Mountains National Park.

horseback rider). National Park Service (NPS)-issued permits were used to count the number of overnight trips during the study period, by entry location, date, and type of visitor (i.e., backpacker, horseback rider, AT-thru hiker). The day use and overnight visitation data were used to parameterize the study model to simulate visitor use levels that correspond to current conditions in May, 2006. Second, information was collected about visitors' travel routes within the study area, by entry location, time of day, and type of visitor. Day use travel routes were obtained via a survey of day use visitors conducted during May, 2006. Overnight travel routes were obtained from the NPS-issued backcountry camping permits and validated with visitor-reported itineraries collected via an overnight visitor survey. The visitation and travel route data were collected in the study area from April 23 to May 31, 2006. This period was selected for the study because it includes the time of year when most AT thru-hikers pass through the study area.

RBSIM MODELING

Input Analysis

The "raw" visitation and travel route data collected in the field and described in the preceding paragraph were formatted into a set of input databases required for the scheduling and routing of simulated visitor groups within RBSim (Itami et al., 2003), the simulation software used to develop the study model. In particular, the visitation data were used to construct databases of daily and normalized weekly arrivals of day use and overnight visitors during the study period. The visitor survey data were used to construct hourly arrival distributions for day use and overnight visitors, and distributions of campsite departure times for overnight visitors. The travel route data from the day use visitor survey and backcountry camping permits were used to construct distributions of travel routes for day use and overnight visitors, by trip starting location, starting time, and model of travel. Due to the programming nature of RBSim, all of the inputs for the simulation model were coded and formatted electronically into Microsoft Access databases. Within each replication of the model, RBSim generates a sequence of random numbers to randomly draw travel routes and departure times from the input distributions. The resultant schedule of simulated trips and associated routes is saved as an Access database and "read" by RBSim to generate a single replication of the model. The process outlined above is repeated for each replication of the model.

SIMULATIONS OF VISITOR USE AND INTER-GROUP ENCOUNTERS

The probabilistic simulation process described above was used to estimate a number of visitor-based and spatially-based measures of visitor use and inter-group encounters, assuming current visitation in May, 2006. Visitor-based outputs generated from model simulations included: 1) the percentage of visitor groups that had at least one period during their hike of 30 minutes or more during which time they did not encounter another group ("temporal encounters"); 2) the percentage of visitor groups that, on average, encountered fewer than 2 other groups per hour ("hourly encounters"); and 3) the percentage of visitor groups that never encountered more than 2 groups per hour in the "interior" of the study area (i.e., 0.5 miles or further from any trailhead; "interior encounters"). Spatially-based outputs generated from model simulations included estimates of average daily visitor use at three particularly popular attraction sites within the study area – Hen Wallow Falls, Midnight Hole, and Mouse Creek Falls ("use_HWF," "use_MH," and "use_MF," respectively). Estimates of average daily hiking use of each trail segment ("trail use") and average nightly camping use of each camping location (i.e., campsites and shelters; "camping use") within the study area were also generated. The outputs were calculated from data generated by the computer simulation model after a warm-up period of 8 days. The warm-up period is designed to "populate" the model with visitor groups representative of the study period (Law & Kelton, 2000), and the 8-day period was selected because all trips observed in the backcountry permit data were 8 days or shorter in duration.

OUTPUT RELIABILITY ANALYSIS

Three alternative methods were used to estimate the number of replications needed to obtain desired levels of precision for the model outputs outlined in the preceding section of this chapter (Centeno & Reyes, 1998; Itami et al., 2005; Law & Kelton, 2000). It should be noted that the output reliability analysis methods used in this study are appropriate only for terminating simulations and do not apply in the case of steady-state simulations (Centeno & Reyes, 1998). The first step within each of the three reliability analysis methods was to define a desired level of precision with which to estimate the model outputs. For this study, the three methods described below were used to determine the number of replications needed to estimate, with 90% confidence, the visitor-based outputs within +/- 5% and the spatially-based outputs within +/- 1 visitor group.

Because the computer simulation model developed in this study was used to generate estimates of multiple outputs simultaneously, the alpha levels for the confidence intervals specified within each of the reliability analysis methods used in this study were adjusted using a Bonferroni Correction. In particular, the specified alpha level was adjusted by dividing it by the number of outputs estimated together (Law & Kelton, 2000). For example, the computer simulation model was used to estimate all three of the visitor-based outputs simultaneously. Thus, the Bonferroni Corrected alpha level for the analysis of visitor-based outputs was equal to 0.033 (0.10 divided by 3).

The next step within each of the three reliability analysis methods was to run the model for a relatively small number of replications, commonly referred to as the "short run." All three of the reliability analysis approaches used in this study require that the short run simulation has been replicated a sufficient number of times that the variances of the outputs of interest have stabilized. The steps that follow from the short run simulation vary across the three reliability analysis methods and are described separately in the following paragraphs.

METHOD OF INDEPENDENT REPLICATIONS

Within the method of independent replications, the following equation is used to compute the confidence interval half-width around the mean of each output of interest resulting from the short run simulation:

$$\pm t_{n-1, 1-\alpha/2} \sqrt{[S^2(n)]/n} \quad (1)$$

Where:
n = number of replications conducted for the short run simulation
$t_{n-1, 1-\alpha/2}$ = $(1 - \alpha / 2)$ percentile of the t-student distribution with n-1 degrees of freedom
$S^2(n)$ = sample variance of the output variable from the short run simulation

If the confidence interval half-width is less than or equal to the user-specified value, then no further replications are needed. For example, if the short run simulation results in a confidence interval half-width of less than or equal to 5% of visitors for each of the visitor-based outputs, then the number of replications performed for the short run simulation is sufficient to generate an estimate of these variables with the specified level of precision. Otherwise, the following equation is needed to compute the number of replications needed to achieve the user-specified level of precision:

$$n^* = \text{Round}\,[\,n \times (h/h^*)^2\,] \quad (2)$$

Where:
n^* = estimated number of replications needed to achieve user-specified level of precision
n = number of replications from short run simulation
h = interval half-width computed using short run results and Equation 1
h^* = user specified confidence interval half-width

The model is then run for n^* replications and the computation process using Equations 1 and 2 is repeated until the desired level of precision is obtained. At each iteration of the method, n^* and $S^2(n^*)$ are substituted into Equation 1.

ITERATIVE METHOD

Law and Kelton (2000) suggest a modification to the method of independent replications referred to as the iterative method. Within the iterative method, Equation 1 is modified such that if the user-specified level of precision is not achieved within the short run simulation, the number of replications within the equation is increased incrementally by a value of one until the desired confidence interval half-width is achieved, as illustrated in the following equation:

$$n^*(\beta) = \min\{i \geq n : t_{i-1,1-\alpha/2}\sqrt{[S^2(n)]/i} \leq \beta\} \quad (3)$$

Where:
$n^*(\beta)$ = estimated number of replications needed to achieve user-specified level of precision
β = the user specified confidence interval half-width
n = number of replications from the short run simulation
$t_{i-1,1-\alpha/2}$ = $(1-\alpha/2)$ percentile of the t-student distribution with i-1 degrees of freedom
$S^2(n)$ = sample variance of the output variable from the short run simulation
i = number of replications at each iteration of the method

The iterative method is more efficient than the method of independent replications because it does not require additional replications of the computer simulation model after the short run simulation has been conducted. However, the iterative method assumes that the population variance of the output of interest will not change significantly as the number of replications is increased.

RELATIVE ACCURACY METHOD

The relative accuracy method is similar to

$$n^*(\lambda') = \min\{i \geq n : (t_{i-1,1-\alpha/2}\sqrt{[S^2(n)]/i})/|\overline{X}(n)| \leq \lambda'\} \quad (4)$$

Where:
$n^*(\lambda')$ = the estimated number of replications needed to achieve a user-specified level of relative accuracy
λ' = the user specified relative accuracy
N = the number of replications from the short run simulation
$t_{i-1,1-\alpha/2}$ = the $(1-\alpha/2)$ percentile of the t-student distribution with i - 1 degrees of freedom
$S^2(n)$ = the sample variance of the output variable from the short run simulation
$\overline{X}(n)$ = the mean of the output variable from the short run simulation
I = the number of replications at each iteration of the method

the iterative method, but is designed to estimate the number of replications needed to achieve a user-specified level of *relative accuracy*, rather than simply a user-specified confidence interval half-width. Within this method, the relative accuracy is calculated as the confidence interval half-width as calculated in Equation 1, divided by the mean of the output variable of interest derived from the short run simulation. Thus, within the relative accuracy method, the equation for computing the number of replications needed to achieve a user-specified level of precision is as follows equation 4.

As with the iterative method, the relative accuracy method assumes that the population variance of the output of interest will not change significantly as the number of replications is increased. The relative accuracy method also assumes that the mean of the output variable of interest will not change significantly as the number of replications is increased.

RESULTS

Results of an iterative, graphical analysis process to select the number of short run replications needed to achieve variance stability among the visitor-based outputs are reported in Figure 3. The visitor-based outputs appear to stabilize between 10 and 20 replications, although the variances increase somewhat as the number of replications increase beyond 20. The same procedure was repeated to select the number of short run replications for the spatially-based outputs. Results for the campsite and trail outputs suggest that population variances for most of the outputs stabilize with relatively few replications (i.e., 15 to 20 replications). While results of the short run replications analysis suggest the population variance for Hen Wallow Falls use stabilizes around 20 or 30 replications, population variances for Midnight Hole and Mouse Creek Falls Use do not appear to stabilize, even with a large number of short run replications. Despite the mixed results with respect to the stability of population variances, 20 replications were used for the short run simulation since many of the outputs of interest report little appreciable change in population variance beyond 20 replications.

Table 1 reports the estimated number of

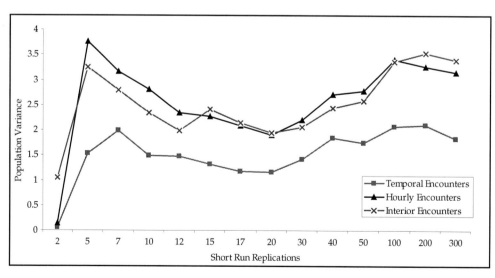

Figure 3. Estimates of population variances for visitor-based outputs with alternative numbers of replications for the "short run" simulation.

replications needed to simultaneously estimate with 90% confidence 1) the three visitor-based outputs within +/- 5% of visitor groups; 2) the three attraction-based outputs within +/- 1 visitor group; and 3) the three visitor-based outputs within +/- 5% of visitor groups and the three attraction-based outputs within +/- 1 visitor group, together. Results of all three reliability analysis methods suggest that only 20 replications are needed to estimate the three visitor-based outputs or the three attraction-based outputs alone. To estimate the visitor-based and attraction-based outputs together, the method of independent replications and iterative method suggest that 20 replications are sufficient to do so with the specified level of precision. However, the relative accuracy method suggests 41 replications are needed to estimate the visitor-based and attraction-based outputs simultaneously with the desired level of precision, with visitor use at Midnight Hole being the constraining factor.

Table 2 reports the results of each of the three reliability analysis methods for estimates of average nightly campsite and shelter use. Since there are 10 campsites and shelters, the Bonferroni Corrected alpha level is 0.01 for each campsite/shelter, with an overall alpha level of 0.1. Results of the method of independent replications and iterative method suggest that a maximum of 58 replications would be needed to estimate, with 90% confidence, average nightly camping use at each camping location, within +/- 1 visitor group. However, results of the relative accuracy method suggest that nearly 300 replications of the model would be needed to estimate average nightly camping use, by location, with the desired level of precision.

Since there are over 100 trail segments within the study area, Table 3 reports the results of the three reliability analysis methods for estimates of daily trail segment use as the percentage of trail segments that require various ranges of replications. Results of both the method of independent replications and iterative method suggest that only 20 replications are needed to estimate with 90% confidence average daily hiking use, by trail segment, within +/- 1 visitor group. The results of the relative accuracy method, however, suggest that while a little over 50% of the trail segments within the study area require only 20 replications to achieve the desired level of precision, nearly 9% of the trail segments require over 100 replications, and over 2% require more than 1,500 replications. Thus, to estimate, with 90% confidence daily hiking use within +/- 1 visitor group for all trail segments in the study area simultaneously, the results of the relative accuracy method suggest that over 1,500 replications of the model would be required.

Using the results of the reliability analyses to guide the number of replications performed with the model, the visitor-based, attraction-based, and camping and hiking use outputs were estimated. Table 4 reports estimates of the visitor-based and attraction-based outputs, estimated separately with 20 replications each, and together, with 50 replications of the model. The number of replications in the experiments produced confidence interval half-widths well within the reliability standards specified for the study (+/- 5% of visitor groups for the visitor-based indicators and +/- 1 visitor group for the attraction-based indicators). Results of the model suggest that about 60% of visitors to the study area have at least one time period of 30 minutes or more during which they encounter no other visitors. Further, the model estimates that approximately 45% of visitors encounter an average of two or fewer people per hour, and about 50% of visitors encounter two or fewer people per hour on trails in

Table 1. Reliability Analysis Results for Visitor-Based and Attraction-Based Outputs.

Model outputs	C.I. half-width	Alpha level (Bonferroni Corrected)	Estimated replications — Method of Independent Replications	Estimated replications — Iterative Method	Estimated replications — Relative Accuracy Method
Visitor-based outputs	-	-	-	-	-
Temporal encounters	± 5% of groups	0.10 (0.033)	20	20	20
Hourly encounters	± 5% of groups	0.10 (0.033)	20	20	20
Interior encounters	± 5% of groups	0.10 (0.033)	20	20	20
Attraction-based outputs	-	-	-	-	-
Hen Wallow Falls – Use	± 1 group	0.10 (0.033)	20	20	20
Midnight Hole – Use	± 1 group	0.10 (0.033)	20	20	20
Mouse Creek Falls – Use	± 1 group	0.10 (0.033)	20	20	20
Visitor and attraction-based	-	-	-	-	-
Temporal encounters	± 5% of groups	0.10 (0.017)	20	20	20
Hourly encounters	± 5% of groups	0.10 (0.017)	20	20	20
Interior encounters	± 5% of groups	0.10 (0.017)	20	20	20
Hen Wallow Falls – Use	± 1 group	0.10 (0.017)	20	20	20
Midnight Hole – Use	± 1 group	0.10 (0.017)	20	20	41
Mouse Creek Falls – Use	± 1 group	0.10 (0.017)	20	20	20

Table 2. Reliability Analysis Results for Estimates of Average Nightly Camping Use, by Camping Location.

Camping Locations	C.I. half-width	Alpha level (Bonferroni Corrected)	Estimated Replications — Method of Independent Replications	Estimated Replications — Iterative Method	Estimated Replications — Relative Accuracy Method
Davenport Gap Shelter	± 1 group	0.10 (0.01)	20	20	88
Cosby Knob Shelter	± 1 group	0.10 (0.01)	20	20	48
Tricorner Knob Shelter	± 1 group	0.10 (0.01)	26	25	112
Laurel Gap Shelter	± 1 group	0.10 (0.01)	20	20	55
Campsite 29	± 1 group	0.10 (0.01)	58	52	116
Campsite 34	± 1 group	0.10 (0.01)	50	45	293
Campsite 35	± 1 group	0.10 (0.01)	20	20	26
Campsite 36	± 1 group	0.10 (0.01)	20	20	60
Campsite 37	± 1 group	0.10 (0.01)	20	20	20
Campsite 38	± 1 group	0.10 (0.01)	20	20	35

Table 3. Reliability Analysis Results for Estimates of Average Daily Hiking Use, by Trail Segment.

Estimated number of replications	Method of Independent Replications	Iterative Method	Relative Accuracy Method
	% of trail segments	% of trail segments	% of trail segments
20 replications	100.0	100.0	53.5
21 – 50 replications	0.0	0.0	28.2
51 – 100 replications	0.0	0.0	6.3
101 – 150 replications	0.0	0.0	6.3
151 – 200 replications	0.0	0.0	1.4
201 – 300 replications	0.0	0.0	2.1
301 – 400 replications	0.0	0.0	0.0
401 – 500 replications	0.0	0.0	0.0
501 – 1000 replications	0.0	0.0	0.0
1001 – 1500 replications	0.0	0.0	0.0
1501 – 2000 replications	0.0	0.0	2.1

n = 111 trail segments

the interior of the study area. The model estimates that average daily use of Hen Wallow Falls and Mouse Creek Falls is about 10 visitor groups per day, while average daily use of Midnight Hole was estimated to be about eight groups per day.

The estimates of camping use reported in Table 5 were generated based on 1) 60 replications of the model, which is approximately the required number of replications estimated from the method of independent replications and iterative method; and 2) 300 replications, which constitutes the number of replications required by the relative accuracy method. While the standard deviations of camping use estimates were generally larger with 300 replications, the confidence interval half-widths were generally smaller compared to the results generated from 60 replications of the model. Although the

Table 4. Visitor-based and Attraction-based Outputs, Estimated Separately and Simultaneously.

Model outputs	Alpha level (Bonferroni Corrected)	Number of replications	Output value	Standard deviation	Confidence interval half-width
Visitor-based indicators			-	-	-
Temporal encounters	0.10 (0.033)	20	61.36	1.078	0.417
Hourly encounters			45.64	1.378	0.533
Interior encounters			51.19	1.395	0.539
Attraction-based indicators			-	-	-
Hen Wallow Falls – Use	0.10 (0.033)	20	10.53	0.115	0.074
Midnight Hole – Use			8.19	0.312	0.199
Mouse Creek Falls – Use			10.73	0.212	0.136
Visitor and attraction based			-	-	-
Temporal encounters	0.10 (0.017)	50	61.53	1.321	0.500
Hourly encounters			45.76	1.669	0.631
Interior encounters			51.23	1.608	0.609
Hen Wallow Falls – Use			10.51	0.140	0.053
Midnight Hole – Use			8.21	0.314	0.119
Mouse Creek Falls – Use			10.76	0.243	0.092

Table 5. Estimates of Average Nightly Camping Use, by Camping Location.

Estimated number of replications	Alpha level (Bonferroni Corrected)	60 Replications			300 Replications		
		Output value	Standard deviation	C.I. half-width	Output value	Standard deviation	C.I. half-width
Davenport Gap Shelter	0.10 (0.01)	4.500	1.780	0.618	4.240	1.739	0.261
Cosby Knob Shelter		3.067	1.520	0.522	2.960	1.649	0.248
Tricorner Knob Shelter		5.417	1.531	0.525	5.503	1.603	0.241
Laurel Gap Shelter		1.267	0.561	0.193	1.373	0.664	0.100
Campsite 29		3.783	2.666	0.915	3.647	2.657	0.399
Campsite 34		2.200	2.165	0.338	2.170	2.259	0.339
Campsite 35		1.050	0.194	0.067	1.110	0.329	0.049
Campsite 36		1.233	0.476	0.163	1.180	0.443	0.066
Campsite 37		1.867	0.891	0.306	1.660	0.818	0.122
Campsite 38		1.533	0.790	0.271	1.573	0.932	0.140

results of the 300-replications experiment produced outputs with higher reliability than the 60-replications experiment, the confidence interval half-widths in each experiment were all within the standard of reliability for the analysis. Furthermore, the mean output values from each of the two experiments did not differ substantively. The results suggest overnight use in the study area is relatively low, with average camping use ranging from a low of one camping group per night at campsite 35 to a high of about five camping groups per night at the Tricorner Knob Shelter. Camping use estimates were highest along the Appalachian Trail, with the Tricorner Knob Shelter receiving the most camping use along the trail and the Cosby Knob Shelter receiving the lowest amount of camping use along the trail.

Table 4.6 reports the percentage of trail segments that fall within various average daily use categories. While the results of the relative accuracy method suggest that over 1,500 replications of the model are necessary to simultaneously estimate, with 90% confidence, average trial use for all of the trail segments in the study area within +/- 1 visitor group, RBSim is unable to process this number of replications with the study model due to file size constraints. Thus, the results reported in Table 4.6 were generated based on 1) 20 replications of the model, which is approximately the required number of replications estimated from the method of independent replications and iterative method; and 2) 1,300 replications, which constitutes the maximum number of replications RBSim is able to process with the study model. It should be noted that this file size constraint is specific to the model developed in this study, and that the total number of replications RBSim can process is a function of the size and complexity of the system being modeled. Further, it is likely that RBSim can be modified in a future release to minimize or eliminate this file size issue. Almost 40% of the trail segments in the study area had an average daily use of less than one visitor group, while 9% of the trail segments had an average daily use of over 20 visitor groups. Confidence interval half-widths for trail use estimates reported in Table 4.6 ranged from 0.005 to 0.575 with 20 replications of the model, and 0.001 to 0.061 with 1,300 replications of the model. Thus, both the 20-replications and 1,300-replications experiments produced confidence interval half-widths for all of the trail use outputs within the reliability standard set for the study.

Table 6. Ranges of Estimated Average Daily Hiking Use of Trail Segments.

Average trail use	20 replications	1300 replications
	% of trails segments	% of trails segments
Less than 1 group	37.8	37.8
1 to less than 2 groups	23.4	23.4
2 to less than 5 groups	11.7	11.7
5 to less than 10 groups	16.2	16.2
10 to less than 15 groups	0.9	0.9
15 to less than 20 groups	0.9	0.9
20 or more groups	9.0	9.0

n = 111 trail segments

Note. Alpha level = 0.10, Bonferroni Corrected alpha level = 0.001, confidence interval half-width = ± 1 visitor group.

DISCUSSION

As noted, previous applications of computer simulation modeling to outdoor recreation management and planning have generally done little to assess the reliability of model estimates. This study demonstrates the application of reliability analysis procedures developed in the broader field of discrete-event simulation to modeling recreational use in a low use area. These same procedures are applicable to visitor landscapes, in general, including those that receive greater levels of use (Itami et al., 2005). Thus, this study serves to document reliability analysis procedures that can be adopted as standard practice for computer simulation modeling of visitor landscapes in general. It is interesting to note, however, that the three reliability analysis methods used in this study produced substantively different results. In particular, while the method of independent replications and iterative method generally yielded similar estimates of the number of replications needed to achieve desired levels of precision for the model outputs, the relative accuracy method typically resulted in much larger estimates of replication requirements. For example, results of both the method of independent replications and iterative method suggest that only 20 replications are needed to estimate the visitor-based and attraction-based outputs at the specified level of precision. In contrast, the relative accuracy method results suggest just over 40 replications would be needed. The differences between the results of the relative accuracy method and the other two methods are even more pronounced for the camping use and trail use estimates, with the relative accuracy method estimating the need for more than 5 times as many replications for the camping use outputs than estimated by the other two methods, and estimating the need for over 1,500, rather than 20, replications to estimate the trail use outputs.

While the relative accuracy method produces results that constitute the most stringent requirements for model replications, it is arguably the preferred reliability analysis method because, within its estimation of replication requirements, the variance of the model outputs is considered in relation to the size of the corresponding mean values of the outputs. This is a particularly important issue within low use recreation environments where it is expected that at least some, if not many, of the model outputs' mean values will be relatively small. That being said, the results of this study suggest that the relative accuracy method may overestimate replication requirements. In particular, experiments based on the less stringent replication requirements of the other two reliability analysis methods produced outputs that met the reliability standards specified for the study. These findings suggest that additional research on the relative merits of the three reliability analysis methods used in this study is warranted.

While the results of the reliability analyses conducted in this study varied depending on the method used, the findings regarding the feasibility of generating precise estimates of visitor use-related outputs from simulation models of low use environments are generally encouraging. The study findings are particularly promising for the visitor-based outputs generated with the study model, with only 20 replications of the model needed to achieve precise estimates. The visitor-based outputs include inter-group encounters, which has been the most commonly adopted indicator of wilderness solitude, as well as indicators that account for the temporal and spatial dimensions of encounters. Thus, the results of this study suggest that computer simulation modeling is a reliable tool for helping to implement visitor-based indicators of wilderness solitude within a VERP or LAC monitoring program.

While the findings from the reliability analyses for the visitor-based outputs were encouraging, results from the reliability analyses for the spatially-based outputs were somewhat mixed. Results suggest that precise estimates of visitor use at the three attractions sites for which outputs were obtained can be generated with just 20 replications of the study model. Further, the results suggest that the study model can reliably estimate the visitor-based and attraction-based outputs simultaneously, with fewer than 50 replications. Results of the method of independent replications and iterative method suggest that no more than 60 replications are needed to reliably estimate average nightly camping use at each of the 10 campsites and shelters in the study area, however, the relative accuracy method suggests nearly 300 replications of the model are needed. While this is a substantially larger number of replications than that required for the visitor-based and attraction-based outputs, it involves relatively inconsequential amounts of computer processing time and capacity. In contrast, results of the reliability analyses suggest that as many as 1,700 replications are needed to simultaneously estimate average daily hiking use on all 111 of the trail segments in the study area. This exceeds the maximum number of replications RBSim can process with the study model, due to file size constraints. Thus, the findings suggest that precise estimates cannot be readily obtained for the lowest use trail segments within the study area. However, as noted, 90% confidence interval half-widths for estimates of average daily hiking use on the 111 trail segments based on 1,300 replications were all well below the target half-width of +/- 1 visitor group. Furthermore, confidence interval half-widths produced from the 20-replications experiment were all within the reliability standard specified for the trail use outputs. As noted above, these finding suggest that the relative accuracy may provide an upper-bound estimate, perhaps even an overestimate, of the number of replications needed to generate outputs at specified levels of precisions. If the relative accuracy method systematically produces overestimates of replication requirements, this is not necessarily an inconsequential issue, as the run time required to produce 1,300 replications of the study model totaled more than seven days. In any case, results of this study suggest the feasibility of simultaneously generating a relatively large number of use-related outputs from simulation models of low use visitor landscapes may be limited due to file size and processing time constraints. However, these are limitations that could likely be addressed with advances in computing technology.

Aside from advances in computing technology, there are several alternative approaches to address the challenges associated with generating precise estimates of use-related outputs in low use recreation environments. One option would be to eliminate the lowest use trail segments from simulation model output analyses. In the case of this study, eliminating the three lowest use trail segments from the output analysis would reduce the number of replications needed to obtain precise estimates of average daily hiking use from over 1,700 to less than 250. This approach might be particularly attractive in cases where areas of concentrated use are of particular concern and interest to managers. However, it could be argued that the lowest use portions of backcountry and wilderness areas are the most important to monitor because they afford rare opportunities for solitude that may be threatened by even small increases in visitor use and inter-group encounters. Thus, eliminating low use areas or zones from simulation model output analyses may not be an acceptable approach.

An alternative approach to address issues of reliability in simulations of low use environments would be to produce estimates of visitor use and use-related indicators aggregated by management zone. For example, modeling could be used to generate estimates of average daily use for "primitive zone trails," "threshold zone trails," and "corridor zone trails." Results using this approach would be less spatially precise, but may be equally or more sufficient for the purposes of helping to monitor and manage visitor use and opportunities for solitude in low use environments. This type of approach could be programmed within RBSim and other simulation software packages with relative ease, and the approach could be tailored to the management zoning of a particular study area.

Alternatively, variance reduction techniques could be used to reduce the number of replications that would be needed to obtain precise estimates of all of the outputs of interest. Variance reduction techniques are characterized as methods used to increase the efficiency and speed of simulating a study environment to estimate the outputs of interest as precisely as desired (Law & Kelton, 2000). A few of the more commonly used variance reduction techniques include: 1) using common random numbers to compare two separate, alternative system simulations; 2) antithetic variates, which introduce negative correlation between separate runs of the same system; 3) importance sampling, where the chance of events of interest are increased to occur more often; and 4) conditioning the model to remove one source of variability. While RBSim is programmed to use common random numbers as a default, selection of the appropriate variance reduction techniques are model specific. That is, one or more variance reduction techniques may work well within one simulation modeling application, but perform poorly when applied to other models. Furthermore, the specific efficiency of each separate technique's ability to reduce the variance of the system is unknown for the model of interest. Future research should explore which variance reduction techniques may work best for low use recreation areas, and whether or not variance reduction techniques improve the ability of computer simulation models to produce estimates at managerially useful levels of precision.

CONCLUSIONS

This study serves to document procedures to assess the reliability of outputs from computer simulation models of visitor landscapes. It is recommended that these procedures be adopted as standard practice within applications of computer simulation modeling to outdoor recreation planning and management. Further, this study provides insights into the feasibility of generating visitor use-related model outputs at managerially relevant levels of precision. The results suggest that precise estimates can be obtained for a small to moderate number of visitor-based and spatially-based outputs. However, there are constraints to generating precise estimates of use-related outputs as the number of outputs estimated simultaneously becomes large. This challenge is particularly pronounced in cases where at least some of the outputs are derived for low use attractions, trails, or camping locations. Future studies should explore variance reduction techniques to enhance the reliability of computer simulation modeling in cases where a large number of outputs are desired and/or low use environments are modeled.

LITERATURE CITED

Aplet, G., Thomson, J., & Wilbert, M. 2000. *Indicators of wildness: Using attributes of the land to assess the context of wilderness*. USDA Forest Service General Technical Report RMRS-P-15-VOL-2.

Centeno, M. A., & Reyes, M. F. 1998. *So you have your model: What to do next. A tutorial on simulation output analysis*. Paper presented at the 1998 Winter Simulation Conference.

Dawson, C. P. 2004. Monitoring outstanding opportunities for solitude. *International Journal of Wilderness, 10*(3), 12-14, 29.

Freimund, W., Peel, S., Bradybaugh, J., & Manning, R. E. 2003, 2004. *The wilderness experience as purported by planning compared with that of visitors to Zion National Park*. Paper presented at the Protecting Our Diverse Heritage: The Role of Parks, Protected Areas, and Cultural Sites (Proceedings of the 2003 George Wright Society / National Park Service Joint Conference), Hancock, MI.

Hall, T. E. 2001. Hikers' perspectives on solitude and wilderness. *International Journal of Wilderness, 7*(2), 20-24.

Hendee, J. C., & Dawson, C. P. 2002. *Wilderness Management* (Third ed.). Golden, CO: Fulcrum Publishing.

Itami, R. M., Raulings, R., MacLaren, G., Hirst, K., Gimblett, H. R., Zanon, D., et al. 2003. RBSim 2: simulating the complex interactions between human movement and the outdoor recreation environment. *Journal for Nature Conservation, 11*(4), 278-286.

Itami, R. M., Zell, D., Grigel, F., & Gimblett, R. 2005. *Generating confidence intervals for spatial simulations - determining the number of replications for spatial terminating simulations*. Paper presented at the MODSIM 2005 International Congress on Modelling and Simulation, Melbourne, Australia.

Law, A. M., & Kelton, W. D. 2000. *Simulation Modeling and Analysis* (Third ed.): McGraw-Hill Higher Education.

Lawson, S., Itami, R., Gimblett, R., & Manning, R. 2006. Benefits and challenges of computer simulation modeling of backcountry recreation use in the Desolation Lake Area of the John Muir Wilderness. *Journal of Leisure Research, 38*(2), 187-207.

Lawson, S. & Manning, R. 2003. Research to inform management of wilderness camping at Isle Royale National Park: Part I – descriptive research. *Journal of Park and Recreation Administration, 21*(3), 22-42.

Manning, R. E. 2001. Visitor experience and resource protection: A framework for managing the carrying capacity of national parks. *Journal of Park and Recreation Administration, 19*, 93-108.

National Park Service. 1997. *VERP: The Visitor Experience and Resource Protection (VERP) Framework - A handbook for planners and managers*. USDI National Park Service Technical Report.

National Park Service. 2005. *Camper nights summary report: Great Smoky Mountains National Park*. USDI National Park Service.

Saarinen, J. 1998. Cultural influences on wilderness encounter responses. *International Journal of Wilderness, 4*(1), 28-32.

Stankey, G. H., Cole, D. N., Lucas, R. C., Peterson, M. E., Frissell, S., & Washburne, R. 1985. *The Limits of Acceptable Change (LAC) System for Wilderness Planning.* USDA Forest Service General Technical Report INT-176.

Stewart, W. P., & Cole, D. N. 2001. Number of encounters and experience quality in Grand Canyon backcountry: consistently negative and weak relationships. *Journal of Leisure Research, 33*(1), 106-120.

Whittaker, D., & Shelby, B. 1992. *Developing good standards: Criteria, characteristics, and sources.* USDA Forest Service General Technical Report PNW-GTR-305.

CHAPTER 21

REPLICATION, RELIABILITY AND SAMPLING ISSUES IN SPATIAL SIMULATION MODELING: A CASE STUDY EXPLORING PATTERNS OF HUMAN VISITATION IN CANADA'S MOUNTAIN PARKS

Robert M. Itami
Randy Gimblett
Frank Grigel
Darrel Zell
David McVetty

Abstract: Parks Canada has the responsibility to improve the visitor experience and ecological integrity of Canada's Mountain National Parks. This chapter addresses the need to improve the understanding of the complex patterns of visitor use with simulation modeling. This chapter explores two major issues in spatial simulation modeling; how to determine the number of replications required that meets user specified reliability and accuracy as well as the spatial and temporal sampling of visitor behavior to develop probabilistic simulations for accurate forecasting and decision making. These issues are explored and discussed in a spatial simulation of travel patterns for Banff, Yoho, Kootenay and Jasper National Parks.

Keywords: Spatial Simulation, Simulation Statistics, Spatial and Temporal Patterns, Replication, Reliability, Sampling, Probabilistic

INTRODUCTION

Parks Canada is the federal agency responsible for managing Canada's vast system of national parks, national historic sites, and national marine conservation areas. The Agency's Corporate Plan lays out a set of performance objectives in three areas: protection, public education, and visitors' experiences. Each of these pillars needs reliable monitoring data to guide management of these special Canadian places. There is a strong need by the Agency to understand and monitor the outcomes of management plans and policies – both positive and negative – not just numbers of visitors. To be successful, the Agency needs ways to forecast the impacts of plans and policies.

Impacts of visitor use are a function of the number of users, the amount of time they spend in protected areas, and their behaviors while visiting. Environmental impacts can be changed with fluctuations in the number of visitors, but also with alterations of behavior of the same number of visitors.

On the border of Alberta and British Columbia lie Banff, Jasper, Kootenay, and Yoho National Parks (See Figure 1). Visitors view and use these parks in the collective, moving back and forth between parks during their visits without concern that they are crossing the borders into different management units. Historically the Parks Canada Agency managed these parks as separate entities. While recognizing the presence of nearby organizational units, management decisions largely focused on lands and visitors in each of the parks. The Agency's increased emphasis on Ecological Integrity initiated a movement to recognize the interconnection of the lands in these parks. Over time, this has resulted in a greater emphasis on co-operative management between the four parks.

While the management of the parks was beginning to become more integrated across park borders, many of the existing monitoring programs (primarily visitation estimates) remained extremely park-specific. This situation highlighted the need to develop a coordinated monitoring approach in the mountain parks, as Parks Canada was unable to report on relatively simple numbers, like overall visitation to the mountain parks. Simply adding the number of visitors reported by each park resulted in a great deal of double-counting as all the visitors who moved between any of the four parks was counted in each park's attendance estimates. Any discussion of visitation at the mountain parks level was largely speculative.

The goal of this study was to capture a statically representative sample of visitor use for an entire calendar year, with reliable samples of park visitors to all four mountain parks from each quarter (January to March; April to June; July to September; and October to December) and then using this sample to develop a probabilistic spatial simulation of tourist flows for each park. Being able to predict mountain park visitation was the primary reason that the 2003 survey of visitors to Banff, Jasper, Kootenay, and Yoho National Parks was originally proposed. Adding complexity to capturing a representative sample is the interception of visitors on two of the three west-east corridors through the Rocky Mountains (the Trans-Canada highway and the more northerly Highway 16). In total almost six million (5.8 million) vehicles enter the mountain national parks of Banff, Jasper, Yoho, and Kootenay each year. About 1.4 million of these vehicle entries qualify as visitor vehicle arrivals to the national parks, with the rest including commercial vehicles, through traffic, park employees, and visitor re-entries.

VISITOR SAMPLING

During the design of this study, it became apparent that this would also provide the perfect opportunity to capture information at a level that could be applicable to visitors to the mountain parks collectively, but also would provide information about visitors to any one of the four parks. In addition to being able to predict parks' attendance, the 2003 survey was seen as an opportunity to collect detailed information about visitors' use of and movement across the landscape of the mountain parks. This information was intended to be able to predict visitor behavior at different levels of resolution: at the mountain parks, the individual parks, and at major nodes within each park. The research design was not intended to predict behavior at the level of individual facilities or trails.

The study used a multi-stage sampling design to collect a representative sample of visit party arrivals from the population of independent vehicle entries in each quarter at each entry point. The multi-stage design helped to ensure that the sample was large enough to provide reliable data for each of the 28 cells. The sur-

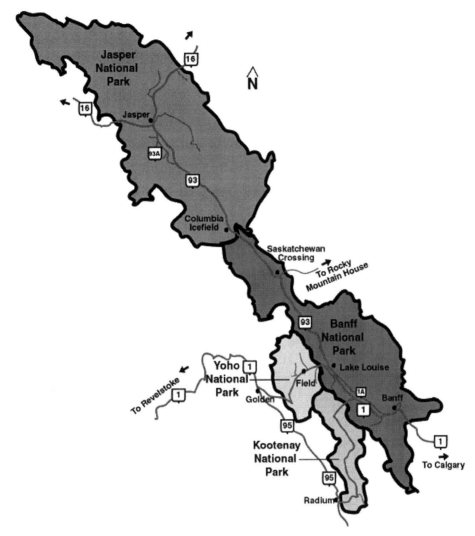

Figure 1. Illustrates the study area and the road system inside the parks.

vey captured visit party entries at the seven entry points to the parks: Banff East Gate (at Canmore); Trans Canada Hwy West (at Field); Yellowhead Hwy East (at Hinton); Yellowhead Hwy West (at Mt. Robson); David Thompson (at Saskatchewan River Crossing); Hwy 93 (at Radium); and at the Calgary Airport Shuttle Terminal. Visitor re-entries result from visitors leaving the parks to visit or access services in nearby communities or in adjacent parks. Re-entries had not been factored out of previous visitor estimates for the individual parks. As parks were visualized as distinct geographic entities, their visitation estimates were produced from stopping samples of entering vehicles and asking if they asked if the party inside planned on stopping in the park. No consideration was given to those parties who may have begun their visit on an earlier day and were returning to the park after accessing services in one of the communities outside the park, or returning after having visited a neighboring park. There are communities within a 45-minute drive at four of the parks six entry gates. Two of the communities (Canmore, AB and Radium, BC) are directly adjacent to the parks. While these communities have always offered services and accommodations to parks visitors, the number of accommodation opportunities in the communities has risen dramatically over the past 15 years, with Canmore experiencing the largest growth. In planning the 2003 survey we recognized that visitor re-entries needed to be measured and we expected that this could result in a drop in new visitation estimates. While we expected that this number could be fairly significant, it was difficult to predict how large it may be.

The move to predicting individual park visitor estimates from estimates developed at the four mountain park level provided the perfect opportunity to begin to think about park visitation as a part of visitor movements across the four mountain park landscape. Recognizing the financial investment required to make this project happen, Parks Canada partnered with a number of other Agencies to secure the necessary funding. These information needs of all these agencies resulted in the development of an extremely long questionnaire (twelve pages) that included demographic, motivational, and satisfaction questions, as well as information about behavior and activities inside and outside the park.

The key feature of the questionnaire was the trip diary. It asked visitors to identify their activities at key locations within the mountain parks. The trip diary was intended to provide detailed information just not about where visitors went, but what they did at each location, and for how long. The initial goal was to analyze this information to identify specific patterns of visitor use that could be used to segment visits into distinct types. These segments were envisioned as distinct patterns of activities and places visited that would describe the visits to the mountain parks, not the visitors. It can become difficult to distinguish between the visit (places and activities) and the visitor (the person doing those things). The key difference between visits and visitors is that a single person may use the parks in different ways on different visits. A summer day trip in the frontcountry of a park will result in a significantly different visit profile than an over-night cross-country ski trip in the backcountry, even when it is the same person doing both. By focusing on the places visited and activities engaged in, visit profiles are distinctly different from the visitor profiles that we have traditionally developed.

The sample size for each cell, based on a confidence level of 95% and a margin of error of ±5%, is 384. In some cases, low traffic volumes suggested that a higher margin of error had to be accepted (e.g., the David Thompson has 68,000 vehicle entries per year), so the study aimed for a larger margin of error (±10%, or a sample

Table 1. Sample Size by Gate

Entry Points	Annual Traffic	Sample	Margin of Error
Banff East Gate (Canmore)	3,100,000	384	±5%
Trans Canada Hwy West (Field)	940,000	384	±5%
Yellowhead Hwy East (Hinton)	670,000	384	±5%
Yellowhead Hwy West (Mt. Robson)	440,000	384	±5%
David Thompson (Sask Crossing)	68,000	96	±10%
Hwy 93 (Radium)	560,000	96	±10%
Total	5,778,000	1,824	

size of 96 returned questionnaires) in those cases (See Table 1).

A multi-stage sampling design was used to obtain the required sample for each gate within each season. With seven entry points and four quarters, this yielded a 7 x 4 design, with each containing a reliable sample. The number of sampling days in each cell was determined by working backward, based on:

- Estimated proportion of visitor traffic in all vehicle entries (varies by entry point and season, based on past research)
- Proportion of visit arrivals (varies by entry point and season, based on past research);
- Estimated participation rate for sampled parties (estimated to be 90% of the vehicles selected); and
- Estimated response rate for those who participate (estimated to be 30%).

To collect a sample of 384 completed questionnaires for each cell (96 in the cells where a higher margin of error is accepted), a total of 1,280 forms must be distributed per quarter per entry point (340 in the smaller cells). The results are presented in Table 2 below.

Data were weighted to the gate traffic information (hourly counts), therefore it was critical to ensure a representative sample is drawn from each gate otherwise the weighting process could compound any errors associated with the gate sample. Parks Canada maintains reliable hourly count data for each of the entry points used in the survey. These data are critical to weighting. Parks Canada developed a traffic partition factor that was applied to the traffic counts in order to arrive at the number of visitors and non-visitors. Vehicles were selected at random at each of the entry points. Party intercept information was collected for each vehicle using a hand held computer. A more detailed survey will be given to the visitor party to be returned in the supplied postage-paid envelope at the end of their trip. The survey data were linked to the intercept data with a unique identifier. A floating random sample was used at each entry point to minimize the impact on visitor traffic. The floating sampled selects the "next available vehicle" once an interview is completed. This created a higher probability of selection for party arrivals in slow traffic periods that was addressed by weighting the survey data to the hourly traffic counts for that entry point. Weighting is required to adjust the sample to reflect the true proportions within the population. For each hourly time block, it will be necessary to apply the traffic partition factor to estimate the total visitor to

Table 2. Number of Survey Days by Quarter

	Quarter 1	Quarter 2	Quarter 3	Quarter 4	Total
Banff East Gate (Canmore)	50 days	50 days	50 days	50 days	**200 days**
Trans Canada Hwy West (Field)	25 days	25 days	25 days	25 days	**100 days**
Yellowhead Hwy East (Hinton)	30 days	30 days	30 days	30 days	**120 days**
Yellowhead Hwy West (Mt. Robson)	20 days	20 days	20 days	20 days	**80 days**
David Thompson (Sask Crossing)	10 days	10 days	10 days	10 days	**40 days**
Hwy 93 (Radium)	25 days	25days	25 days	25 days	**100 days**
Calgary Airport Shuttle Terminal		10 days	10 days	10 days	**30 days**
Quarterly Total	160 days	170 days	170 days	170 days	**670 days**

non-visitor volumes that have entered the gate.

Survey records contain a linking variable to each hourly time block; therefore, it is possible to determine the number of returned surveys in relation to the number of visitors that entered the park during that time block. This will be the first level of weighting applied to the data.

Post stratification of data may be necessary if there is significant bias noted between the survey responses and the baseline data. For example, if there is a significant local or non-local bias or a temporal bias associated with those who are on longer trips responding at a lower rate (perhaps due to losing surveys while on a long trip). Statistical tests will be used to compare he baseline data to the returned sample

SURVEYS TO TRIP ITINERARIES FOR SIMULATION

The 2003 Mountain Parks Study provided the trip itinerary data for a trace simulation. Out of the 13,373 first time arrivals to the Mountain parks, 9348 respondents agreed to complete the questionnaire. Out of the 9348 respondents, 2383 questionnaires were returned, resulting in a 25.5 percent return rate. Out of those 2383 questionnaires, 1982 respondents actually completed the trip diary component (necessary to develop the trip itineraries for the trace simulation), resulting in a 21.2 percent return rate. An evaluation of the trip diary section of the 1982 responses, 1620 actually completed the diary correctly.

For purposes of this study the 1620 diaries became our sample population to both build and run a valid simulation as well as to develop a procedure to determine if the pool of responses accurately represented the spatial and temporal variability necessary to build a probabilistic model. To use a pattern of use simulation for management proposes, a probabilistic model of travel patterns must be developed from analysis of survey trips. Trip itineraries vary within the constraints of seasonal patterns, which will constrain certain activities and access to destinations because of weather. Trips must be grouped or clustered in order to classify trip itineraries according to seasonal variations in pattern of use. In order to do this, a number of different cluster analysis techniques were applied to the 1620 survey trips using a number of different clusters and against different definitions of a "season". The strongest clustering results came from a technique called "kmeans". In this technique, the analyst must nominate the number of clusters. 3 and 4 clusters were tested with 3 clusters differentiating the winter season trips most distinctly. The other trips were clustered on the basis of the type of activity: either active (hiking, biking etc) or inactive.

Based on analysis of the survey trips looking at sample size during each week of the year and the results of the cluster analysis, two periods of the year were selected to develop a probabilistic simulation from survey trip itineraries and traffic counts. The winter period selected is the month of January 2003. The summer period selected is 4 weeks beginning June 23, 2003 and ending July 24, 2003. We report on the January results in this paper.

PATTERN OF USE SIMULATION FOR CANADA'S FOUR MOUNTAIN PARKS

A probabilistic spatial simulation of tourist flows in Banff, Jasper, Kootenay and Yoho National Parks in the Canadian Rocky Mountains was developed using RBSim software (Cole 2005; Gimblett et al. 2000; Gimblett 2002; Itami et al. 2004; Itami et al. 2005). The next step in the development of the probabilistic simulation is to randomly assign, for each day of the simulation, the correct number of trip arrivals to each day of the simulation and then scheduling the exact minute of arrival based on the hourly arrival distribution. A trip itinerary must be randomly selected from a pool of trips specified from the cluster analysis for the entry gate and time of year. This is done for each gate by proceeding from the first day of the simulation to the last day of the simulation according to the following procedure:

For each day,

1. The week of arrival is determined (1 through 53)
2. The week day is determined (Monday through Sunday).
3. The total number of arrivals for the week for the current gate is determined from the weighted traffic count data for the week.
4. The weekly arrival distribution for the current week is selected
5. The total number of arrivals for the week is multiplied by each day in the weekly arrival distribution. This determines the number of arrivals for each day of the week.
6. The hourly distribution is then selected
7. The exact time of arrival for the current trip is selected from the hourly arrival distribution by generating a random number and calculating the exact minute of arrival by piecewise linear interpolation from the hourly arrival distribution.
8. Finally, the trip clusters that occur in this week are looked up and then all trips that fall in these clusters for the current entry gate are selected to create the pool of trips. Each trip in the pool has an equal chance of being selected. A random number is generated and a single trip is selected from the pool
9. The trip selection is now complete and the process is repeated for the next trip until all trips for the current day are selected. If all trips are scheduled, the process progresses to the next day of the simulation and the process is repeated until the last day of the simulation.

The first two weeks of the simulation outputs must be discarded because the simulation starts with no visitors so any outputs during an initial "warm up" period are not representative of system behavior. To correct this problem, the simulation is allowed to run until the system comes to full capacity. In this case the period that was selected was two weeks since the longest visits are 11 days. The warm up period gives the simulation time to populate the system to capacity before collecting statistics from the simulation. For the winter simulation 1 week of simulation was obtained for analysis.

Two performance indicators were measured for links: Link Use and Link Encounters. Link use is a frequency count for the number of parties visiting each link for each day of the simulation. Link Encounters is the number of direct contacts between parties along a link. There were 658 links visited and 493 links with encounters recorded in the simulation. Since daily link use and link encounters were generated, there were 3799 link use days and 2083 link encounter days recorded with confidence intervals calculated for each.

STATISTICAL METHODS FOR DETERMINING THE NUMBER OF REPLICATIONS FOR TERMINATING SIMULATIONS

The method of independent replications requires the model is run for a "small" number of replications. In the case of probabilistic simulations, "small" may mean 10 to 15 replications.

The next step is to calculate the (1-α) confidence interval using equation 1. (See Centeno and Reyes, 1998 and Law and Kelton, 2000 pp 253-259)

$$\overline{X}(n) \pm t_{1-\alpha/2}\sqrt{\frac{S^2(n)}{n}} \quad (1)$$

WHERE $\overline{X}$ is the mean of the performance indicator for the current replication
$t_{1-\alpha/2}$ is the (1-α) percentile of the t-student distribution with n-1 degrees of freedom
$S^2(n)$ is the sample variance

In equation 1 the expression: $t_{1-\alpha/2}\sqrt{\frac{S^2(n)}{n}}$

is referred to as the confidence interval half width. If this value is less than the user specified accuracy after the initial n replications for the "short run" then there is no need for further replications. However if this value is larger then the user specified accuracy, then n can be estimated using Equation 2.

$$n^* = \text{Round}\left[n \times \left(\frac{h}{h^*}\right)^2\right] \quad (2)$$

WHERE n^* is the estimated number of replications needed
h is the half width from the sample run
h* is the desired half width or absolute accuracy specified by the user.

Law and Kelton (2000, p512) suggest a modification of the above estimate in a method they call the iterative method. In this case the number of replications is increased by 1 each time and the confidence interval is recomputed after each iteration, until the desired accuracy is achieved. This method assumes that the population variance will not change (appreciably) as the number of replications increase.

$$n_a^*(\beta) = \min\left\{i \geq n : t_{i-1,1-\alpha/2}\sqrt{\frac{S^2(n)}{i}} \leq \beta\right\} \quad (3)$$

WHERE $n_a^*(\beta)$ is the estimated number of replications needed with absolute accuracy β
n is the number of replications from the "short run"
$t_{i-1,1-\alpha/2}$ is the is the (1-α) percentile of the t-student distribution with i-1 degrees of freedom
$S^2(n)$ is the sample variance from n replications
i is the iteration (greater than n)

A third method described by Law and Kelton (2000, p. 513) uses a measure called "relative accuracy". Relative accuracy is the Confidence Interval Half Width from Equation 1 divided by $\overline{X}$ is an estimate of the actual relative error. In this method the user specifies a desired relative error λ.

$$n_r^*(\lambda) = \min\left\{i \geq n : \frac{t_{i-1,1-\alpha/2}\sqrt{S^2(n)/i}}{|\overline{X}(n)|} \leq 1\right\} \quad (4)$$

WHERE $n_r^*(\lambda)$ is the estimated number of replications needed with relative accuracy λ
N is the number of replications from the "short run"
λ is the user-specified relative accuracy
$t_{1-\alpha/2}$ is the is the (1-α) percentile of the t-student distribution with n-1 degrees of freedom
$S^2(n)$ is the sample variance from n replications
i is the iteration (greater than n)

MULTIPLE PERFORMANCE INDICATORS AND THE BONFERRONI CORRECTION

In a typical spatial simulation, there is normally more than one performance indicator being measured. If simulations are viewed as experiments where we are testing hypotheses about the system under study, then the alpha levels of the statistical tests applied in the previous section must be adjusted using the Bonferroni Correction (MathWorld, 2005). The Bonferroni Correction is used when several tests are being performed simultaneously. Where a given alpha level for a single performance indictor may be appropriate, it is not for the set of all comparisons. The simplest form of the Bonferroni Correction is to take the desired alpha level and divide by the number of performance indicators being tested. Thus, if the desired alpha level is 0.10 and there are five performance indicators, the adjusted alpha level would be 0.10/5 or 0.02 for each test.

Law and Kelton (2000) suggest another approach in which the sum of the alpha levels for each test equals the desired alpha level, suggesting that each performance indicator can have a unique alpha level in the statistical test. They also suggest that more than 10 performance indicators is impractical and that, given the stochastic nature of simulation it may be impractical to meet statistical requirements for all performance indicators simultaneously and that one must just have to accept that some indicators may not be used reliably in drawing conclusions from the simulation.

DETERMINING THE NUMBER OF REPLICATIONS FOR SPATIAL SIMULATIONS

In traditional industrial applications of simulation such as manufacturing and queuing simulations a single mean for each performance indicator is all that is needed. In spatial simulations however, the problem is more complex as performance indicators can vary spatially as in the case of travel simulations where performance indicators for each destination must be analysed simultaneously. Essentially, the approach required is to apply the same statistical methods described in section 2 to each and every location in the spatial simulation where performance indicators are to be measured. For instance, in travel simulations we may be interested in:

- the total visits per destination,
- average visit duration per destination, and
- average queuing times at parking facilities.

In this case, we have three performance indicators. Suppose our network has 10 destinations we wish to evaluate. We must first determine the alpha level we wish to test for. If the overall alpha is 0.10 then, according to the Bonferroni Correction we must use an alpha level of 0.10/3 or 0.03.

Next we decide which method will be used to determine the number of replications either by specifying the confidence interval half width for equations 2 and 3 or by specifying the relative accuracy for equation 4.

The simulation is then replicated for a "short run" of say, 10 replications and the outputs from each of the 10 destinations for the 3 performance indicators are gathered. Using this output we then apply the corresponding method (equations 2, 3 or 4) for each of the 10 destinations for the 3 performance indicators using an alpha level of 0.03.

For purposes of understanding the implications of the three methods for estimating the number of replications needed to obtain user specified measures of confidence and reliability from the simulation, the results of two links are shown in the tables below. Confidence intervals for the 7 replications for daily link use and daily link encounters were calculated using different alpha values and different user-specified confidence half-width values to investigate the impact these values have on the three methods for estimating the number of replications.

Table 3 shows the means and standard deviations for link use and link encounters for two links. Link 116 and Link 26. Link 116 is typified by low number of encounters as compared to Link 126. Note that standard deviations for link encounters are relatively high compared to link use. This is because link encounters are much more sensitive to random variations than link use since small changes in arrival times and volumes can change the number of encounters on a link.

Table 4 shows the confidence interval half width (CI Half) and relative precision (Relative Precision) from confidence intervals calculated with an alpha of 0.10 (90% confidence). Law and Kelton (2000) suggest that an alpha of 0.10 is a reasonable level of confidence given that random numbers are used to generate the performance indicators. Note the large values for relative precision for link encounters as compared to link use. Remember that the relative precision is the ratio of the confidence interval half width with the mean. This shows the value of this measure to indicate the amount of variation between simulation runs.

Table 5 shows the estimated number of replications for link use using the three different methods described in this paper. For equations 2 and 3 an absolute accuracy (the user specified confidence interval half width) of 1 is used. For Equation 4 a user specified relative accuracy of 0.15 is used. For link use equation 4 shows that there is no need for further replications whereas equations 2 and 3 show that many more replications are needed to reduce the confidence interval to a half width of 1. This is instructive, because it may indicate that the criteria for absolute accuracy may be unrealistic given the high standard deviations for link use. Note that equation 3 (the iterative method) shows a consistently lower estimate for the number of replications required to generate confidence intervals with a half width of 1.

For equations 2 and 3 an absolute accuracy (the user specified confidence interval half width) of 1 is used. For Equation 4 a user specified relative accuracy of 0.15 is used. For link use equation 4 shows that there is no need for further replications whereas equations 2 and 3 show that many more replications are needed to reduce the confidence interval to a half width of 1. This is instructive, because it may indicate

Table 3. Means and Standard Deviations for Link Use and Link Encounters

	Date	Link Use		Link Encounters	
Link	Jan 03	Mean	StDev	Mean	StDev
116	15	380.71	20.18	2.86	3.23
116	16	387.00	15.00	3.86	5.64
116	17	401.57	25.55	3.71	2.86
116	18	432.86	21.79	3.71	5.31
116	19	448.71	20.88	6.29	6.82
126	15	559.14	18.41	32.86	16.23
126	16	580.14	19.73	48.71	31.44
126	17	595.43	28.58	32.29	35.96
126	18	679.00	16.41	86.71	74.31
126	19	670.71	17.73	25.14	28.74

Table 4. Confidence Half Intervals and Relative Precision for Link Use and Link Encounters, Alpha = 0.10

	Date	Link Use		Link Encounters	
Link	Jan-03	CI Half	Rel Prec	CI Half	Rel Prec
116	15	14.82	0.04	2.37	0.83
116	16	11.02	0.03	4.14	1.07
116	17	18.76	0.05	2.10	0.57
116	18	16.00	0.04	3.90	1.05
116	19	15.33	0.03	5.01	0.80
126	15	13.52	0.02	11.92	0.36
126	16	14.49	0.02	23.09	0.47
126	17	20.99	0.04	26.41	0.82
126	18	12.05	0.02	54.57	0.63
126	19	13.02	0.02	21.10	0.84

Table 5. shows the estimated number of replications for link use using the three different methods described in this paper.

	Date	Link Use			Link Encounters		
Link	Jan 03	Eq. 2	Eq. 3	Eq. 4	Eq. 2	Eq. 3	Eq. 4
116	15	1583	1113	7	39	31	156
116	16	850	616	7	120	88	260
116	17	2465	1784	7	31	25	74
116	18	1792	1297	7	106	79	249
116	19	1646	1192	7	176	128	144
126	15	1280	927	7	994	720	32
126	16	1470	1064	7	3731	2701	152
126	17	3084	2232	7	4883	3534	152
126	18	1016	736	7	20848	15089	91
126	19	1187	860	7	3118	2257	159

that the criteria for absolute accuracy may be unrealistic given the high standard deviations for link use. Note that equation 3 (the iterative method) shows a consistently lower estimate for the number of replications required to generate confidence intervals with a half width of 1.

In Table 5 the number of replications required for link encounters is greater for Equation 4 than equations 2 and 3. This is because there are relatively few encounters per link (see Table 6) for link 116. However for relative accuracy, we see that we would require anywhere from 74 to 260 replications in order to reach a relative precision of 0.15 for link 116. This reflects the high standard deviations as compared to the means for link encounters for link 116 and the resulting high relative precision. For link 126 there were many more encounters per day than link 116. The result is that we have a larger confidence interval half width, which requires many more replications to achieve our user specified absolute accuracy of 1. This shows how important it is to carefully select appropriate standards of accuracy for each performance indicator.

To examine the impact of the Bonferroni Correction, we now assume that in order to achieve an overall alpha of 0.10 for the simulation, we need to estimate replications using an alpha level of 0.05 for each of our two performance indicators.

Table 6 shows, as we would expect, that the confidence interval half widths and relative precisions have increased because of the lower alpha value.

In Table 7 we see a significant drop in the number of replications required from equations 2 and 3 for link use because we have now increased the user specified half width from 1 to 10. Equation 4 for link use shows no need for more replications because of the low relative precision values obtained as the result of the high use levels and small confidence interval half widths. For link encounters no further replications are required from equations 2 and 3 because of the low number of encounters and the high user specified confidence interval half width of 10. Equation 4 for node 116 still shows a large number of replications due to the high standard deviations as compared to the small confidence interval half widths (see Table 4). For Node 126 we see a large reduction in the number of replications as compared to table 3 primarily because of the increase in the user specified confidence interval half width of 10.

Now we look at the estimated replications for the entire network. Tables 8 and 9 show the summary of results for all links, for link use and link encounters for an alpha of 0.05, and a confidence interval

Table 6. Confidence Half Intervals and Relative Precision for Link Use and Link Encounters, Alpha = 0.05

Link	Date Jan-03	Link Use		Link Encounters	
		CI Half	Rel Prec	CI Half	Rel Prec
116	15	18.67	0.05	2.98	1.04
116	16	13.88	0.04	5.22	1.35
116	17	23.63	0.06	2.65	0.71
116	18	20.15	0.05	4.91	1.32
116	19	19.31	0.04	6.31	1.00
126	15	17.03	0.03	15.01	0.46
126	16	18.25	0.03	29.07	0.60
126	17	26.43	0.04	33.26	1.03
126	18	15.17	0.02	68.73	0.79
126	19	16.40	0.02	26.58	1.06

Table 7. Estimated replications using Equations 2, 3 and 4 for Link Use and Link Encounters, Alpha = 0.05, User CI Half Width = 10, User Relative Precision = 0.15

Link	Date Jan-03	Link Use			Link Encounters		
		Eq. 2	Eq. 3	Eq. 4	Eq. 2	Eq. 3	Eq. 4
116	15	24	19	7	7	7	221
116	16	13	12	7	7	7	370
116	17	39	28	7	7	7	104
116	18	28	21	7	7	7	354
116	19	26	20	7	7	7	204
126	15	20	16	7	16	13	44
126	16	23	18	7	59	41	74
126	17	49	34	7	77	53	215
126	18	16	13	7	331	200	128
126	19	19	15	7	49	35	226

half width of 10. Note in table 6 it is possible to achieve the user desired accuracy for 90% of the links with less than 60 replications for link use whereas in table 7 it takes over 180 replications to achieve the same coverage for link encounters.

Tables 8, 9 & 10 show the results of equation 4 with a relative accuracy of 0.15. A similar pattern is seen here with less than 40 replications needed for link use to achieve the relative precision for 0.15 for link use and over 500 replications required for link encounters. Since both link use and link encounters are generated in the same simulation the obvious method for selecting the required simulations is simply to take the maximum value from all estimates, which is this case is over 500 replications. However this ignores the costs of processing time. In this simulation, each replication takes around an hour to run. If we run the simulation for 500 replications this means 500 hours of computer time and with limitations on file sizes for output databases, it is likely the simulation will fail from reaching file size limits.

These results show a number of important characteristics of performance indicators and the impact on the number of replications required to meet user-specified reliability measures. First the more sensitive a performance measure is to random variation, the higher the variances and the wider the confidence interval and therefore the greater number of replications required. Second, if using equations 2 and 3 for estimating the number of replications, it is important to carefully select the absolute accuracy (desired CI half width) for each performance indicator – it may be helpful to use the output analysis for the "short run" to help determine reasonable values for each performance indicator. Third, absolute accuracy and relative precision are two very different measures yielding very different results in terms of calculating the number of replications. It may be useful to evaluate both measures when estimating the number of replications for terminating simulations using the methods described in this paper.

The results of this analysis showed that for node use, link use, and overnight node encounters the required 95% confidence intervals could be achieved within the 5 replications. For link encounters however, because of the higher impact of small variations on encounters, 95.7% of the links produced 95% confidence intervals within 5 replications, another 2.82% requiring an estimated 6 to 10 replications and the remaining 1.5% requiring an estimated 11 to 104 replications to achieve 95% confidence intervals (See Table 11)

Table 8. Number of replications for link use using equation 3 for alpha = 0.05, CI Half Width = 10 for entire network

Link Use Absolute Accuracy = 10			
Reps	Count	%Links	Accum%
1: 20	2926	77.02%	77.02%
21: 40	428	11.27%	88.29%
41: 60	111	2.92%	91.21%
61: 80	12	0.32%	91.52%
81:100	16	0.42%	91.95%
101:120	6	0.16%	92.10%
121:140	17	0.45%	92.55%
141:160	16	0.42%	92.97%
181:200	267	7.03%	100.00%
Total	3799	100.00%	

Table 9. Number of replications for link encounters using equation 3 for alpha = 0.05, CI Half Width = 10 for entire network

Link Encounters Absolute Accuracy = 10			
Reps	Count	%Links	Accum%
1: 20	1120	53.77%	53.77%
21: 40	169	8.11%	61.88%
41: 60	95	4.56%	66.44%
61: 80	66	3.17%	69.61%
81:100	61	2.93%	72.54%
101:120	45	2.16%	74.70%
121:140	36	1.73%	76.43%
141:160	33	1.58%	78.01%
161:180	26	1.25%	79.26%
181:200	432	20.74%	100.00%
Total	2083	100.00%	

Table 10. Number of replications for link use using equation 4 for alpha = 0.05 relative accuracy = 0.15 for entire network

Link Use - Relative Precision = 0.15			
Reps	Count	%	Accum%
1: 40	3493	91.95%	91.95%
41: 80	87	2.29%	94.24%
81:120	44	1.16%	95.39%
121:160	23	0.61%	96.00%
201:240	14	0.37%	96.37%
281:320	10	0.26%	96.63%
401:440	30	0.79%	97.42%
481:500	23	0.61%	98.03%
501:	75	1.97%	100.00%
Total	3799	100.00%	

DEMONSTRATION OF PROCEDURES FOR DEVELOPING PROBABILISTIC SIMULATIONS FOR WINTER AND SUMMER MONTHS

After analysis of the baseline trips looking at sample size during each week of the year and the results of the cluster analysis described in section 7.2, two periods of the year were selected to demonstrate techniques developed to generate probabilistic trips from survey trip itineraries and traffic counts. The winter period selected is the month of January 2003. The summer period selected is 4 weeks beginning June 23, 2003 and ending July 24, 2003. The first two weeks of the simulation outputs must be discarded because the simulation starts with no visitors so any outputs during an initial "warm up" period are not representative of system behavior. To correct this problem, the simulation is allowed to run until the system comes to full capacity. In this case the period that was selected was two weeks since the longest visits are 11 days. The warm up period gives the simulation time to populate the system to capacity before collecting statistics from the simulation. For the winter simulation 2 weeks of simulation was obtained for analysis, for the summer simulation period 1 week of simulation outputs were obtained.

Table 11. Number of replications for link encounters using equation 4 for alpha = 0.05 relative accuracy = 0.15 for entire network

Link Encounters - Relative Precision = 0.15			
Reps	Count	%	Accum%
1: 40	57	2.74%	2.74%
41: 80	92	4.42%	7.15%
81:120	127	6.10%	13.25%
121:160	139	6.67%	19.92%
161:200	117	5.62%	25.54%
201:240	124	5.95%	31.49%
241:280	117	5.62%	37.11%
281:320	97	4.66%	41.77%
321:360	54	2.59%	44.36%
361:400	28	1.34%	45.70%
401:440	147	7.06%	52.76%
441:480	131	6.29%	59.05%
481:500	48	2.30%	61.35%
501:	805	38.65%	100.00%
Total	2083	100.00%	

RESULTS OF TRAFFIC COUNT ANALYSIS

Traffic count data for the six entry gates from which surveys were obtained were adjusted to reflect the survey population. These weighted counts were then analyzed to determine if there were common weekday arrival distributions through the year for each gate, and then re-analyzed to determine if there were common hourly arrival distributions during 2003. The detailed results for weekly and hourly distributions have been provided to Parks Canada in the form of a Microsoft Access database.

WEEKDAY ARRIVAL DISTRIBUTIONS AT GATES

In this analysis, a week is defined as a 7-day period beginning on Mondays. In 2003, there are 53 weeks. A Chi Square test was performed for each gate comparing each weekly arrival distribution against each other week in the year. Table 12 shows the results of this analysis. 3 gates: Entry 1(Banff East), entry 4 (Jasper East) and entry 7 (Kootenay) showed the highest variation in weekday arrival distributions, with entry 1 having no two weeks that had the same weekday arrival distribution. Gate 6 had lowest number of unique distributions with almost 55% of the weeks sharing the same distribution.

HOURLY ARRIVAL DISTRIBUTIONS AT GATES

The hourly distributions were examined for the 365 days of the year in 2003. Table 13 shows a similar result to the weekly distributions in that entries 1, 4 and 7 had the greatest variation in hourly arrival distributions.

CLUSTER ANALYSIS OF SURVEY ITINERARIES BY SEASON AND BY ACTIVITY.

The survey trips represent a "single realization" of the full variation of travel patterns in the four parks. In reality, the selection of trip itineraries will vary within the constraints of seasonal patterns, which will constrain certain activities and access to destinations because of weather. In probabilistic simulations, trip itineraries are drawn at random from a "pool" of trips that have been sampled from the population of all possible trips. Table 14 shows the number of trip diaries collected at each entry gate.

Because it is expected that trip itineraries change during the season from different entry points, the trips must be grouped or clustered in order to classify trip itineraries according to seasonal variations in pattern of use. In order to do this, a number of different cluster analysis techniques were applied to the 1620 survey trips using a number of different clusters and against different definitions of a "season".

The strongest clustering results came from a technique called "kmeans". In this technique, the analyst must nominate the number of clusters. 3 and 4 clusters were tested with 3 clusters differentiating the winter season trips most distinctly. The other trips were clustered on the basis of the type of activity: either active (hiking, biking etc) or inactive. Figure 2 shows the results of the kmeans cluster analysis with 3 clusters. Table 15 illustrates the results of the analysis where cluster 3 is the winter trips cluster 1 and 2 are the clusters based on the type of activity (active or inactive).

METHODOLOGY FOR INTEGRATING ARRIVAL DISTRIBUTIONS WITH CLUSTERED TRIP ITINERARIES

The next step in the development of the probabilistic simulation is to randomly assign, for each day of the simulation, the correct number of trip arrivals to each day of the simulation and then scheduling the exact minute of arrival based on the hourly arrival distribution. A trip itinerary must be randomly selected from a pool of trips specified from the cluster analysis for the entry gate and time of year. This is done for each gate by proceeding from the first day of the simulation to the last day of the simulation according to the following procedure:

For each day,

1. The week of arrival is determined (1 through 53)

Table 12. Results of Chi Square test to determine common weekday arrival distributions for 53 weeks in 2003 for each entry gate.

Entry Gate Location Description	Unique Weekly Distributions
Entry 1 1.6 KM E OF BANFF PARK GATES	53
Entry 3 2.3 KM E OF BANFF PARK GATES	33
Entry 4 EAST JASPER PARK GATES	52
Entry 5 19.1 KM W OF 16 & 93 W JASPER TOWNSITE Entry 6 ALTA-BC BORDER Entry 7 0.3 KM S OF KOOTENAY PARK GATES	35 29 49

Table 13. Results of Chi Square test to determine common hourly arrival distributions for 365 days in 2003 for each entry gate

Entry	Location Description	Hourly Distributions Unique
Entry 1	1.6 KM E OF BANFF PARK GATES	133
Entry 3	2.3 KM E OF BANFF PARK GATES	39
Entry 4	EAST JASPER PARK GATES	168
Entry 5	19.1 KM W OF 16 & 93 W JASPER TOWNSITE	68
Entry 6	ALTA-BC BORDER	79
Entry 7	0.3 KM S OF KOOTENAY PARK GATES	127

Table 14. Survey trip diaries collected at each entry Gate

Gate	Location Description	Number of Trip diaries
Gate 1	1.6 KM E OF BANFF PARK GATES	878
Gate 3	2.3 KM E OF BANFF PARK GATES	94
Gate 4	EAST JASPER PARK GATES	385
Gate 5	19.1 KM W OF 16 & 93 W JASPER TOWNSITE ALTA-BC	173
Gate 6	BORDER	34
Gate 7	0.3 KM S OF KOOTENAY PARK GATES	65

2. The weekday is determined (Monday through Sunday).

3. The total number of arrivals for the week for the current gate is determined from the weighted traffic count data for the week.

4. The weekly arrival distribution for the current week is selected

5. The total number of arrivals for the week is multiplied by each day in the weekly arrival distribution. This determines the number of arrivals for each day of the week.

6. The hourly distribution is then selected

7. The exact time of arrival for the current trip is selected from the hourly arrival distribution by generating a random number and calculating the exact minute of arrival by piecewise linear interpolation from the hourly arrival distribution.

8. Finally, the trip clusters that occur in this week are looked up and then all trips that fall in these clusters for the current entry gate are selected to create the pool of trips. Each trip in the pool has an equal chance of being selected. A random number is generated and a single trip is selected from the pool

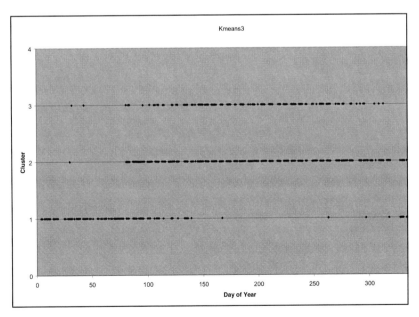

Figure 2. Results of cluster analysis of destinations and activities for 1620 survey trips using kmeans and 3 clusters.

Table 15. Number of trips allocated to each cluster at each entry gate from the kmeans cluster analysis with 3 clusters. Note that 6 trips were not allocated to clusters.

Gate	Location Description	Cluster	Num Trips
Gate 1	1.6 KM E OF BANFF PARK GATES		5
Gate 1	1.6 KM E OF BANFF PARK GATES	1	223
Gate 1	1.6 KM E OF BANFF PARK GATES	2	550
Gate 1	1.6 KM E OF BANFF PARK GATES	3	100
Gate 3	2.3 KM E OF BANFF PARK GATES	1	6
Gate 3	2.3 KM E OF BANFF PARK GATES	2	80
Gate 3	2.3 KM E OF BANFF PARK GATES	3	8
Gate 4	EAST JASPER PARK GATES		1
Gate 4	EAST JASPER PARK GATES	1	87
Gate 4	EAST JASPER PARK GATES	2	250
Gate 4	EAST JASPER PARK GATES	3	47
Gate 5	19.1 KM W OF 16 & 93 W JASPER TOWNSITE	1	12
Gate 5	19.1 KM W OF 16 & 93 W JASPER TOWNSITE	2	97
Gate 5	19.1 KM W OF 16 & 93 W JASPER TOWNSITE	3	64
Gate 6	ALTA-BC BORDER	1	5
Gate 6	ALTA-BC BORDER	2	23
Gate 6	ALTA-BC BORDER	3	6
Gate 7	0.3 KM S OF KOOTENAY PARK GATES	1	10
Gate 7	0.3 KM S OF KOOTENAY PARK GATES	2	46
Gate 7	0.3 KM S OF KOOTENAY PARK GATES	3	9

9. The trip selection is now complete and the process is repeated for then next trip until all trips for the current day are selected. If all trips are scheduled, the process progresses to the next day of the simulation and the process is repeated until the last day of the simulation.

DEMONSTRATION OF PROBABILISTIC SIMULATION FOR WINTER ARRIVALS

The above methodology was implemented for a two-week period in winter (after deleting a two week "warm up" period). Table 16 shows the weekday arrivals for the 4 weeks of the winter simulation (weeks 1-4 2003). A single replication of the simulation was run since this was for demonstration purposes only. However, the statistical methods for determining the number of replications required to establish confidence intervals from the simulation outputs are identical to those describe for the baseline simulation. Figure 9 and Figure 10 show examples of mapped output from the simulation for node use and link use. Other possible outputs could include number of encounters at nodes and links per day, use at busiest hour of the day, the busiest link or node, and other summaries of use which can be made by time, by location, or by traveler.

DEMONSTRATION OF PROBABILISTIC SIMULATION FOR SUMMER ARRIVALS

The above methodology was implemented for a one-week period in summer (after deleting a two week "warm up" period). Table 17 shows the weekday arrivals for the 4 weeks of the summer simulation (weeks 25-28 2003). A single replication of the simulation was run since this was for demonstration purposes only. However, the statistical methods for determining the number of replications required to establish confidence intervals from the simulation outputs are identical to those describe for the baseline simulation. Figure 4 and 5 show examples of mapped output from the simulation for node use and link use.

CONCLUSIONS

The purpose of this study was to develop a baseline model of patterns of use in the four Mountain parks based on the 2003 Visitor Patterns of Use Study. This study used visitor patterns of use data in conjunction with simulation to evaluate spatial and temporal patterns of use and explore the data to determine if there was a sufficient sample to construct a probabilistic model. Using simulation in conjunction with visitor use data has many benefits. It provides:

- A comprehensive and dynamic understanding of visitor behavior, interactions between visitors and interactions between visitors and the resource base.

- A framework for a more holistic and comprehensive way of incorporating visitor information into the planning and management process.

- A way of measuring visitor interactions that is difficult or expensive to do in the field.

- A way to test alternative management scenario's and putting management into an exploratory and experimental framework

- An effective way to communicate complex inter-related issues in recreation management to the public and decision makers.

Table 16. Weekday arrivals for each entry gate for winter simulation from weighted traffic count 2003

Entry Gate	Week	Weekly Total	Mon	Tue	Wed	Thu	Fri	Sat	Sun
Entry 1	1	5091	0	0	967	1094	1104	1037	889
Entry 1	2	5257	669	688	680	695	867	915	743
Entry 1	3	5558	625	649	648	708	962	1061	905
Entry 1	4	4963	612	587	586	664	884	824	806
Entry 3	1	80	0	0	17	21	18	9	15
Entry 3	2	71	15	15	6	5	6	9	15
Entry 3	3	67	15	8	8	7	6	10	13
Entry 3	4	80	13	11	10	8	11	10	17
Entry 4	1	1969	0	0	334	504	466	357	308
Entry 4	2	1906	265	280	267	266	350	257	221
Entry 4	3	2024	238	228	240	281	444	294	299
Entry 4	4	1746	221	247	220	244	337	244	233
Entry 5	1	171	0	0	25	38	36	37	35
Entry 5	2	129	0	0	19	29	27	28	26
Entry 5	3	124	0	0	18	28	26	27	25
Entry 5	4	102	0	0	15	23	22	22	21
Entry 6	1	463	0	0	81	96	97	98	91
Entry 6	2	356	0	0	62	74	75	75	70
Entry 6	3	350	0	0	61	73	73	74	69
Entry 6	4	329	0	0	58	68	69	70	65
Entry 7	1	778	0	0	198	167	136	128	149
Entry 7	2	448	60	52	55	56	66	56	103
Entry 7	3	434	50	44	50	50	59	55	126
Entry 7	4	402	51	43	44	46	61	47	110

While simulation has been shown to an effective tool to evaluation spatial and temporal patterns of use, there are still some inherent issues that must be addressed before reliable and accurate simulations can be developed. This chapter addresses two of those major issues: *How to determine the number of simulation replications required that meets some user specified degree of reliability and accuracy?* and *How to develop methods and collect accurate data that encompasses the spatial and temporal patterns of visitor behavior necessary to develop probabilistic simulations for accurate forecasting and decision making?* These two issues are critical in any spatial simulation project that is focused on outputs that are directly used for management decision making. This chapter has shown that accurate and reliable spatial simulations can be developed using the procedures outlined above for evaluating patterns for Banff, Yoho, Kootenay and Jasper National Parks. Results clearly indicate that for three selected indicators; node (destination) use, link (route) use, and overnight destination encounters, the required 95% confidence intervals could be achieved within the 5 replications. For link encounters however, because of the higher impact of small variations on encounters, 95.7% of the links produced 95% confidence intervals within 5 replications, another 2.82% requiring an estimated 6 to 10 replications and the remaining 1.5% requiring an estimated 11 to 104 replications to achieve 95% confidence intervals.

With regards to visitor sampling, this study has concluded that since conventional social/economic sampling assumes a single population for the whole season of use, that spatial and temporal simulation modeling has a much higher degree of complexity and data requirements. Since these types of spatial simulations require sampling at multiple entry points, over a variety of seasons that there is a much higher need for systematic random sampling for weekday and hourly arrivals. The sample size acquired in the 2003 Pattern of Visitor Use study certain met the requirements of any well designed social science study, but the data collection design falls short of the requirements needed to develop valid travel simulation models for management decision making. Such methods do not take into account the spatial and temporal variations. This study clearly illustrated that based on the minimum requirements to do a chi-square test on daily arrival distributions a minimum of 5 arrivals per day for 80% of the cases being compared are required. This means that for weekdays 6 out of 7 weekdays with a minimum of 5 arrivals or 30 arrivals per week (realistically something closer to 50 surveys a week) are required in order to make any statistical generalizations or comparisons between the results of the trace simulation to the probabilistic simulation.

Based on the minimum requirements to do a chi-square test on daily arrival distributions a minimum of 5 arrivals per day for 80% of the cases being compared is required. This means that for weekdays 6 out of 7 weekdays with a minimum of 5 arrivals or 30 arrivals per week (realistically something closer to 50 surveys a week) are required in order to make any statistical generalizations or comparisons between the results of the trace simulation to the probabilistic simulation. An evaluation of the 2003 Pattern of Visitor use data revealed only four weeks that met this standard that could be used to develop a probabilistic simulation. While this was adequate for demonstration purposes, but it falls far short of the data required to build an accurate probabilistic simulation necessary for decision making. Much more work needs to be done to ensure that data collection and field based methodologies in particular be developed to capture the data necessary for building accurate and

Table 17. Weekday arrivals for each entry gate for summer simulation from weighted traffic count 2003

Entry Gate	Week	Week Total	Mon	Tue	Wed	Thu	Fri	Sat	Sun
Entry 1	25	9970	1341	1253	1306	1431	1751	1523	1365
Entry 1	26	14054	1375	1368	1485	1804	3023	2691	2308
Entry 1	27	18829	2085	3037	2452	2692	3007	2629	2927
Entry 1	28	20696	2865	2525	2564	2804	3500	3256	3182
Entry 3	25	276	41	31	27	38	48	43	48
Entry 3	26	448	38	35	43	53	91	96	92
Entry 3	27	878	82	136	115	127	141	137	140
Entry 3	28	903	119	103	103	116	136	165	161
Entry 4	25	4523	610	583	586	645	800	611	688
Entry 4	26	6567	683	635	700	885	1329	1245	1090
Entry 4	27	6805	1022	951	890	924	1095	947	976
Entry 4	28	6569	958	855	843	913	1103	955	942
Entry 5	25	1501	209	209	203	220	228	211	221
Entry 5	26	1964	220	221	242	287	368	318	308
Entry 5	27	3509	338	608	446	553	563	475	526
Entry 5	28	3408	465	444	423	491	507	494	584
Entry 6	25	1386	182	222	193	181	184	212	212
Entry 6	26	1695	218	258	207	206	232	269	305
Entry 6	27	3433	276	427	761	497	466	498	508
Entry 6	28	3295	492	554	438	424	431	467	489
Entry 7	25	2655	305	287	302	326	366	344	725
Entry 7	26	2796	347	309	320	335	379	391	715
Entry 7	27	5488	683	1170	641	637	642	658	1057
Entry 7	28	4851	652	576	617	621	638	614	1133

reliable simulation models that decision makers can use and obtain defensible and credible results.

LITERATURE CITED

Centeno, M. and R. M. Florencia. 1998, "So you have your model: What to do next. A tutorial on simulation output analysis" in Proceedings of the 1998 winter Simulation conference, D.J. Medeiros, E.F. Watson, J.S. Carson and M.S. Manivannan, editors. http://www.informs-cs.org

Cole, D. (compiler). 2005. Computer Simulation Modeling of Recreation Use: Current Status, Case Studies, and Future Directions. Gen. Tech. Rep. RMRS-GTR-143. Ogden, UT: U.S. Department of Agriculture, Forest Service, Rocky Mountain Research Station. September 2005. Pgs 75. http://www.fs.fed.us/rm/pubs/rmrs_gtr143.html.

GeoDimensions Pty Ltd. 2005. Pattern of Human Use Simulation in Canada's Mountain Parks. Final Report - Parks Canada. March, 2005.

Gimblett, H. R., M. Richards & R. Itami. 2000. RBSim: Geographic simulation of wilderness recreation behavior. Journal of Forestry, 99(4), 36-42.

Gimblett, H.R. (Ed.). (2002). Integrating GIS and Agent Based Modeling Techniques for Understanding Social and Ecological Processes. Oxford University Press. 2002. 329pgs.

Itami, R., Raulings, R., MacLaren, G., Hirst, K., Gimblett, R., Zanon, D., & Chladek, P. 2004. RBSim 2: Simulating the complex interactions between human movement and the outdoor recreation environment. Journal of Nature Conservation, 11(4), 278-286.

Itami, R.M., D. Zell, F. Grigel & H.R. Gimblett. 2005. Generating Confidence Intervals for Spatial Simulations: Determining the Number of Replications for Spatial Terminating Simulations. In Zerger, A. and Argent, R.M. (eds) MODSIM 2005 International Congress on Modelling and Simulation. Modelling and Simulation Society of Australia and New Zealand, December 2005, pp. 141-148. ISBN: 0-9758400-2-9. http://www.mssanz.org.au/modsim05/papers/itami.pdf

Law, Averill M. & D. W. Kelton. 2000. Simulation Modeling and Analysis, Third Edition, McGraw Hill, Boston.

McVetty, D. 2003. Understanding Visitor Flows In Canada's Mountain National Parks: The Patterns Of Visitor Use Studies In Banff, Jasper, Kootenay And Yoho National Parks. Paper presented at the 5th International SAMPAA Conference. Victoria, British Columbia. Canada. May 11-16, 2003.

CHAPTER 22

WHAT CAN BE LEARNED FROM MULTI-AGENT SYSTEM MODELING?

Danielle J. Marceau

Abstract: Considering the increasing popularity of multi-agent system (MAS) modeling applied to environmental systems, this paper aims at investigating how this simulation approach fits within the generally accepted scientific framework and how useful are these models to scientists, managers and decision makers concerned by environmental resource management. To do so, MAS modeling is situated within the context of Complexity Theory and Post-Normal Science. The key role of MAS modeling is highlighted and emerging research trends that incorporate advanced concepts required for the study of complex environmental systems within a MAS architecture are identified.

Key Words: multi-agent system, complexity theory, post-normal science, exploratory approach

INTRODUCTION

Multi-agent systems (MAS) originate from computer science, more specifically from the field of distributed artificial intelligence (Moulin and Chaib-Draa 1996). Since its very beginning in the 1950s, artificial intelligence aimed at the creation of an intelligent machine capable of performing complex tasks similar to those achieved by human beings, such as reasoning, communication and learning. The development of MAS results from the desire and need of building computer systems able to operate independently, without direct human intervention, and to interact efficiently, i.e. to cooperate and reach agreement with other systems and the external environment when necessary (Wooldrige 2002; Conte et al., 1998). In that context, an agent is defined as a computer system, situated in some environment, and capable of its own action in order to satisfy its design objectives (Wooldrige 2002). By extension, a multi-agent system is a community of agents that work together to solve problems that are beyond their individual capabilities (Durfee et al. 1989), and which possesses the following characteristics. They are autonomous, i.e. they act independently of any controlling intelligence; they are goal-driven and tries to fulfill specific objectives; they are aware of and can respond to changes in their environment, defined as the space that supports their activities; they can also move within that environment; they are social and communicative: they interact with one another by exchanging messages using some language; they have the ability to cooperate, coordinate, and negotiate with each other, and finally, they can be designed to learn and adapt their state and behavior in response to stimuli from other agents and their environment. This emphasis on interactions that exist between agents is what distinguishes MAS from other systemic modeling approaches (Ferber 1999; Hare and Deadman 2004; Bousquet and Trébuil 2005).

Two main categories of agents can be distinguished: artificial or software agents, and agents that interact in the real world, which are the subject of this paper. The independent studies of Schelling (1971) and Sakoda (1971) are often referred to as the first real applications of a MAS conducted in the field of social sciences (Benenson and Torrens 2004; Axtell 2006). Their models aim at understanding households' migration behavior using a population of human agents interacting with each other. Another pioneer paper published by Hogeweg and Hesper (1983) describes an application of agent-based modeling (then called individual-oriented modeling) to study the emergent behavior of bumble bees within a colony. They illustrate that the population dynamics of a bumble bee colony combined with simple behavior of adult bees can generate the social interaction structure of the colony and its ontogeny. In 1984, Axelrod developed a well-known multi-agent model called the iterated prisoner's dilemma, a game in which each participant has to choose his/her strategy (typically in the form of cooperate or betray) to win a certain benefit. His work generated considerable interest in social and biological sciences and contributed to improve the understanding of the evolution of social norms and behaviors (Anderies 2000), as well illustrated by the work of Batten (2007) on self-defeating systems. However, it is only at the end of the 1980s, with the advent of object-oriented programming languages, that several research groups interested in testing the potential of MAS in a broad range of applications emerged. Ferber (1999) distinguishes five main areas of application of MAS, namely problem solving, building artificial worlds, collective robotics, program design, and multi-agent simulation. In this paper, we will only focus on the later, emphasizing the purpose and contribution of multi-agent system modeling in natural and social sciences, mostly for environmental resource management.

Over the last fifteen years, MAS have been increasingly developed to simulate different aspects of natural and social systems. In ecology, they have been implemented to study the relationships between predation, competition dynamics and ecosystem indicators (Bousquet et al. 1994), as well as population dynamics (Parry et al., 2004). They have also been used to simulate the migratory and foraging behavior of animal species and to understand how animals perceive, learn and adapt to their environment (Dumont and Hill, 2001; Robinson and Graniero 2004; Topping et al. 2003). MAS have been designed to simulate human/wildlife interactions (Ahern et al. 2001; Mathevet et al. 2003; Anwar et al. 2007; An et al. 2005) and human/landscape interactions (Gimblett et al. 2001; Itami and Gimblett 2001; Gimblett et al. 2002). Several authors have investigated the potential of MAS to simulate land-use changes, capture urban dynamics, and test alternative planning scenarios (Lim et al. 2002; Irwin and Bockstael 2002; Barros 2003; Kii and Doi 2005; Saarloos et al. 2005; Benenson, 1998; Omer 2005; Batty 2005; Benenson and Torrens 2004; Brown et al. 2004). MAS have also been intensively used to study human behavior and to simulate the decision making of stakeholders in order to assess the impact of their decisions on ecosystems

(Ligtenberg et al. 2001; Valkering et al. 2004; Kaufmann and Gebetsroither 2004; Monticino et al. 2007). MAS have been developed to simulate pedestrian movement in a city (Batty 2001; Kerridge et al. 2001; Haklay et al. 2001; Helbing et al. 2001). Finally, MAS have been applied to address a broad range of issues related to environmental resource management, such as water management (Feuillette et al. 2003; Lacombe and Naivinit 2005), forest management (Guizol and Purmono 2005; Moreno et al. 2007; Hoffmann et al. 2000), and agro-ecosystem management (Suphanchaimart et al. 2005; Lynam, T. 2000; Trébuil et al. 2005; Sengupta et al. 2005; Gross et al. 2006).

Considering this increasing popularity of MAS and the fact that it is a field of scientific investigation that has reached a certain level of maturity, this paper attempts to answer the generic question: What can be learned from MAS? More specifically, how does the MAS simulation approach fit within the generally accepted scientific framework to achieve knowledge? How can scientists as well as managers and decision makers use MAS models to achieve a greater understanding of the behavior of environmental systems and propose relevant management strategies and policies? To answer these questions, MAS modeling will be situated within the context of Complexity Theory and Post-Normal Science, and its key scientific role will be identified. The paper will conclude with the presentation of emerging research trends that need to be pursued to achieve the full potential of MAS modeling.

MAS AND COMPLEXITY THEORY

Complexity Theory is a new field of interdisciplinary research that emerged mainly over the last two decades from the interplay of physics, mathematics, biology, economy, and computer science, to study a category of systems referred to as complex systems for which conventional science approaches have been found inappropriate. Complex systems display the following characteristics. They are dynamic and consist of a large number of components that interact in a non-linear way, creating recognizable patterns of organization across spatial and temporal scales. The richness of these interactions generates emergent structures and behaviors described as global outcomes that cannot be predicted by examining the components of the system in isolation. Complex systems are open systems that undergo spontaneous self-organization; they use available energy and material to naturally organize and maintain themselves into a state that is at the threshold between complete disorder and complete order. These self-organizing systems are adaptive, i.e. they do not passively respond to events, but display remarkable adaptation to changing circumstances using feedback mechanisms. Examples of complex systems are provided by the stock market, the weather, earthquakes, traffic jams, ant colonies, ecosystems, and all living and intelligent organisms (Waldrop 1992; Chu et al. 2003; Parwini 2005; Kay et al. 1999; Parker et al. 2003).

These unique characteristics of complex systems make them intrinsically unpredictable; conventional reductionist science and traditional methods of analysis are unable to predict their behavior. In response to such a challenge, a novel scientific investigation approach has emerged referred to as simulation modeling, which has become the *de facto* methodology of much research on complexity (Batty 2005; Crawford et al. 2005). Simulation involves the use of digital computers as a laboratory for controlled experimentation. A simulation model can be compared to a box having several knobs, which allow a user to change the parameters, explore the links between components, tests various conditions and scenarios, and display the outputs in order to achieve understanding of the dynamics of the system under investigation (Grimm 1999). Computer simulation modeling is increasingly used to assist in environmental management and decision-making (Batten 2007).

MAS are considered by several scientists as an excellent simulation tool for the study of complex systems for several reasons. First, MAS rely on a bottom-up approach that starts by explicitly considering the components of a system (i.e. individual agents) and tries to understand how the system's properties emerge from the interactions among these components (Grimm 1999; Grimm et al. 2005). They therefore represent a powerful modeling tool to capture the dynamics of a system through agents' behavior and interactions. An et al. (2005) developed an agent-based spatial model to simulate the impact of the growing rural population on the forests and panda habitat in a natural reserve in China. These authors emphasize the necessity of the bottom-up approach that allows the decomposition of the population dynamics into individual life histories and the characterization of the dynamics of all households in the study area to really capture the importance of demographic factors in their model. Batty (2005) also advocates the necessity of a bottom-up modeling approach for the study of cities and urban development. He mentions that the spatial order we can observe in city patterns can only be explained by representing the basic elements of the city through cells and agents, where cells represent the physical and spatial structure of the city and where agents correspond to the human and social units that make the city evolve. He considers that agent-based modeling is one of the cornerstones of complexity theory where individual agents and their interactions are essential in explaining system behavior at a more aggregate level. This significance of the interactions between system's components, which is at the core of MAS, is central to the study of complex systems. Rather than focusing on the components of a system and their properties as in more traditional science, it is the relationship between these components that really matters when attempting to capture the dynamics of a system and the emergence of global structures (Parrott and Kok, 2000).

Another key aspect inherent to the study of complex systems that can be implemented within a MAS model is the capacity for the agents to perceive and adapt to their evolving environment. Portugali et al. (1997) and Benenson (1998) have conducted MAS simulation experiments in which agents adapt their residential behavior depending on the properties of their neighborhood and the whole city. Similarly, Barros (2003) has designed a MAS to simulate the locational decision of different economic groups in Latin American cities in order to better understand the process of 'peripherization', characterized by the formation of low-income residential areas in the peripheral ring of these cities. Adaptation requires a memory and some knowledge needed by the agents to make decisions. One example of such implementation is given by the study of Anwar et al. (2007) who built a MAS model to simulate whale-watching activities in the Saguenay-St. Lawrence Marine Park in Quebec, Canada. The 'boat operator' agents are designed with a memory and the capacity to plan and modify their itinerary based on messages received by other agents on whale location, and the success in encountering whales during their trips.

MAS allow the coupling of ecological and social systems such that the role of adaptive disaggregated human decision-making and its impact on the ecological system can be modeled. This is particularly useful to address problems of access and use of limited natural and renewable environmental resources (Hare and Deadman 2004; Parker et al. 2003, Bousquet and LePage 2004). A clear illustration of this approach is provided by Trébuil et al. (2005) in a study undertaken to explore the relationship between crop diversification and the risk of land degradation in northern Thailand. They build a MAS model to investigate simultaneously the interaction between the agronomic and socioeconomic components of the system, and to represent the diversity of the farming community and farmers' decision-making processes driving land-use changes at the village catchment level. Their model considerably improved the understanding of the functioning of this complex agricultural system. Another good example is provided by Monticino et al. (2007) who implemented a MAS to represent how various interacting stakeholders make decisions that affect land-use change and forest ecosystem sustainability in Texas, USA. A similar demonstration is given by several studies conducted to simulate the complex interactions between humans and outdoor recreation environments (Gimblett et al. 2001; Gimblett et al. 2002; Itami et al. 2003). The integration of both the physical properties of the environment and the characteristics of the human agents along with the planning of their activities and the way they make decisions is critical to fully capture the dynamics of such systems.

Understanding a complex system involves investigating interactions at a given level as well as among various connected levels. According to Holling (2001) in its recently proposed theory of Panarchy, the complexity of human and natural living systems emerges from a small number of controlling processes, referred to as critical self-organized variables, which take place in nested sets observable at various spatial and temporal scales. MAS can be effectively employed to understand the multi-scale organization of a system and how interactions are structured at each level of the hierarchy. Servat et al. (1998) have developed a multi-scale MAS to simulate runoff and infiltration processes in a watershed. The first level of their model consists of a population of individual water entities called waterball agents, which move according to their environment. The second level is composed of hydrological agents such as ponds and ravines, which are dynamically created when the speed of a waterball becomes null in a local minimum of topography. These authors advocate the necessity of integrating different scales in MAS simulations to capture scale transfer processes that are characteristics of complex systems. When applied in a social context, MAS can allow the investigation of the interactions among stakeholders at different scales and the emergence of collective responses that can be translated into environmental management policies. One illustration is provided by Ligtenberg et al. (2001) who built a spatial multi-actor model to simulate land-use changes as the result of actor-based decision-making. The agents have the ability to build their own image of how the spatial environment should be organized and to participate in a collective decision making procedure to allocate the new land use in the territory.

MAS AND POST-NORMAL SCIENCE

The concept of post-normal science has been first introduced in 1985 by Funtowicz and Ravetz, in response to the challenges of coping with many uncertainties in environmental policy issues, and has continuously evolved since then (Funtowicz and Ravetz 1993; Ravetz 1999; Ravetz and Funtowicz 1999; De Marchi and Ravetz

1999; Ravetz 2004). It is a new conception of science, which links epistemology and policy that focuses on problems typically neglected in traditional science and that are characteristics of complex systems. According to post-normal science, the reductionist analytical worldview that attempts to understand systems by dividing them into smaller elements is being replaced by a systemic, synthetic and humanistic approach. Systems are recognized as dynamic and complex and the scientific approach appropriate to deal with them is based on the assumptions of unpredictability, incomplete control, uncertainty, the influence of value, and a plurality of legitimate perspectives (Funtowicz and Ravetz 1993). It is generally acknowledged that post-normal science encompasses three essential elements: those related to uncertainty analysis and management, the integration of different sources of knowledge, and the participation of all the stakeholders, ranging from the expert scientist to the lay person referred to as the extended peer community, who has the desire to contribute to the resolution of the issue.

James Kay is among environmental scientists who strongly advocated the link between complexity and post-normal science as the only appropriate way to adequately deal with environmental systems. In one of his papers (Kay et al. 1999), he and his collaborators state: 'The dynamics of ecosystems and human systems need to be addressed in the context of post-normal science grounded in complex systems thinking'. In a series of publications, Kay recognizes that conventional science approaches to modeling and forecasting ecosystems are inappropriate and clearly promote both the necessity of a new understanding based on the notions of complex system theory and a radically different role the scientists should play when investigating such systems. That role should shift from being the expert inferring what will happen (i.e. making predictions) to providing decision makers and the community with an appreciation of how the future might unfold based on the description of how the system under investigation might evolve (Kay 1991; 1994; Kay and Schneider 1994; Kay et al. 1999).

The strongest example of design, implementation and application of MAS models within the post-normal science paradigm is provided by the companion modeling approach developed by François Bousquet and his collaborators. The main principle of this approach is to develop MAS simulation models that integrate various stakeholders' points of view, including the scientist's one, and to use them as a platform to facilitate dialogue, share learning, and improve collective decision making (Bousquet and Trébuil 2005; Bousquet et al. 2003). Stakeholders are considered more than simple users; they are actively involved in the construction of the model to ensure its relevance. A key aspect of companion modeling is the emphasis on the interactions among social agents regarded as actors who achieve decisions that affect the evolution of ecosystems through agreement, learning, conflict resolution, and collective action. One premise of this approach is that while the future of an ecosystem is unpredictable, it is partially decidable and desirable pathways of development can be encouraged through appropriate regulations and policies. Therefore, the objective of the modeling exercise is not prediction but rather interactive learning to identify desirable options based on various management scenarios.

Companion modeling typically involves three main steps. First, the stakeholders are identified. Then, techniques such as interviews and role-playing games are applied to gather information from these stakeholders about how they perceive and make decisions about the environment. Such information is used to validate hypotheses and design the MAS. Finally, simulations are run to illustrate the system's dynamics that emerges from the interactions among the agents and between the agents and their environment (Gurung 2005; Bousquet et al. 2003). An application of this methodology is provided by Campo (2005) who implemented a MAS model to simulate how individual actions of stakeholders collectively affect fish stock in coastal Bohol in the Philippines. Scenarios were developed to illustrate the potential of the MAS model as a tool for increasing the fishermen awareness of the current state of the coastal environment and natural resources of their municipality. A second example is given by the study conducted by Guizol and Purnomo (2005) who used a multi-agent simulation to explore scenarios of collaboration between stakeholders for forest plantation management in Malaysia. The purpose of their modeling approach was to create a virtual world that allows stakeholders to learn about the effects that their decisions might have on themselves and the environment. Promising solutions for increasing economic benefits of many stakeholders were identified through the MAS simulations. A similar demonstration of a participatory agent based modeling is done by Valkering et al. (2004). Their model incorporates the goals and beliefs of stakeholders and simulates the negociation process among them to identify which strategy receives the greatest support in the context of river management.

MAS AND THE SCIENTIFIC ENDEAVOR

Reflecting on the above contributions, how does MAS modeling fit within the general scientific endeavor of achieving knowledge? From the perspective of traditional science, two paths can be used in the pursuit of knowledge: the deductive and the inductive approaches. According to the former, a scientist states a problem that is linked to a specific body of knowledge or a recognized theory, formulates hypotheses, collects relevant observations, and designs an appropriate experiment to either verify or falsify the hypotheses. The scientist is expected to provide a formal proof of his/her results in support of the theory that represents the foundation of his/her work. If consistent experiments contradict the accepted theory, then a paradigm shift can occur in favor of the establishment of a new theory. This process has been well described by Kuhn (1970) and is referred to as 'normal science'. When using the inductive approach, a scientist starts with a set of observations that are in some ways classified in order to identify meaningful patterns. The scientist is seeking to achieve a certain degree of generalization that can be used as evidence in support of an existing theory (Skidmore 2002).

When scrutinized through these two traditional pathways, it might be argued that MAS modeling fails to provide satisfactory scientific answers. While a MAS modeler can effectively formulate relevant hypotheses about a macroscopic phenomena and systematically tests alternative explanations of this phenomena, he/she cannot prove the results using formal mathematics or logic. This statement has raised a strong debate on the credibility of MAS results and the necessity of applying appropriate validation techniques (Manson 2000; Amblard et al. 2006). Several strategies have been proposed including: providing a rigorous presentation of the structure of the model, comparing the results obtained from a MAS with other types of models, and testing the accuracy of the assumptions regarding the behavior and interactions of agents through experimental approaches (Bousquet and Le Page

2004). Other commonly applied validation procedures consist in comparing the simulated data with independent observed data using various metrics, performing sensitivity analysis to test the relative sensitivity of the model outputs to different parameters, and providing a qualitative assessment of the realism of the model outcomes (Ginot and Monod 2006). While the use of one of several of these techniques adds to the credibility of the simulation experiment, it still does not constitute a formal proof of the results.

When a MAS modeler conducts a series of simulation experiments in search of patterns or trends to understand the dynamics of a system, he/she follows an inductive approach. However, in contrast to pure induction, the set of observation data collected about the system in question is used to parameterize the model and, through simulations, generate a set of outputs. The risk in using these model outcomes to make generalization is that they might reflect some artifacts of the model rather than being meaningful patterns related to real-world phenomena. Another limitation of the inductive approach is that it does not provide a clear understanding of the exact mechanism linking the observed, or in this case simulated, facts and the conclusion drawn from them (Skidmore 2002).

In attempts to answer the question: What kind of science do we practice when using MAS models, Parker et al. (2003) and Couclelis (2001) reply that the role of MAS models can be described according to an explanatory and a predictive approach. Within the former context, MAS models are used to test candidate or alternative explanations of an empirical macroscopic phenomenon that can add support to an existing theory. They can also help uncovering novel hypotheses that might open a fruitful research avenue. When used within a descriptive approach, MAS models are then designed to mimic real-world entities and their interactions as closely as possible. Such models are parameterized using empirical data and can be employed to forecast the evolution of a wide range of phenomena. Both approaches have potential drawbacks. In the first case, it remains difficult to evaluate the utility of the simulations and to establish clearly what the model is telling about the real world. In the second case, because of the impossibility to formally prove the predictions achieved by the model, there is a risk of erroneously interpreting the results.

When using the combined paradigms of Complexity Theory and Post-Normal Science to assess the contribution of MAS modeling, a third perspective can be formulated: exploration. As elegantly expressed by Bradbury (2000), the true role of MAS models is to explore the universe of possibilities of a system. To do so, he suggests a strategy consisting in a series of stepping stones, called the adjacent possibilities, to discover the paths a system can take in the future, starting from the realized possibility of its present state. This perspective is increasingly advocated by several authors concerned by the acute problem of environmental resource management. Since human and natural systems defined as complex adaptive systems are inherently unpredictable, the role of scientists should shift from attempting to predict and control these systems to understanding their possible evolution, keeping in mind that humans are actors affecting them (Kay 1991; Pahl-Wostl 2007). The emergence of biocomplexity over the last decade as an analysis paradigm also supports the idea of developing new classes of models that incorporates concepts of self-organization, adaptation, resilience, multi-scale interactions and multiple actors to better simulate complex environmental systems and explore their alternative futures (Colwell 1998; Bolte et al. 2006).

According to Bradbury (2000), to fully achieve its goal, a MAS model should be itself a complex adaptive system and such a purposeful model has not yet been built. However, a look at the emerging research trends in MAS modeling reveals a clear interest in integrating more advanced concepts and approaches that are required for the study of complex environmental systems. As an example, Giret and Botti (2004) have proposed a methodology based on the concept of abstract agent to build a large-scale multi-agent system. Such a system includes multiple types of agents and may encompass multiple MASs, each of them having distinct properties. It is designed to take into consideration hierarchical and multi-scale requirements that are characteristics of complex systems. Similarly, Adam and Mandiau (2005) have developed some functionalities to incorporate the notions of role and hierarchy, using a platform dedicated to hierarchical MAS, in which agents have different roles and levels of responsibility. Techniques have been proposed to allow changes in agents' interactions and agents' roles such that agents operating in an open environment can react to unpredictable events (Picard et al. 2006). Considerable scientific effort is also being made to fully incorporate the environment as an essential component within a MAS, which gave rise to the domain of situated multi-agent systems (Weyns et al. 2005a; Hassas 2006). Such a system is a computational framework for the definition of complex systems characterized by the presence of autonomous entities interacting in an environment whose spatial structure is a determinant factor affecting their actions and their possible interactions (Bandini et al. 2006). The environment is defined as an active entity with its own processes that can change its own state, independent of the activity of the embedded agents (Weyns et al. 2005b). Research is also conducted to increase the autonomy of the agents and provide them with some intelligent skills, so that they can acquire new knowledge, transfer their capabilities and acquire new ones, and adapt their structure according to the changes in the environment (Suna and Seghrouchni 2005). Context-based multi-agent architectures are also being developed in which agents are able to learn how to distinguish different contexts and make appropriate decisions based on them (Bucur et al. 2005). A large part of these recent developments have been accomplished in the context of artificial or software agents, but considering their relevance and potential one can assume that they will progressively be incorporated within MAS models specifically designed to investigate the dynamics of environmental systems.

CONCLUDING REMARKS

This paper has explored the large domain of MAS modeling applied to environmental systems through the paradigms offered by Complexity Theory and Post-Normal Science. This exercise provides a strong illustration of the unique contribution of MAS models in the understanding of these systems considered as complex adaptive systems, whose behavior is inherently unpredictable. This scientific contribution is well summarized through the explorative approach, which aims not at predicting and controlling these systems, but rather at exploring the envelope of their possible alternative futures, taking into consideration the human influence on that evolution. Therefore, MAS models can be useful to scientists, managers, decision makers and even the general public in providing a conceptual and computational framework to simulate the behavior of environmental systems under various scenarios and at multiple scales, to reproduce aspects of the human decision process and the feedback mechanism between natural and human systems, and to help identifying appropriate management strategies. However, the full potential of MAS modeling still remains to be fulfilled and more advanced concepts required to capture the complexity of environmental systems still need to be incorporated within MAS architectures.

LITERATURE CITED

Adam, E. and R. Mandiau. 2005. Roles and hierarchy in multi-agent organizations. *4th International Central and Eastern European Conference on Multi-Agent Systems*, Budapest, Hungary, September 2005, Springer, vol. 3690, pp. 539-542.

Ahearn, S., J. L. D. Smith, A.R. Joshi, and J. Ding, J. 2001. TIGMOD: an individual-based spatially explicit model for simulating tiger/human interaction in multiple use forests. *Ecological Modelling* 140(81-97).

Amblard, F., J. Rouchier, and P. Bommel. 2006. Évaluation et validation de modèles multi-agents. *Modélisation et simulation multi-agents*, Amblard F. and D. Phan, eds, Hermes-Lavoisier: 103-140.

An, L., M. Linderman, J. Qi, A. Shortridge, and J. Liu. 2005. Exploring complexity in a human-environment system: An agent-based spatial model for multidisciplinary and multiscale integration. *Annals of the Association of American Geographers* 95(1): 54-79.

Anderies, J. M. 2000. The transition from local to global dynamics: A proposed framework for agent-based thinking in social-ecological systems. *Complexity and Ecosystem Management*. M. A. Janssen, ed., Edward Elgar: 13-34.

Anwar, M., C. Jeanneret, L. Parrott, and D.J. Marceau. 2007. Conceptualization and implementation of a multi-agent model to simulate whale-watching activities in the St. Lawrence estuary in Quebec, Canada. *Environmental Modeling and Software*, Vol. 22 pp. 1775-1787.

Axelrod, R. 1984. *The evolution of cooperation*. Basic Books.

Axtell, R. 2006. La fin des débuts pour les systèmes multi-agents en sciences sociales. *Modélisation et simulation multi-agents*. F. Amblard and D. Phan, eds, Hermes-Lavoisier: 161-171.

Bandini, S., M. L. Federici, S. Manzoni, and G. Vizzari. 2006. Towards a methodology for situated cellular agent based crowd simulations. *6th International Workshop on Engineering Societies in the Agents World*, Kusadasi, Turkey, October 2005, Springer, vol. 3963, pp. 203-220.

Barros, J. 2003. Simulating urban dynamics in Latin American Cities. *Proceedings of the 7th International Conference on Geocomputation*, University of Southampton, UK, 8-10 Sept.

Batten, D. 2007. Are some human ecosystems self-defeating? *Environmental Modelling and Software* 22: 649-655.

Batty, M. 2001. Agent-based pedestrian modeling. *Environment and Planning B* 28: 321-326.

Batty, M. 2005. *Cities and complexity*. The MIT Press.

Benenson, I. and P. Torrens. 2004. *Geosimulation: Automata-based modeling of urban phenomena*, Wiley and Sons.

Benenson, I. 1998. Multi-agent simulations of residential dynamics in the city. *Computers, Environment and Urban Systems* 22(1): 25-42.

Bolte, J. P., D. W. Hulse, S.V. Gregory, and C. Smith. 2007. Modeling biocomplexity - actors, landscapes and alternative futures. *Environmental Modelling and Software* 22: 570-579.

Bousquet, F. and C. Le Page. 2004. Multi-agent simulations and ecosystem management: A review. *Ecological Modelling* 176: 313-332.

Bousquet, F. and G. Trébuil. 2005. Introduction to companion modeling and multi-agent systems for integrated natural resource management in Asia. *Companion modeling and multi-agent systems for integrated natural resource management in Asia*. F. Bousquet, G. Trébuil, and B. Hardy, eds, International Rice Research Institute (IRRI): 1-17.

Bousquet, F., C. Cambier, and P. Morand, 1994. Distributed artificial intelligence and object-oriented modeling of a fishery. *Mathematical and Computer Modeling* 20: 97-107.

Bousquet, F., O. Barreteau, P. D'Aquino, M. Etienne, S. Boissau, S. Aubert, C. Le Page, D.

Babin, and J.C. Castella. 2003. Multi-agent systems and role games: Collective learning processes for ecosystem management. In *Complexity and ecosystem management: The theory and practice of multi-agent approaches*, M.A. Janssen, ed., Edward Elgar Publishers, pp. 248-285.

Bradbury, R. 2000. Futures, predictions and other foolishness. *Complexity and Ecosystem Management*. M. A. Janssen, ed., Edward Elgard: 48-62.

Brown, D. G., S. E. Page, R. Riolo, and W. Rand. 2004. Agent-based and analytical modeling to evaluate the effectiveness of greenbelts. *Environmental Modelling and Software* 19: 1097-1109.

Bucur, O., P. Beaune, and O. Boissier. 2005. What is context and how can an agent learn to find and use it when making decisions? *4th International Central and Eastern European Conference on Multi-Agent Systems*, Budapest, Hungary, September 2005, Springer, vol. 3690, pp. 112-121.

Campo, P. C. 2005. Integrating multi-agent systems and geographic information systems modeling with remote-sensing data for participatory natural resource management in coastal Bohol, Philippines. *Companion modeling and multi-agent systems for integrated natural resource management in Asia*. F. Bousquet, G. Trébuil, and B. Hardy, eds, International Rice Research Institute (IRRI): 255-274.

Chu, D., R. Strand, and R. Fjelland. 2003. Theories of complexity. *Complexity* 8(3): 19-30.

Colwell, R. 1998. Balancing the biocomplexity of the planet's living systems: A twenty-first century task for science. *Bioscience* 48(10): 786.

Conte, R., N. Gilbert, and J.S. Sichman. 1998. MAS and social simulation: A suitable commitment. *Multi-agent systems and agent-based simulation. Proceedings of the first international workshop on MABS*, Paris, France, July 1998. J. S. Sichman, R. Conte, and N. Gilbert, eds, Springer: 1-9.

Couclelis, H. 2001. Why I no longer work with agents: A challenge for ABMs of human-environment interactions. In *Meeting the challenge of complexity: Proceedings of a Special Workshop on Land-use/Land-cover Change*, Parker, D.C., T. Berger, and S.M. Manson, eds, Irvine, California.

Crawford, T. W., J. P. Messina, S.M. Manson, and D. O'Sullivan. 2005. Complexity science, complex systems, and land-use research. *Environment and Planning B* 32: 792-798.

De Marchi, B. and J. R. Ravetz. 1999. Risk management and governance: A post-normal science approach. *Futures* 31: 743-757.

Dumont, B. and D. R. C. Hill. 2001. Multi-agent simulation of group foraging in sheep: Effects of spatial memory, cospecific attraction and plot size. *Ecological Modelling* 141: 201-215.

Durfee, E.H., V.R. Lesser, and D.D. Corkill. 1989. Trends in cooperative distributed problem solving. *IEEE Transactions on Knowledge and Data Engineering*, KOE-11(1): 63-83

Ferber, J., 1999. *Multi-agent systems: An introduction to distributed artificial intelligence*. Addison-Wesley.

Feuillette, S., F. Bousquet, and P. Le Goulven. 2003. SINUSE: a multi-agent model to negotiate water demand management on a free access water table. *Environmental Modelling and Software* 18: 413-427.

Funtowicz, S. and J. R. Ravetz. 1993. Science for the post-normal age. *Futures* 25: 739-755.

Funtowicz, S. and J. Ravetz. 1985. Three types of risk assessment: A methodological analysis. In *Risk analysis in the private sector*, C. Whipple and V.T. Covello, eds, Plenum, pp. 217-232.

Gimblett, H. R., C. A. Roberts, T.C. Daniel, M. Ratliff, M.J. Meitner, S. Cherry, D. Stallman, R. Bogle, R. Allred, D. Kilbourne, and J. Bieri. 2002. An intelligent agent-based model for simulating and evaluating river trip scenarios along the Colorado river in Grand Canyon National Park. In *Integrating geographic information systems and agent-based modeling techniques*. H. R. Gimblett, ed., Oxford University Press, pp. 245-275.

Gimblett, R., T. C. Daniel, S. Cherry, and M.J. Meitner. 2001. The simulation and visualization of complex human-environment interactions. *Landscape and Urban Planning* 54: 63-78.

Ginot, V. and H. Monod. 2006. Explorer les modèles par simulation: application aux analyses de sensibilité. *Modélisation et simula-*

tion multi-agents, Amblard, F. and D. Phan, eds, Hermes Lavoisier: 75-102.

Giret, A. and A. Botti. 2004. On the definition of meta-models for analysis of large-scale MAS. *Second German Conference on Multi-Agent System Technologies*, Erfurt, Germany, September 2004, Springer, vol. 3187, pp. 273-286.

Grimm, V. 1999. Ten years of individual-based modelling in ecology: What have we learned and what could we learn in the future? *Ecological Modelling* 115: 129-148.

Grimm, V., E. Revilla, F. Jeltsch, W. M. Mooij, S.F. Railsback, H.H. Thulke, J. Weiner, T. Wiegand, and D. L. DeAngelis. 2005. Pattern-oriented modeling of agent-based complex systems: Lessons from ecology. *Science* 310: 987-991.

Gross, J. E., R. R. J. McAllister, N. Abel, D.M. Stafford Smith, and Y. Maru. 2006. Australian rangelands as complex adaptive systems: A conceptual model and preliminary results. *Environmental Modelling and Software* 21: 1264-1272.

Guizol, P. and H. Purmono. 2005. Modeling multi-stakeholder forest management: The case of forest plantations in Sabah. *Companion modeling and multi-agent systems for integrated natural resource management in Asia*. F. Bousquet, G. Trébuil, and B. Hardy, eds, International Rice Research Institute (IRRI): 275-291.

Gurung, T. R. 2005. Companion modeling to examine water-sharing arrangements among rice-growing villages in west-central Bhutan: Preliminary results. *Companion modeling and multi-agent systems for integrated natural resource management in Asia*. F. Bousquet, G. Trébuil, and B. Hardy, eds, International Rice Research Institute (IRRI): 101-120.

Haklay, M., D. O'Sullivan, and M. Thurstain-Goodwin. 2001. So go downtown: Simulating pedestrian movement in town centres. *Environment and Planning B* 28: 343-359.

Hare, M. and P. J. Deadman. 2004. Further towards a taxonomy of agent-based simulation models in environmental management. *Mathematics and Computers in Simulation* 64: 25-40.

Hassas, S. 2006. Engineering complex adaptive systems using situated multi-agents. *6th International Workshop on Engineering Societies in the Agents World*, Kusadasi, Turkey, October 2005, Springer, vol. 3963, pp. 125-141.

Helbing, D., P. Molnar, I.J. Farkas and K. Bolay. 2001. Self-organizing pedestrian movement. *Environment and Planning B* 28: 361-383.

Hoffmann, M., H. Kelley, and T. Evans. 2000. Simulating landcover change in South-Central Indiana: An agent-based model of deforestation and afforestation. In *Complexity and ecosystem management: The theory and practice of multi-agent approaches*, ed. M.A. Janssen, Edward Elgar Publishers, pp. 218-247.

Hogeweg, P. and B. Hesper. 1983. The ontogeny of the interaction structure in bumble bee colonies: A MIRROR model. *Behavioural Ecology and Sociobiology* 12: 271-283.

Holling, C. S., 2001. Understanding the complexity of economic, ecological and social systems. *Ecosystems* 4: 390-405.

Irwin, E. G. and N. E. Bockstael. 2002. Interacting agents, spatial externalities and the evolution of residentail land use patterns. *Journal of Economic Geography* 2(1): 31.

Itami, R. M. and H. R. Gimblett. 2001. Intelligent recreation agents in a virtual GIS world. *Complexity International* 8.

Itami, R. M., R. Raulings, G. MacLaren, K. Hirst, H.R. Gimblett, D. Zanon, and P, Chladek. 2003. RBSim2: Simulating the complex interactions between human movement and the outdoor recreation environment. *Journal for Nature Conservation* 11: 278-286.

Kaufmann, A. and E. Gebetsroither. 2004. Modelling self-organization processes in socio-economic and ecological systems for supporting the adaptive management of forests. *Transactions of the 2nd Biennial Meeting of the International Environmental Modelling and Software Society*, University of Osnabrück, Germany, 14-17 June 2004, pp. 283-288.

Kay, J. and E. Schneider. 1994. Embracing complexity: The challenge of the ecosystem approach. *Alternatives* 20(3): 32-38.

Kay, J. J., H. A. Regier, M. Boyle, and G. Francis. 1999. An ecosystem approach for sustainability: Addressing the challenge of complexity. *Futures* 31: 721-742.

Kay, J., 1991. A non-equilibrium thermodynamic framework for discussing ecosystem integrity. *Environmental Management* 15(4): 483-495.

Kay, J. 1994. Some notes on: the ecosystem approach, ecosystems as complex systems and state of the environment reporting. Montreal, Canada, North American Commission for Environmental Cooperation, State of the North American Ecosystem Meeting: 42 p.

Kerridge, J., J. Hine, and M. Wigan. 2001. Agent-based modelling of pedestrian movements: The questions that need to be asked and answered. *Environment and Planning B* 28: 327-341.

Kii, M. and K. Doi. 2005. Multi-agent land-use and transport model for the policy evaluation of a compact city. *Environment and Planning B* 32: 485-504.

Kuhn, T. 1970. *The structure of scientific revolution*. University of Chicago Press.

Lacombe, G. and W. Naivinit. 2005. Modeling a biophysical environment to better understand the decision-making rules for water use in the rainfed lowland rice ecosystem. *Companion modeling and multi-agent systems for integrated natural resource management in Asia*. F. Bousquet, G. Trébuil, and B. Hardy, eds, International Rice Research Institute (IRRI): 191-209.

Ligtenberg, A., A. K. Bregt, and R. Van Lammeren. 2001. Multi-actor-based land use modelling: Spatial planning using agents. *Landscape and Urban Planning* 56: 21-33.

Lim, K., P. J. Deadman, E. Moran, E. Bondizio, and S. MacCraken. 2002. Agent-based simulations of household decision making and land use change near Altamira, Brazil. In *Integrating geographic information systems and agent-based modeling techniques for simulating social and ecological processes*. H. R. Gimblett, ed., Oxford University Press, pp. 277-310.

Lynam, T. 2000. Complex and useful but certainly wrong: A multi-agent agro-ecosystem model from the semi-arid areas of Zimbabwe. In *Complexity and Ecosystem Management: The theory and practice of multi-agent approaches*, ed. M.A. Janssen, Edward Elgar Publishers, pp. 188-217.

Manson, S. M. 2000. Validation and verification of multi-agent systems. *Complexity and Ecosystem Management*. M. A. Janssen, ed., Edward Elgar: 63-74.

Mathevet, R., F. Bousquet, C. Le Page, and M. Antona. 2003. Agent-based simulations of interactions between duck population, farming decisions and leasing of hunting rights in the Camargue (Southern France). *Ecological Modelling* 165: 107-126.

Monticino, M., M. Acevedo, B. Callicott, T. Cogdill, and C. Lindquist. 2006. Coupled human and natural systems: A multi-agent approach. *Environmental Modelling and Software* 22(5): 656-663.

Moreno, N., R. Quintero, et al. 2007. Biocomplexity of deforestation in the Caparo tropical forest reserve in Venezuela: An integrated multi-agent and cellular automata model. *Environmental Modelling and Software* 22: 664-673.

Moulin, B. and B. Chaib-Draa 1996. An overview of distributed artificial intelligence. *Foundations of distributed artificial intelligence*. G. M. P. O'Hare, and N.R. Jennings, eds, Wiley and Sons: 3-55.

Omer, I. 2005. How ethnicity influences residential distributions: An agent-based simulation. *Environment and Planning B* 32: 657-672.

Pahl-Wostl, C., 2007. The implications of complexity for integrated resources management. *Environmental Modelling and Software* 22: 561-569.

Parker, D. C., S. M. Manson, M.A. Janssen, M.J. Hoffman, and P. J. Deadman 2003. Multi-agent systems for the simulation of land-use and land-cover change: A review. *Annals of the Association of American Geographers* 93(2): 314-337.

Parrott, L. and R. Kok 2000. Incorporating complexity in ecosystem modelling. *Complexity International* 7: 1-19.

Parry, H., A. J. Evans, and D. Morgan 2004. Aphid population dynamics in agricultural landscapes: An agent-based simulation model. *Transactions of the 2nd Biennial Meeting of the International Environmental Modelling and Software Society*, University of Osnabrück, Germany, 14-17 June 2004, pp. 914-919.

Parwani, R. R. 2005. Complexity: An introduction. *Complexity International*.

Picard, G., S. Mellouli, and M.-P. Gleizes, 2006. Techniques for multi-agent system reorganization. *6th International Worshop on Engineering Societies in the Agents World*, Kusadasi, Turkey, October 2005, Springer, vol. 3963, pp. 142-152.

Portugali, J., I. Benenson, and I. Omer 1997. Spatial cognitive dissonance and sociospatial emergente in a self-organizing city. *Environment and Planning B* 24(2): 263-285.

Ravetz, J. R. and S. Funtowicz, 1999. Post-normal science: An insight now maturing. *Futures* 31: 641-646.

Ravetz, J. R., 1999. What is post-normal science? *Futures* 31(647-653).

Ravetz, J. R. 2004. The post-normal science of precaution. *Futures* 36: 347-357.

Robinson, V. B. and P. A. Graniero. 2004. Spatially explicit individual-based ecological modeling with mobile fuzzy agents. *Fuzzy modeling with spatial information for geographic problems*. F. E. Petry, V. B. Robinson and M. A. Cobb, eds, Springer: 299-334.

Saarloos, D., T. A. Arentze, A. Borgers, and H. Timmermans. 2005. A multi-agent model for alternative plan generation. *Environment and Planning B* 32: 505-522.

Sakoda, J.M. 1971. The checkerboard model of social interaction. *Journal of Mathematical Sociology* 1: 119-132.

Schelling, T.S., 1971. Dynamic models of segregation. *Journal of Mathematical Sociology* 1: 143-186.

Sengupta, R., C. Lant, S. Kraft, J. Beaulieu, W. Peterson, and T. Loftus. 2005. Modelling enrollment in the Conservation Reserve Program by using agents within spatial decision support systems: An example from Southern Illinois. *Environment and Planning B* 32: 821-834.

Servat, D., E. Perrier, J.-P. Treuil, and A. Drogoul. 1998. When agents emerge from agents: Introducing multi-scale viewpoints in multi-agent simulations. *First International Workshop on Multi-Agent Systems and Agent-Based Simulation*. J. S. Sichman, R. Conte, and N. Gilbert, eds., Paris, France, July 1998, Springer, vol. 1534, pp. 183-198.

Skidmore, A. 2002. Taxonomy of environmental models in the spatial sciences. *Environmental modelling with GIS and remote sensing*. A. Skidmore, ed., Taylor and Francis: 8-25.

Suna, A. and A. E. F. Seghrouchni. 2005. Adaptive mobile multi-agent systems. *4th International Central and Eastern European Conference on Multi-Agent Systems*, Budapest, Hungary, September 2005, Springer, vol. 3690, pp. 41-50.

Suphanchaimart, N., C. Wongsamun, and P. Panthong. 2005. Role-playing games to understand farmers' land-use decisions in the context of cash-crop price reduction in upper northeast Thailand. *Companion modeling and multi-agent systems for integrated natural resource management in Asia*. F. Bousquet, G. Trébuil, and B. Hardy, eds, International Rice Research Institute (IRRI): 121-139.

Topping, C., T. S. Hansen, T.S. Jensen, J.U. Jepsen, F. Nikolajsen, and P. Odderskaer. 2003. ALMASS, an agent-based model for animals in temperate European landscapes. *Ecological Modelling* 167: 65-82.

Trébuil, G., F. Bousquet, B. Ekasingh, C. Baron, and C. Le Page. 2005. A multi-agent model linked to a GIS to explore the relationship between crop diversification and the risk of land degradation in northern Thailand highlands. *Companion modeling and multi-agent systems for integrated natural resource management in Asia*. F. Bousquet, G. Trébuil, and B. Hardy, eds, International Rice Research Institute (IRRI): 167-190.

Valkering, P., J. Krywkow, J. Rotmans, and A. Van Der Veen. 2004. Simulating stakeholder support for river management. *Transactions of the 2nd Biennial Meeting of the International Environmental Modelling and Software Society*, University of Osnabrück, Germany, 13-17 June, pp. 184-189.

Waldrop, M.M. 1992. *Complexity: The emerging science at the edge of order and chaos*. Touchstone.

Weyns, D., H. V. D. Parunak, F. Michel, T. Holvoet, and J. Ferber. 2005a. Environments for multi-agent systems: State-of-the art and research challenges. D. Weyns, ed., Springer-Verlag: 1-47.

Weyns, D., M. Schumacher, A. Ricci, M. Viroli, and T. Holvoet. 2005b. Environments in multi-agent systems. *The Knowledge Engineering Review* 20(2): 127-141.

Wooldridge, M. 2002. *An introduction to multi-agent systems*. Wiley and Sons.

CHAPTER 23

CONCLUSION: MONITORING, SIMULATION AND MANAGEMENT OF VISITOR LANDSCAPES

Randy Gimblett
Hans Skov-Petersen

There is a growing body of theoretical and applied research focused within the context of human-environment interactions. This book has explored the current state of spatial/temporal simulations that integrate human behavior and environmental factors as applied to decision-making in spatially referenced dynamic environments. Visitor landscapes are those that humans inhabit temporarily and explore, seeking experience opportunities in a natural outdoor setting. These experience opportunities such as solitude or adventure, come with the potential to destroy the very nature of the experience sought through crowding, overuse and degradation to the landscape. Frequency of use, duration of stay, type of encounter, recreation activity visitors are engaged in, season of use, numbers of visitors in the setting and the fragility of the landscape, all have a significant impact on social and ecological values and require extensive management.

While there are many important points that you have encountered by authors who submitted chapters to this book, it's main purpose was to provide a glimpse of field methods and tools that can aid decision makers in planning and managing lands. Those are the lands where there is a great need to accommodate increasing human use while maintaining their ecological integrity. One important discovery made by the coeditors prior to the writing of this book is that conventional methods used in planning and management of human-landscape interactions fall far short of the needs of decision-makers who must evaluate the cascading impacts of humans in visitor landscapes. Many public land agencies, local governments and international organizations are exploring tools such as multi-agent simulations coupled with social science data to develop long-term strategies for evaluating human-landscape interactions. In particular, spatial agent-based models are being explored with some success to provide a better understanding of the spatial and temporal patterns of human-landscape interactions and to predict how distributions of this use are likely to change in response to policy implementation and proactive management actions. While the application of simulation to study human-landscape interactions is in its infancy, there is need to develop a comprehensive and empirically based framework for linking the social, biophysical and geographic disciplines across space and time. This is what this volume has attempted to accomplish.

Chapters in this book were organized in such as way as to demonstrate the necessity of careful integration of monitoring data using statistically valid sampling and data collection techniques, with a growing number of modeling platforms (ie. Agent-based, State preference/Choice models and Baysian belief systems) to specifically answer natural resource based management questions. Several chapters in this book provide good examples of how decision makers can work together with modelers and field staff to identify specific management questions and develop monitoring and data collection techniques, coupled with the appropriate modeling platform to directly answer those questions and implement results to manage visitor use. Without such a planning framework, a model can be only minimally useful, under utilized in predicting future trends and lead to misrepresentative or speculative results. If decision makers are our intended audience, who use data and model outputs to aid decision making, then it is our ethical responsibility as professionals to ensure that current monitoring techniques are consistently explored and statistically representative and valid methods are used to capture spatial and temporal patterns of human use that meet social science standards. In addition, spatial simulation models need to be constructed using appropriate simulation standards including degrees of reliability and confidence intervals.

This volume explored all of these issues and provided excellent examples of where coupling monitoring techniques with spatial simulations and management questions can lead to policy that dramatically improves decision-making in visitor landscapes. The following sections provide a glimpse of some of the significant findings of this volume.

EXPERIENCES AND BEHAVIOR IN VISITOR LANDSCAPES

Much is written about the importance of affordance theory in how humans select and interact in landscapes. Gibson (1979) coined the term "affordance" that has evolved into the foundation of ecological psychology and direct perception. Gibson's notion of the affordance is based on a unified relationship between a human and its' environment. Salient, functional features that are present in the environment lead to direct sensation, perception, and action by the human. Any notion of information is functional or situational, involves both human and environment, and the affordance-based relationship between the two.

Cornwell et al. (2003) suggest that: "the use of affordance theory in multi-agent systems arose from engineering constraints rather than a theoretical predisposition." Agents created in many software implementations (agent-based simulations) of modeling human-landscape interactions contained a functional representation of every other object in its environment and made its decisions by consulting some type of internal schema. Cornwell et al. (2003) suggest that "if people each carry within themselves a functional representation of the objects in their environment – a robust mental model of the world maintained by perceptual information – shouldn't artificial intelligence (agent-based simulations) intended to traverse and manipulate a virtual landscape possess a similar structure?" They suggest that in order to begin to capture the subtleties of social interaction or simulate human emotionality, simulated agents must act based on their own unique socio-cultural background and personal experience. Affordance theory then provides a framework for developing methods for capturing and encapsulating data about human or visitor experience as well as formalizing within simulation models the way in which this data is used to represent human cognition, action, wayfinding etc.

The distinction of affordances for purposes of wayfinding, task-performance, social territories, cultural expression and visual/non-visual aesthetics can initially be useful in prompting visitors to recall the variety of different kinds of experiences that occurred in a landscape. Doxtater (Chapter 2) suggests that this taxonomy can then be used to structure information about environmental experiences in subsequent processes of computer simulation, most of which involve the management of inputs from multiple visitor groups. Doxtater (Chapter 2) suggests that most applications of categories of affordances in visitor landscapes will begin with a close evaluation of what people are actual doing in the setting. Instead of more open-ended methods of gathering information about experiences, one can envision separate mapping of many type of experiences visitors have. Considering individual (demographic) differences between kinds of visitors, methods must enable the researcher to capture information about object, space and movement relevant to each category of experience. The development of simulation models of experience has occurred first in landscape oriented adjacent fields. It is here that ideas about categories of experience might be expected to bear first fruits as part of applied simulation to make better decisions about changes in these physical settings.

Steven's (Chapter 3) echoes some of the same issues that Doxtater (Chapter 2) outlined. He suggests that little is known of the capacity for informal learning, or sense making, which affects visitor transactions within natural environments. What stimuli are encountered in the biophysical landscape, and how do cognitive and affective responses to those stimuli affect visitor experiences? Using a multi-layer cognitive mapping instrument, *Environmental N*, visitor experiences on the Bardedjilidji Sandstone Track in Australias Kakadu National Park were recorded on site. A typology of visitor experiences grounded in their responses provided the basis for characterizing experiences at Kakadu. Three core experience themes encapsulating fifteen sub-themes emerged from the data: Wayfaring; Seeking, discovering, encountering; and Making sense of place. All these are shown to be important indicators of how a visitor experiences landscape.

The above indicators are all extremely important concepts that provide a fundamental understanding of visitor experience. Even though these concepts are relevant they are seldom operationalized into management of visitor landscapes. Elands and Marwijk (Chapter 4) are adamant that ecologists and social scientists have focused on behavioral practices of visitors in nature areas, but less on the existential basis of 'what makes people visit nature areas?' In their chapter they conceptualize recreation quality by means of different experiential worlds of those in visitor landscapes. They link these experiential worlds conceptually with characteristics of nature areas that can be described in use, perception, narrative form and appropriation value. These values are subsequently operationalized into experiential parameters important for management of visitor landscapes. They strongly suggest that the relation between experience and time-space behavior is complicated; besides experiential factors, other factors such as familiarity, influence the movement patterns of visitors. Not every environmental value – and its subsequent operationalization into experiential parameters – can be and should be integrated into simulation models. Their ideas are extremely compelling and offer much to the theme of this book. They argue that that simulation models are not the solution for investigating interactions between visitors and landscape. Simulation models can anticipate the influence of visitor use of a landscape, however, stakeholder acceptation of management strategies should also be taken into account. Qualitative research work, discussions and stakeholder deliberations (e.g. with participation groups) should be part of the planning and management of visitor landscapes.

MONITORING SPATIAL AND TEMPORAL PATTERNS OF VISITOR USE

While the case studies presented in this chapter demonstrate the potential utility of computer simulation modeling for planning and management of visitor landscapes, there are several challenges that need to be addressed. A primary challenge has to do with collecting reliable, representative data with which to construct computer simulation models. Due to the dispersed nature of outdoor recreation in many visitor landscapes, it is often difficult to capture travel itineraries and other data needed to construct a computer simulation. Research and coordination with land managers is needed to identify methods for collecting trip and visitor characteristics data in a reliable and consistent manner. Several chapters in this book address this issue. Many years of research have focused in the study of human impacts on social and ecological systems (Cole 2004) and the impacts that human activities have on wildlife and their associated habitats (Steidl & Powell 2006). Cole (2004), Blahna & McCool (2004) and Manning & Lime (2000) also emphasize that dispersed use patterns make it extremely difficult to uniformly manage for visitors. However this work does emphasize the importance of collecting spatial and temporal data to empirically derive distributions across a visitor landscape. They have identified that the critical management concerns are the density of visitors, the destination and duration of visit, the mode of travel, the type and desired level of encounters of other visitors. Understanding how these factors can lead to a decline in visitor experience or displacement to some other location of similar quality is an essential management goal for recreational areas, which according to Cole (2004) may cause local impacts to be more severe and widely distributed. Given this situation, Nelson et al. (Chapter 17) analyzed the degree and location of encounters and evaluating wilderness quality standards occurring in Misty Fjords National Monument and Wolf et al. (Chapter 19) undertaken in Prince William Sound, Alaska to measure the potential conflict between visitors that could be detrimental to the level of wilderness experience. Both of these studies used primary and secondary data sources to analyze, simulate and map locations of high use and where potential conflicts could potentially occur within the each area.

A primary challenge then has to do with collecting reliable, representative data with which to construct computer simulation models. Chapters (5 & 7) by Xia, Arrowsmith and Skov-Petersen et al. clearly reveal that reliable data can be strategically collected using automatic traffic counters or aerial surveys and feed directly into the simulation models to construct trip itineraries. Chapters (16 & 17) by Lace et al., Nelson et al. reveal that diaries, counters and survey methods are labor intensive but can capture essential visitor use numbers, direction of movement, characteristics of visitors and experiential data from which to construct artificial visitor agents and simulate real visitor behavior. Assessment of the number of tickets or licenses granted (see for instance Lace et al. in the present volume), interviews with visitors which are frequently based on structured questionnaires (see for instance Taczanowska et al. in the present volume) or the evaluation of preferences, norms, experiences, and choices including ranking statements and photographs (e.g Jensen 1999), manipulated photos displaying different levels of visitation (Manning 2007), and in situ self-registration of the linkage between stimuli/thoughts/feeling Stevens (Chapter 3) all are useful methods for capturing data for agent based models. Preparation of sketch maps Taczanowska et al. (Chapter 9) and more automated sports timing devices, and application of GPS technology for registrations itineraries (Chapters 15, 5, 6) for instance by Jochem et al., Xia and Arrowsmith, and Loiterton and Bishop) are capable of capturing continuous data about visitor movement patterns which has immense benefits to the modeler. Automating these procedures and capturing larger sample sizes over longer periods of time with much lower labor cost, should be the goal of any field manager or researcher. It is apparent from the chapters in this section of the book that additional research is required to explore, adapt and in many cases develop alternative data collection methods, including GPS and satellite, automatic timing systems (i.e., race technology), mechanical counters and radio frequency technology as well as related technologies is warranted. Also apparent from most of the chapters presented in this book is that no one technique alone can capture all the essential spatial and temporal characteristics of visitor behavior in order to build models. A combination of these techniques is frequently required to successfully build models.

In addition, the challenge is not just exploring methods for acquiring data, nor is it automating the process to collect vast amounts of continuous data, but how to collect a statistically valid and representative sample of the spatial and temporal patterns of use across a visitor landscape. This issue remains a challenge and holds data collected with conventional methods suspect. Many of the methods to date, employed by researchers, field staff and managers in visitor landscapes have failed to provide the data required to build accurate and reliable models of visitor landscapes. While conventional social/economic sampling assumes a single population for the whole season of use, spatial and temporal simulation modeling has a much higher degree of complexity and data requirements. Since these types of spatial simulations require sampling at multiple entry points, over a variety of seasons, there is a much higher need for systematic random sampling to capture seasonal variations and more specifically for weekday and hourly arrivals.

Itami et al. (Chapter 18) continue to investigate methods for collecting accurate data that encompasses the spatial/temporal patterns of visitor behavior necessary to develop probabilistic simulations for accurate forecasting and decision-making. In a simulation developed using visitor data from the 2003 Pattern of Visitor Use study (McVetty 2003), the sample size acquired met all the requirements of any well-designed social science study. However as Itami et al. reports the data collection design fell short of the requirements needed to develop valid travel simulation models for management decision-making. Such methods do not take into account the spatial and temporal variations of visitor behavior. This study clearly illustrated how important research design is in acquiring a spatially and temporally representative sample that is statistically valid and concludes that based on the minimum requirements to do a chi-square test on daily arrival distributions a minimum of 5 arrivals per day for 80% of the cases being compared is required. This means that for weekdays 6 out of 7 weekdays with a minimum of 5 arrivals or 30 arrivals per week (realistically something closer to 50 surveys a week) are required in order to make any statistical generalizations. An evaluation of the 2003 Pattern of Visitor use data (McVetty 2003) revealed only four weeks that met this standard that could be used to develop a probabilistic simulation. While this was adequate for demonstration purposes, it falls far short of the data required to build an accurate probabilistic simulation necessary for decision-making. This study strongly illustrates how important population sampling is to developing accurate and reliable simulation models that decision makers can use and obtain defensible and credible results.

Finally, Kisler et al. (Chapter 20) recommends that statistical procedures for determining the number of replications be adopted as standard practice within applications of computer simulation modeling to planning and management of visitor landscapes. Further, this study provides insights into the feasibility of generating visitor use-related model outputs at managerially relevant levels of precision. As identified above, there are some minimum requirements that need to be met if a probabilistic simulation for accurate forecasting and prediction is the desired outcome. This requirement is less problematic where the frequency of visits is high and consistent throughout the season. But what about the effectiveness of models to forecast or predict outcomes for areas of low visitor use? In diverse, vast visitor landscapes, such as the Alaskan Landscape, Misty Fjords National Monument Nelson et al., (Chapter 17), Prince William Sound Wolfe et al., (Chapter 19), where federal regulation requires regular planning and management, visitor numbers of not high enough to obtain enough data that is spatially and temporally representative of the patterns of use. For example, in visitor landscapes such as the Gates of the Arctic National Refuge, encounters and encounter rates have been a standard to measure the quality of a wilderness experience, where encounters are the antithesis of solitude. If solitude is a measure of a high quality wilderness experience, then increasing encounter rates should be a measure of diminishing that quality. That seems to be a reasonable measure however planners are now questioning this measure in large, remote visitor landscapes that have low use. Gates of the Arctic is over eight and a half million acres that receives less then fifteen hundred visitors annually. Those are pretty low encounter rates. So the question is can reliable estimates of encounters or solitude-related indicators be generated for low use recreation environments, such as backcountry, remote and wilderness areas? Is it possible to generate estimates at a level of precision that is useful for management purposes for some, but not all locations within a low use recreation area (i.e., selected trails/trail segments and camping locations)? This is one example of where a model would be most useful for estimating and predicting seasonal patterns of use across a visitor landscape. The research by Kisler et al. (Chapter 20) suggests that precise estimates can be obtained even for a small to moderate number of visitor and spatially based outputs. Their study clearly demonstrates the application of reliability analysis procedures developed in the broader field of discrete-event simulation to modeling recreational use in a low use area. While the results of the reliability analyses conducted in this study varied depending on the method used, the findings regarding the feasibility of generating precise estimates of visitor use-related outputs from simulation models of low use environments are generally encouraging. The study findings are particularly promising for the visitor-based outputs generated with the study model, with only 20 replications of the model needed to achieve precise estimates.

The visitor-based outputs include intergroup encounters, which has been the most commonly adopted indicator of wilderness solitude, as well as indicators that account for the temporal and spatial dimensions of encounters. Thus, the results of this study suggest that computer simulation modeling is a reliable tool for helping to implement visitor-based indicators of wilderness solitude. Future studies should explore variance reduction techniques to enhance the reliability of computer simulation modeling in cases where a large number of outputs are desired and/or low use environments are modeled.

SIMULATION AND MANAGEMENT OF VISITOR LANDSCAPES

Cole et al. (2005) have suggest that "as the demographics of public land recreational visitors change, planners and managers of public lands face the challenge of protecting resources while providing high quality visitor experiences." They go on to say, "because our political environment demands ever more reliance on scientific data and transparent decision making, planners and managers need better tools to help them understand current visitor use, analyze potential alternatives for future use, and communicate the implications of various alternative decisions in ways that are meaningful to the stakeholder. Computer simulation is one such tool that is gaining popularity in this arena. Lawson et al. (Chapter 10) have noted that recent research has identified at least five ways in which computer simulation can facilitate more informed planning and management of protected natural areas, including 1) describing existing visitor use conditions; 2) monitoring "hard to measure" indicator variables; 3) "proactively" managing carrying capacity; 4) testing the effectiveness of alternative visitor use management practices; and 5) guiding the design of research on stakeholders attitudes.

Lawson (Chapter 10) suggests that an underlying assumption of computer simulation models is that visitor headways and travel route distributions will remain the same, even as use levels rise in the study areas modeled. Since a visitors' recreation behavior can be driven by the amount of time they have and the location of key attractions (e.g., water bodies, vistas, mountain tops), these assumption are probably reasonable in many contexts. However, these assumptions may become more problematic for dramatic increases in overall use levels in which case use densities may override the attractiveness of specific landscape features and, to a lesser extent, time constraints in determining people's recreation behavior. In cases where the scenarios to be simulated deviate dramatically from existing conditions, it may be more appropriate to use a rule-based simulation approach. With this approach, simulated visitor behavior is driven by a set of rules that take into account such factors as the number of other visitors in selected locations, time constraints, and the attractiveness of landscape features. A primary challenge of rule-based simulation lies in defining rules that accurately represent human behavior. If rules of human behavior are not specified correctly, results of rule-based simulations will not be valid. Lace et al (Chapter 16) demonstrate that accurate representations of spatial and temporal patterns of use in visitor landscapes can be captured with very simple rules and using a rule-based simulation. More research is needed to develop empirical bases for defining context-specific and generalizable rules of recreation behavior.

There is no doubt that agent-based models are an important tool for analyzing social and ecological systems. Marceau (Chapter 22) eloquently summarizes the unique contribution of agent-based models for understanding complex adaptive systems, whose behavior is inherently unpredictable. She suggests that this type of modeling should be placed into an explorative framework, which aims not at predicting and controlling these systems, but rather at exploring the envelope of their possible alternative futures, taking into consideration the human influence on that evolution. She goes on to suggest that agent based models can be useful to scientists, managers, decision makers and even the general stakeholder in providing a conceptual and computational framework to simulate the behavior of environmental systems under various scenarios and at multiple scales. In addition to reproduce aspects of the human decision process and the feedback mechanism between natural and human systems, and to help identifying appropriate management strategies. It is apparent that the full potential of agent based modeling still remains to be fulfilled and more advanced concepts required to capture the complexity of environmental systems still need to be incorporated within agent based models architectures.

Exploratory frameworks such as described by Marceau for modeling outdoor recreation behavior in visitor landscapes can be extremely powerful analytical tools. In this volume Hunt (Chapter 11) explains that simulation modeling of outdoor recreation behaviors can benefit from the tremendous amount of work that has developed and applied theories of decision-making on outdoor recreation topics. Choice models are showing promise to guide behaviors of agents. Hunt suggests that first, choice models provide a theoretical rationale for the behavioral rules used to guide agent choices. Rather than dictating the behavioral rules through expert opinion, these behavioral rules are revealed to researchers through empirical data. Second, choice models allow one to test the statistical significance of preferences for various attributes. This statistical estimation gives confidence to researchers about the importance that various attributes have at influencing choice decisions of agents. Third, choice models forecast probabilities of behaviors that are ideally suited for simulation modeling. Finally, choice models, which are consistent with random utility theory, allow one to estimate changes in economic value of behaviors from different management scenarios. He suggests that hybrid choice and agent based models may benefit from the statistical rigor and theoretical foundation from choice modeling and the flexibility and ability to uncover emergent behaviors from agent based models. These hybrid models should continue to receive attention by both choice model and agent based model researchers.

While choice models are revealing important contributions to agent modeling, Sinay et al. (Chapter 12) presents comparable framework and a modeling approach for assessing cultural impact and predicting the efficacy of cultural change management strategies. A Bayesian Belief Network modeling approach has been applied to cultural change management on a visited protected area inhabited by traditional people. From empirical study, practical indicators and measurements to forecast cultural change were identified. The greatest contribution of this research is its innovative approach to understanding, managing, forecasting and studying cultural change. The continuation of this research may lead to the identification of general theories relating pressures, responses, weights the importance of indicators of cultural change.

Itami (Chapter 18) reveals another framework the Limits of Acceptable Activity (LSA) has proven to be a robust way of obtaining user-based definitions of river capacity, quality of service and the consequences of increasing traffic. The quantitative definition of LSA developed for river traffic was useful in interpreting the results of river traffic simulation. The use of focus groups to illicit current and maximum acceptable LSA levels for mixed traffic was efficient, easy to employ and stimulated positive and constructive feedback on the current conditions, desired conditions and the consequences of exceeding desired LSA levels. Itami concludes that while the LSA technique coupled with stakeholder participation shows great promise, further research needs to be done to get better coincidence between LSA levels identified in workshops and traffic densities generated in simulations. However the flexibility of the framework for handling mixed and competing uses in a wide range of environments supports its further refinement and development. This framework reveals great promise for multilevel decision-making and demonstrated effective implementation of strategic management objectives.

Elands et al. (Chapter 4) concur with Itami in that they conclude that simulation models can anticipate the influences of visitors but they strongly suggest that stakeholder acceptance of management strategies should also be taken into account. They suggest like Itami that qualitative research work, discussions and stakeholder deliberations (e.g. with participation groups) should be part of management practices as well. Pouwels et al. (Chapter 14) also echo the fact that simulation models are useful in communication with stakeholders. They suggest that managers can show clearly what the effects of different scenarios are for attainment of both habitat goals and recreation goals. Communicative action among participating stakeholder groups allows eventual construction of a consensus policy on recreational access.

This idea of incorporating the stakeholders into the planning process is echoed by Jochem et al. (Chapter 15) who describe the development and application of MASOOR (Multi Agent Simulation of Outdoor Recreation), an interactive decision-making framework for evaluating front country situations. This chapter is unique in that it places the multi-agent tool into a decision-making framework. Their chapter concludes with an application of the model to examine patterns of visitor use in Dwingelderveld National Park in the Netherlands. Pouwels et al. (Chapter 14) and Taczanowska et al. (Chapter 9) both employ MASOOR to evaluate a variety of scenarios for ranking the effects of disturbance on individuals and populations. It can be seen as part of a conceptual planning/managing framework. This framework incorporates three dimensions for managing multifunctional land use problems: goals, monitoring and designs. Nature management can only be meaningful if all dimensions of the planning framework for different functions are elaborated and dealt with at the same time. Taczanowska et al. (Chapter 9) use MASOOR to used to undertake an empirical study of the Danube Floodplains National Park, Austria, where main emphasis was put on exploring human-environment interactions as well as on the characteristics of the individual routes These three chapters together emphasize the importance of coupling monitoring and simulation within a decision-making framework that includes stakeholder input for evaluating a range of strategic alternatives related to human-landscape interactions.

While MASOOR, RBSim (Chapter 16), and other software that aids the modeler in building spatial agent-based models will continual to be enhanced, Skov-Petersen (Chapter 8) demonstrates the potential of incorporating probabilistic visibility into these models. He demonstrates how probabilistic visibility (PV) can be measured, modeled, and applied to simulation models. He suggests that when evaluating whether one agent can see another, the probability can be revealed from Probabilistic Visibility Graphs (PVG) for all relevant pairs of locations (e.g. raster cells). For each agent the probability is juxtaposed with a random number (between 0 and 1), to determine if visual contact is actually made. The relevance of applying probabilistic visibility to spatial agent-based models is amplified if more detailed phenomena's, in spatial or temporal terms, are modeled. The finer the details of the landscape anticipated, the shorter the model time step, the more heterogeneous the landscape elements (e.g. vegetation), the denser the path- and road network, the higher the number of visitors; the greater the need for finer detail in the perceptive abilities of the model agents. In order to capture the finer details in the perceptive abilities of model agents the classical Boolean realization of visibility needs to be broadened to include probabilistic visibility. This technique shows great promise for aiding agents in evaluating their environment and others within it.

While simulation has been shown to an effective tool to evaluation spatial and temporal patterns of use, there are still some inherent issues that must be addressed before reliable and accurate simulations can be developed. Validation of computer simulation model outputs presents another significant challenge to be addressed in future research. Existing applications of computer simulation to recreation management and planning, including the examples presented in the paper by Lawson et al., (Chapter 10) have generally lacked or been limited in the use of quantitative validation techniques such as those outlined by Law & Kelton (2000). Furthermore, while there is a relatively extensive body of literature describing validation techniques for simulation models of manufacturing systems, there is a lack of recent research concerning the appropriateness of alternative statistical techniques for validation of spatial computer simulation models of visitor landscapes.

The issue of reliability is particularly important because computer simulation modeling uses random numbers to generate input variables (e.g., visitor arrival times, durations at destinations) and therefore the estimates from a model vary across replications of the model. Consequently, conclusions should not be drawn from a single replication of a model (Law & Kelton 2000). So the question that is frequently asked and not so frequently practiced is how to determine the number of simulation replications required that meets some user specified degree of reliability and accuracy. This issue is critical in any spatial simulation project that is focused on outputs that are directly used for management decision-making. Itami et al. (Chapter 22) have shown that accurate and reliable spatial simulations can be developed using the procedures outlined above for evaluating patterns of use in Canada's Mountain Parks (Banff, Yoho, Kootenay and Jasper). Results clearly indicate that for three selected indicators, travel routes, day and and overnight destination encounters, the required 95% confidence intervals could be achieved within the 5 replications. For travel route encounters however, because of the higher impact of small variations on encounters, 95.7% of the routes produced 95% confidence intervals within 5 replications, another 2.82% requiring an estimated 6 to 10 replications and the remaining 1.5% requiring an estimated 11 to 104 replications to achieve 95% confidence intervals. This study concludes that a relatively high degree of precision and reliability in results can be achieved through implementation of proper simulation techniques.

Likewise, Kiser et al. (Chapter 20) and

Lawson et al. (Chapter 10) carefully document procedures to assess the reliability of outputs from computer simulation models of visitor landscapes. They recommend that these procedures be adopted as standard practice within applications of computer simulation modeling to outdoor recreation planning and management. Further, this study provides insights into the feasibility of generating visitor use-related model outputs at managerially relevant levels of precision. The results suggest that precise estimates can be obtained for a small to moderate number of visitor and spatially-based outputs. However, there are constraints to generating precise estimates of use-related outputs as the number of outputs estimated simultaneously becomes large. This challenge is particularly pronounced in cases where at least some of the outputs are derived for low use attractions, trails, or camping locations. Future studies should explore variance reduction techniques to enhance the reliability of computer simulation modeling in cases where a large number of outputs are desired and/or low use environments are modeled.

Finally, while many of the previous applications of computer simulation modeling to park and protected area planning and management have generally done little to assess the reliability of model estimates, more research is warranted to develop standardized methods and procedures to assess the validity of computer simulation models designed for outdoor recreation management. Chapters (10, 18, 20) by Lawson et al., Itami and Kisler et al. have demonstrated the effectiveness of using appropriate statistical methods for identifying the number of simulation replications needed to estimate model outputs at a level of precision that is useful for management purposes. Future applications of computer simulation modeling to parks and protected areas need to continue on this trend.

It should be apparent as one reads this volume that visitor use is one of the primary "agents of change" in visitor landscapes. Thus, management and planning requires information about how visitors use of these areas. In many cases decision makers have had to rely on intuition, secondary data sources about visitor use or data collected with little foresight as to how it might be used to answer specific management questions. Many of the questions that decision makers need to know to proactively manage their landscapes are spatial and temporal in nature. How do visitors disperse across a landscape? Where are the major concentration areas? How does visitor use respond or is influenced by seasonal change? What are the short and long term impacts of use in visitor landscapes? There are many examples presented in this volume that demonstrate, computer simulation has significant potential to provide a more informed basis for planning and management of visitor landscapes.

This volume examines the need to develop a comprehensive and empirically based framework for linking the social, biophysical and geographic disciplines across space and time. Simulation modeling has the potential to bridge a significant social science knowledge gap to improve the ability of decision making to positively and proactively manage and promote long-term protection visitor landscapes (Gimblett 2005). Spatial agent-based simulations coupled with accurate monitoring techniques provide:

- A comprehensive and dynamic understanding of human behavior, interactions between humans and their environment;
- A framework for a more holistic and

comprehensive way of incorporating human-landscape information into the planning and management process;

- A way of measuring human interactions that are difficult or expensive to do in the field;
- A way to test alternative management scenario's and place planning and management into an exploratory and experimental framework;
- Communicating complex inter-related issues in human-landscape management to stakeholders and decision makers;
- A comprehensive framework for human monitoring, understanding human use patterns, and as a decision support system for defining and testing alternative management responses to changing condition.

As with the introduction of any new methods, they do not come without growing pains. While the spatial agent-based modeling community has come a long way in adopting and adapting techniques from many disciplines and developing new ones specific to agent modeling, it is apparent that there is much more research and development work that needs to occur before the decision making community can have full confidence in the results these models provide. To ensure that computer simulation models provide valid and reliable information in a cost-effective manner, it is important that state-of-the-art knowledge, procedures, and methods be adopted for simulation of use in visitor landscapes. A primary product of future research should be a set of technical standards and procedures that guide future applications of data collection, long-term monitoring and simulation within a decision-making framework that included stakeholder involvement. Itami (Chapter 18) provides an excellent example of how this can be achieved. Critical to all the technical pieces of the simulation that need to be worked on is the integration of spatial agent-based simulation modeling into a planning and management framework. This volume presents some excellent examples that place simulation into a useful and effective position as an integral part of decision-making. When this occurs on a more regular based, the true value of simulation will be realized as an integral and essential component of a decision-making process. Like the explosion of the development and use of Geographic Information Systems in the 70's and early 80's, the early beginnings of spatial agent-based simulations will become common practice in the near future.

LITERATURE CITED

Blahna, D. & S. McCool. 2004. Managing Visitors in Wildland Settings. Paper presented at a Recreation Simulation Workshop. Anchorage, Alaska. May 17-18, 2004.

Cole, D. N. 2004. Monitoring and Management of Recreation in Protected Areas: the Contributions and Limitations of Science. In: Sievänen, Tuija, Erkkonen, Joel, Jokimäki, Jukka, Saarinen, Jarkko, Tuulentie, Seija & Virtanen, Eija (eds.). Policies, methods and tools for visitor management – proceedings of the second International Conference on Monitoring and Management of Visitor Flows in Recreational and Protected Areas, June 16–20, 2004, Rovaniemi, Finland.

Cole, D., Cahill, K. & M. Hof. 2005. Why Model Recreation Use?. In: Cole, David N. (compiler). Computer Simulation Modeling of Recreation Use: Current Status, Case Studies, and Future Directions. Gen. Tech. Rep. RMRS-GTR-143. Ogden, UT: U.S. Department of Agriculture, Forest Service, Rocky Mountain Research Station. September 2005. Pgs 1-2.

Cornwell, J. K. O'Brien, B. Silverman, J, Toth. 2003. Affordance Theory for Improving the Rapid Generation, Composability, and Reusability. Conference on Behavior Representation in Modeling and Simulation. May 12-15, 2003.

Gibson, J. J. 1979. The ecological approach to visual perception. Boston: Houghton Mifflin, 1979.

Gimblett, H. R. 2005. Human-Landscape Interactions in Spatially Complex Settings: Where are we and where are we going? In Zerger, A. and Argent, R.M. (eds) MODSIM 2005 International Congress on Modelling and Simulation. Modelling and Simulation Society of Australia and New Zealand, December 2005, pgs.11-20. ISBN: 0-9758400-2-9.

Law, A. & Kelton, W. 2000. Modeling simulation and analysis (3rd ed.). Boston: McGraw Hill.

Manning, R.E. and D. W. Lime. 2000. Defining and Managing the Quality of Wilderness Recreation Experiences. USDA Forest Service Proceedings RMRS-P-15 Vol-4. 2000.

McVetty, D. 2003. Understanding Visitor Flows In Canada's Mountain National Parks: The Patterns Of Visitor Use Studies In Banff, Jasper, Kootenay And Yoho National Parks. Paper presented at the 5th International SAMPAA Conference. Victoria, British Columbia. Canada. May 11-16, 2003.

Steidl, R. & B. Powell. 2006. Assessing the Effects of Human Activities on Wildllife. George Wright Society Forum. Volume 23, Number 2 (2006).

LIST OF CONTRIBUTORS

Arne Arnberger
Institute of Landscape Development,
Recreation and Conservation Planning,
BOKU - University of Natural Resources
and Applied Life Sciences in Vienna,
Peter Jordan-Str.82
A-1190 Vienna
Austria
arne.arnberger@boku.ac.at

Colin A. Arrowsmith
RMIT University
School of Mathematical & Geospatial
Sciences
124 La Trobe Street Melbourne, Victoria
3000
Australia
colin.arrowsmith@rmit.edu.au

Ian Bishop
Department of Geomatics
The University of Melbourne
Parkville, Victoria
Australia 3010
i.bishop@unimelb.edu.au

R.W. (Bill) Carter
Faculty of Science, Health and Education
University of the Sunshine Coast
Maroochydore DC Qld 4558
Australia

Dave Crowley
Division of Wildlife Conservation
Alaska Department of Fish and Game
Cordova, Alaska 99574
USA
dcrowley@fs.fed.us

Dennis Doxtater
School of Architecture
University of Arizona
Tucson, Arizona 85712
USA
doxtater@email.arizona.edu

Birgit H.M. Elands
Forest and Nature Conservation Policy
Group, Landscape Centre,
Wageningen University, P.O. Box 47,
6700AA
The Netherlands
birgit.elands@wur.nl

Randy Gimblett
School of Natural Resources
University of Arizona
BSE Bldg. Rm. 325
Tucson, Arizona 85712
USA
Gimblett@ag.arizona.edu

Frank Grigel
Social Science Specialist, Monitoring
Parks Canada Agency
Western & Northern Service Centre
#1550, 635 - 8 Avenue S.W.
Calgary, Alberta
Canada
Frank.Grigel@pc.gc.ca

Jeffrey C. Hallo
Department of Parks, Recreation and
Tourism Management
College of Health, Education, and Human
Development
Clemson University,

263 Lehotsky Hall, Box 340735
Clemson, SC 29634-0735
jhallo@clemson.edu

Len M. Hunt
Centre for Northern Forest Ecosystem Research
Ontario Ministry of Natural Resources
955 Oliver Road
Thunder Bay, ON
Canada
P7B 5E1
len.hunt@ontario.ca

Robert M. Itami
GeoDimensions Pty. Ltd.
16 Tullyvallin Crescent
Sorrento, Victoria
Australia 3943
Bob.Itami@geodimensions.com.au

Frank Søndergaard Jensen
Senior Researcher
Forest & Landscape Denmark
University of Copenhagen
Copenhagen, Denmark
fsj@life.ku.dk,

Rene Jochem
Wageningen University and Research Centre
P.O. Box 47, Wageningen, 6700AA,
The Netherlands
rene.jochem@wur.nl

Laura Kennedy
School of Natural Resources
BSE Bldg. Rm. 325
University of Arizona
Tucson, Arizona
USA
lakenne@gmail.com

Brett C. Kiser
Department of Forestry
College of Natural Resources
Virginia Polytechnic Institute and State University
305 Cheatham Hall
Blacksburg, VA 24061
bkiser@vt.edu

Spencer Lace
School of Natural Resources
BSE Bldg. Rm. 325
University of Arizona
Tucson, Arizona 85712
USA
slace@email.arizona.edu

Steven R. Lawson
Department of Forestry
College of Natural Resources
Virginia Polytechnic Institute and State University
310-A Cheatham Hall
Blacksburg, VA 24061
lawsons@vt.edu

Daniel Loiterton
Department of Geomatics
The University of Melbourne
Parkville, Victoria
Australia 3010
d.loiterton@pgrad.unimelb.edu.au

Danielle J. Marceau
Department of Geomatics Engineering,
University of Calgary
2500 University Drive
Calgary, Alberta
Canada
Email: marceau@geomatics.ucalgary.ca

Robert E. Manning
Rubenstein School of Environment and Natural Resources
The University of Vermont
George D. Aiken Center
81 Carrigan Drive
Burlington, VT 05405-0088
Robert.Manning@uvm.edu

Ramona van Marwijk
Forest and Nature Conservation Policy Group, Landscape Centre,
Wageningen University, P.O. Box 47, 6700AA
The Netherlands
ramona.vanmarwijk@wur.nl

David McVetty
Parks Canada
Western & Northern Service Centre
145 McDermot Avenue
Winnipeg Canada
dave.mcvetty@pc.gc.ca

Henrik Meilby,
Associate Professor
Forest & Landscape Institute
University of Copenhagen
Copenhagen, Denmark
heme@life.ku.dk,

Andreas Muhar
Institute for landscape development, recreation and conservation planning
BOKU - University of Natural Resources and Applied Life Sciences
Peter Jordan-Str.82
A-1190 Vienna
Austria
andreas.muhar@boku.ac.at

Rachel M. Nelson
School of Natural Resources
University of Arizona
BSE Bldg. Rm. 325
Tucson, Arizona 85712
USA
nelsonrachelmarie@hotmail.com

Aaron Poe
Chugach National Forest
Glacier Ranger District
Chugach National Forest
Girdwood, AK 99587
USA
apoe@fs.fed.us

Chris Prew
Pike & San Isabel National Forests
Cimarron & Comanche National Grasslands
South Park Ranger District
320 HWY 285 - Box 219
Fairplay, CO 80440 USA
cprew@fs.fed.us

David. G. Pitt
Department of Landscape Architecture,
College of Design,
University of Minnesota,
89 Church St. SE,
Minneapolis, MN
USA.
pittx001@umn.edu.

Rogier Pouwels
Wageningen University and Research Centre
P.O. Box 47, Wageningen, 6700AA,
The Netherlands.
rogier.pouwels@wur.nl

Ulrike Pröbstl
Leader of Working Group 3
AGL Arbeitsgruppe für Landnutzungsplanung
Institut für ökologische Forschung
St.Andrä-Str. 8
DE-82398 Etting-Polling
Germany
office@agl-proebstl.de

Chris Sharp
School of Natural Resources
University of Arizona
BSE Bldg. Rm. 325
Tucson, Arizona 85712
USA
csharp@email.arizona.edu

Laura Sinay
School of Natural and Rural System Management
The University of Queensland

Australia
lsinay@ig.com.br

Hans Skov-Petersen
Senior Researcher
Forest & Landscape Denmark
University of Copenhagen
Copenhagen, Denmark
hsp@life.ku.dk

Carl Smith
School of Natural and Rural System Management
The University of Queensland
Australia
csmith@ig.com.br

Bernhard Snizek
metascapes.org
Refshalevej 110a
DK-1432 Copenhagen K
bs@metascapes.org

Michael L Steven
Honorary Associate
Faculty of Architecture, Design and Planning
University of Sydney, NSW, Australia
&
C/- MWH New Zealand Ltd
PO Box 9624, Te Aro
Wellington
New Zealand
michael.l.steven@mwhglobal.com

Karolina Taczanowska
Institute of Landscape Development, Recreation and Conservation Planning,
BOKU - University of Natural Resources and Applied Life Sciences in Vienna,
Austria
karolina.taczanowska@boku.ac.at

Jana Verboom
Wageningen University and Research Centre
P.O. Box 47, Wageningen, 6700AA,
The Netherlands.
jana.verboom@wur.nl

Peter Visschedijk
Alterra Green World Research,
The Netherlands
Peter.Visschedijk@wur.nl

Phillip Wolfe
College of Forestry, Oregon State University
Corvallis, Oregon
USA
phillip.wolfe@gmail.com

Brian Garber-Yonts
NOAA Fisheries, Alaska Fisheries Science Center
Alaska Fisheries Science Center, NOAA
7600 Sand Point Way N.E., Building 4
Seattle, Washington
USA
brian.garber-yonts@noaa.gov

Jianhong Xia
RMIT University
School of Mathematical & Geospatial Sciences
124 La Trobe Street, Melbourne, Victoria 3000
Australia
Jianong.Xia@rmit.edu.au

Darrel Zell
Parks Canada
Banff National Park
Banff, Canada
darrel.zell@pc.gc.ca

EXTERNAL REVIEWERS AND COLLABORATORS

Many of the authors and coauthors above served as reviewers of the chapters included in this volume. But we would like to personally thank the following evaluators for their hard work in reviewing a significant number of these chapters. Your evaluations have contributed greatly to the success of this book.

Mette Termansen
Lecturer
School of Earth and Environment,
University of Leeds. Uk
m.termansen@see.leeds.ac.uk..

Christopher John Topping
Senior Researcher
Department of Wildlife Ecology and Biodiversity,
National Environmental Research Institute.
University of Århus. Denmark
cjt@dmu.dk..

Ole Hjort Caspersen
Senior Researcher
Department of Urban and Landscape Studies. Forest & Landscape.
University of Copenhagen, Denmark
ohc@life.ku.dk..

Lasse Møller-Jensen
Associate Professor
Institute of Geography and Geology,
University of Copenhagen. Denmark.
lmj@geogr.ku.dk.

Further the editors wants to express their gratitude to Inger Grønkjær Ulrich (Forest & Landscape Denmark) for the hard work on the lay out and desktop publishing of the volume and Bernhard Snizek (metascapes.org) for developing and managing the content management system used during the production of the book.

INDEX

A

Aboriginal 39, 49, 50
Acadia National Park 184
Acceptance of management decisions 246
Accidental sampling 40
Accuracy (of output) 321
Adaptive management 234
Aesthetics 27
Affordance 13, 426
Age and group composition 67
Agent-Based Models (ABM) 3
Agent environment 113
Agent methods 110
Agent parameters 109
Aggregating and simplifying data 100
Aggregating choice alternatives 194
Alaska 295, 318, 349
ALEX 259
ALMASS 259
Alternative management practices 176
Amsterdamse Waterleidingduinen 257
Amusement, rapture, dedication 61
Appalachian Trail 373
Appreciating aspects of place 48
Attitudes towards simulation 242
Attraction 68
Australia 39, 51, 108, 332
Austria 163, 240
Automatic counters 124, 127
Automatic traffic counters 428
Autonomous agents 3, 411

B

Backcountry 317, 392
Bald eagle 350, 362
Banff National Park 390
Bardedjilidji Sandstone Track 39
Baseline simulation 178
Bayesian Belief Network (BBN) 220
Bears 296, 318
Behaviour and behavioural rules 38, 68
Beliefs, values and norms 214
Billabong 48, 51
Biodiversity 59
Biodiversity and recreation 254
Biophysical impacts 350
Blackboard 282
Black oystercatcher 350, 363
Boats 53, 300
Bonferroni Correction 376, 397
Botanic Garden Melbourne 108
Brazil 217
Browsing visitors 67

C

Camera-based counting systems 90
Campsites 300
Canada 390
Carrying capacity 181
Choice modeling 190
Chunking 277
Closed circuit television (CCTV) 95
Cluster analysis 73
Cluster analysis of survey itineraries 403
Co-learning environment 234
Cognitive agents 274
Cognitive and affective responses 41
Cognitive map 40, 47, 277
Cognitive representation 63
Commercial flights 318
Communication with stakeholders 433
Communicative approach to models 271
Comparing model results to observation 322
Complex adaptive systems 418
Complexity Theory 413
Computer games 147
Confidence Intervals 301

Crocodiles 39, 49, 54
Crowding 22, 69, 280
Crowding norms 72
Cultural change 214
Cultural expression 25
Cultural impact 432
Cultural preferences 64
Cultural traditions and expressions 214
Cutthroat trout 350, 363

D

Dangerous space 21
Dangers 49
Danube Floodplains National Park 163
Data availability and costs 245
Data regarding visitor movements 86
Data requirements for choice models 192
Day of the week 125, 164
Decision-making 425
Defensible space 21
DEM (Digital Elevation Model) 116
Denmark 125
Densely populated countries 79
Design of more realistic research 176
Deterministic time table 135
Diaries 394
Dimensions of the transaction 272
Direct observation 89, 124
Discrete Event Simulator 278
Distance-decay 145
Drinking water 257
Dwingelderveld National Park 66, 282
Dynamic models 177

E

Ecological disaster 350
Ecological footprint 271
Ecological footprint modeling 286
Ecological Integrity 390
Ecological management 59
Emotional 61
Encounter 53
Encounters 60, 300, 317, 375, 428
Environmental N (data collection method) 40
Environmental values 62
Estimating preferences 199
European Commission 240
Event types 109

Exclusive Space 22
experience of managers, (GIS-) 246
Experiential themes 47
Extend (software package) 177

F

Factor analysis 73
Factors influencing visitor behavior 244
Feelings (stimuli accompanied by...) 40
First law of Geography 144
France 240
Frontcountry 69, 392

G

Germany 240
Goal-driven 411
GPS (Global position system) 68, 86, 93, 115, 178, 240, 273, 319, 429
Grand Canyon, Arizona 22
Great Britain 240
Great Smoky Mountains National Park 373
Grounded theory 44

H

Harbor seal 350, 363
Hard to measure variables 179
Hazards 49
Heterogeneity in preferences 200
History of place 52
Homing (path integration) 277
Human-environment interactions 159, 425
Human/wildlife interactions 352
Human impact on wildlife 256, 350, 361
Hunters 296

I

Indicators of cultural change 218, 224
Indigenous people 218
Indirect observations 123
Inductive loop detectors 92
Induvidually-Based Models (IBM) 3, 273
Infrared sensors 91
Inter-agent visibility 146
iRAS 108
Isovist 143
Italy 240

J

Jasper National Park 390
John Muir Wilderness 178
Juatinga Ecological Reserve 217

K

Kakadu National Park 39
Kayaks 296, 318
Kootenay National Park 390

L

Level of detail of the models 7
Level of familiarity 78
Level of Sustainable Activity (LSA) 334
Limits of Acceptable Change (LAC) 371
Lobau 163
Location-sensitive questionnaires 116

M

Magnetic detectors 91
Maintenance of cultural integrity 231
Manager-oriented Model 287
Marked paths 18
Marked trail visitors 67
Markov Chain 101
MASOOR 67, 160, 240, 257, 270
Melbourne 332
METAPHOR 257
Methods for tracking 92
Minimum acceptable conditions 176
Misty Fjords National Monument 315
Mobile phone tracking 95
Model validation 78
Moment of experience 41
Motorized Vessels 336
Mountain biking 20
Mountain goats 318
Movement decisions 108
Muir Woods National Monument 181
Multi-stage sampling design 390
Multi Agent Systems (MAS) 3, 273, 411
Multi Criteria Analysis 279

N

Narrative value 65
Native American groups 25
Natura 2000 sites 240

Netherlands 60
Netherlands, the 240, 257, 282
Network 60, 113, 162, 177, 305, 320
Nietzsche 62
Non-Motorized vessels 336
Non-visual aesthetics 18, 21
Number of replications 376, 395, 397, 430, 435

O

Oil tanker 350
Out-there-ness 61

P

Participation and site choice 205
Participatory planning 248
Path choices 119
Paths, signage and marked trails 63
Path segment history 280
Pattern of use 394
PDA (personal digital assistant) 94, 116
Pedestrian models 7
People at One Time (PAOT) 183
Perception, comprehension, reaction 5
Perception of cultural change 230
Perception value 65
Person-environment interaction 62
Phenomenological 61
Photographs 28
Physically disabled 20
Pictograms 40
Pigeon Guillemot 350
Piloting 277
Planning/managing framework 254
Planning and management of visitor use 176
Planning process complexity 247
Post-normal science 415
Pressure pads 90
Prince William Sound 295, 349
Prisoner's dilemma 412
Probabilistic simulation 394
Probabilistic visibility 144, 433
Probabilistic Visibility Graphs 144
PROGRESS 240
Public acceptance 78
Public attitudes toward management 184
Purists and Urbanist 62
Purposive sampling 40

Q

Quality of visitor experiences 271
Questionnaires 394

R

RAMAS 259
Randomness 125
Random Utility Models 3, 191
Random variation 124
Ranked Space 22
Raster data 163
RBSim 296, 298, 320, 353, 375
Recreational data acquisition 88
Recreational experience 38, 44
Recreational quality 60, 427
Recreation Opportunity Spectrum (ROS) 316
Reliability analysis 376
Reliability of outputs 321, 435
Religion 232
Representative sampling 393, 429
Reproduction of birds 256
Residuals 125
Restroom 19
River capacity 333
River management 335
River traffic 342
RouteCalculator (method) 114
Rule-based simulation 431

S

Scientists (role of...) 79
Self-administered questionnaires 93
Sensemaking 38
Sensitivity analysis 120
Sensory experiences 27
Ships 318
Shoreline 298, 317
Signs 18, 39
Similarities of behavioral models 207
Skateboarding 20, 24
Skiing 20
Skylark 257
Slovenia 240
Snake 54
Social agents 416
Social territorial 20
Space Syntax 17, 24, 285

Spatial/temporal patterns of visitor use 352
Spatial behavior 160
Spatial clustering technique 355
Spatial quality 60
Spatial Scales 287
Spiritual overtone (landscapes with...) 25
Spontaneous Space 23
Staff agents 109
Stakeholder acceptance 433
Stakeholder conversation 288
Stakeholder diversity 247
Standards of quality 60
State-Pressure-Response (PSR) 219
Stimuli, thoughts and feelings 41
Stochastic time tables 135
Substitution among alternatives 201
Symbolic landscape 68
Symbolic representations 64

T

Task performance 18, 21
Temporal profiles 125
Termites 55
Territorial space 23
Testing alternative management practices 183
Thoughts (stimuli accompanied by...) 40
Time-space behavior 67
Time of the year 132
Time tables 124
Tongass National Forest 315
Tourists 232, 318, 342
Traditional communities 217
Tragedy of the commons 1
Trail design alternatives 271
Transparancy parameters 147
Transparency decay of vegetation 144
Trip itineraries 161, 298, 357, 394
Typology of recreational experiences 44

U

Universal choice set 193
Universe of possibilities of a system 418
Urban parks 54
USA 178
User interface 113
Utility maximization 191

V

Verification of simulation outputs 321
Viewing angle 144
Viewshed Calculation Application 150
Virtual environments 117
Visitor agent 108
Visitor count 89
Visitor distribution 60
Visitor landscapes 17, 27
Visitor preferences 28
Visitor types 168
Visitor use day, a 305
Visual and non-visual aesthetics 27
Visual contact 144

W

Water taxi services 296
Wayfinding 16, 53

Weather 52, 125
Wetlands 52
Wilderness 316
Wilderness Act 371
Wilderness quality 428
Wildlife 39, 50, 318
Woodlark 256

Y

Yoho National Park 390

Z

Zion National Park 179
Zoning scenario 262

About the Editors

HANS SKOV-PETERSEN is Senior Researcher (GIS and recreation) at Forest & Landscape Denmark, University of Copenhagen. He holds a Masters Degreein Horticulture (1989) and a Ph.D. in Human Geography and GIS (2002).His main research interests include spatial modeling of human behavior andwayfinding, Spatial decision support systems in physical planning, Monitoring of recreational behavior and preferences, and communicativeaspects of GIS. Hans is editor in charge of the Danish GIS- and geodata-magazine Perspektiv.

RANDY GIMBLETT is a professor at the University of Arizona in the School of Natural Resources. He received his Ph.D. from the University of Melbourne (1998), a Masters in Landscape Architecture from the University of Guelph (1984) and a Diploma in Landscape Architecture from Ryerson Polytechnical University (1977) in Toronto. He is the editor of a book titled Integrating GIS and Agent Based Modeling Techniques for Understanding Social and Ecological Processes. His main research interests include Spatial Dynamic Ecosystem Modeling, Cognition and Environmental Perception, Modeling and Simulating Human movement and behavior in Visitor Landscapes.